## 李文超教授和研究团队

1997年,李文超教授和周国治院士在与来访的瑞典皇家工学院 Seshadri Seetharaman 教授、杜嗣琛教授进行学术交流后合影

左起:文洪杰、黄向东、李文超、王习东、周国治、张海军、Seshadri Seetharaman、甄 强、杜嗣琛、
　　　张 梅、王福明、滕立东、黄绵亮

1996年,李文超教授与俄罗斯科学院结构宏观动力学研究所斯多林研究员合影,陪同人员为张海军博士

1996年,李文超教授在与俄罗斯科学院结构宏观动力学研究所斯多林研究员进行学术交流后合影

左起:文洪杰、斯多林、李文超、张海军、
　　　甄 强

1998年,适逢1994级本科生毕业,在实验室合影
前排左起:李文超、王福明
后排左起:张 梅、李山丹、郭宇艳、杜雪岩、秦晓军、甄 强、丁保华、
　　　　　张海军、黄绵亮、黄向东

1999年,同学聚会后合影
　前排左起:张 梅、刘静波
　后排左起:刘国华、滕立东、文洪杰、杜雪岩、王习东、魏志峰

1997年，文洪杰博士学位论文答辩后留念

左起：黄绵亮、张海军、黄向东、
　　　文洪杰、甄　强

1997年，师兄弟在北京科技大学西校门前合影

左起：黄绵亮、张海军、甄　强、
　　　杜雪岩、黄向东、丁保华

1997年，研究团队登上北京香山合影留念

左起：张　梅、杜雪岩、甄　强、张海军、黄绵亮

2006年10月28日,北京科技大学物理化学研究团队成员从各地回到母校,共同庆祝李文超教授70华诞
前排左起:刘克明、岳昌盛、王海娟、张 慜、赵斯琴、梁子福、刘晓丹、董鹏莉、唐续龙、李玉祥、马 刚
中排左起:苏鑫明、张 辉、金 烁、刘丽丽、郭 敏、孙贵如、李文超、王亚丽、周 媛、盖鑫磊、马 腾、黄剑雄、张艳君、张 梅
后排左起:赛音巴特尔、杨修春、刘建华、张海军、杜雪岩、高立春、甄 强、王习东、赵海雷、叶 超、李 超、杨传钰、王 荣、张作泰

谨以本书祝贺

李文超教授八十寿辰！

# 冶金与材料物理化学专题文集

Symposium on Physical Chemistry
in Metallurgy and Materials

本书编委会 编

北 京
冶金工业出版社
2015

## 内 容 提 要

本文集简要介绍了冶金与材料物理化学、资源综合利用、耐火材料基础理论上取得的研究成果，收录了冶金物理化学、钢铁冶金、有色冶金、陶瓷材料、耐火材料和工业固体废弃物高效综合利用等方面的学术论文50篇，分为物理化学应用基础研究、新型高温结构陶瓷和功能陶瓷材料三个部分。这些论文反映出该领域的最新研究进展和成果，代表了我国目前冶金与材料物理化学领域的研究水平。

本文集可供物理化学研究者和冶金与材料工作者阅读。

**图书在版编目(CIP)数据**

冶金与材料物理化学专题文集/《冶金与材料物理化学专题文集》编委会编. —北京：冶金工业出版社，2015.10
ISBN 978-7-5024-7048-7

Ⅰ.①冶… Ⅱ.①冶… Ⅲ.①材料科学—物理化学—文集 Ⅳ.①TB3-53

中国版本图书馆CIP数据核字(2015)第242029号

出 版 人 谭学余
地　　址　北京市东城区嵩祝院北巷39号　邮编　100009　电话　(010)64027926
网　　址　www.cnmip.com.cn　电子信箱　yjcbs@cnmip.com.cn
责任编辑　刘小峰　美术编辑　彭子赫　版式设计　孙跃红
责任校对　李　娜　责任印制　李玉山
ISBN 978-7-5024-7048-7
冶金工业出版社出版发行；各地新华书店经销；三河市双峰印刷装订有限公司印刷
2015年10月第1版，2015年10月第1次印刷
787mm×1092mm　1/16；28印张；2彩页；889千字；431页
180.00元

冶金工业出版社　投稿电话　(010)64027932　投稿信箱　tougao@cnmip.com.cn
冶金工业出版社营销中心　电话　(010)64044283　传真　(010)64027893
冶金书店　地址　北京市东四西大街46号(100010)　电话　(010)65289081(兼传真)
冶金工业出版社天猫旗舰店　yjgycbs.tmall.com

(本书如有印装质量问题，本社营销中心负责退换)

# 本书编委会
（以拼音为序）

| 丁保华 | 董　倩 | 杜雪岩 | 樊成才 | 高立春 |
| 黄绵亮 | 黄向东 | 李兴康 | 连　芳 | 刘国华 |
| 刘建华 | 刘静波 | 刘克明 | 秦建武 | 赛音巴特尔 |
| 滕立东 | 王海川 | 王金淑 | 王习东 | 文洪杰 |
| 张海军 | 张　辉 | 张　梅 | 张作泰 | 甄　强 |
| 赵海雷 | 仲维斌 | 庄又青 | | |

# 编者的话

李文超教授是北京科技大学教授、博士生导师，一直从事冶金物理化学领域的科研与教学工作。作为建国初期赴苏联留学并回国的专家，直接参与和见证了我国当代冶金和耐火材料工业以及相关科学的发展。李文超教授在冶金及材料物理化学教育和研究上的贡献，及其治学、科研和育人之道，深得师生的尊敬。

李文超教授于1936年11月生于山东招远；1959年毕业于莫斯科钢铁学院金属物理化学系；1986年赴莫斯科钢与合金学院作为高级访问学者；1991~1996年曾任北京科技大学理化系主任；1991年受聘为北京科技大学理化系教授、博士生导师，兼职应聘为重庆大学、安徽工业大学教授；1987年中国金属学会恢复后，即任冶金物化学术委员会委员兼秘书，并先后担任过冶金物化学术委员会（分会理事长）副主任、主任，多次被评为中国科协先进工作者；1991~2006年担任中国金属学会第七届常务理事、第五、六、七届中国金属学会理事，曾担任第一届有色学会理事，历任多届稀土火法冶金专业委员会委员，之后被授予中国金属学会名誉理事和荣誉会员，并被授予"银质终身成就奖"；曾任国家科委冶金学科组秘书、国家自然科学奖评委、杰出青年基金评委和国家自然科学基金会评专家等；长期担任《钢铁》《稀有金属（中、英文版）》《耐火材料（中、英文版）》等期刊编委。

李文超教授为北京科技大学第一届（也是全国第一届）冶金物化本科生讲授了"冶金物化研究方法"专业课，并出版了全国第一本《冶金物化研究方法》讲义；指导了首届4名冶金物化本科毕业生的毕业论文；此外，还先后为本科生讲授过冶金热力学、冶金动力学、冶金物化研究方法、冶金学、示踪原子应用、金属扩散等专业课；为硕士研究生讲授过冶金热力学、无机非金属材料动力学、近代物理化学研究方法、相图热力学分析、高温陶瓷等学位课及选修课；为博士研究生讲授过陶瓷原理与方法、物理化学原理与方法、冶金物理化学进展等课程。1997年被评为北京市优秀教师。从1959年留学回国在北京钢铁学院任教起,从事教育工作45年,培养了一大批冶金和材料领域的科技和管理人才,为我国的冶金和材料领域的高等教育做出了重要的贡献。先后培养博士研究生21名、硕士研究生13名。其中11人成为了国内外的大学教授,2人成为国家杰出青年基金获得者。

李文超教授的研究领域包括:网络工艺数据库在材料合成中应用、近代耐火材料理论基础研究、功能陶瓷、结构陶瓷和古陶瓷材料物理化学。他从1985年获得了首项科学院基金的资助,到国家自然科学基金建立,先后承担了11项自然科学基金项目(1985年成立科学院基金会,后正式改为国家自然科学基金委);冶金部攻关课题等4项(包括:"六五"、"七五"、"八五"、"九五"科技攻关项目各1项)。荣获国家自然科学三等奖1项、国家教委科技进步一等奖1项、冶金部科技进步一等奖1项、国家教委科技进步二等奖1项、冶金部科技进步二等奖1项、省部级科技进步三等奖3项;获得国家发明专利4项。在国内外学术刊物和会议上共发表学术论文260余篇;出版专著1部、教材3种,其中《冶金热力学》获北京市优秀教材一等奖。

李文超教授在冶金及材料物理化学、资源综合利用、耐火材料基础理论上取得了一系列重大的研究成果,对促进冶金及材料物理化学的发展具有重要意义。通过建立冶金及高温陶瓷材料体系内部各个组元热力学参数的预报和评估的方法,完善了热力学理论体系,并从材料热力学和相图计算的角度指导多元多相体系材料的化学成分设计;利用模式识别技术优化材料制备过程中的多因素技术参数;提出了高温材料制备和应用过程中的反应动力学建模方法和理论;通过对不同的熔渣体系和金属体系的扩散动力学研究,发现了液相化学扩散也存在宏观流动的现象。

李文超教授及其团队所取得的研究成果对我国今后的科研工作有所启发和帮助,为此,我们特以"冶金与材料物理化学"为专题,收集了冶金物理化学、钢铁冶金、有色冶金、陶瓷材料、耐火材料和工业固体废弃物高效综合利用方面的部分代表性学术论文50篇,编辑成本文集出版。由于时间和水平所限,编委会组成仓促,未能做到更加全面地反映研究成果,在此表示歉意。

"**文**风高白雪,品格**超**青云"。李文超教授虽已80高龄,但精神矍铄,身体健康,实乃品格修养所致。李文超教授治学严谨,为人师表,平易近人;在科学研究工作中,勤奋努力,求实进取,开拓创新,持之以恒,为我国冶金及陶瓷材料高温物理化学的发展做出了非常重要的贡献。李文超教授的"做人之本、治学之道、科研之风、育人之德"可谓年轻学者们的学习楷模。他教育我们严谨治学，勤奋刻苦，求实进取，以科技创新报效祖国。

感谢李文超教授多年来对学生们的指导与关心！钟香崇院士、周国治院士、朱元凯教授、王俭教授、董元篪教授、谢志鹏教授、王福明教授、唐清教授、包宏教授、王迎军教授、徐利华教授等以及瑞典皇家工学院的教授曾经和李文超教授一起给予我们指导与帮助，在此表示衷心的感谢！我们目前所在的工作单位，特别是北京科技大学、北京大学、北京工业大学、上海大学、武汉科技大学、兰州理工大学、安徽工业大学，还有挪威 DNV GL 公司、ABB 公司、得克萨斯州农工大学—金斯维尔校区、中国钢研科技集团有限公司、北京交通大学、首钢技术研究院、广州白云区发改局、北京仪尊时代科技有限公司、南京亿达高科环保技术有限公司等，为我们提供了良好的工作平台，在此对这些单位和同事表示感谢！

谨以本书，祝贺李文超教授八十华诞！

本书编委会
2015 年 6 月

# 目 录

## 第一部分 物理化学应用基础研究

钢中稀土夹杂物生成的热力学规律 ………………………………………… 李文超　3

物理化学在古陶瓷研究中的应用 …………………………………………… 李文超　19

Prediction of Thermodynamic Properties for Multicomponent System with Chou Model
………… Zhen Qiang（甄强）　Bao Hong（包宏）　Wang Fuming（王福明）　等　25

由稳定化合物熔化焓提取二元系活度 ………………… 李兴康　刘四俊　王 俭　等　30

Statistical Mechanics Model of Liquid Binary Alloy and Its Parameters
…………………………………… Fan Chengcai　Wang Jian　Li Wenchao　et al.　35

Does Nitrogen Transport in Vitreous Silica only Take Place in Molecular Form?
………………………………………………………………… Dong Qian　G. Hultquist　47

Influence of Additives on Kinetic Behavior of $SiO_2$-C-$N_2$ System
………………………………………………… Zhuang Youqing　Li Wenchao　56

Silica Photonic Crystals with Quasi-full Band Gap in the Visible Region Prepared in Ethanol
………………………… Zhang Hui　Wang Xidong　Zhao Xiaofeng　et al.　60

Fe-C-$j$($j$ = Ti、V、Cr、Mn)熔体的热力学性质规律 ……… 王海川　王世俊　乐可襄　等　66

A Data Treatment Method of the Carbon Saturated Solubility in Fe-C-Cr Melt
………………………… Wang Haichuan　Wang Shijun　Yue Kexiang　et al.　74

低碳 FeMnSiAl 系 TWIP 钢冶炼技术研究 ………………… 刘建华　庄昌凌　李世琪　等　81

Influence of Vanadium on Microstructure and Properties of Medium-chromium White Cast Iron
………………………… Liu Keming　Wang Fuming　Li Changrong　et al.　88

Phase Relationships and Thermodynamic Properties in the Mn-Ni-C System
………………………… Teng Lidong　Ragnhild Aune　Li Wenchao　et al.　95

Experimental Investigation and Modeling of Cooling Processes of High Temperature Slags
………………………… Sun Yongqi　Shen Hongwei　Wang Hao　et al.　108

Thermodynamic Investigation of Synthesizing Metastable β-Sialon-Alon Composite Ceramic
………………………… Huang Xiangdong　Li Wenchao　Wang Fuming　et al.　121

高炉渣合成 Ca-α-Sialon-SiC 粉的热力学分析及工艺优化
………………………………………… 刘克明　王福明　李文超　等　127

## 第二部分 新型高温结构陶瓷

Kinetic Studies of Oxidation of MgAlON and a Comparison of the Oxidation Behaviour of AlON, MgAlON, O'-SiAlON-ZrO$_2$ and BN-ZCM Ceramics
………………………… Wang Xidong  Li Wenchao  Seshadri Seetharaman  139
Thermal Diffusivity/Conductivity of MgAlON-BN Composites
………………………… Zhang Zoutai  Li Wenchao  S. Seetharaman  155
热压合成 AlON-VN 复相陶瓷的研究 …………… 赛音巴特尔  张作泰  李文超  167
Synthesis of TiN/AlON Composite Ceramics
………………………… Wang Xidong  Gao Lichun  Li Guobao  et al.  173
The Effect of Al$_2$O$_3$(Mul.) on Phase Compositions of O'-Sialon Ceramics
………………………… Zhong Weibin  Li Wenchao  Zhong Xiangchong  181
氮化硼对锆刚玉莫来石材料力学性能及显微结构的影响
………………………… 赵海雷  李文超  钟香崇  等  186
ZrO$_2$-CaO-BN 复合材料的研制 ……………… 黄绵亮  李文超  钟香崇  194
Synthesis Mechanism of Silicon Nitride Obtained from Silica Reduction
…… Zhuang Youqing（庄又青）  Wang Jian（王俭）  Li Wenchao（李文超）  等  198
O'-Sialon-ZrO$_2$-SiC 复合材料的摩擦磨损性能研究
………………………… 张海军  李文超  姚熹  等  203
合成 β-SiAlON-AlON 复相材料的热力学分析的研究
………………………… 黄向东  李文超  王福明  等  211
热压烧结 Ta/β'-Sialon 系梯度功能材料的残余热应力分析 ……… 丁保华  李文超  217
石英向 α-方石英转化率研究 ……………………… 王金淑  王俭  李文超  等  223
刚玉强化日用瓷的理论分析 ……………………… 李文超  王俭  刘建华  等  228
用穆斯堡尔谱和吸收光谱研究汝瓷天青釉呈色机理
………………………… 秦建武  李国桢  李文超  等  237
MgO-SiO$_2$-Al$_2$O$_3$ 体系用后耐火材料合成新材料的研究
………………………… 赛音巴特尔  廖洪强  岳昌盛  等  242
Recent Development of Andalusite Refractories in China
………………………… Wen Hongjie  Li Wenchao  Wang Jinxiang  et al.  248
红柱石分解过程的分形研究 …………………… 文洪杰  李文超  王金相  等  257
Fractal Calculation of Mo/β'-Sialon Functionally Gradient Materials by Powder Metallurgy
………………………… Ding Baohua  Li Wenchao  Wang Fuming  等  261

## 第三部分 功能陶瓷材料

Effects of Preparing Conditions on Controllable One-step Electrodeposition of ZnO Nanotube Arrays
………………………… Lu Hui  Zheng Feng  Zhang Mei  et al.  267
Oxygen Sensitivity of Nano-CeO$_2$ Coating TiO$_2$ Materials

·················· Zhang Mei　Wang Xidong　Wang Fuming　et al.　280

Thermodynamic Analysis of Combustion Synthesis of $Al_2O_3$-TiC-$ZrO_2$ Nanoceramics

·················· Dong Qian　Tang Qing　Li Wenchao　et al.　287

Microwave Plasma Sintered Nanocrystalline $Bi_2O_3$-$HfO_2$-$Y_2O_3$ Composite Solid Electrolyte

·················· Zhen Qiang　Girish M. Kale　He Weiming　et al.　296

一种新型湿化学方法合成 Ba($Mg_{1/3}Ta_{2/3}$)$O_3$ 纳米粉末的研究

·················· 连　芳　徐利华　王福明　等　311

Effect of Ti Content on the Martensitic Transformation in Zirconia for Ti-$ZrO_2$ Composites

·················· Teng Lidong　Li Wenchao　Wang Fuming　318

Preparations and Characterizations of New Mesoporous $ZrO_2$ and $Y_2O_3$-stabilized $ZrO_2$ Spherical Powders ·················· Zhang Hui　Lu Hu　Zhu Yawei　et al.　325

Catalytically Highly Active Top Gold Atom on Palladium Nanocluster

·················· Zhang Haijun　Tatshuya Watanabe　Mitsutaka Okumura　et al.　340

X-Ray Photoelectron Spectrascopy Investigation of Ceria Doped with Lanthanum Oxide

·················· Du Xueyan(杜雪岩)　Li Wenchao(李文超)　Liu Zhenxiang(刘振祥)　et al.　349

A New Highly Selective $H_2$ Sensor Based on $TiO_2$/PtO-Pt Dual-Layer Films

·················· Du Xueyan　Wang Yuan　Mu Yongyan　et al.　352

Mesoporous $TiO_2$ Thin Films Exhibiting Enhanced Thermal Stability and Controllable Pore Size: Preparation and Photocatalysed Destruction of Cationic Dyes

·················· Wang Jinshu　Li Hui　Li Hongyi　et al.　360

Evaluation of $La_{0.3}Sr_{0.7}Ti_{1-x}Co_xO_3$ as Potential Cathode Material for Solid Oxide Fuel Cells

·················· Du Zhihong　Zhao Hailei　Shen Yongna　et al.　374

Platinum Decorated Aligned Carbon Nanotubes: Electrocatalyst for Improved Performance of Proton Exchange Membrane Fuel Cells

·················· Yuan Yuan　Joshua A. Smith　Gabriel Goenaga　et al.　392

Preparation and Characterization of $Li^+$-modified $Ca_xPb_{1-x}TiO_3$ Film for Humidity Sensor

·················· Liu Jingbo　Li Wenchao　Zhang Yanxi　et al.　408

摩托车尾气催化净化技术原理与应用 ·················· 秦建武　416

基于 WEB 的人工神经网络材料设计系统 ·················· 刘国华　包　宏　李文超　427

# 第一部分

## 物理化学应用基础研究
Applied Foundamental Research of Physical Chemistry

物理化学是诸多过程学科的重要基础，诸如冶金过程、材料科学与工程、化学工程、石油工程等。如何利用物理化学的理论和方法分析工程科学研究中出现的问题，解决工程生产实践中具体问题是科技人员和工程技术人员所关注的。

本文集通过冶金和材料研究中一些实例，利用物理化学作为理论工具，分析科研和生产中出现的问题和现象，为物理化学研究者和冶金与材料工作者搭建桥梁，促进相关学科的发展。

# 钢中稀土夹杂物生成的热力学规律

李文超

(北京科技大学理化系，北京　100083)

**摘　要**：本文叙述了钢中稀土夹杂物生成的热力学分析方法，并以含砷低碳钢为例讨论了稀土夹杂物生成的热力学规律，与试验结果作了比较。

## Thermodynamics of the Formation of Rare-Earth Inclusions in Steel

Li Wenchao

(University of Science and Technology Beijing, Beijing 100083)

**Abstract**: In this paper the thermodynamic calculation method for the formation of rare-earth inclusions in steel is discussed, and the calculation results are compared with the experimental data for as-containing low carbon steel.

稀土元素加入钢液后，由于它们对氧、硫、砷、锡、铋、锑、铅、镉等有害杂质有较强的亲和力，因此稀土元素可以脱氧、脱硫、去除有害杂质，从而达到了净化钢液、净化晶界，改善杂质分布规律和夹杂物的形态，有效地提高了钢材的性能。

稀土元素脱氧、脱硫、去除杂质的产物，一方面助于近代实验手段，诸如探针、扫描电子显微镜、透射电镜、X射线衍射等进行鉴定和分析；另一方面，可利用无机热力学数据库进行物理化学分析，预测稀土夹杂物生成的可能性热力学条件及生成顺序。

已发表的文章对钢液中[Ce]-[O]-[S]平衡研究较多，崴海德(A. Vahed)和恺(A. R. Kay)等人建立了[Ce]-[O]-[S]沉淀图，可预测钢液中稀土夹杂物生成的顺序，但有局限性。它仅适用于稀土—氧—硫的产物，对其他杂质产物不能应用。因此，掌握稀土元素与钢液作用的热力学分析方法具有普遍的意义。首先从热力学上分析，稀土元素加入钢液中会生成哪些稀土夹杂物，它们的生成条件和顺序如何？

## 1　稀土夹杂物生成的可能性

决定稀土夹杂物生成的可能性是在相同的条件下（同一标准态下，取一个摩尔稀土为比较标准），计算各类稀土夹杂物的生成反应自由能，进行比较。

### 1.1　计算生成稀土夹杂物的标准吉布斯自由能二项式

$$[RE] + \frac{y}{x}[Imp] = \frac{1}{x}(RE)_x(Imp)_y(s)$$

$$\Delta_r G^\ominus = A + BT$$

式中，[RE] 为溶解于钢液中的各种稀土金属元素；[Imp] 代表溶解于钢液中的各种杂质元素，诸如：[O]、[S]、[As]、[N]、[Sn]、[C] 等。

### 1.2 利用化学反应等温式计算实际条件下反应的吉布斯自由能

$$\Delta_r G = \Delta_r G^\ominus + RT \ln Q = \Delta_r G^\ominus + RT \ln \frac{a_{(RE)_x(Imp)_y}^{\frac{1}{x}}}{a_{[RE]} a_{[Imp]}^{\frac{y}{x}}}$$

式中，$Q$ 为产物的活度积与反应物活度积之比。

活度的计算公式为：

$$a_i = f_i [\%i]$$

活度系数为：

$$\lg f_i = \sum_{j=1}^{n} e_i^j [\%j]$$

式中，$e_i^j$ 为钢液中 $j$ 元素对 $i$ 组元的相互作用系数。

### 1.3 计算实例

现以我国某含砷铁矿冶炼低碳钢为例，对其夹杂物进行热力学计算。其化学成分为：0.20% C；0.017% Si；0.45% Mn；0.067% P；0.044% S；0.03% Al；0.22% Cu；0.014% O；0.21% As；0.002% N；0.1% Sn；0.090% Ce。计算在电渣重熔时，能够生成哪些稀土夹杂物。

由活度相互作用系数（见表1），计算钢液中各组元的活度（见表2）。

**表1 1873K 时钢液中各组元的活度相互作用系数**

| 元素 | O | C | Si | Mn | P | S | N | Al | Ce | As | Sn |
|---|---|---|---|---|---|---|---|---|---|---|---|
| O | -0.20 | -0.45 | -0.131 | -0.021 | 0.07 | -0.133 | 0.057 | -3.9 | -0.57 | 0.07 | 0.011 |
| S | -0.27 | 0.11 | 0.063 | -0.026 | 0.029 | -0.28 | 0.01 | 0.035 | -1.91 | 0.0041 | |
| N | 0.05 | 0.13 | 0.047 | -0.02 | 0.045 | 0.007 | 0.0 | -0.028 | | 0.018 | 0.007 |
| C | -0.34 | 0.14 | 0.08 | -0.012 | 0.051 | 0.046 | 0.11 | 0.043 | | 0.043 | 0.041 |
| Al | -6.6 | 0.091 | 0.0056 | | | 0.03 | -0.058 | 0.045 | -0.43 | | |
| Ce | -5.03 | | | | | -8.36 | | -2.25 | -0.003 | | |
| As | | 0.25 | | | | 0.0037 | 0.077 | | | 0.296 | |
| Sn | -0.11 | 0.37 | 0.057 | | 0.036 | -0.028 | 0.027 | | | | 0.0016 |

**表2 溶解于钢液中各组元的活度**

| 活度 | $f_i$ | | | | | | | | $a_i$ | | | | | | |
|---|---|---|---|---|---|---|---|---|---|---|---|---|---|---|---|
| | $f_O$ | $f_S$ | $f_N$ | $f_{Al}$ | $f_{Ce}$ | $f_C$ | $f_{As}$ | $f_{Sn}$ | $a_O$ | $a_S$ | $a_N$ | $a_{Al}$ | $a_{Ce}$ | $a_C$ | $a_{As}$ | $a_{Sn}$ |
| 加铈前 | 0.62 | 0.998 | 1.059 | 0.848 | | 1.095 | 1.296 | 1.188 | $0.868 \times 10^{-2}$ | $4.39 \times 10^{-2}$ | $0.212 \times 10^{-2}$ | $2.54 \times 10^{-2}$ | | 0.219 | 0.272 | 0.119 |
| 加铈后 | 0.55 | 0.672 | 1.059 | 0.776 | 0.312 | 1.059 | 1.296 | 1.188 | $0.77 \times 10^{-2}$ | $2.96 \times 10^{-2}$ | $0.212 \times 10^{-2}$ | $2.33 \times 10^{-2}$ | $2.81 \times 10^{-2}$ | 0.219 | 0.272 | 0.119 |

由已知反应的标准自由能,根据化学反应等温式计算实际条件下夹杂物生成的自由能,计算结果示于表3。

表3 炼钢温度下夹杂物生成的自由能

| 反 应 | $\Delta_r G^\ominus$/kJ·mol$^{-1}$[Ce] | $\Delta_r G$/kJ·mol$^{-1}$[Ce] | $\Delta_r G_{1873K}$/kJ·mol$^{-1}$ |
|---|---|---|---|
| [Ce]+[N]=CeN(s) | −172.89+0.0811T | −172.89+0.162T | +130.46 |
| [Ce]+2[C]=CeC$_2$(s) | −131.00+0.250T | −131.00+0.145T | +141.07 |
| [Ce]+1.5[C]=0.5Ce$_2$C$_3$(s) | −112.00+0.103T | −112.00+0.124T | +120.29 |
| [Ce]+2[O]=CeO$_2$(s) | −852.72+0.361T | −852.72+0.361T | +177.38 |
| [Ce]+1.5[O]=0.5Ce$_2$O$_3$(s) | −714.38+0.180T | −714.38+0.270T | −208.45 |
| [Ce]+[O]+0.5[S]=0.5Ce$_2$O$_2$S(s) | −675.70+0.166T | −675.70+0.250T | −206.92 |
| [Ce]+[Al]+3[O]=CeAlO$_3$(s) | −1366.46+0.364T | −1366.46+0.547T | −342.65 |
| [Ce]+[S]=CeS(s) | −422.10+0.120T | −422.10+0.179T | −86.20 |
| [Ce]+1.5[S](s)=0.5Ce$_2$S$_3$ | −536.42+0.164T | −536.42+0.237T | −91.68 |
| [Ce]+4/3[S]=1/3Ce$_3$S$_4$(s) | −497.67+0.146T | −497.67+0.215T | −94.96 |
| [Ce]+[As]=CeAs(s) | −302.04+0.237T | −302.04+0.278T | +218.13 |
| [Ce]+0.5[S]+0.5[As]=0.5(CeAs·CeS)(s) | −352.27+0.179T | −352.27+0.229T | +76.72 |
| [Ce]+[O]+[F]=CeOF(s) | −904.30+0.226T | −904.30+0.297T | −348.86 |
| [Ce]+0.5[Sn]=0.5Ce$_2$Sn(s) | −199.92+0.102T | −199.92+0.119T | +103.30 |
| [Ce]+3[Sn]=CeSn$_3$(s) | −190.20+0.280T | −190.20+0.316T | +401.60 |

由上述计算结果可以看出,CeN、CeO$_2$、CeAs、(CeAs·CeS)(s)等均不能生成。对稀土砷化物进一步计算表明,只有当 $\frac{[RE]+[As]}{[O]+[S]} \geq 6.7$ 时方可生成(CeAs·CeS)(s);而本实验条件为 $\frac{[RE]+[As]}{[O]+[S]} = 5.17$,故不能生成稀土砷化物。

## 2 稀土夹杂物生成的热力学条件与顺序

仅计算实际条件下的生成自由能还不够,还必须考虑各类稀土氧化物和稀土硫化物之间的相互转换,才能最终决定它们的生成热力学条件和顺序。

为此,由稀土夹杂物生成的标准自由能计算不同稀土夹杂物的活度积(见表4),再依据稀土夹杂物转换反应式,计算自由能,确定它们热力学相互转换条件。

表4 稀土夹杂物的活度积

| 反 应 | 1873K时活度积 $\Pi_i$ |
|---|---|
| [Ce]+2[O]=CeO$_2$(s) | $\Pi_{a1} = 0.188 \times 10^{-10}$ |
| [Ce]+1.5[O]=0.5Ce$_2$O$_3$(s) | $\Pi_{a2} = 0.291 \times 10^{-10}$ |
| [Ce]+[O]+0.5[S]=0.5Ce$_2$O$_2$S(s) | $\Pi_{a3} = 0.63 \times 10^{-10}$ |
| [Ce]+[S]=CeS(s) | $\Pi_{a4} = 0.328 \times 10^{-5}$ |
| [Ce]+4/3[S]=1/3Ce$_3$S$_4$(s) | $\Pi_{a5} = 0.578 \times 10^{-6}$ |
| [Ce]+1.5[S]=0.5Ce$_2$S$_3$(s) | $\Pi_{a6} = 0.397 \times 10^{-6}$ |

## 2.1 计算稀土氧化物夹杂相互转换的热力学条件

$CeO_2(s)$ 与 $Ce_2O_3(s)$ 转换的热力学条件:

$$Ce_2O_3(s) + [O] = 2CeO_2(s) \quad \Delta_r G^\ominus = RT\ln\frac{\Pi_{a1}^2}{\Pi_{a2}^2}$$

由化学反应等温方程式计算实际条件下的标准自由能:

$$\Delta_r G = \Delta_r G^\ominus + RT\ln\frac{1}{a_O} = RT\ln\frac{\Pi_{a1}^2}{\Pi_{a2}^2 a_O}$$

当 $\Delta_r G < 0$,即 $\frac{\Pi_{a1}^2}{\Pi_{a2}^2 a_O} < 1$ 或 $a_O > 0.417$,才能生成 $CeO_2(s)$ 夹杂物,而本实验条件下,$a_O < 0.417$(见表2),故不可能生成 $CeO_2(s)$ 夹杂。

$$Ce_2O_3(s) + [S] = Ce_2O_2S(s) + [O] \quad \Delta_r G^\ominus = RT\ln\frac{\Pi_{a3}^2}{\Pi_{a2}^2}$$

于是

$$\Delta_r G = \Delta_r G^\ominus + RT\ln\frac{a_O}{a_S} = RT\ln\frac{\Pi_{a3}^2 a_O}{\Pi_{a2}^2 a_S}$$

当 $\Delta_r G < 0$,即 $\frac{\Pi_{a3}^2 a_O}{\Pi_{a2}^2 a_S} < 1$,本实验条件下 $\frac{a_O}{a_S} < 0.213$ 或写为 $a_O < 0.213 a_S$(参见表2),显然 $Ce_2O_3(s)$ 夹杂转换为 $Ce_2O_2S(s)$ 夹杂物。

## 2.2 稀土硫化物夹杂物转换的热力学条件

$Ce_3S_4(s)$ 与 $CeS(s)$ 转换的热力学条件:

由反应 $\quad Ce_3S_4(s) = 3CeS(s) + [S] \quad \Delta_r G^\ominus = RT\ln\frac{\Pi_{a4}^3}{\Pi_{a5}^3}$

因而 $\quad \Delta_r G = RT\ln\frac{\Pi_{a4}^3 a_S}{\Pi_{a5}^3}$

当 $\Delta_r G < 0$,$a_S = 0.0055$,则生成 $CeS(s)$ 夹杂物;而本实验条件 $a_S > 0.0055$(参见表2),故不能生成 $CeS(s)$ 夹杂物。

$Ce_2S_3(s)$ 与 $Ce_3S_4(s)$ 转换的热力学条件:

由反应 $\quad 3Ce_2S_3(s) = 2Ce_3S_4(s) + [S] \quad \Delta_r G^\ominus = RT\ln\frac{\Pi_{a5}^6}{\Pi_{a6}^6}$

于是 $\quad \Delta_r G = RT\ln\frac{\Pi_{a5}^6 a_S}{\Pi_{a6}^6}$

当 $\Delta_r G < 0$,即 $a_S < 0.105$ 时,生成 $Ce_3S_4(s)$ 夹杂物的条件,由表2可以看出,$a_S = 0.0296$,满足生成 $Ce_3S_4(s)$ 夹杂物的热力学条件。

通过上述热力学分析,得知在本实验条件下应该存在:$Ce_3S_4(s)$、$Ce_2O_2S(s)$、$CeAlO_3(s)$ 等稀土夹杂物。

## 2.3　实验验证

在热力学计算的基础上，对钢中夹杂物进行分析鉴定。为防止硫化物等一些微量溶于水的夹杂物的流失，采用非水电解质溶液电解法把夹杂物从钢中萃取分离出来。试样为阳极，不锈钢片为阴极，阳极电流密度不大于 $100mA/cm^2$，电解液的 pH = 8，电解时控制电解液的温度为 $-5 \sim +5℃$，在电解分离过程中采用氩气保护，以防止稀土夹杂物的二次氧化，实验装置示于图 1。

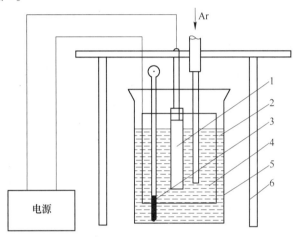

图 1　钢中夹杂物电解分离示意图
1—阳极；2—阴极；3—温度计；4—电解液；5—电解槽；6—阳极支架

经电解分离夹杂物后，在 SEM 上进行观测，其结果见图 2。

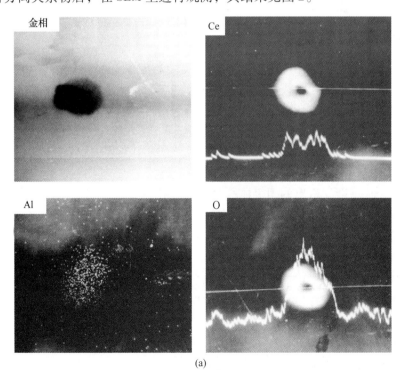

(a)

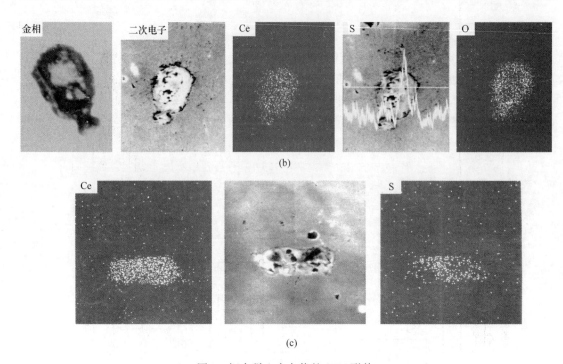

图 2　钢中稀土夹杂物的 SEM 形貌

(a)（CeAlO$_3$(s)）600×；(b)（Ce$_2$O$_2$S(s)）600×；(c)（Ce$_3$S$_4$(s)）600×

## 3　凝固过程稀土夹杂物生成的热力学计算

凝固过程中由于杂质元素在固—液两相的分配不同，因而形成化学偏析，杂质偏析的热力学计算式为：

$$C_l = \frac{C_0}{1 - g(1-k)}$$

式中，$C_0$ 未凝固前金属液中所含杂质的浓度；$C_l$ 同一温度下金属液所含溶质的浓度；$g$ 为凝固率，即某温度下析出固体的重量和溶液未开始凝固前重量之比；$k$ 为偏析系数，又称平衡分配比。

$$k = \frac{C_s}{C_l}$$

式中，$C_s$ 为某温度下凝固析出的固溶体所含溶质的浓度。

由已知的数据利用表 5 计算 $C_l^{Ce}$、$C_l^{Mn}$、$C_l^{S}$、$C_l^{N}$、$C_l^{C}$，得知，当金属液凝固 90% 时，钢液中还不可能生成 MnS、CeN、Ce$_x$C$_y$ 等夹杂物。

表 5　元素的偏析系数

| 元素 | C($\delta$-Fe) | C($\gamma$-Fe) | N | S | As | Mn($\delta$-Fe) | Mn($\gamma$-Fe) | O($\delta$-Fe) | O($\gamma$-Fe) | Si | Cu |
|---|---|---|---|---|---|---|---|---|---|---|---|
| $k$ | 0.2 | 0.35 | 0.25 | 0.045 | 0.33 | 0.9 | 0.75 | 0.02 | 0.03 | 0.84 | 0.9 |

由图 3 可以看出，凝固过程在 1743K（1470℃）生成，EDS 分析结果见表 6，且在固相氧化物夹杂物的表面析出了 CeN、Ce$_x$C$_y$ 等夹杂物。

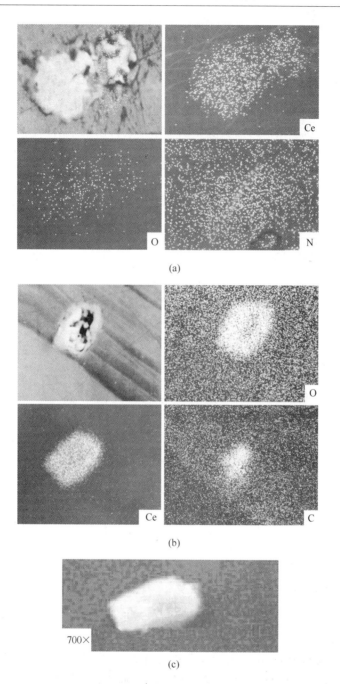

图3 凝固过程析出 MnS，及在铝酸稀土夹杂物表面析出 CeN 和 $Ce_xC_y$ 夹杂的 SEM 形貌

(a) (CeN) 500×; (b) ($Ce_xC_y$) 400×; (c) (MnS)

表6 MnS 的 EDS 分析结果

| 单 位 | S | Mn |
| --- | --- | --- |
| mol% | 50.42 | 49.58 |
| wt% | 37.24 | 62.76 |

## 4 关于稀土砷化物夹杂物的热力学计算及实验验证

### 4.1 稀土砷化物夹杂物的热力学计算

由于在文献中缺少砷在铁液中的活度和溶解自由能的数据,于是根据 Fe-As 相图(见图 4)用熔化自由能法求出铁的活度及其活度系数,再用吉布斯—杜亥姆方程,并引入 $\alpha$ 函数对摩尔浓度进行图解积分求出 $\gamma_{As}$、$\gamma_{As}^0$、$\gamma_{As}^{As}$,而后计算砷在铁液中的溶解自由能。

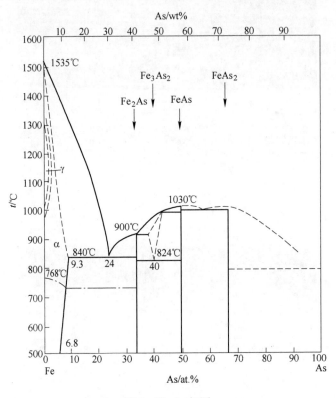

图 4  Fe-As 相图

用熔化自由能法求铁的活度和活度系数,选取纯液态 δ-Fe 为标准态,则纯固态铁的自由能为:

$$\Delta_{fus}G^{\ominus} = G_{(s)} - G_{(1)}^{\ominus} = -RT\ln a'_{(s)}$$

式中,$\Delta_{fus}G^{\ominus}$ 为熔化自由能;$a'_{(s)}$ 为纯固态铁的活度。

已知 δ-Fe 的熔点为 1808K,熔化焓为 13.08kJ/mol,熔化熵可由热容进行计算 $\Delta_{fus}C_p = C_{p(1)} - C_{p(s)} = 1.26 \text{J/(mol·K)}$,于是熔化自由能:

$$\Delta_{fus}G^{\ominus} = 18080 + 2.17T - 1.26T\ln T$$

在液相线上固态铁与溶液处于平衡,即

$$G_{(s)} = \overline{G}_{(1)}$$

式中,$\overline{G}_{(1)}$ 为溶液中溶剂铁的偏摩尔自由能。

$$G_{(1)}^{\ominus} + RT\ln a'_{(s)} = G_{(1)}^{\ominus} + RT\ln a_{(1)}$$

所以，$a'_{(s)} = a_{(l)}$。

由于砷与铁形成 α 固溶体，因此液态铁并非与固体平衡，而是与 α 固溶体平衡。又由于砷在铁液中固溶两相对较低，可近似地认为符合拉乌尔定律，即与固相线上铁的摩尔分数成正比，即有：

$$a_{(s)} = a'_{(s)} \cdot x'_{Fe}$$

式中，$x'_{Fe}$ 为固相线上铁的摩尔浓度。

于是熔化自由能可写为：

$$\Delta_{fus}G^{\ominus} = -2.303RT\lg a'_{(s)} = -2.303RT\lg a_{(s)} + 2.303RT\lg x'_{Fe}$$

而

$$a_{(s)} = \gamma_{Fe} \cdot x_{Fe}$$

式中，$x_{Fe}$ 为液相线上铁的摩尔浓度。

于是

$$\lg\gamma_{Fe} = \lg a_{(s)} - \lg x_{Fe}$$

再将由熔化自由能与活度的关系式计算得到的代入，得到：

$$\lg\gamma_{Fe} = \frac{\Delta_{fus}G^{\ominus}}{RT} + \lg x'_{Fe} - \lg x_{Fe}$$

将已知的数据代入，得到：

$$\lg a_{Fe} = 0.152\lg T - \frac{684}{T} - 0.113 + \lg x'_{Fe}$$

$$\lg\gamma_{Fe} = 0.152\lg T - \frac{684}{T} - 0.113 + \lg x'_{Fe} - \lg x_{Fe}$$

由此两个式子可以计算液相线温度下 Fe-As 二元系溶剂铁的活度和活度系数。

假定 Fe-As 为正规溶液，则满足

$$RT\ln\gamma_{Fe} = bx_{As}^2$$

式中，$b = (\Delta_{vap}H_{Fe}^{0.5} - \Delta_{vap}H_{As}^{0.5})^2$ 与温度无关的常数，因为 $\Delta_{vap}H_{Fe}$ 和 $\Delta_{vap}H_{As}$ 分别为铁和砷的汽化焓不随温度改变。

当溶液成分一定时，$\ln\gamma_{Fe}$ 与 $T$ 成反比，即已知液相线上铁的活度和活度系数，即可求出任何温度下铁的活度和活度系数。

引入 α 函数，利用 Gibbs-Duhem 方程可以计算砷的活度和活度系数。

α 函数：

$$\alpha_{Fe} = \frac{\lg\gamma_{Fe}}{(1-x_{Fe})^2}$$

分部积分得到

$$\lg\gamma_{As} = -\alpha_{Fe}x_{Fe}x_{As} + \int_{x_{Fe}=0}^{x_{Fe}=1}\alpha_{Fe}dx_{Fe}$$

当 $x_{Fe} \to 1$，$x_{As} \to 0$ 时，则 $\gamma_{As} \to \gamma_{As}^0$，则 $\gamma_{As} \to \gamma_{As}^0$，即有

$$\lg\gamma_{As}^0 = \int_0^1\alpha_{Fe}dx_{Fe}$$

由文献查得，在 1573K，$x_{Fe} = 0.76$ 时，$\gamma_{As} = 0.12$。代入砷的活度公式，并换算到冶炼温度 1873K 时砷的活度系数，于是

$$\lg\gamma_{As} = -1.13 - \alpha_{Fe}x_{Fe}x_{As} + \int_{0.76}^{x_{Fe}}\alpha_{Fe}dx_{Fe}$$

作 $\alpha_{Fe}$ 与 $x_{Fe}$ 关系图（见图5），图解积分后，计算 $\gamma_{As}$。

根据 $\ln\gamma_{As}$ 与 $x_{As}$ 的关系图（见图6）外插得到 $\gamma_{As}^0 = 0.0062$。

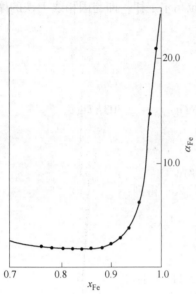 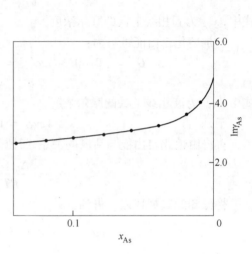

图5　Fe-As 二元系 $\alpha_{Fe}$ 与 $x_{Fe}$ 的关系　　　图6　$\ln\gamma_{As}$ 与 $x_{As}$ 的关系

已知 $\gamma_{As}^0 = 0.0062$，可以计算砷在铁液中的溶解自由能。

$$0.5As_2(g) = [As]$$

$$\Delta_{sol}G_{As}^{\ominus} = RT\ln\frac{\gamma_{As}^0 M_{Fe}}{100 M_{As}}$$

式中，$M_{Fe}$、$M_{As}$ 分别为铁和砷的原子量。

在冶炼温度1873K时，$\Delta_{sol}G_{As}^{\ominus} = -155.4\text{kJ}\cdot\text{mol}^{-1}[As]$。

对正规溶液，其熵变为：

$$\Delta_{sol}S^{\ominus} = -19.142\lg\frac{55.85}{100\times74.92} = 0.0407\text{kJ}\cdot\text{mol}^{-1}\cdot\text{K}^{-1}$$

$$\Delta_{sol}H^{\ominus} = 19.142T\lg\gamma_{As}^0 = -79.15\text{kJ}\cdot\text{mol}^{-1}$$

于是得到溶解自由能与温度的关系式：

$$\Delta_{sol}G_{As}^{\ominus} = -79.15 - 0.0407T, \quad \text{kJ}\cdot\text{mol}^{-1}$$

由图6可以计算砷的自相互作用系数：

$$e_i^j = \left(\frac{\partial\ln f_i}{\partial[\%j]}\right)_{[\%j]}$$

在 $x_{As}\to0$ 处曲线的斜率求出 $\varepsilon_{As}^{As}$，进而得到 $e_{As}^{As} = 0.296$。

在此基础上，便可以计算砷化物夹杂生成的热力学。又由于文献中缺乏 CeAs 的标准生成自由能数据，于是用晶体离子熵公式计算熵变，用绝对熵法计算其标准生成自由能。

根据卡普斯津斯基（А. Ф. Капустинский）提出的公式：

$$S_c^i = \frac{3}{2}R\ln M_i - 1.5\frac{Z_i^2}{r_i}$$

式中，$R$ 为气体常数；$M_i$ 为元素的原子量（$M_{Ce}=140.12$，$M_{As}=74.92$）；$Z_i$ 为 $i$ 离子的价数（$Z_{Ce^{3+}}=3$，$Z_{As^{3-}}=3$）；$r_i$ 为 $i$ 离子半径（$r_{As^{3-}}=0.222nm$，$r_{Ce^{3+}}=0.118nm$）。

计算结果：

$$S_{298,Ce^{3+}}^{\ominus}=0.0138 kJ\cdot mol^{-1}\cdot K^{-1}$$

$$S_{298,As^{3-}}^{\ominus}=0.0284 kJ\cdot mol^{-1}\cdot K^{-1}$$

$$S_{298,As}^{\ominus}=0.0352 kJ\cdot mol^{-1}\cdot K^{-1}$$

$$S_{298,Ce}^{\ominus}=0.064 kJ\cdot mol^{-1}\cdot K^{-1}$$

所以，$\Delta_f S_{298,CeAs}^{\ominus}=S_{298,Ce^{3+}}^{\ominus}+S_{298,As^{3-}}^{\ominus}-S_{298,Ce}^{\ominus}-S_{298,As}^{\ominus}=-0.057 kJ\cdot mol^{-1}\cdot K^{-1}$

文献查得：$\Delta_f H_{298,CeAs}^{\ominus}=-288.3 kJ\cdot mol^{-1}$

于是反应 $As(s)+Ce(s)=CeAs(s)$ 的标准生成自由能为：

$$\Delta_f G_{CeAs}^{\ominus}=\Delta_f H_{CeAs}^{\ominus}-\Delta_f S_{CeAs}^{\ominus}=-288.3+0.057T,\ kJ\cdot mol^{-1}$$

已知：

$$As(s)=0.5As_2(g)\quad \Delta G^{\ominus}=100.42-0.0845T, kJ\cdot mol^{-1}$$

$$Ce(s)=Ce(l)\quad \Delta G^{\ominus}=9.21-0.0086T, kJ\cdot mol^{-1}$$

于是 $0.5As_2(g)+Ce(l)=As(s)+Ce(s)\quad \Delta G^{\ominus}=-109.63+0.0931T, kJ\cdot mol^{-1}$

根据已知的数据可以计算反应 $0.5As_2(g)+Ce(l)=CeAs(s)$ 的标准自由能：

$$\Delta_f G^{\ominus}=-397.93+0.15T,\ kJ\cdot mol^{-1}$$

因为溶解反应 $0.5As_2(g)=[As]$ 的溶解自由能为：$\Delta_{sol}G^{\ominus}=-79.15-0.041T$，$kJ\cdot mol^{-1}$；反应 $Ce(l)=[Ce]$ 的溶解自由能为：$\Delta_{sol}G^{\ominus}=-16.74-0.0464T$，$kJ\cdot mol^{-1}$；所以

$$[As]+[Ce]=CeAs(s)$$

$$\Delta_f G^{\ominus}=-302.04+0.237T,\ kJ\cdot mol^{-1}$$

由此可以算出在冶炼温度 1873K 下 $\Delta G^{\ominus}=142.2 kJ\cdot mol^{-1}$ 不可能生成 CeAs 夹杂物，但当 $T\leq 1273K$ 时，$\Delta G^{\ominus}\leq 0$，故在凝固过程中有可能析出。

若硫化稀土和砷化稀土生成固溶体，且满足理想溶液，则 $\Delta_{mix}H=0$。取 $x_{CeAs}=0.5$，$x_{CeS}=0.5$，于是反应：

$$CeS(s)+CeAs(s)=(CeAs\cdot CeS)_{ss}$$

$$\Delta_{mix}G^{\ominus}=0.5RT\ln x_{CeS}\cdot x_{CeAs}=-5.76T$$

查得：$0.5S_2(g)=[S]\quad \Delta_{sol}G^{\ominus}=-135.0+0.0234T, kJ\cdot mol^{-1}$

$$[Ce]+[S]=CeS(s)\quad \Delta G^{\ominus}=-402.5+0.127T, kJ\cdot mol^{-1}$$

反应 $2[Ce]+[S]+[As]=(CeAs\cdot CeS)_{ss}\quad \Delta G^{\ominus}=-704.54+0.359T,\ kJ\cdot mol^{-1}$

热力学计算表明：在冶炼温度 1873K 下，$\Delta G^{\ominus}=-32.88 kJ\cdot mol^{-1}$，可以生成该复合夹杂物。

## 4.2 实验验证

在光学显微镜定性观测的基础上，进行了 SEM 观测和 EDS 分析，发现了不规则块状的稀土砷化物与硫化物复合夹杂，见图 7。

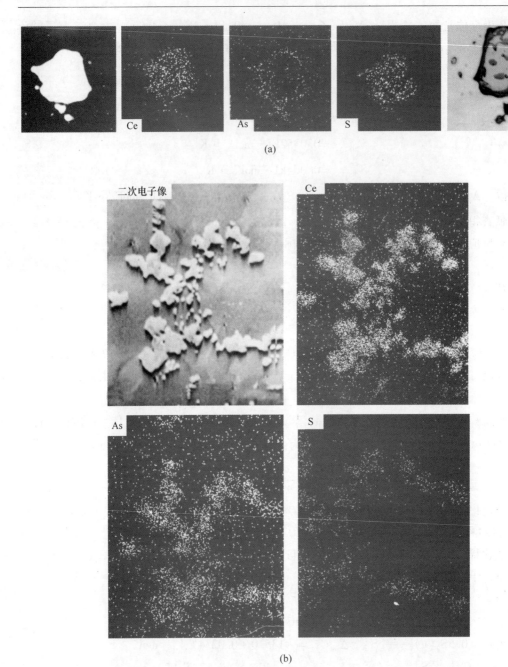

图 7　稀土砷化物与稀土硫化物的复合夹杂物

## 5　氟氧化稀土夹杂物的热力学计算及实验验证

### 5.1　热力学近似计算

1941 年合成了 LaOF，但至今文献中没有其热力学数据。因此，只能用近似的方法进行计算。可以用自键焓法、哈伯—伯恩（Haber-Born）热化学循环与卡普斯金斯基

(Капустинский) 晶格能计算法、哈伯—伯恩热化学循环与菲列曼 (фереман) 点阵能计算法三种近似的方法计算标准生成焓，计算结果在三级误差范围内吻合较好。这里仅以哈伯—伯恩热化学循环与卡普斯金斯基晶格能计算法为例进行详细计算，LaOF 的生成焓的热化学循环见图 8。

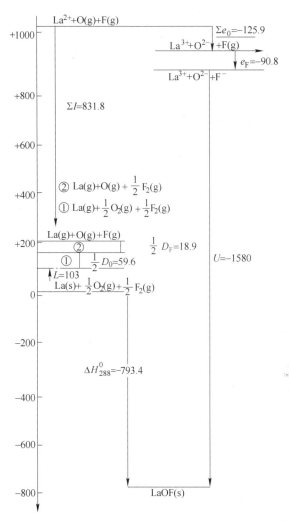

图 8 LaOF 的 Haber-Born 热化学循环

使金属镧变为镧蒸气，需要获得升华焓 $\Delta L_{sub}^{\ominus} = 430.95 \text{kJ} \cdot \text{mol}^{-1}$；使镧蒸气变为镧离子需要电离能，需打掉 3 个电子，总电离能为 $\sum I = I_1 + I_2 + I_3 = 3480.25 \text{kJ} \cdot \text{mol}^{-1}$；使双原子气体 $F_2$、$O_2$ 分别变为单原子气体 F、O 需要解离能，$0.5D_F = 79.08 \text{kJ} \cdot \text{mol}^{-1}$，$0.5D_O = 249.37 \text{kJ} \cdot \text{mol}^{-1}$；使单原子气体 F、O 变为气态离子需要电子亲和能，$e_{F^-} = -413.38 \text{kJ} \cdot \text{mol}^{-1}$，$\sum e_{O^{2-}} = -562.77 \text{kJ} \cdot \text{mol}^{-1}$；使气态离子 $La^{3+}$、$F^-$ 和 $O^{2-}$ 从无穷远集拢到一个晶格中形成 LaOF 离子晶体需要晶格能，表示为：

$$U = 12.02 \frac{Z_a Z_c \sum n}{r_a + r_c} \left(1 - \frac{0.345}{r_a + r_c}\right) = 6610.72 \text{kJ} \cdot \text{mol}^{-1}$$

式中,$r_a$、$r_c$ 分别为阳离子和阴离子半径（$r_{La^{3+}} = 0.115\text{nm}$, $r_{O^{2-}} = 0.14\text{nm}$, $r_{F^-} = 0.136\text{nm}$，算得 $r_{OF^{3-}} = 0.167\text{nm}$）；$Z_a$、$Z_c$ 分别为阳离子和阴离子的电价；$\sum n$ 为离子数总和。

依据 LaOF 的哈伯—伯恩热化学循环求得氟氧化镧的标准生成焓：

$$\Delta_f H^{\ominus}_{298,LaOF} = L_{La} + 0.5 D_O + 0.5 D_F + \sum I_{La^{3+}} + \sum e_{O^{2-}} + e_{F^-} + U_{LaOF} = -3319.59 \text{kJ} \cdot \text{mol}^{-1}$$

根据卡普斯金斯基等人给出的单一气体离子熵的计算公式计算离子熵：

$$S_i^{\ominus} = 1.5 R \ln M_i - 1.5 \frac{Z_i^2}{r_i}$$

式中,$R = 8.314 \text{J} \cdot \text{mol}^{-1} \cdot \text{K}^{-1}$ 为气体常数；$r_i$ 为 $i$ 离子半径（$r_{La^{3+}} = 0.122\text{nm}$）；$Z_i$ 为 $i$ 离子的价数；$M_i$ 为 $i$ 元素的原子量（$M_{La} = 138.9$）。

于是得到：$S^{\ominus}_{298,La^{3+}} = 15.06 \text{J} \cdot \text{mol}^{-1} \cdot \text{K}^{-1}$。

文献查得：$S^{\ominus}_{298,O^{2-}} = 20.5 \text{J} \cdot \text{mol}^{-1} \cdot \text{K}^{-1}$, $S^{\ominus}_{298,F^-} = 24.27 \text{J} \cdot \text{mol}^{-1} \cdot \text{K}^{-1}$, $S^{\ominus}_{298,La} = 56.9 \text{J} \cdot \text{mol}^{-1} \cdot \text{K}^{-1}$, $0.5 S^{\ominus}_{298,O_2} = 102.5 \text{J} \cdot \text{mol}^{-1} \cdot \text{K}^{-1}$, $0.5 S^{\ominus}_{298,F_2} = 101.67 \text{J} \cdot \text{mol}^{-1} \cdot \text{K}^{-1}$。

对反应

$$\text{La(s)} + 0.5 \text{O}_2(\text{g}) + 0.5 \text{F}_2(\text{g}) = \text{LaOF(s)}$$

所以

$$\Delta_f S^{\ominus}_{298,LaOF} = S^{\ominus}_{298,La^{3+}} + S^{\ominus}_{298,O^{2-}} + S^{\ominus}_{298,F^-} - S^{\ominus}_{298,La} - 0.5 S^{\ominus}_{298,O_2} - 0.5 S^{\ominus}_{298,F_2} = -0.201 \text{kJ} \cdot \text{mol}^{-1} \cdot \text{K}^{-1}$$

用绝对熵法计算氟氧化镧的标准生成自由能：

$$\Delta_f G^{\ominus}_{298,LaOF} = \Delta_f H^{\ominus}_{298,LaOF} - T \Delta_f S^{\ominus}_{298,LaOF} = -3319.59 + 0.201 T, \text{ kJ} \cdot \text{mol}^{-1}$$

在温度 1873K 电渣重熔铁铬铝合金时，已知合金中 [La] = 0.030%，[O] = 0.002%，渣中含氟化钙，计算 LaOF 的生成自由能。

反应为

$$[\text{La}] + 1.5[\text{O}] + 0.5 \text{CaF}_2(\text{l}) = \text{LaOF(s)} + 0.5 \text{CaO(s)}$$

已知：

$\text{La(s)} + 0.5 \text{O}_2(\text{g}) + 0.5 \text{F}_2(\text{g}) = \text{LaOF(s)}$  $\Delta_f G^{\ominus} = -1445.57 + 0.201 T, \text{kJ} \cdot \text{mol}^{-1}$

$[\text{La}] = \text{La(s)}$  $\Delta_{sul} G^{\ominus} = 20.50 + 0.067 T, \text{kJ} \cdot \text{mol}^{-1}$

$1.5[\text{O}] = \frac{3}{4} \text{O}_{2(\text{g})}$  $\Delta_{sul} G^{\ominus} = 175.73 + 0.004 T, \text{kJ} \cdot \text{mol}^{-1}$

$0.5 \text{CaF}_2(\text{l}) = 0.5 \text{F}_2(\text{g}) + 0.5 \text{Ca(g)}$  $\Delta_f G^{\ominus} = 73.43 - 0.138 T, \text{kJ} \cdot \text{mol}^{-1}$

$0.5 \text{Ca(g)} + 0.25 \text{O}_2(\text{g}) = 0.5 \text{CaO(s)}$  $\Delta_f G^{\ominus} = -389.45 + 0.092 T, \text{kJ} \cdot \text{mol}^{-1}$

于是由上述 5 个反应之和，即可计算 LaOF 的生成自由能：

$$\Delta_f G^{\ominus}_{LaOF} = -904.29 + 0.226 T, \text{ kJ} \cdot \text{mol}^{-1}$$

根据化学反应等温方程计算实际条件下的反应自由能：

$$\Delta_r G = \Delta_r G^{\ominus} + RT \ln \frac{a_{LaOF} a_{CaO}^{0.5}}{a_{[La]} a_{[O]}^{1.5} a_{CaF_2}^{0.5}}$$

以纯物质作标准态,故 $a_{LaOF} = a_{CaO} = a_{CaF_2} = 1$。需要计算 La 和 O 的活度值:
$$a_{[La]} = f_{La}[La]$$
$$\lg f_{La} = e_{La}^{La}[La] + e_{La}^{O}[O] = -0.01 ; f_{La} = 0.98$$

所以,$a_{[La]} = 2.9 \times 10^{-2}$;同理计算 O 的活度 $a_{[O]} = 6.8 \times 10^{-6}$。

将相关的数据代入等温方程式,于是得到:
$$\Delta_r G_{LaOF} = -904.29 + 0.404T, \text{kJ·mol}^{-1}$$

在冶炼温度下,$\Delta_r G_{1873,LaOF} = -147.6 \text{kJ·mol}^{-1}$,故 LaOF 夹杂物可以生成。

应当指出,用三种方法计算了实际条件下 LaOF 夹杂物生成的自由能均为负值,热力学计算表明其可以生成。

## 5.2 实验验证

SEM、XRD 和离子探针的观测和分析也都发现了 LaOF 夹杂物,SEM 的结果见图 9。

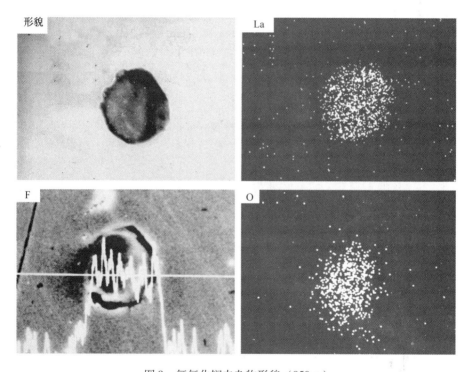

图 9　氟氧化镧夹杂物形貌 (850×)

### 参 考 文 献

[1] 李文超. 钢中夹杂物生成的热力学规律 [J]. 钢铁,1986,21 (3):7.
[2] Vahe A., Kay D. A. R. Met. Trans., 1976, 7B: 375.
[3] 魏寿昆. 冶金过程热力学 [M]. 上海:上海科技出版社,1980.
[4] 李文超,林勤,叶文,等. 含砷低碳钢中稀土夹杂物形成的热力学计算 [J]. 稀有金属(国外版),1983,2 (1):53.
[5] 李文超,林勤,叶文. 钢中 (RE) OF 夹杂物生成的热力学计算 [J]. 稀有金属,1981,(1):13.
[6] А. Ф. Капустинский. Журнал общей химии,1943,T13:497.

[7] A. E. Фереман,Теохимия,1937,T3:1.
[8] 李文超,林勤,叶文,等.稀土元素在铁铬铝合金中的作用[J].中国稀土学报,1983,1(1):47.
[9] 李文超,林勤,叶文,等.35CrNi3MoV钢中稀土夹杂物生成热力学计算[J].中国稀土学报,1984,2(2):57.
[10] 李钒,李文超.冶金与材料热力学[M].北京:冶金工业出版社,2012.

(原载于《钢铁》,1986,21(3):7-12)

# 物理化学在古陶瓷研究中的应用

李文超

（北京科技大学理化系，北京 100083）

**摘 要**：本课题组用物理化学的原理和方法对宋代几类古陶瓷作了一些研究，取得了有益的结果。本文就本课题组对古陶瓷的热力学分析和动力学探讨进行扼要的综述。

**关键词**：古陶瓷；热力学；动力学

## Application of Physico-Chemistry to Study on Ancient Ceramics

Li Wenchao

(University of Science and Technology Beijing, Beijing 100083)

**Abstract**: Our research group has investigated some kinds of ancient ceramics in the Song Dynasty by means of theories and methods in Physico-Chemistry, and obtained some significant results. In the present paper, the analyses by thermodynamics and the approach to kinetics in the study on ancient ceramics are briefly summarized based on our investigation.

**Keywords**: ancient ceramics; thermodynamics; kinetics

我国宋代陶瓷工艺技术突出，无论在数量和质量上都超过了历史上任何时期，出现了驰名国内外的五大名瓷"汝、钧、官、哥、定"，尤以汝为魁。汝瓷釉似天青，钧瓷釉瑰丽如朝霞，官瓷釉厚如堆脂，哥瓷釉以开片为见长，定瓷刻花而取胜，其传世精品至今仍是各国考古学家的研究对象[1,2]。历代仿制较多，但均未达到宋代水平，至今只有汝瓷于1988年通过鉴定。

随着科学技术的发展，科技工作者借助于现代科技手段，深化了对古陶瓷的研究[3,4]。本文在本课题组研究工作的基础上，着重叙述了物理化学原理和方法在研究古陶瓷中的应用。

## 1 热力学分析方法在古陶瓷中的应用

从古陶瓷釉的化学成分可知，无论是钧釉还是汝釉中变价元素（Fe、Ti、Mn、Cu 等）是其成色的主要基础。钧釉中主要是铜离子、汝釉中主要是铁离子成色，而其他变价元素的离子含量较少，应为辅助成色。然而，缺乏理论和实验依据。现以汝瓷天青釉为例，结合物理化学热力学分析和配位场理论，讨论其成色机理，再通过实验验证。

### 1.1 热力学分析

根据天青釉烧成的温度1513K（1240℃）和气氛（4%~7% CO 和 $H_2$，因古代用木材烧成）绘制热力学参数状态图，见图1。由图可以看出，在1373K（1100℃）釉料尚未熔化时，$Fe_2O_3$ 已完全转化为 $Fe_3O_4$，随着温度升高只有部分 $Fe_3O_4$ 还原转化为 $FeO$。因此在釉中以 $Fe^{2+}$ 为主，还有一定量的 $Fe^{3+}$；即由热力学分析可以认为，天青釉中以 $Fe^{2+}$ 成色为主，而 $Fe^{3+}$ 辅助成色。同理可以分析钧釉成色主要是由于 $Cu/Cu^+$ 比值变化，而出现葡萄红、玫瑰紫等色泽。

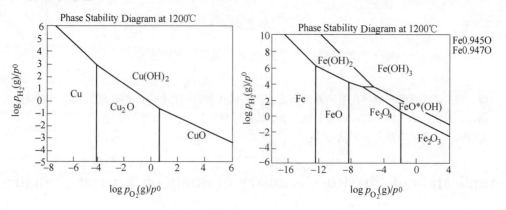

图1　铁（铜）氧化物还原的热力学参数状态图

Fig. 1　Phase stability diagram of Fe(Cu)-O-H

## 1.2　实验验证[5]

将宝丰清凉寺出土的汝瓷残片的釉层分离制成穆斯堡尔谱（Mössbauer）的试样，分析铁的价态与配位状态。以 γ-Fe 为标样，在室温下测谱。按洛仑兹（Lorentz）函数，用最小二乘法高斯—牛顿法与不同矩阵法在计算机上拟合，最后得到穆斯堡尔曲线与参数，见图 2 和表 1。由图可以看出，拟合曲线由两套双峰和一套单峰组成，从而得到穆斯堡尔参数。

依据相关文献可以确定，双峰 1 为 $Fe^{2+}$ 八面体配位的吸收；双峰 2 为 $Fe^{2+}$ 四面体配位的吸收；单峰 3 为 $Fe^{3+}$ 四面体配位的吸收。根据吸收峰的面积可以计算其相对含量：$Fe_{tet}^{2+}$ 为 19.3%；$Fe_{oct}^{2+}$ 为 67.7%；$Fe_{tet}^{3+}$ 为 13.0%。根据配位场理论，天青釉

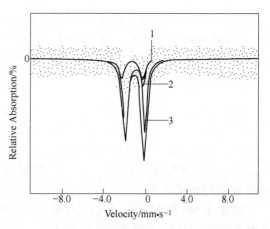

图2　汝瓷天青釉穆斯堡尔谱曲线及计算机拟合

Fig. 2　Mössbauer spectrum of sky blue glaze and computer fitted curves

1，2—双峰；3—单峰

表1　计算机拟合的穆斯堡尔参数

Table 1　Mössbauer parameters of computer fitted spectrum

| Subspec. No | I. S. /mm·s$^{-1}$ | Q. S. /mm·s$^{-1}$ | Area/% | $\Gamma a$/mm·s$^{-1}$ |
|---|---|---|---|---|
| 1 | 1.304 | 2.006 | 67.7 | 0.258 |
| 2 | 0.967 | 1.879 | 19.3 | 0.258 |
| 3 | 0.281 | — | 13.0 | 0.129 |

中四面体和八面体配位的铁离子,其外层电子 $d$ 轨道发生不同方式的分裂。当低能级轨道电子向高能级轨道跃迁时,伴有光的吸收;吸收光的能量等于轨道能级差,即配位场强度。因此,可以计算吸收光的波长。$Fe_{oct}^{2+}$ 应在近红外区 1051nm 处产生吸收;由于天青釉为石灰碱釉,使氧离子的极化率增加,吸收峰向短波方向移动,延伸到可见光区,从而釉呈蓝色;$Fe_{tet}^{2+}$ 在远红外区 4000nm 引起吸收,对釉的成色影响不大;$Fe_{tet}^{3+}$ 在 280nm 处产生吸收,由于其含量较少,对釉的着色影响不大,为辅助成色。后又用吸收光谱进行了检验,实验结果与由穆斯堡尔谱计算的结果相吻合。

## 2 动力学研究方法及应用

气泡在古陶瓷中起着重要的作用,由于其对光的散射,产生晶莹四射的效果。釉中气泡的大小,可用动力学理论进行分析和计算[6]。

由图 3 可以看出,胎釉界面存在着气泡层。这是由于在 1473K(1200℃)烧成温度下,釉中有尚未熔化的残余石英,胎釉界面存在着大量的活性孔隙,这为碳酸盐分解产生的 $CO_2$ 异相生核、长大提供了有利条件。其长大机理可分为三步:(1)碳酸盐分解的化学反应;(2)气体产物($CO_2$)向微气孔扩散;(3)扩散到边界层的气体进入 $CO_2$ 气泡中。显然第(2)步是控速环节,于是可以得到气泡长大的动力学公式(微分表达式和积分表达式):

$$r = \left\{ \frac{7RT\Delta P_{CO_2} D_{CO_2} \varepsilon \eta \xi}{g\Delta\rho} \left[ \ln\left(\delta + \frac{p_g}{\rho g}\right) - \ln\left(\frac{p_g}{\rho g}\right) \right] \right\}^{\frac{1}{3}}$$

式中,$T$ 为烧成温度(1523K);$\eta$ 为釉的黏度(200Pa·s);$\xi$ 为复杂扩散路径系数;$\delta$ 为釉层厚度(2mm);$\Delta\rho$ 为釉与 $CO_2$ 的密度差;$g$ 为重

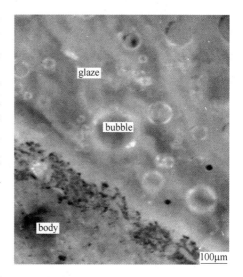

图 3  汝瓷釉中气泡
Fig. 3  The bubble layer in Ru glaze

力加速度;$D_{CO_2}$ 为 $CO_2$ 的扩散系数($1 \times 10^{-7} cm^2/s$);$p_g$ 为炉气压力(0.1MPa);$\varepsilon$ 为胎的孔隙度。如果把已知的数据代入上式,可得到釉中气泡的半径为 0.045mm,即当釉中的气泡大于该值时,将穿过釉层进入炉气。经 SEM 观测,在古青瓷釉中未发现半径大于 0.05mm 的气泡。由此可见,动力学计算与实验观测的结果基本吻合。

## 3 古陶瓷中过渡层形成机理分析[7]

在古陶瓷胎与釉之间存在着过渡层,其化学成分介于胎釉之间,层内有析晶和大小不一的气泡。这是由于在烧成过程中胎釉界面上,液固相互作用,发生溶解、扩散和化学反应的结果。在汝、钧和耀州瓷中均观察到过渡层,见图4。

从结构上看,铝、硅离子以[$AlO_4$]、[$SiO_4$]配位体进入网络,形成釉的骨架。由化学成分计算表明,三种古陶瓷的过渡层的成分代表点均落在 $SiO_2$-$CAS_2$(钙长石)-$KAS_4$(钾长石)三元系中,在釉的软化温度和烧成温度下做其双等温截面图,见图5。在烧成

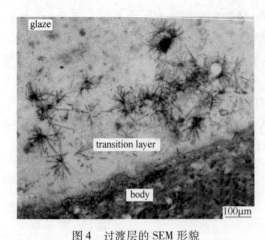

图 4 过渡层的 SEM 形貌

Fig. 4 Microstructure of glaze-body interlayer

温度 1513K 得到 $SiO_2$-$CAS_2$-$KAS_4$-$L_3$-$L_2$-$L_1$-$L_4$ 等温截面图，在釉的凝固点 1373K 得到局部等温截面图 $L_5$-$L_6$-$L_7$-$L_8$（阴影区）。根据罗策布规则，在冷却过程中应析出 $CAS_2$、$SiO_2$，但由于熔体黏度较大，$SiO_2$ 不可能析出。组成沿 $SiO_2$-$CAS_2$ 共熔线变化，从 $L_2$ 到 $L_6$，直到熔体全部凝固。从此间结构参数的变化可以看出，过渡层中的桥氧数远小于釉中的桥氧数，即过渡层的黏度小于釉层的黏度，故气泡向过渡层迁移。气泡的尺寸为 2~50μm，$CAS_2$ 为 0.1~2μm，与可见光的波长相当，且钙长石的折射率大于釉的折射率，对可见光有放射作用，再加上釉的镜面反射分数较高，故使古陶瓷具有"乳光晶莹、釉汁浑厚"的艺术效果。

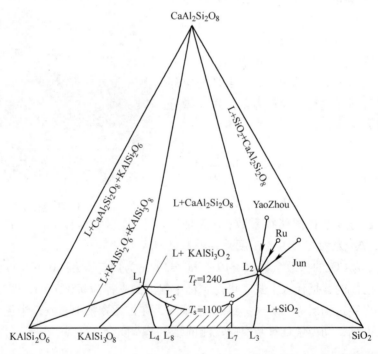

图 5 在 1513K（1240℃）和 1373K（1100℃）下 $SiO_2$-$CAS_2$-$KAS_4$ 三元系的双等温截面图

Fig. 5 Double-isothermal cut in the $SiO_2$-$CAS_2$-$KAS_4$ diagram at 1513K（1240℃）and 1373K（1100℃）

## 4 统计模式识别在古陶瓷研究中的应用

随着计算机信息科学的发展，逐步建立了统计模式识别[8]、神经网络、遗传算法等应用程序，并在材料科学中得到应用。本研究曾用统计模式识别优化了青花瓷的工艺，对古白瓷釉（定、景德镇、邢、德化、巩县等窑）的成分分布作了分析（见图 6），并利用逆

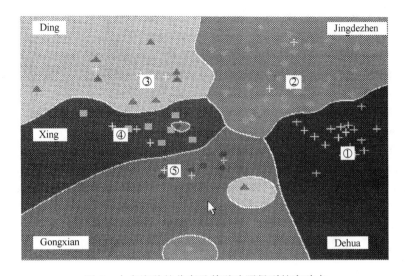

图 6 古白瓷釉的分布及其逆映照得到的实验点

Fig. 6 NLM analysis of ancient white porcelain glaze in various dynasties

① to ⑤ are simulating experiment points

映照选择了五个点在实验室进行了仿制,其釉色、白度均与古白瓷的相近,效果较好[9]。

此外,金格瑞(Kingery)根据釉的相图认为钧釉处于分相区。于是本课题组还用分形数学对釉的分相进行了描述,见图7。可以看出模拟的结果(右上角图)与 TEM 观察的结果相吻合。分形维数为1.85,属分相区内的亚稳态分解。

## 5 小结

物理化学的原理和方法可以深化古陶瓷烧成机理的研究,为仿制古陶瓷的工艺提供理论依据。

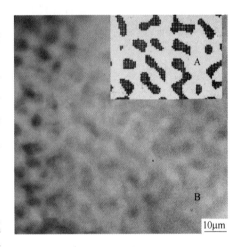

图 7 钧釉分相及其模拟

Fig. 7 The phase separation of Jun glaze and results of simulation

## 参 考 文 献

[1] Ye Zhemin. The elementary introduction to China ancient ceramics science [M]. Beijing: Light industry publishing house, 1982.
叶哲民. 中国古陶瓷科学浅说 [M]. 北京:中国轻工业出版社,1982.

[2] Qin Jianwu. Study on the Ru ware in the Song Dynasty by Physico-Chemistry [D]. Beijing: University of Science and Technology, 1989.
秦建武. 中国宋代汝瓷的物理化学研究 [D]. 北京:北京科技大学,1989.

[3] W. David Kingery. Technical insights into Chinese Material Culture [C]. Science and Technology of Ancient

Ceramics 3$^{rd}$ of Proceedings of The International Symposium. Shanghai, 1995: 261-266.

[4] Li Shenghua, Miao Jianmin. The effect of light on a chronology of ancient ceramics with thermoluminescence dating technique [J]. Sci. China-D, 2001, 31 (5): 1-6.
李盛华, 苗建民. 光照作用对古陶瓷热释光年代的影响 [J]. 中国科学 (D辑), 2001, 31 (5): 1-6.

[5] Li Wenchao, Wang Jian, Qin Jianwu. Investigation of Colouration Mechanism of Ru Ware's Sky Blue Glaze by Mössbauer Spectrum [J]. China's Ceramics, 1991, (3): 59-62.
李文超, 王俭, 秦建武. 用穆斯堡尔谱和吸收光谱研究汝瓷天青釉呈色机理 [J]. 中国陶瓷, 1991, (3): 59-62.

[6] Li Guozhen, Wang Jian, Li Wenchao. Discussion on the Technology on YaoZhou Ware Glaze of Successive Dynasties, Scientific and Technological achievement in Chinese Ancient Ceramics [M]. Shanghai: Shanghai Science and Technology Publisher, 1992.
李国桢, 王俭, 李文超. 历代耀州青瓷工艺探讨, 古陶瓷科学技术 [M]. 上海: 上海科技文献出版社, 1992.

[7] Zhuang Youqing, Wang Jian, Li Wenchao. Emulsification of Sky Blue Glaze of Ru Ware [C]. Proceeding of The Third China-Japan Bilateral Conference on Molten Salt Chemistry and Technology. Beijing, 1990: 202-205.

[8] Li Lei, Li Wenchao. Application of Pattern Recognition in Ancient White Porcelain [C]. Science and Technology of Ancient Ceramics 3$^{rd}$ of Proceedings of The International Symposium. Shanghai, 1995: 433-438.

(原载于《金属学报》, 2002, 38 (6): 613-616)

# Prediction of Thermodynamic Properties for Multicomponent System with Chou Model

Zhen Qiang（甄强）　　Bao Hong（包宏）
Wang Fuming（王福明）　　Li Wenchao（李文超）

(Department of Physical Chemistry, University of Science and
Technology Beijing, Beijing 100083)

**Abstract:** A new geometrical model, Chou model, was developed as a computer program for calculating thermodynamic properties of multicomponent system. Calculated results show that the new model is more reasonable and convenient than other symmetrical (or asymmetrical) models. With this model, the excess Gibbs free energies of Ga-Al-As-In compound semiconductor system were calculated.

**Keywords:** thermodynamic properties; new geometrical model; Ga-Al-As-In alloy

## 1　Introduction

The group Ⅲ-Ⅴ compound semiconductors and their solid solutions are important materials for photoelectronic and high speed electronic applications. The fabrication of these components involves many complex processes. For example, the solid and liquid phases coexist at near equilibrium condition including crystal growth and liquid phase epitaxy, the solid and vapour phases coexist including chemical vapour deposition. Therefore, the phase diagram and thermodynamic properties for these systems are important for providing boundary condition in analysis of processes.

Though databases were previously published and additional experimental measurements were reported for many systems to modify the conclusions of the previous databases, it is still not enough. Because different models and reference states are used, it is difficult to supply systematically thermodynamic data for multicomponents.

Recently, Chou[1] established a new geometric model for calculating the thermodynamic properties of liquid multicomponent systems. It overcomes the defects of previous traditional geometric models[2,3]. And it is expected to apply in some liquid multicomponent systems, in which it is difficult to select symmetrical (or asymmetrical) geometric model and asymmetrical component. In this article, the new geometric model was applied in prediction for the thermodynamic properties of Ga-Al-As-In compound semiconductor system. And the results were compared with that of previous typical geometrical models.

---

Foundation item: Project (5967428) supported by the National Natural Science Foundation of China.

甄强，1996～2000年于北京科技大学师从李文超教授攻读博士学位。目前在上海大学工作，现任科技发展研究院书记兼主持工作院长职务。获得"赣鄱英才"省级特聘教授荣誉称号，发表论文60余篇，荣获省部级科技进步奖4项。

## 2  Chou Model

Firstly, a quantity $\eta_{(ij,ik)}$ is defined as deviation sum of squares:

$$\eta_{(ij,ik)} = \int_0^1 (\Delta G_{ij}^E - \Delta G_{ik}^E)^2 dX_{i(ij)} \tag{1}$$

where $\Delta G_{ij}^E$ and $\Delta G_{ik}^E$ represent the excess Gibbs free energies of binary systems $i-j$ and $i-k$, respectively; $X_{i(ij)}$ indicates the mole fraction of component $i$ in $i$-$j$.

Then another quantity $\xi_{i(ij)}^k$, similarity coefficient, is introduced, which is defined as

$$\xi_{i(ij)}^k = \frac{\eta_{(ij,ik)}}{\eta_{(ij,ik)} + \eta_{(ji,jk)}} \tag{2}$$

On basis of the above definition, the following binary compositions are selected for this new model:

$$X_{i(ij)} = x_i + \sum_{\substack{k=1 \\ k \neq i,j}}^{n} x_k \xi_{i(ij)}^k \tag{3}$$

where $x_i$ and $x_k$ denote the mole fractions of components $i$ and $k$ in a multicomponent system, respectively.

This new model differs from other geometrical models in its special selection of binary compositions that are of close relation with the multicomponent system considered. For instance, the selection of binary composition in $i$-$j$ system depends on the characteristics of systems $k$-$i$ and $j$-$k$ in a $i$-$j$-$k$ ternary system. The coefficient $\xi_{i(ij)}^k$, which expresses the symmetrical characteristics of components quantitatively, represents similarity of $k$ with $i$ and $j$. When two components are exactly same in a ternary system, the ternary system will become a binary one with this model, it overcomes the defect of symmetrical geometric models.

In multicomponent system, $\Delta G_{ij}^E$ is expressed as the following Ridlich-Kister[4] multinominal:

$$\Delta G_{ij}^E = X_{i(ij)} X_{j(ij)} \sum_{o}^{n} L_{ij}^v [X_{i(ij)} - X_{j(ij)}]^v \tag{4}$$

where $X_{j(ij)}$ denotes mole fraction of component $j$ in the binary $i$-$j$. $L_{ij}^v$ represents the interaction parameter for binary $i$-$j$, which is independent of compositions but dependent on temperature.

Combining Eqns. (1) ~ (4), the excess Gibbs free energy $\Delta G^E$ can be obtained for a multicomponent system from relative binaries:

$$\Delta G^E = \sum_{i=1}^{n-1} \sum_{j=i+1}^{n} \frac{x_i x_j}{X_{i(ij)} X_{j(ij)}} \Delta G_{ij}^E \tag{5}$$

## 3  Results and Discussion

The new model (Chou model), other symmetrical and asymmetrical models were developed as computer application program, the mixing molar Gibbs free energies for the asymmetrical system Ag-Sn-Zn (Ag as the asymmetrical component) at 900K are calculated, as shown in Fig. 1. For comparison, the results of some typical symmetrical models (Kohler, Colinet, Muggianu)[5-7] or asymmetrical model (Toop model)[8] as well as the experimental data are also included. From the results, it can be seen that the calculated data from new model is much more similar to the experimental data[9] than those predicted by other models.

In multicomponent Ga-Al-As-In system, the molar excess Gibbs free energies of all sub-binaries in this system at 2100K are shown in Fig. 2. The similarity coefficients for all sub-binaries in this system at 2100K are shown in Table 1. And the molar excess Gibbs free energies of liquid Ga-Al-As-In four component system were calculated, as shown in Fig. 3.

By comparing the molar excess Gibbs free energies of its sub-binaries at 2100K, it can be seen that Al-As-Ga-In system is an asymmetrical one. So the asymmetrical model should be used rather than the symmetrical one. But it is not clear which component is the asymmetrical one. From Fig. 3, when As or In acts as the a-

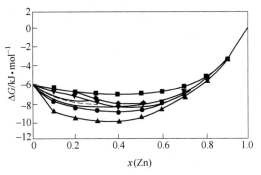

Fig. 1 Comparison between measured and calculated molar Gibbs free energies in ternary Ag-Sn-Zn (Ag/Sn = 1)
—New model; ▼—Kohler model;
+—Colinet model; ▲—Muggianu model;
■—Toop model (Sn as asymmetrical component);
∗—Toop model (Zn as asymmetrical component);
- - -Toop model (Ag as asymmetrical component);
●—Experimental data

symmetrical component in this case meet the new model very well, and when Ga or Al acts as the a-symmetrical component will lead to deviation due to the wrong selection for asymmetrical component.

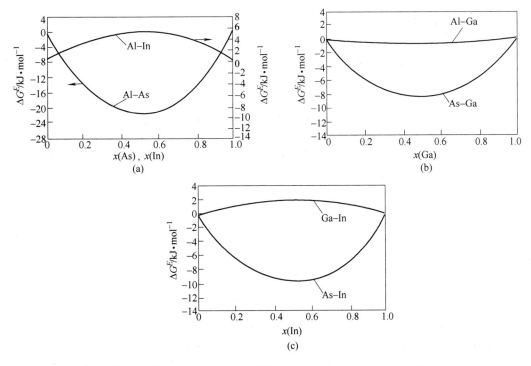

Fig. 2 Comparison of excess Gibbs free energies of sub-binaries in Al-As-Ga-In system at 2100K
(a) Al-As, Al-In; (b) Al-Ga, As-Ga; (c) As-In, Ga-In

Table 1 Similarity coefficient in Al-As-Ga-In system at 2100K

| Similarity coefficient | Al-As-Ga-In (1-2-3-4) | As-Ga-In-Al (1-2-3-4) | Ga-In-Al-As (1-2-3-4) | In-Al-As-Ga (1-2-3-4) |
|---|---|---|---|---|
| $\xi_{12\text{-}3}$ | 0.71547 | 0.01532 | 0.38981 | 0.23507 |
| $\xi_{12\text{-}4}$ | 0.83511 | 0.74320 | 0.44416 | 0.23613 |
| $\xi_{13\text{-}2}$ | 0.87919 | 0.01228 | 0.16491 | 0.60734 |
| $\xi_{13\text{-}4}$ | 0.83893 | 0.39226 | 0.12081 | 0.98772 |
| $\xi_{14\text{-}2}$ | 0.76403 | 0.28453 | 0.98468 | 0.61019 |
| $\xi_{14\text{-}3}$ | 0.75982 | 0.16575 | 0.25680 | 0.55584 |
| $\xi_{23\text{-}1}$ | 0.74320 | 0.44416 | 0.23613 | 0.83425 |
| $\xi_{23\text{-}4}$ | 0.01532 | 0.38981 | 0.23507 | 0.71547 |
| $\xi_{24\text{-}1}$ | 0.38998 | 0.12081 | 0.98772 | 0.83509 |
| $\xi_{24\text{-}3}$ | 0.01228 | 0.16491 | 0.60734 | 0.87919 |
| $\xi_{34\text{-}1}$ | 0.37788 | 0.23507 | 0.71547 | 0.01532 |
| $\xi_{34\text{-}2}$ | 0.44416 | 0.23613 | 0.83425 | 0.74320 |

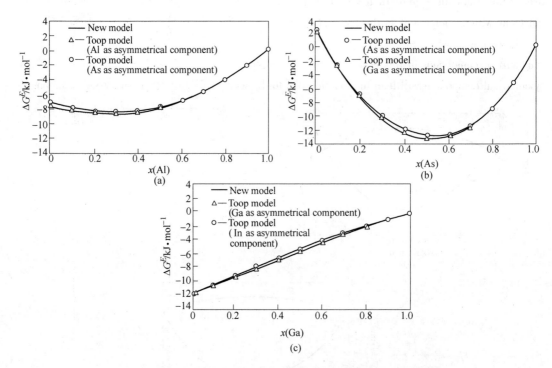

Fig. 3 Excess Gibbs free energies of Ga-Al-As-In system at 2100K
(a) $x(As) = x(Ga) = x(In)$; (b) $x(Ga) = x(In) = x(Al)$; (c) $x(In) = x(Al) = x(As)$

Obviously, it is very important to select an appropriate model or asymmetrical component in calculation. For those systems whose symmetry is not obvious, neither symmetrical model or asymmetrical model can meet the condition. The new model properly indicates the relationship of the components in systems by using similarity coefficient and can be used in most of practical systems than symmetric or asymmetric models do.

## References

[1] Chou K C. A general solution model for predicting ternary thermodynamic properties [J]. CAL PHAD, 1987, 11 (2): 293-297.

[2] Chou K C, Wei S K. New generation solution model for predicting thermodynamic properties of multicomponent system from binaries [J]. Metallurgical and Materials Transactions, 1997, 28B (3): 439-445.

[3] Chou KC, Li W C, Li F S, et al. Formalism of new ternary model expressed in terms of binary regular-solution type parameters [J]. CAL PHAD, 1996, 20 (3): 395-399.

[4] Redlich O, Kister A J. Thermodynamics of nonelectrolyte solutions: algebraic representation of thermodynamic properties and the classification of solution [J]. Ind Eng Chem, 1948, 40: 345-348.

[5] Hultgren R, Desal P D, Hawkins D T, et al. Selected values of the thermodynamic properties of binary alloys [M]. Ohio: American Society for Metals, Metals Park, 1973.

[6] Moser Z, Dutkiewicz J, Gasior W, et al. The Sn-Zn system [J]. Bulletin of alloy phase diagrams, 1985, 6 (4): 330-334.

[7] Kamada K. Activities of liquid gold-zinc and silver-zinc binary alloys by EMF measurememts using zirconia solid electrolyte cells [J]. Transaction of the Japan Institute of Metals, 1987, 28: 41-44.

[8] Karlhuber S, Komarek K L, Mikula A. Thermodynamic properties of liquid Ag-Sn-Zn alloy [J]. Z Metallic, 1994, 85 (5): 307-401.

[9] Ansara I, Chatillon C, Lukas H L, et al. A binary database for III-V compound semiconductor system [J]. CALPHAD, 1994, 18 (2): 177-180.

(原载于 *Transactions of Nonferrous Metals Society of China*, 2000, 10 (5): 642-644)

# 由稳定化合物熔化焓提取二元系活度

李兴康　刘四俊　王　俭　李文超

（北京科技大学理化系，北京　100083）

**摘　要**：本文从二元体系中稳定化合物与液相平衡关系入手，导出了由稳定化合物的熔化焓计算二元系的新公式，通过对 In-Sb 二元系的活度计算验证了本方法的可行性，并用这一新方法预报了半导体 Ga-As 二元系的活度。

**关键词**：熔化焓；二元系；活度；相图

## Calculation the Activities from Binary Systems Containing Congruently Melting Compounds Using Their Melting Enthalpy

Li Xingkang　Liu Sijun　Wang Jian　Li Wenchao

(University of Science and Technology Beijing, Beijing 100083)

**Abstract**: A new method of calculating activities from binary systems has been developed. The new formulae require that the phase diagrams must contain congruently melting compound and its melting enthalpy is known. Their feasibility is tested by applying the new formulae to In-Sb system. By this method, the activities of an important semi-conductor system Ga-As have also been predicted.

**Keywords**: melting enthalpy; binary phase diagram; activity; phase diagram

由相图求活度是获取活度数据的一条重要途径。1964 年，有人改进了前人的工作，提出了一种由化合物的标准生成自由能求活度的方法，但它只能适用于 1∶1 型的化合物。1965 年周国治[1]将这种方法推广到任意 $\eta:\xi$ 型（$A_\eta B_\xi$）化合物中，从而扩大了该方法的应用范围。1987 年周国治、王建军[2]又提出了由化合物的标准生成熵求活度的方法。鉴于化合物熔化焓的数据较之标准生成自由能、标准生成熵更易于获得，本文推导了由稳定化合物的熔化焓提取二元系活度的新方法，并预报了 Ga-As 二元系的活度。

## 1　公式推导

设在 A-B 二元系中有一种稳定的化合物（$A_\eta B_\xi$）（见图 1），可将相图分为 2 个简单

---

本课题由国家自然科学基金资助。

李兴康，1987~1991 年于北京科技大学师从李文超教授攻读博士学位。博士论文题目为《由二元相图提取活度的新方法及化合物稳定性规律的研究》。

共晶二元系。现以化合物左侧的共晶系为例来推导由化合物的熔化焓（$\Delta H_M$），进而求体系活度的计算公式。

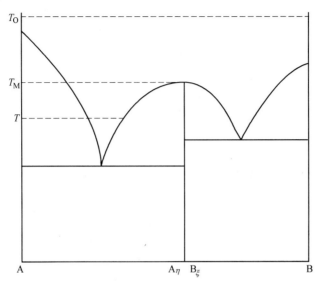

图 1　包含一种稳定化合物的二元系

Fig. 1　A binary system including one congruently melting compound

假定在 $A\text{-}A_\eta B_\xi$ 体系中，组元 $A_\eta B_\xi$ 的成分与活度分别用 $x_{A_\eta B_\xi}$ 及 $a'_{A_\eta B_\xi}$（沿液相线）来表示，则由冰点下降公式可得到：

$$\mathrm{d}\ln a'_{A_\eta B_\xi} = \frac{\Delta H_M}{RT^2}\mathrm{d}T \tag{1}$$

两边积分有：

$$\ln a'_{A_\eta B_\xi} = \int_{T_M}^{T} \frac{\Delta H_M}{RT^2}\mathrm{d}T \tag{2}$$

又由在温度 $T$ 下液相与化合物 $A_\eta B_\xi$ 平衡关系可得到下列两式：

在 $A\text{-}A_\eta B_\xi$ 系中：

$$RT\ln a'_{A_\eta B_\xi} = G^{(s)}_{A_\eta B_\xi} - G^{(l)}_{A_\eta B_\xi} \tag{3}$$

在 A-B 系中：

$$RT(\eta \ln a'_A + \xi \ln a'_B) = G^{(s)}_{A_\eta B_\xi} - \eta G^{0(l)}_A - \xi G^{0(l)}_B \tag{4}$$

其中 $a_A$、$a_B$ 为 A-B 二元系中沿液相线的活度。结合式（3）、式（4）并考虑到了：

$$G^{(l)}_{A_\eta B_\xi} = \eta G^{0(l)}_A - \xi G^{0(l)}_B + RT(\eta \ln a'_A + \xi \ln a'_B)$$

得到：

$$\ln a'_{A_\eta B_\xi} = \ln a_A^{\eta_1} a_B^{\xi_1} - \ln(a_A^{\eta_1} a_B^{\xi_1})(x_B) \tag{5}$$

将式（5）代入式（2）得到：

$$\ln a_A^{\eta_1} a_B^{\xi_1} - \ln(a_A^{\eta_1} a_B^{\xi_1})(x_B) = \int_{T_M}^{T} \frac{\Delta H_M}{RT^2}\mathrm{d}T \tag{6}$$

上式可化为：

$$\ln\gamma_A^\eta \gamma_B^\xi - \ln(\gamma_A^\eta \gamma_B^\xi) x_B$$

$$= \int_{T_M}^{T} \frac{\Delta H_M}{RT^2} dT - \left( \eta \ln \frac{\eta+\xi}{\eta} x_A + \xi \ln \frac{\eta+\xi}{\eta} x_B \right) \tag{7}$$

两边同乘以 $T$，并假定活度系数与温度的关系符合正规溶液的规律，同时对两边微分可得到：

$$T_0 d\ln\gamma_A^\eta \gamma_B^\xi = \frac{\Delta H_M}{RT} dT + \left( \int_{T_M}^{T} \frac{\Delta H_M}{RT^2} dT \right) dT - \left( \eta \ln \frac{\eta+\xi}{\eta} x_A + \xi \ln \frac{\eta+\xi}{\eta} x_B \right) -$$

$$T\left( \frac{\eta x_B - \xi x_A}{x_A x_B} \right) dx_A \tag{8}$$

式中，$\gamma_A$、$\gamma_B$ 为温度 $T$ 下组元 A、B 的活度系数。

再由变通的 Gibbs-Duhem 方程[2]得到：

$$d\ln\gamma_A = \frac{1}{(\eta x_B - \xi x_A) T_0} \left[ \frac{x_B \Delta H_M}{RT} + x_B \int_{T_M}^{T} \frac{\Delta H_M}{RT^2} dT - \xi \sum x_i \ln x_i - \right.$$

$$\left. x_B \left( \eta \ln \frac{\eta+\xi}{\eta} + \xi \ln \frac{\eta+\xi}{\xi} \right) \right] dT - d\left( \frac{T}{T_0} \ln x_A \right) \tag{9}$$

$$d\ln\gamma_B = \frac{1}{(\xi x_A - \eta x_B) T_0} \left[ \frac{x_A \Delta H_M}{RT} + x_A \int_{T_M}^{T} \frac{\Delta H_M}{RT^2} dT - \eta \sum x_i \ln x_i - \right.$$

$$\left. x_A \left( \eta \ln \frac{\eta+\xi}{\eta} + \xi \ln \frac{\eta+\xi}{\xi} \right) \right] dT - d\left( \frac{T}{T_0} \ln x_B \right) \tag{10}$$

其中，$\sum x_i \ln x_i = x_A \ln x_A + x_B \ln x_B$。

积分式（9）、式（10）就能求得组元 A、B 的活度系数。

## 2 可行性的验证

为了对本文提出的计算方法的可行性进行验证，利用相图对 In-Sb[4] 体系的活度进行了计算。根据文献 [6]、[7] 中的表所列数据可以得到：

$$\Delta H_{M(InSb)} = 17.6T - 7.5 \times 10^{-3} T^2 + 40141, J/mol$$

$$\Delta H_{M(Sb)} = 0.90779T + 7.6923 \times 10^{-3} T^2 - 2.000 \times 10^5 T^{-1} -$$

$$5.979 \times 10^{-6} T^2 + 1741, J/mol$$

由上面的熔化焓的表达式和本文导出的式（9）和式（10），经计算得到表1所示的结果。为便于比较，将实验值一并列入。

表1 在973K，实验测定与计算活度的比较
Table 1 Comparison of the calculated with the experimental activities at 973K

| $x_{In}$ | $a_{In}$ | | | $a_{Sb}$ | | |
|---|---|---|---|---|---|---|
| | 计算值 | 实验值 | 偏差 | 计算值 | 实验值 | 偏差 |
| 0.00 | 0.000 | 0.00 | 0.000 | 1.000 | 1.00 | 0.000 |
| 0.10 | 0.031 | 0.03 | 0.001 | 0.900 | 0.90 | 0.000 |
| 0.20 | 0.065 | 0.07 | 0.005 | 0.792 | 0.80 | 0.008 |

续表1

| $x_{In}$ | $a_{In}$ | | | $a_{Sb}$ | | |
|---|---|---|---|---|---|---|
| | 计算值 | 实验值 | 偏差 | 计算值 | 实验值 | 偏差 |
| 0.30 | 0.111 | 0.11 | 0.001 | 0.662 | 0.69 | 0.028 |
| 0.40 | 0.172 | 0.17 | 0.002 | 0.587 | 0.53 | 0.057 |
| 0.50 | 0.271 | 0.28 | 0.009 | 0.361 | 0.35 | 0.011 |
| 0.60 | 0.391 | 0.42 | 0.029 | 0.232 | 0.21 | 0.022 |
| 0.70 | 0.555 | 0.57 | 0.015 | 0.122 | 0.12 | 0.002 |
| 0.80 | 0.725 | 0.73 | 0.005 | 0.055 | 0.06 | 0.005 |
| 0.90 | 0.874 | 0.88 | 0.006 | 0.018 | 0.02 | 0.002 |
| 1.0 | 1.000 | 1.00 | 0.000 | 0.000 | 0.00 | 0.000 |

由表1可以看出，计算值与实验值相符得很好，证明本方法是可行的。计算误差主要取决于现有相图、热力学数据的准确性及计算时所引入的积分误差。

## 3 预报 Ga-As 体系的活度

Ga-As 体系的活度数据是半导体研究中的重要参数。由于 As 元素的某些特性，其活度数据很难通过实验获得。又由于 As 元素在通常条件下得不到液体，因而其标准生成熵、标准生成焓、标准自由能等热力学参数也很难获得。这就会给文献中已有的活度计算方法带来不可克服的困难，而用本文提出的方法计算组元的活度不受此限制。

Ga-As 体系有一个稳定化合物，它的相图示于图 2[5]。有关的热力学数据可以从手册[7]中得到，

$$\Delta H_{M(GaAs)} = 13.8T - 3.033 \times 10^{-3}T^2 + 73198, \text{J/mol}$$

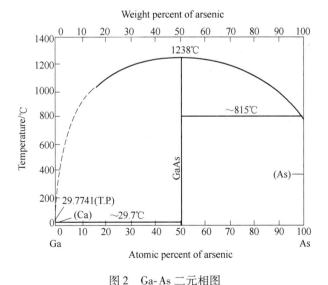

图 2　Ga-As 二元相图
Fig. 2　The phase diagram of Ga-As

所得到的计算结果示于图 3。

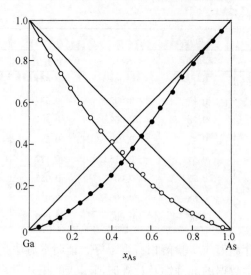

图 3　1573K 时 Ga-As 二元系的活度

Fig. 3　Activities of Ga-As system at 1573K

## 4　结论

（1）本文导出的由稳定化合物的熔化焓计算二元系活度的公式是可行的。

（2）由于熔化焓数据较易获得，用此计算体系的活度适用性强，使用范围较大。

### 参 考 文 献

[1] 周国治. 金属学报，1965，8：545.

[2] Kuo-Chih Chou, Jian-Jun Wang. Metallurgical Transaction, 1987, 18 (A): 323.

[3] Hanson M. Constitution of Binary Alloys (2nd Ed) [M]. New York: Mcgyaw-Hill Book Company, 1958.

[4] Massalsski T B, et al. Binary AlloyPhase Diagram [M]. Ohio: Metals Park, 1986, 2: 1395.

[5] Barin I, Knacke O. Thermodynamical Properties of Inorganic Substances [M]. Berlin: Springer-Verlag, 1973.

[6] Barin I, Knacke O, Kubaschewski O. Thermodynamical Properties of Inorganic Substances [M]. Berlin: Springer-Verlag, 1977.

[7] Hultgren R. Selected Values of the Thermodynamic Properties of Binary Alloy [M]. Ohio: American Society for Metals, Metals Park, 1973.

（原载于《北京科技大学学报》，1992，14（6）：607-611）

# Statistical Mechanics Model of Liquid Binary Alloy and Its Parameters

Fan Chengcai[1,2]　Wang Jian[1]　Li Wenchao[1]　Qian Jiaxu[2]

(1. University of Science and Technology Beijing, Beijing 100083;
2. General Research Institute for Non-ferrous Metals, Beijing 100088)

**Abstract:** From the view of chemical short range order and incomplete random mixing existing in liquid binary alloy, absorbing the rational part of past statistical mechanics a statistical mechanics model of liquid binary alloy has been proposed. From the model, the expressions of component activity were derived. Some parameters have been investigated, and their calculation formulae were obtained. According to the formulae of calculating coordination number, the coordination numbers of Cu-Zr, Fe-B, Ni-B, Co-P, Fe-P systems were calculated, and the results agreed well with those in literature and in experiments.

**Keywords:** statistical mechanics model; liquid binary alloy; coordination number

## 1 Introduction

The past statistical mechanics models have provided possibility for thermodynamical calculation of some liquid alloy solutions. However, it can not yet be used for some liquid alloy solutions in which chemical short-range order and incomplete random mixing exist. In most of these models, the chemical short-range order was neglected, complete random distribution was adopted. But most binary or more component alloy solutions are dealt with regular solution model[1-3], the past statistical mechanics models are required to be improved and developed further. Thus, a statistical mechanics model based on chemical short-range order and incomplete random mixing existing in liquid binary alloy has been developed. Simultaneously, this paper gives the expressions of component activity $a_i$ of liquid binary alloy and the expressions of the parameters, such as co-ordination number, probability of pairwise forming chemical bond $i$-$j$.

## 2 Statistical mechanics model of liouid binary alloy

A statistical mechanics model based on chemical short-range order and incomplete random mixing existing in liquid binary alloy is described as follows:

(1) Pairwise interactive energies in solution is symbolized as $\varepsilon_{ij}$ ($i, j = A, B$), which is varied with compound $A_X B_Y$ forming or not in the solution. When two atoms are combined into chemical bond, the devotion to the interaction energies is symbolized as $\Delta\varepsilon_{ij}$ ($i, j = A, B$). The probability of forming chemical bond is symbolized as $P_{ij}$, and the pairwise interactive energies in pure

---

Supported by National Natural Science Foundation of China.

substance is symbolized as $\varepsilon_{ii}^0$ ($i$ = A, B).

(2) Inner and translational degrees of freedom of atom in pure substance are independent, and equal to that in pure substance, therefore only the near pairwise interactive energies are considered.

(3) Because of difference between two atoms and between pairwise interactive energies, when A and B are mixed into solution, the mixing is incomplete random and obeys some rules; and there is chemical short-range order in the system.

(4) The co-ordination number of A is not equal to that of B. The co-ordination numbers of A, B and pure substance $i$ ($i$ = A, B) are symbolized as $Z_A$, $Z_B$ and $Z_i^0$ ($i$ = A, B).

(5) Because of existing pair interaction, the solution system can be regarded as a system composed of dependent particles, and its Hamiltonian is

$$H = \sum_i \varepsilon_i + E \tag{1}$$

where $\varepsilon_i$ is the energy level of the atom $i$ (associated with inner and translational degrees of freedom), $E$ is the sum of pairwise interactive energies in the solution, named as configurational energy.

The canonical partition function of the liquid binary alloy is as follows:

$$Z_{AB} = \sum \frac{(N_{AA} + N_{BB} + N_{AB})!}{N_{AA}! N_{BB}! N_{AB}!} Q_A^{N_A}(T) Q_B^{N_B}(T) \exp(-E_{ABj}/kT) \tag{2}$$

where $Q_i(T) = \sum_j \exp(-\varepsilon_{ij}/kT)$ is partition function associated with inner and translational degrees of freedom of atom $i$ ($i$ = A, B), $N_{ij}$ ($i,j$ = A, B) is pair number in the solution.

## 3 Activity expressions

### 3.1 Component activity and canonical partition function

Statistical mechanics shows that macroscopic quantity is assembly average of microscopic quantity, hence it is also an ensemble average. According to above, formula of Helmholtz free energy can be given as follows:

$$F = -kT\ln Z \tag{3}$$

where $Z$ is canonical partition function of the system. And the relation of chemical potential and canonical partition function is

$$\mu_i = -RT\left(\frac{\partial \ln Z}{\partial N_i}\right)_{T,V} \tag{4}$$

Comparing with thermodynamical relation formula of chemical potential and activity, $\mu_i = \mu_i^0 + RT\ln a_i$, relation formula of component activity and canonical partition function is obtained:

$$\ln a_i = \left[\frac{\partial \ln Z(T, V, N_i)}{\partial N_i}\right]_{T,V} - \left[\frac{\partial \ln Z(T, V, N_i, N_j)}{\partial N_i}\right]_{T,V,N_j} \tag{5}$$

## 3.2 Activity expressions

According to existent pattern of compound in the solution, liquid binary alloy solutions could be divided into two styles: non-compound system and compound system.

### 3.2.1 Activity expressions of non-compound system

Because of no compound existing in the solution, the devotion of chemical bond formed by two atoms is unnecessary to be considered, the configurational energy of the system is

$$E_{AB} = N_{AB}\varepsilon_{AB} + N_{AA}\varepsilon_{AA} + N_{BB}\varepsilon_{BB} \tag{6}$$

For system composed of pure substance A, its configurational energy is

$$E_A = N_A Z_A^0 \varepsilon_{AA}^0 \tag{7}$$

Define $\{X_{ij}\} = \{N_{ij}\} / \sum_{i,j}\{N_{ij}\}$, where $\{X_{ij}\}$ is called pair fraction; substitute Eqns. (2), (6) and (7) into Eqn. (5) and according to the principle of lowest energy, activity expression of component A is obtained as:

$$\ln a_A = \left(\frac{\partial\{N_{AB}\}}{\partial N_A}\right)_{T,V,N_B} \ln\{X_{AB}\} + \left(\frac{\partial\{N_{AA}\}}{\partial N_A}\right)_{T,V,N_B} \ln\{X_{AA}\} + \left(\frac{\partial\{N_{BB}\}}{\partial N_A}\right)_{T,V,N_B} \ln\{X_{BB}\} + \frac{\{N_{AB}\}}{kT}\left(\frac{\partial\varepsilon_{AB}}{\partial N_A}\right)_{T,V,N_B} + \frac{\{N_{AA}\}}{kT}\left(\frac{\partial\varepsilon_{AA}}{\partial N_A}\right)_{T,V,N_B} + \frac{\{N_{BB}\}}{kT}\left(\frac{\partial\varepsilon_{BB}}{\partial N_A}\right)_{T,V,N_B} - \frac{Z_A^0 \varepsilon_{AA}^0}{2kT} \tag{8}$$

According to the same deduction, activity expression of component B is also given by

$$\ln a_B = \left(\frac{\partial\{N_{AB}\}}{\partial N_B}\right)_{T,V,N_A} \ln\{X_{AB}\} + \left(\frac{\partial\{N_{AA}\}}{\partial N_B}\right)_{T,V,N_A} \ln\{X_{AA}\} + \left(\frac{\partial\{N_{BB}\}}{\partial N_B}\right)_{T,V,N_A} \ln\{X_{BB}\} + \frac{\{N_{AB}\}}{kT}\left(\frac{\partial\varepsilon_{AB}}{\partial N_B}\right)_{T,V,N_A} + \frac{\{N_{AA}\}}{kT}\left(\frac{\partial\varepsilon_{AA}}{\partial N_B}\right)_{T,V,N_A} + \frac{\{N_{BB}\}}{kT}\left(\frac{\partial\varepsilon_{BB}}{\partial N_B}\right)_{T,V,N_A} - \frac{Z_B^0 \varepsilon_{BB}^0}{2kT} \tag{9}$$

Neglecting the relation of pairwise interactive energy and component concentration, activity expressions of component A, B become as follows:

$$\ln a_A = \left(\frac{\partial\{N_{AB}\}}{\partial N_A}\right)_{T,V,N_B} \ln\{X_{AB}\} + \left(\frac{\partial\{N_{AA}\}}{\partial N_A}\right)_{T,V,N_B} \ln\{X_{AA}\} + \left(\frac{\partial\{N_{BB}\}}{\partial N_A}\right)_{T,V,N_B} \ln\{X_{BB}\} - \frac{Z_A^0 \varepsilon_{AA}^0}{2kT} \tag{10}$$

$$\ln a_B = \left(\frac{\partial\{N_{AB}\}}{\partial N_B}\right)_{T,V,N_A} \ln\{X_{AB}\} + \left(\frac{\partial\{N_{AA}\}}{\partial N_B}\right)_{T,V,N_A} \ln\{X_{AA}\} + \left(\frac{\partial\{N_{BB}\}}{\partial N_B}\right)_{T,V,N_A} \ln\{X_{BB}\} - \frac{Z_B^0 \varepsilon_{BB}^0}{2kT} \tag{11}$$

### 3.2.2 Particle number of compound system

Let $[N_{A\text{-}B}]$, $[N_A]$ and $[N_B]$ be particle number of compound $A_XB_Y$, component A and component B in the solution respectively; $Q_A$, $Q_B$ and $Q_{A-B}$ represent partition function associated with inner and translational degrees of freedom of particles A, B and $A_X B_Y$. When compound $A_X B_Y$ exists in the solution, chemical equilibrium established among component $A_X B_Y$, component A and component B is as follows:

$$X_A(1) + Y_B(1) = A_X B_Y(1)$$

According to the calculation formula of equilibrium constant in statistical mechanics[4], particle number of compound $A_X B_Y$, component A and component B are related with

$$\frac{[N_{A-B}]}{[N_A]^X [N_B]^Y} = \frac{Q_{A-B}}{Q_A^X Q_B^Y} \exp(-\Delta E^0/kT) \tag{12}$$

where $\Delta E^0$ is the difference of dissociative energy of reactant and product in ground state, it is also the remaining sum of ground state of reactant molecula and product molecula.

The following relation exists among component particle number:

$$[N_A] = N_A - X[N_{A-B}] \tag{13}$$

$$[N_B] = N_B - Y[N_{A-B}] \tag{14}$$

$$\frac{[N_{A-B}]}{(N_A - X[N_{A-B}])^X (N_B - Y[N_{A-B}])^Y} = \frac{Q_{A-B}}{Q_A^X Q_B^Y} \exp(-\Delta E^0/kT) \tag{15}$$

In Eqn. (15), at given temperature, partition function $Q_{A-B}$, $Q_A$, $Q_B$ and $\Delta E^0$ are independent of particle number, they are constant.

Let

$$AA = \frac{Q_{A-B}}{Q_A^X Q_B^Y} \exp(-\Delta E^0/kT) \tag{16}$$

hence

$$\frac{[N_{A-B}]}{(N_A - X[N_{A-B}])^X (N_A - X[N_{A-B}])^Y} = AA \tag{17}$$

In Eqn. (17), $N_A$, $N_B$, $X$, $Y$ and $AA$ are fixed value in a given system, therefore particle number of component $A_X B_Y$ is easily obtained. According to Eqn. (13) and Eqn. (14), particle number of component A, B are also easily obtained.

3.2.3 Activity expression of compound system

Considering the effect of compound existing in the solution on particle number, canonical partition function is

$$Z_{AB} = \sum_j \frac{(N_{AA} + N_{BB} + N_{AB})!}{N_{AA}! N_{BB}! N_{AB}!} Q_A^{N_A}(T) Q_B^{N_B}(T) Q_{A-B}^{N_{AB}}(T) \exp(-E_{ABj}/kT) \tag{18}$$

When compound exists in the solution, the devotion of bond-forming to configurational energy is needed to be considered. Thus, configurational energy is

$$E_{AB} = N_{AB}(\varepsilon_{AB} + P_{AB}\Delta\varepsilon_{AB}) + N_{AA}(\varepsilon_{AA} + P_{AA}\Delta\varepsilon_{AA}) +$$
$$N_{BB}(\varepsilon_{BB} + P_{BB}\Delta\varepsilon_{BB}) \tag{19}$$

where $P_{ij}$ ($i, j = $ A, B) is the probability of pairwise forming chemical bond $i$-$j$.

According to the similar deduction of section 3.2.1, activity expressions of component A, B in compound system are

$$\ln a_A = \left(\frac{\partial \{N_{AB}\}}{\partial N_A}\right)_{T,V,N_B} \ln\{X_{AB}\} + \left(\frac{\partial \{N_{AA}\}}{\partial N_A}\right)_{T,V,N_B} \ln\{X_{AA}\} +$$

$$\left(\frac{\partial \{N_{BB}\}}{\partial N_A}\right)_{T,V,N_B} \ln\{X_{BB}\} + \frac{\{N_{AB}\}}{kT}\left[\left(\frac{\partial \varepsilon_{AB}}{\partial N_A}\right)_{T,V,N_B} +\right.$$

$$\left.\left(\frac{\partial P_{AB}}{\partial N_A}\right)_{T,V,N_B} \Delta\varepsilon_{AB} + \left(\frac{\partial \Delta\varepsilon_{AB}}{\partial N_A}\right)_{T,V,N_B} P_{AB}\right] +$$

$$\frac{\{N_{AA}\}}{kT}\left[\left(\frac{\partial \varepsilon_{AA}}{\partial N_A}\right)_{T,V,N_B} + \left(\frac{\partial P_{AA}}{\partial N_A}\right)_{T,V,N_B} \Delta\varepsilon_{AA} +\right.$$

$$\left.\left(\frac{\partial \Delta\varepsilon_{AA}}{\partial N_A}\right)_{T,V,N_B} P_{AA}\right] + \frac{\{N_{BB}\}}{kT}\left[\left(\frac{\partial \varepsilon_{BB}}{\partial N_A}\right)_{T,V,N_B} +\right.$$

$$\left.\left(\frac{\partial P_{BB}}{\partial N_A}\right)_{T,V,N_B} \Delta\varepsilon_{BB} + \left(\frac{\partial \Delta\varepsilon_{BB}}{\partial N_A}\right)_{T,V,N_B} P_{BB}\right] -$$

$$\left(\frac{\partial [N_{A-B}]}{\partial N_A}\right)_{T,V,N_B} \ln Q_{A-B} - \left(\frac{\partial [N_A]}{\partial N_A}\right)_{T,V,N_B} \ln Q_A -$$

$$\left(\frac{\partial [N_B]}{\partial N_A}\right)_{T,V,N_B} \ln Q_B - \ln Q_A - \frac{Z_A^0 \varepsilon_{AA}^0}{2kT} \tag{20}$$

$$\ln a_B = \left(\frac{\partial \{N_{AB}\}}{\partial N_B}\right)_{T,V,N_A} \ln\{X_{AB}\} + \left(\frac{\partial \{N_{AA}\}}{\partial N_B}\right)_{T,V,N_A} \ln\{X_{AA}\} +$$

$$\left(\frac{\partial \{N_{BB}\}}{\partial N_B}\right)_{T,V,N_A} \ln\{X_{BB}\} + \frac{\{N_{AB}\}}{kT}\left[\left(\frac{\partial \varepsilon_{AB}}{\partial N_B}\right)_{T,V,N_A} +\right.$$

$$\left.\left(\frac{\partial P_{AB}}{\partial N_B}\right)_{T,V,N_A} \Delta\varepsilon_{AB} + \left(\frac{\partial \Delta\varepsilon_{AB}}{\partial N_B}\right)_{T,V,N_A} P_{AB}\right] +$$

$$\frac{\{N_{AA}\}}{kT}\left[\left(\frac{\partial \varepsilon_{AA}}{\partial N_B}\right)_{T,V,N_A} - \left(\frac{\partial P_{AA}}{\partial N_B}\right)_{T,V,N_A} \Delta\varepsilon_{AA} -\right.$$

$$\left.\left(\frac{\partial \Delta\varepsilon_{AA}}{\partial N_B}\right)_{T,V,N_A} P_{AA}\right] + \frac{\{N_{BB}\}}{kT}\left[\left(\frac{\partial \varepsilon_{BB}}{\partial N_B}\right)_{T,V,N_A} +\right.$$

$$\left.\left(\frac{\partial P_{BB}}{\partial N_B}\right)_{T,V,N_A} \Delta\varepsilon_{BB} + \left(\frac{\partial \Delta\varepsilon_{BB}}{\partial N_B}\right)_{T,V,N_A} P_{BB}\right] -$$

$$\left(\frac{\partial [N_{A-B}]}{\partial N_B}\right)_{T,V,N_A} \ln Q_{A-B} - \left(\frac{\partial [N_A]}{\partial N_B}\right)_{T,V,N_A} \ln Q_A -$$

$$\left(\frac{\partial [N_B]}{\partial N_B}\right)_{T,V,N_A} \ln Q_B + \ln Q_B - \frac{Z_B^0 \varepsilon_{BB}^0}{2kT} \tag{21}$$

Neglecting the relation of pairwise interactive energy and component concentration, activity expressions of component A, B in the solution are as follows:

$$\ln a_A = \left(\frac{\partial \{N_{AB}\}}{\partial N_A}\right)_{T,V,N_B} \ln\{X_{AB}\} + \left(\frac{\partial \{N_{AA}\}}{\partial N_A}\right)_{T,V,N_B} \ln\{X_{AA}\} +$$

$$\left(\frac{\partial \{N_{BB}\}}{\partial N_A}\right)_{T,V,N_B} \ln\{X_{BB}\} + \frac{\{N_{AB}\}}{kT}\left(\frac{\partial P_{AB}}{\partial N_A}\right)_{T,V,N_B} \Delta\varepsilon_{AB} +$$

$$\frac{\{N_{AA}\}}{kT}\left(\frac{\partial P_{AA}}{\partial N_A}\right)_{T,V,N_B} \Delta\varepsilon_{AA} + \frac{\{N_{BB}\}}{kT}\left(\frac{\partial P_{BB}}{\partial N_A}\right)_{T,V,N_B} \Delta\varepsilon_{BB} -$$

$$\left(\frac{\partial [N_A]}{\partial N_A}\right)_{T,V,N_B} \ln Q_A - \left(\frac{\partial [N_B]}{\partial N_A}\right)_{T,V,N_B} \ln Q_B + \ln Q_A -$$

$$\left(\frac{\partial [N_{A-B}]}{\partial N_A}\right)_{T,V,N_B} \ln Q_{A-B} - \frac{Z_A^0 \varepsilon_{AA}^0}{2kT} \tag{22}$$

$$\ln a_B = \left(\frac{\partial \{N_{AB}\}}{\partial N_B}\right)_{T,V,N_A} \ln\{X_{AB}\} + \left(\frac{\partial \{N_{AA}\}}{\partial N_B}\right)_{T,V,N_A} \ln\{X_{AA}\} +$$

$$\left(\frac{\partial \{N_{BB}\}}{\partial N_B}\right)_{T,V,N_A} \ln\{X_{BB}\} + \frac{\{N_{AB}\}}{kT}\left(\frac{\partial P_{AB}}{\partial N_B}\right)_{T,V,N_A} \Delta\varepsilon_{AB} +$$

$$\frac{\{N_{AA}\}}{kT}\left(\frac{\partial P_{AA}}{\partial N_B}\right)_{T,V,N_A} \Delta\varepsilon_{AA} + \frac{\{N_{BB}\}}{kT}\left(\frac{\partial P_{BB}}{\partial N_B}\right)_{T,V,N_A} \Delta\varepsilon_{BB} -$$

$$\left(\frac{\partial [N_A]}{\partial N_B}\right)_{T,V,N_A} \ln Q_A - \left(\frac{\partial [N_B]}{\partial N_B}\right)_{T,V,N_A} \ln Q_B + \ln Q_B -$$

$$\left(\frac{\partial [N_{A-B}]}{\partial N_B}\right)_{T,V,N_A} \ln Q_{A-B} - \frac{Z_B^0 \varepsilon_{BB}^0}{2kT} \tag{23}$$

## 4 Pair number

### 4.1 Pair number and Cowley-Warren chemical short range order parameter

Because of the difference between two atoms and between pairwise interactive energies, when different metals A and B are mixed into solution, the mixing is incomplete random, the chemical concentration around each atom is not equal to their average concentration, and there exists chemical short range order in the system. In general, the probability of finding atom A as a nearest neighbour of a given atom A is not equal to that of a given atom B. Similarly, the probability of finding atom B as a nearest neighbour of a given atom A is not equal to that of a given atom B.

Suppose $(i/j)$ to represent the probabilities of finding atom $i$ as a nearest neighbour of a given atom $j$ ($i, j = A, B$); and $(i, j)$ to represent the probabilities of atoms $i, j$ becoming nearest neighbour ($i, j = A, B$), obviously,

$$(B/A) + (A/A) = 1 \tag{24}$$

$$(A/B) + (B/B) = 1 \tag{25}$$

The probabilities (A, A), (A, B), (B, B) and (B, A) are related to the conditional probabilities (A/B), (A/A), (B/B) and (B/A) through the relations:

$$(A, B) = X_A(B/A) \tag{26}$$

$$(A, A) = X_A(A/A) \tag{27}$$

$$(\text{B, B}) = X_B(\text{B/B}) \tag{28}$$

$$(\text{B, A}) = X_B(\text{A/B}) \tag{29}$$

where $X_i$ ($i =$ A, B) are the molar fraction of atom $i$.

The total nearest neighbour pair number is then:

$$N_{\text{total}} = (Z_A N_A + Z_B N_B)/2 \tag{30}$$

where $Z_i$ ($i =$ A, B) are the co-ordination number of atom $i$ in the solution, $N_i$ ($i =$ A, B) are the number of atom $i$.

According to the definition of Cowley-Warren chemical short range order parameter $\alpha_1$ for the first co-ordination shell[5-7], conditional probabilities (A/B), (B/A) are related to $\alpha_1$ as follows:

$$(\text{A/B}) = X_A(1 - \alpha_1) \tag{31}$$

$$(\text{B/A}) = X_B(1 - \alpha_1) \tag{32}$$

Thus, above-mentioned relations are readily reduces to the relations of nearest neighbour pair number and Cowley-Warren chemical short range order parameter, namely

$$\begin{aligned} \{N_{AB}\} &= (\text{A, B})N_{\text{total}} + (\text{B, A})N_{\text{total}} \\ &= (Z_A N_A + Z_B N_B)X_A X_B(1 - \alpha_1) \end{aligned} \tag{33}$$

$$\begin{aligned} \{N_{AA}\} &= (\text{A, A})N_{\text{total}} \\ &= (Z_A N_A + Z_B N_B)X_A(X_A + X_B \alpha_1)/2 \end{aligned} \tag{34}$$

$$\begin{aligned} \{N_{BB}\} &= (\text{B, B})N_{\text{total}} \\ &= (Z_A N_A + Z_B N_B)X_B(X_B + X_A \alpha_1)/2 \end{aligned} \tag{35}$$

## 4.2 Cowley-Warren chemical short range order parameter

When the solution is considered to be an open system, its grand partition function is

$$\Xi = \sum_j \omega Q_A^{N_A}(T) Q_B^{N_B}(T) \exp[(\mu_A N_A + \mu_B N_B - E_j)/kT] \tag{36}$$

where $\mu_i$ ($i =$ A, B) denote the chemical potentials of component $i$, $Q_i(T)$ ($i =$ A, B) are partition functions associated with inner and translational degrees of freedom of atom $i$, $E_j$ are the configurational energy when the system is in level $j$.

The solution system can be considered to be composed of lots of small domains, and each domain is also an open system, assuming that the solution system is divided into two domains, domain 1 and domain 2. This enables us to write down the grand partition function as the product of the grand partition of the two domains:

$$\Xi = \Xi_1 * \Xi_2 \tag{37}$$

where

$$\Xi_1 = \sum_j \omega_1 \zeta_A^{N_{1A}} \zeta_B^{N_{1B}} \exp[-(E_j + \overline{E}_{12})/kT] \tag{38}$$

$$\Xi_2 = \sum_j \omega_2 \zeta_A^{N_{2A}} \zeta_B^{N_{2B}} \exp[-(E_j + \overline{E}_{12})/kT] \tag{39}$$

$$\zeta_A = Q_A(T)\exp(\mu_A/kT) \tag{40}$$

$$\zeta_B = Q_B(T)\exp(\mu_B/kT) \tag{41}$$

$$N_{1A} + N_{2A} = N_A \quad (42)$$

$$N_{1A} + N_{2B} = N_B \quad (43)$$

$N_{ij}$ ($j$ = A, B, $i$ = 1, 2) are the particle number of atoms $j$ in domain $j$. $E_{ij}$ ($i$ = 1, 2) are the configurational energy of domain 1, 2 in level $j$. $\overline{E}_{12}$ is the average value of the interaction between atoms of domain 1 and domain 2. Estimation of exp ($-\overline{E}_{12}/kT$) of equations (38) and (39) is quite difficult, $\overline{E}_{12}$ is related to concentration, coordination number, atomic volume and so on. Instead of making a rigorous approach, we resort to a simple approximation for $\overline{E}_{12}$ which is gained by Fowler and Guggenheim[8]:

$$\exp(-\overline{E}_{12}/kT) \approx \Phi_A^{f_A} \Phi_B^{f_B} \quad (44)$$

where $f_A$ are the mumbers of atoms in domain 2 which are the nearest neighbours of atom A in domain 1, $f_B$ is similar to $f_A$, $\Phi_A$, $\Phi_B$ are constant.

Therefore,

$$\Xi = \sum_j \omega_1 \zeta_A^{N_{1A}} \zeta_B^{N_{1B}} \Phi_A^{f_A} \Phi_B^{f_B} \exp(-E_j/kT) \quad (45)$$

When there is only one atom in domain 1, no pair (AA, AB or BB) is existing in the system, the configurational energy $E_1$ is zero. The nearest neighbours of atom A, B in domain 2 are equal to their coordination number, thus grand partition $\Xi$ of the system composed of domain 1 can be written as

$$\Xi_1^{(1)} = \zeta_A \Phi_A^Z + \zeta_B \Phi_B^Z \quad (46)$$

When there are two atoms in domain 1, pair (AA, AB or BB) is existing in the system, the configurational energy $E$ is not equal to zero, and the devotion of pair which becomes a part of compound $A_xB_y$ to interaction energy must also be considered, therefore its grand partition can be obtained as

$$\Xi_1^{(2)} = \zeta_A^2 \Phi_A^{2(Z_A-1)} Q_{AA} + \zeta_B^2 \Phi_B^{2(Z_B-1)} Q_{BB} + \zeta_A \Phi_A^{Z_A-1} \cdot \zeta_B \Phi_B^{Z_B-1} Q_{AB} \quad (47)$$

and

$$Q_{ij} = \exp\left[(-\varepsilon_{ij} + P_{ij}\Delta\varepsilon_{ij})/kT\right] \quad (i, j = A, B) \quad (48)$$

where $\varepsilon_{ij}$ ($i, j$ = A, B) are the interaction energy between $i, j$, $\Delta\varepsilon_{ij}$ ($i, j$ = A, B) are the devotion of pair forming chemical bond to interaction energy. By employing Eqn. (47), we can immediately express the probabilities of two atom $i, j$ ($i, j$ = A, B) becoming the nearest neighbour.

$$(A, A) = \zeta_A^2 \Phi_A^{2(Z_A-1)} Q_{AA} / \Xi_1^{(2)} \quad (49)$$

$$(B, B) = \zeta_B^2 \Phi_B^{2(Z_B-1)} Q_{BB} / \Xi_1^{(2)} \quad (50)$$

$$(B, A) = (A, B)$$
$$= \zeta_A \Phi_A^{Z_A-1} \zeta_B \Phi_B^{Z_B-1} Q_{AB} / \Xi_1^{(2)} \quad (51)$$

Therefore

$$\frac{(A, A) * (B, B)}{(A, B)^2} = Q_{AA} * Q_{BB} / Q_{AB}^2 \quad (52)$$

By setting

$$\eta^2 = Q_{AA} * Q_{BB} / Q_{AB}^2 \quad (53)$$

Equation (50) provides

$$(A, A) * (B, B)/(A, B)^2 = \eta^2 \quad (54)$$

Equations (36) – (39) and (54) provide

$$(A, B) = \frac{2X_A X_B}{\beta + 1} \quad (55)$$

where

$$\beta = \sqrt{1 + 4X_A X_B (\eta^2 - 1)} \quad (56)$$

Equations (36), (41) and (56) further yield

$$\alpha_1 = (\beta - 1)/(\beta + 1) \quad (57)$$

## 4.3 Nearest neighbour pair number

By making use of the relations of nearest neighbour pair number to chemical short range parameter to the relations of chemical short range parameter and interaction energy between atoms, we readily obtain the relation of nearest neighbour pair number and interaction energy as follows.

$$\{N_{AB}\} = 2(Z_A N_A + Z_B N_B) X_A / (\beta + 1) \quad (58)$$

$$\{N_{AA}\} = (Z_A N_A + Z_B N_B) X_A (\beta - 1 + 2X_A) / 2(\beta + 1) \quad (59)$$

$$\{N_{BB}\} = (Z_A N_A + Z_B N_B) X_B (\beta + 1 - 2X_A) / 2(\beta + 1) \quad (60)$$

From the above equations, we can know that if the interaction energy between atoms $\varepsilon_{ij}$ ($i, j = A, B$), devotion of forming chemical bond to interaction energy $\Delta\varepsilon_{ij}$ ($i, j = A, B$) and the probabilities of forming chemical bond are given, neighbour pair number can be obtained.

## 5 Probabilities of forming chemical bond

Assuming that if $(X - 1)$ A, $(Y - 1)$ B are the nearest neighbour of pair AB, compound $A_x B_y$ is formed, thus the probabilities of pairwise $(i, j)$ becoming a part of compound $A_x B_y$ are

$$P_{AB} = [(B/B)^Y - (B/A)^Y][(A/A)^X - (A/B)^X] / [(A/A) - (A/B)]^2 \quad (61)$$

$$P_{AA} = (X - 1)(Y + 1)(A/A)^{X-2}(B/A) \quad (62)$$

$$P_{BB} = (X + 1)(Y - 1)(A/B)^X (B/B)^{Y-2} \quad (63)$$

When $X$ or $Y$ equals to one, the chemical bond AA or BB does not exist in solution, in this instance, $P_{AA}$ or $P_{BB}$ is zero, the above equations (62) and (63) are also correct.

## 6 Theoretic formulae for calculating coordination number

According to the relations of conditional probabilities, concentration and interaction energies in binary system, theoretic formulae of calculating coordination number can be obtained. Generally, atoms in liquid binary alloy are close to each other as shown in Fig. 1. Obviously, *EF* is at right angles with *AB*, *EG* is at right angles with *CD*. Relying on theorem about triangle, triangle *ABF*, *ABE*, *GDC* and *EDC* are right triangles.

In right triangles *ABF* and *ABE*,

$$AB^2 + (R_A - H_{A2})^2 = R_A^2 \tag{64}$$

$$AB^2 + (R_A + H_{A2})^2 = (\min(R_A, R_B) + R_A)^2 \tag{65}$$

Equations (64) and (65) provide

$$H_{A2} = [(\min(R_A, R_B))^2 + 2R_A \min(R_A, R_B)]/4R_A \tag{66}$$

$$H_{A1} = \min(R_A, R_B) - H_{A2} = [-(\min(R_A, R_B))^2 + 2R_A \min(R_A, R_B)]/4R_A \tag{67}$$

Similarly,

$$H_{B2} = [(\min(R_A, R_B))^2 + 2R_A \min(R_A, R_B)]/2(R_A + R_B) \tag{68}$$

$$H_{B1} = [2R_B \min(R_A, R_B) - (\min(R_A, R_B))^2]/2(R_A + R_B) \tag{69}$$

From equations (66), (67), (68) and (69), $H_{A2}$, $H_{A1}$, $H_{B2}$, $H_{B1}$ can be calculated. According to the volume formula, volume of single coordination atom A or B in the globe whose radius is $R_A + \min(R_A, R_B)$ and whose center is $E$ can easily be obtained,

$$V_{AA} = \frac{\pi}{3}[H_{A1}^2(R_A + \min(R_A, R_B)) - H_{A1} + H_{A2}^2(3R_A - H_{B2})] \tag{70}$$

$$V_{AB} = \frac{\pi}{3}[H_{B1}^2(R_A + \min(R_A, R_B)) - H_{B1} + H_{B2}^2(3R_B - H_{B2})] \tag{71}$$

$V_{AA}$, $V_{AB}$ denote volume of the single coordination atom A and B. Simultaneously, coordination numbers $Z_{AA}$, $Z_{AB}$ are related to conditional probabilities (B/A), (A/A) and coordination numbers $Z_A$ through

$$Z_{AB} = (B/A)Z_A \tag{72}$$

$$Z_{AA} = (A/A)Z_A \tag{73}$$

where $Z_A$ denotes coordination numbers of atom A, $Z_{AB}$ and $Z_{AA}$ denote number of atom B and A around atom A. Thus the volume occupied by the coordination atoms A and B can be expressed as

$$\begin{aligned}V_{co-A} &= Z_{AA}V_{AA} + Z_{AB}V_{AB} \\ &= \frac{\pi}{3}Z_A\{(A/A)[H_{A1}^2(R_A + \min(R_A, R_B) - H_{A1}) + H_{A2}^2(3R_A - H_{A2})] + (B/A)[H_{B1}^2(3(R_A + \min(R_A, R_B)) - H_{B1}) + H_{B2}^2(3R_B - H_{B2})]\}\end{aligned} \tag{74}$$

The volume occupied by the coordination atoms A and B can also be expressed as

$$V'_{co-A} = 4\pi/3[R_A + \min(R_A, R_B)^3 - R_A^3] \tag{75}$$

Combine Eqn. (74) and Eqn. (75), co-ordination numbers $Z_A$ of atom A are obtained as:

$$Z_A = 4\{[\overline{R}_A + \min(\overline{R}_A, \overline{R}_B)]^3 - \overline{R}_A^3\}/\{(A/A)[H_{A1}^2(3(R_A + $$

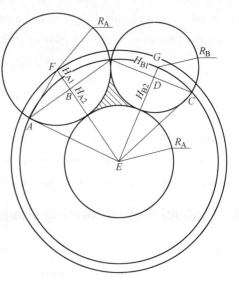

Fig. 1  Sketch of atom distribution

$$\min(R_A, R_B)) - H_{A1}) + H_{A2}^2(3R_A - H_{A2})] +$$
$$(B/A)[H_{B1}^2(3(R_A + \min(R_A, R_B)) - H_{B1}) +$$
$$H_{B2}^2(3R_B - H_{B2})]\} \qquad (76)$$

Similarly,
$$Z_B = 4\{[\overline{R}_B + \min(\overline{R}_A, \overline{R}_B)]^3 - \overline{R}_B^3\}/\{(A/B)[H_{A1}^2(3(R_B + \min(R_A', R_B)) - H_{A1}) + H_{A2}^2(3R_A - H_{A2})] +$$
$$(B/B)[H_{B1}^2(3(R_B + \min(R_A, R_B)) - H_{B1}) +$$
$$H_{B2}^2(3R_B - H_{B2})]\} \qquad (77)$$

In the above formulae, the atom radius is denoted with different symbol, because the interstitial volume is a part of the volume measured by its radius from equation (74), but it is not a part of that from equation (75); although using the same radius, the error will be very great in equation (74). Therefore we use density radius calculated from mass density instead, the density radius is as follows

$$R = (3M/4\rho\pi N)^{1/3} \qquad (78)$$

where $M$ is molar weight, $\rho$ is density.

## 7 Numerical values of co-ordination number

In general, the interaction energy between atoms increases with bond energy. Because of datum shortage of interaction energy, bond energy was substituted for interaction energy, calculated results are shown in Table 1.

**Table 1 Coordination mumber**

| Alloys (A – B) | $Z_A$ (calculated) | $Z_A$ (ref.) | relative error /% | references |
| --- | --- | --- | --- | --- |
| $Fe_{0.83}B_{0.17}$ | 12.52 | 13.2 | 5.15 | [9] |
| $Ni_{0.64}B_{0.34}$ | 12.6 | 13.0 | 3.00 | [9] |
| $B_{0.36}Ni_{0.64}$ | 9.93 | 9.0 | 10.33 | [9] |
| $Cu_{0.60}Zr_{0.40}$ | 11.4 | 10.3 | 7.18 | [10] |
| $Co_{0.80}P_{0.20}$ | 11.61 | 12.3 | 5.61 | [11] |
| $Co_{0.75}P_{0.25}$ | 11.69 | 13.0 | 10.76 | [11] |
| $Fe_{75.7}P_{24.3}$ | 11.76 | 12.78 | 7.98 | [12] |
| $Fe_{0.75}P_{0.25}$ | 11.7 | 13.0 | 9.46 | [13] |
| $P_{0.25}Fe_{0.75}$ | 10.15 | 11.9 | 14.71 | [13] |
| $Fe_{0.82}P_{0.18}$ | 11.64 | 12.54 | 7.12 | experiment |
| $P_{0.18}Fe_{0.82}$ | 10.15 | 11.90 | 14.71 | experiment |

## 8 Conclusions

(1) From the view of chemical short range and incomplete random mixing existing in liquid binary alloy solution, absorbing the rational part of past statistical mechanics model, a new model of liquid binary alloy solution has been proposed in this paper. According to the model, expressions of com-

ponent activity of liquid binary alloy solution are obtained.

(2) According to the model, the formulae of calculating probabilities of forming chemical bond $i$-$j$, nearest neighbour pair number, Cowley-Warren chemical short range order etc. are obtained.

(3) According to the relation of co-ordination number and condition probabilities, using geometrical relation of atoms distribution, formulae of co-ordination number in liquid binary alloy are obtained.

(4) According to the formulae and reference data, numerical values of co-ordination number are obtained. The results agree well with the reference data and experiment data within the range of error, the key problem about coordination number is solved.

## References

[1] Richardson F D. The Physical Chemistry of Metals [M]. London: Academic Pr, 1953.
[2] Chou K C, Wang J J. Metallurgical Transactions, 1987, 18A (2): 323.
[3] Chou Kouchih. CALPHAD, 1990, 14 (1): 41; 1990, 14 (3): 275.
[4] Tang Youqi. Statistical Mechanics and its Application in physicochemistry. (in Chinese) [M]. Beijing: Science Pub, 1964.
[5] Warren B E. X-Ray diffraction [M]. Addison-Wesley Pub Company Inc., 1969.
[6] Cowley J M. Phys Rev, 1950, 77: 667.
[7] Cowley J M. J Appl Phys, 1950, 21: 24.
[8] Fowler R H, Guggenheim E A. Statistical Thermodynamics [M]. London: Combridge University Press, 1939.
[9] Wu Guoan. Journal of physics, (in Chinses), 1984, 33: 645.
[10] Chip D R, Jennings L D, Giessu B C. Bull Am Phys Soc, 1978, 23: 467.
[11] Sadoc J F, Diamer J. Mat Sci Engr, 1976, 23: 137.
[12] Takeo Fujiwara, Yasushi Ishii. J Phy s F: Meta Phys, 1980, 10: 1901.
[13] Yashi Wasedo, Hideo Okazaki, Tsayoski Masunuoto. J Mater Sci, 1977, 12: 202.

（原载于 *Transactions of Nonferrous Metals Society of China*, 1997, 7 (3): 140-148）

# Does Nitrogen Transport in Vitreous Silica only Take Place in Molecular Form?

Dong Qian　G. Hultquist

(Division of Corrosion Science, Department of Materials and
Engineering, Royal Institute of Technology,
Dr. Kristinas väg 51, Stockholm SE10044, Sweder)

**Abstract:** It is generally believed that nitrogen transport in vitreous silica exclusively takes place in molecular form, although no evidence for this is found in the literature. Actually, an analysis of literature data of transport of nitrogen and noble gases in vitreous silica at 900℃ suggests it may not be the case. In order to clarify the operative species of nitrogen transport in this material, experiments of permeation and uptake/release have been performed with the use of gas phase analysis and isotopic labeling of nitrogen. By comparing the relative distributions of $^{14,14}N_2$, $^{14,15}N_2$, and $^{15,15}N_2$ in exposure gas, permeated gas and released gas with distribution of equilibrated nitrogen molecules, the percentage of dissociated nitrogen in the transport has been evaluated at different temperatures. It has then been found that nitrogen undergoes dissociation not only on the surface of vitreous silica but also in its bulk and that the overall dissociation of nitrogen increases with temperature. It is concluded that nitrogen diffuses both in molecular and atomic forms with approximately 15% atomic nitrogen transport at 900℃. The observed transport rates are explained by diffusion of molecular nitrogen combined with a retardation of dissociated nitrogen in reversible traps.

## 1 Introduction

Vitreous silica ($v$-$SiO_2$) is one of a few commercially produced glasses with a simple chemical composition. It is an important technological material with use in several applications such as electronic devices (vitreous silica has been used for envelopes of vacuum sealed electronic devices, such as halogen incandescent and discharge lamps and as dielectric thin films in integrated circuit devices), optical elements such as dielectric layers in microelectronics, and as catalyst supports. In all these applications, the transport properties of vitreous silica play an important role in process control and ultimate properties. This has stimulated many investigations of this material. Hence, the transport of gases in vitreous silica is an old subject of both scientific and technological importance. The relatively high diffusion rate of gases in this glass makes it possible to investigate a variety of gases, and the high glass transformation temperature allows measurements over a wide temperature range.

---

董倩，1998~2001年于北京科技大学师从李文超教授攻读博士学位。目前在挪威DNV GL公司工作，任高级工程师。

Several measurements of the transport of noble gases, hydrogen, and oxygen gas in a variety of vitreous silica have been published. However, only two studies of transport of nitrogen in silica have been reported, namely, by Barrer[1,2] and Johnson and Burt.[3] Shelby has summarized permeation and diffusion data of a number of gases including the noble gases He, Ne, Ar, and Kr as well as the diatomic gases $H_2$, $N_2$, and $O_2$ in Ref. [2] Shelby commented that values reported by Barrer and by Johnson for nitrogen permeation appeared to be somewhat "suspect" due to the extremely high perme-ability values as compared to those for other gases. It is generally believed that nitrogen transport in vitreous silica exclusively takes place in molecular form, but no evidence for this opinion is found in the literature. This situation has motivated us to conduct the work presented here, where the transport of nitrogen in vitreous silica has been studied by using gas phase analysis and isotopic labeling of nitrogen.

## 2 Theory: isotopic equilibration

An efficient way to identify the operative species during gas transport in materials is to expose the materials to isotopic labeled gas and measure the exchange of isotopes between gas molecules.

In a closed system with the two isotopes $^{14}N$ and $^{15}N$, there are three possible diatomic nitrogen molecules: $^{14,14}N_2$, $^{14,15}N_2$, and $^{15,15}N_2$. The fractions of $^{14}N$ and $^{15}N$, $f_{14_N}$ and $f_{15_N}$, in the system are

$$f_{14_N} = \frac{[^{14,14}N_2] + 0.5[^{14,15}N_2]}{[^{14,14}N_2] + [^{14,15}N_2] + [^{15,15}N_2]}$$

$$f_{15_N} = \frac{[^{15,15}N_2] + 0.5[^{14,15}N_2]}{[^{14,14}N_2] + [^{14,15}N_2] + [^{15,15}N_2]}$$

where $[^{14,14}N_2]$, $[^{14,15}N_2]$, $[^{15,15}N_2]$ are the concentrations of $^{14,14}N_2$, $^{14,15}N_2$, and $^{15,15}N_2$, respectively, and $f_{14_N} + f_{14_N} = 1$.

The statistically equilibrated fractions, $^\infty f_{14,14N_2}$, $^\infty f_{14,15N_2}$, and $^\infty f_{15,15N_2}$, are calculated from $f_{14_N}$ and $f_{15_N}$:

$$^\infty f_{14,14N_2} = \frac{(f_{14_N})^2}{(f_{14_N} + f_{15_N})^2} = (f_{14_N})^2$$

$$^\infty f_{14,15N_2} = \frac{2f_{14_N}f_{15_N}}{(f_{14_N} + f_{15_N})^2} = 2f_{14_N}f_{15_N}$$

$$^\infty f_{15,15N_2} = \frac{(f_{15_N})^2}{(f_{14_N} + f_{15_N})^2} = (f_{15_N})^2$$

Statistically equilibrated fractions of $^{14,14}N_2$, $^{14,15}N_2$, and $^{15,15}N_2$ are plotted as a function of $f_{14_N}/f_{15_N}$ in Fig. 1. It can be noted that the highest equilibrium fraction of $^{14,15}N_2$ is 0.5 and this is obtained in a gas mixture where $f_{14_N} = f_{15_N}$.

In a nonequilibrated nitrogen gas, an increase of the mixed molecule, $^{14,15}N_2$ results from dissocia-

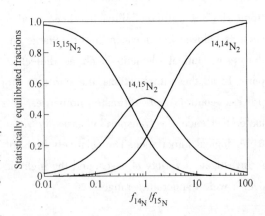

Fig. 1 Statistically equilibrated fractions of $^{14,14}N_2$, $^{14,15}N_2$, and $^{15,15}N_2$ as a function of $f_{14_N}/f_{15_N}$

tion of $^{14,14}N_2 \rightarrow {}^{14}N + {}^{14}N$ and $^{15,15}N_2 \rightarrow {}^{15}N + {}^{15}N$ followed by association of $^{14}N$ and $^{15}N$ to $^{14,15}N_2$.

The effect of exclusive molecular transport and exclusive atomic transport of nitrogen through a membrane on distribution of $N_2$ molecules is illustrated in Fig. 2, where the nitrogen gas with 50% $^{14,14}N_2$ and 50% $^{15,15}N_2$ at the highpressure side is nonequilibrated. Exclusive molecular transport results in the same distribution at the high-pressure side. On the other hand, if all nitrogen molecules dissociate on the surface or inside the membrane, the equilibrated concentration of $^{14,15}N_2$ is found at the low-pressure side. Consequently, exclusively atomic transport produces equilibrated gas at the low-pressure side.

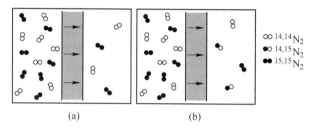

Fig. 2 (a) Exclusive molecular transport in permeation;
(b) Exclusive atomic transport in permeation

The effect of exclusive molecular transport and exclusive atomic transport of nitrogen during uptake/release on distribution of $N_2$ molecules is illustrated in Fig. 3, where the material surface is exposed to nitrogen gas with 50% $^{14,14}N_2$ and 50% $^{15,15}N_2$. If no dissociation takes place, the same distribution of nitrogen molecules is released. On the other hand, if exclusive atomic transport takes place, a composition with 25% $^{14,14}N_2$, 50% $^{14,15}N_2$, and 25% $^{15,15}N_2$ is released.

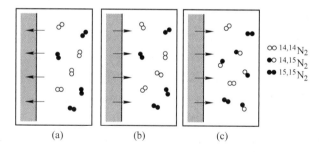

Fig. 3 (a) Uptake; (b) Exclusive molecular transport in release;
(c) Exclusive atomic transport in release

## 3 Experiment

### 3.1 Gas phase analysis

Experiments of permeation and uptake/release were performed with a technique named gas phase analysis (GPA). The GPA apparatus consists of three main parts separated by ultrahigh vacuum valves: a gas handling system, an enclosed volume of 100 cm$^3$, and a mass spectrometer (MS) in ultra-high vacuum. The gas handling system supplies nitrogen isotopes in the enclosed volume. The enclosed volume consists of a stainless steel part and a vitreous silica tube, which is heated by a tube furnace. The isotopic composition of nitrogen molecules in the enclosed volume is measured by the MS. In order to quantify the MS signal into flux data, calibration has been done with a known amount of gas determined by pressure, volume, and temperature in the enclosed reaction chamber. This gas was pumped away with the ion pump where the integrated pressure versus time curve corresponds to the amount of gas. Details of GPA and its calibration are found in Refs. [4] and [5].

## 3.2 Permeation experiment

A vitreous silica tube was used with an outer diameter of 1.2 cm, inner diameter of 1.0 cm, and heated length of 25 cm. The impurity levels in wt ppm were Al-30, B-0.5, Ca-3.5, Cu-0.1, Fe-2.5, Li-4, Mn-0.1, Mg-1, P-0.005, K-2.5, Ti-0.7, and Na-4.

The silica tube was heated at 900℃. The inner side was kept at $<5\times10^{-9}$ mbar by continuous evacuation by an ion pump and the outer side was kept at $<10^{-3}$ mbar for 24 h by continuous pumping with a rough pump. After 24h the outer side of the silica tube was exposed to 225 mbar nitrogen gas with a composition 57.1% $^{14,14}N_2$, 3.6% $^{14,15}N_2$, and 39.3% $^{15,15}N_2$. The permeation of nitrogen through the silica wall was measured by the MS until a steady state flux was obtained.

## 3.3 Uptake/release experiment

A vitreous silica tube with similar dimensions as the one used in permeation experiment was used. The impurity levels in wt ppm were Al-16, Ca-0.4, Fe-0.3, K-0.7, Li-0.7, Na-0.9, and Ti-1.4.

Approximately 400 mbar $N_2$ gas with $^{14,14}N_2 \approx {}^{15,15}N_2$ was exposed to the inner side of the silica tube for 20 h at 25, 200, 550, and 900℃, respectively. The exposures were preceded, by an outgassing at 900℃ for 16-24h until no significant decrease of nitrogen was observed. After 20h of exposure, the tube was quickly brought to room temperature and evacuated for 1min with a rough pump, whereby the free gas phase nitrogen was removed. The gas released from the silica tube was then monitored by the MS. The release process started at room temperature, followed by release at exposure temperature until virtually all $^{14,15}N_2$ was released from the tube wall. The amount of nitrogen released from the tube was obtained by integration of release rate over time.

# 4 Results and discussion

## 4.1 Permeation experiment

Fig. 4 shows the transient flux of nitrogen molecules through the silica wall at 900℃ during permeation experiment. It is seen that $^{14,14}N_2$, $^{14,15}N_2$, and $^{15,15}N_2$ increase during the first 20h. During 20-100h, $^{14,15}N_2$ continues to increase while $^{14,14}N_2$ $^{15,15}N_2$ decrease. Steady state fluxes are found after approximately 150h of exposure. The increase in fraction of $^{14,15}N_2$ is due to formation of $^{14,15}N_2$ by combination of $^{14}N$ and $^{15}N$, which are produced by dissociation of $^{14,14}N_2$ and $^{15,15}N_2$ on the surface and bulk of the silica wall. A possible reason for the $^{14,14}N_2$ and $^{15,15}N_2$ decrease as well

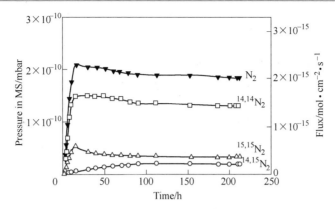

Fig. 4  Flux upon exposure of 225 mbar nitrogen to 1mm thick silica wall at 900 ℃

as the decrease in total nitrogen, $N_2$, after 20h is that approximately 25 mbars were taken out for analysis after approximately 15h. On the other hand, $^{14,15}N_2$ is increasing monotonously up to about 100h. The reason for that is not clear but may be due to an initial filling of relatively deep traps by dissociated nitrogen and formation of $^{14,15}N_2$ in consumption of $^{14,14}N_2$ and $^{15,15}N_2$. It can be calculated that the steady state flux of nitrogen through a silica wall is $2 \times 10^{-15}$ mol·cm$^{-2}$·s$^{-1}$, corresponding to $1.2 \times 10^9$ mol·cm$^{-2}$·s$^{-1}$ with approximately 200 mbar $N_2$ at 900 ℃.

By using the method explained in Sec. Ⅱ, the equilibrated distributions of $^{14,14}N_2$, $^{14,15}N_2$ and $^{15,15}N_2$ have been calculated based on the $^{14}N/^{15}N$ ratio in gas composition at the low-pressure side. The relative distributions of the nitrogen molecules at high-pressure side, at low-pressure side, and at statistical equilibrium are shown in Fig. 5. The distribution in statistical equilibrium was calculated from those at low-pressure side. The equilibrated fraction of $^{14,15}N_2$ is calculated to be 0.37. The actual fraction of $^{14,15}N_2$ at highpressure side is approximately 0.03 and the measured fraction of $^{14,15}N_2$ at steady state at low-pressure side is 0.10. A significant increase of $^{14,15}N_2$ at the low-pressure side is therefore present. Some atomic transport of nitrogen during permeation at 900 ℃ has taken place, although a statistical equilibrium among isotopic molecules is not reached. The reason why a statistical equilibrium among isotopic molecules is not reached even at steady state after 150h is due to the fact that nitrogen transport in silica mainly takes place in molecular form.

In Fig. 5, a higher $^{14}N/^{15}N$ ratio at passage through the wall can also be observed. This is likely due to residual $^{14}N$ in spite of the preceded outgassing of the tube.

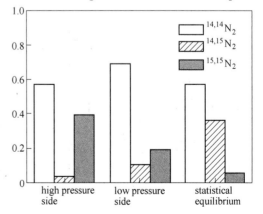

Fig. 5  Relative distributions of $^{14,14}N_2$, $^{14,15}N_2$, and $^{15,15}N_2$ at steady state in permeation experiment with 225 mbar nitrogen at 900 ℃

## 4.2 Uptake/release experiment

To obtain $^{14,15}N_2$ fractions, analysis of data has been done in uptake/release experiment at 25, 200, 550, and 900℃. The released nitrogen after exposure at 900℃ is shown in Fig. 6 during outgassing at 25℃ followed by 900℃. The fraction of $^{14,15}N_2$ after 3.5h outgassing is 0.06, which is higher than the fraction in exposure (<0.02). Relative distributions of $^{14,14}N_2$, $^{14,15}N_2$, and $^{15,15}N_2$ in exposure gas, released gas, and statistical equilibrium in released gas can be compared in Fig. 7. The distribution in statistical equilibrium was calculated from those in release. The increase of $^{14,15}N_2$ must result from dissociation of $^{14,14}N_2$ and $^{15,15}N_2$ on the surface and in the silica during the uptake and release process.

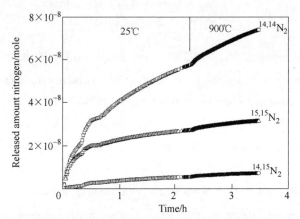

Fig. 6　Nitrogen release from silica after 20h exposure in 400 mbar nitrogen at 900℃

The overall results in uptake/release experiment together with the $^{14,15}N_2$ fraction in the permeation experiment at 900℃ are summarized in Fig. 8. A slight increase in $^{14,15}N_2$ upon temperature can be seen in the figure.

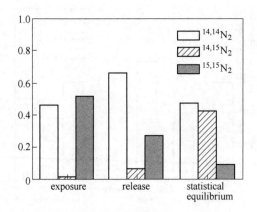

Fig. 7　Relative distribution of $^{14,14}N_2$, $^{14,15}N_2$ and $^{15,15}N_2$ in uptake/release experiment with 400 mbar nitrogen at 900℃ in exposure

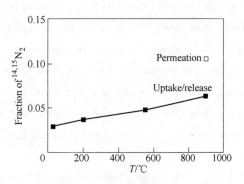

Fig. 8　Fraction of $^{14,15}N_2$ vs temperature at different temperatures

## 4.3 Percentage of atomic and molecular transports

The percentage of atomic and molecular transports can be calculated in the following way.

In permeation experiment,

$$\% \text{ atomic transport} = \frac{f_{\text{low pressure}} - f_{\text{high presure}}}{f_{\text{equilibrium}} - f_{\text{high pressure}}} \times 100$$

where $f_{\text{low pressure}}$, $f_{\text{high pressure}}$, and $f_{\text{equilibrium}}$ are fractions of $^{14,15}N_2$ at low-pressure side, at high-pressure side, and at statistical equilibrium, respectively.

In uptake/release experiment,

$$\% \text{ atomic transport} = \frac{f_{\text{release}} - f_{\text{exposure}}}{f_{\text{equilibrium}} - f_{\text{exposure}}} \times 100$$

where $f_{\text{release}}$, $f_{\text{exposure}}$, and $f_{\text{equilibrium}}$ are fractions of $^{14,15}N_2$ during release, during exposure, and at statistical equilibrium, respectively,

$$\% \text{ molecular transport} = 100\% - \% \text{ atomic transport}$$

The percentage of atomic and molecular transports at different temperatures is shown in Fig. 9. The dissociation of molecular nitrogen in the silica tube facilitates nitrogen transport in atomic form. It can be seen that there are 12% atomic transport in uptake/release experiment and 20% atomic transport in permeation experiment at 900℃. The percent of atomic nitrogen transport is higher in permeation experiment than that in uptake/release experiments. It is likely due to an additional dissociation of nitrogen in the bulk. Considering the average of above-mentioned percent-ages of atomic transport in the two experiments, a value of 15% is used for atomic nitrogen transport in vitreous silica at 900℃.

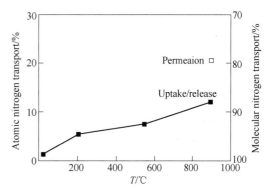

Fig. 9  Percentage of atomic and molecular nitrogen transports vs temperature

## 4.4 Dissociation rate of nitrogen

Dissociation rate of nitrogen on the surface of silica can be measured[6] During the 20h of exposure in 400 mbar isotopic nitrogen, a small but significant increased fraction of $^{14,15}N_2$ in exposure gas was detected at 900℃, which corresponds to approximately $3 \times 10^{-13}$ mol $(N_2) \cdot cm^{-2} \cdot s^{-1}$. As expected, this is much higher than the steady state flux of $^{14,15}N_2$ at 900℃, which is $2 \times 10^{-15}$ mol $(N_2) \cdot cm^{-2} \cdot s^{-1}$ as seen in Fig. 4.

## 4.5 Analysis of literature data and comparison with present data

Fig. 10 summarizes the permeability of the noble gases He, Ne, Ar, and Kr and the diatomic gases hydrogen, oxygen, and nitrogen in vitreous silica at 900 ℃. These data are obtained from Refs. [2] and [7]. The permeability of nitrogen in silica at 900 ℃ is measured to be $6 \times 10^8$ mol · cm$^{-1}$ · s$^{-1}$ atm$^{-1}$ in this study and is also plotted in Fig. 10 (▲). It is seen in Fig. 10 that the logarithm of the permeabilities versus molecular collision diameter of the noble gases falls on a straight line, here called the noble gas line.

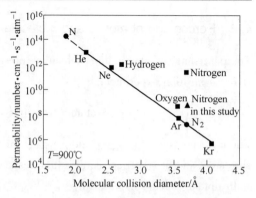

Fig. 10 Permeability of noble gases and some diatomic gases in vitreous silica (■) at 900 ℃ from Ref. [2] vs molecular collision diameter data from Ref. [7]. Permeability for nitrogen measured in this study is also plotted in the figure (▲). Permeability of $N_2$ (●) is determined by the noble gases but is found to be higher in this study. Permeability of N (●) is taken from extrapolation of solid line determined by the noble gases. Diameter of N is assumed to be half that of $N_2$

The diffusion of noble gases in vitreous silica takes place with a single mode diffusion process and the gas has virtually zero chemical interaction with the glass network[8,9] Their permeability, or rather diffusivity, is strongly size dependent. The experimentally obtained permeabilities of the diatomic gases hydrogen, oxygen, and nitrogen are significantly higher—40 times for hydrogen, 35 times for oxygen, and 60 times for nitrogen—than expected from values given by the noble gas line, as seen in Fig. 10. We consider two possibilities to explain these high permeabilities. One possibility is that a very small fraction (approximately $10^{-4}$) of diatomic molecules dissociates and permeates freely in atomic form at a rate determined by a size half that of molecules and in accordance with the noble gas line. In a previous study[4] it was found that more than 95% of hydrogen had undergone dissociation in silica at 900 ℃, whereas in the present work of nitrogen the corresponding value is approximately 15%. We exclude freely permeating atoms due to the high degree of dissociation observed compared to $10^{-4}$. The second possibility is that several percent of diatomic molecules dissociate into diffusing atoms, which are retarded due to an interaction with the host material. In effect, this means a reversible trapping of atoms. (An irreversible trapping does not have any influence on the permeability.) A retardation of permeating atoms due to reversible trapping should be influenced by the concentration of traps and their interaction strength. In this description, the observed total permeabilities of diatomic gases are the sums of a molecular permeability, roughly indicated by the noble gas line, and a permeability due to reversibly trapped atoms.

The difference in size between molecules and atoms is greater for nitrogen than for hydrogen and oxygen which, at least partly, explains the experimental fact that nitrogen permeates with the highest rate relative to its expected rate among these gases. The observed relatively high permeability of diatomic gases compared with that of monatomic gases (noble gases) could be explained by a transport of diatomic molecules combined with a retarded permeation of atoms.

In addition, it is interesting to compare our measured nitrogen permeability (▲) with the one given by Ref. [2] (■) shown in Fig. 10. The permeability measured in Ref. [2] is approximately

400 times higher than our value, which is likely due to a higher dissociation fraction of molecular nitrogen in silica used in Ref. [2]. In conclusion, it has been found that a significant fraction of nitrogen has undergone dissociation in this study and nitrogen transport does not only take place in molecular form.

## 5  Summary

Experiments of permeation and uptake/release in vitreous silica have been performed with the use of gas phase analysis and isotopic labeling of nitrogen. It has been shown that some molecules dissociate on the surface and in the bulk of vitreous silica. The dissociation rate of nitrogen on the surface of silica in 400 mbar nitrogen is measured to be $3 \times 10^{-13}$ mol($N_2$) $\cdot$ cm$^{-2}$ $\cdot$ s$^{-1}$ at 900℃. Nitrogen diffuses both in molecular and atomic forms in the silica. The percentage of atomic transport of nitrogen in silica increases with the temperature. At 900℃, approximately 15% atomic nitrogen transport is present. Our measured steady state flux corresponds to a permeability of nitrogen through a silica wall at 900℃ being $5.5 \times 10^{8}$ mol $\cdot$ cm$^{-1}$ $\cdot$ s$^{-1}$, which is 400 times lower than the value measured by Barrer and Johnson. However, also our measured permeability is higher than the expected permeability of molecular nitrogen. An explanation for these experimental results is suggested to include temporarily trapping of atomic nitrogen.

**Acknowledgements:** The authors gratefully acknowledge the financial support from Swedish Foundation for Strategic Research [for one of the authors (Q. D.)] and from Center of Competence in High-Temperature Corrosion [for the other author (G. H.)]. M. J. Graham is also acknowledged for the interest he showed in our work.

### References

[1] R. M. Barrer. Diffusion In and Through Solids [M]. England: Cambridge University Press, Cambridge, 1941.
[2] J. E. Shelby. Handbook of Gas Diffusion in Solids and Melts [M]. Ohio: ASM international, Metals Parks, 1996.
[3] J. Johnson, R. Burt. J. Opt. Soc. Am, 1922, 6: 734.
[4] E. Hörnlund, G. Hultquist. J. Appl. Phys., 2003, 94: 4819.
[5] T. Åkermark, G. Hultquist., Q. Lu. J. Mater. Eng. Perform, 1996, 5: 516.
[6] E. Hörnlund. Appl. Surf. Sci, 2002, 199: 195.
[7] D. R. Lide. CRC Handbook of Chemistry and Physics, 80th ed [M]. Florida: CRC, Boca Raton, 1999.
[8] J. E. Shelby. Mass Transport Phenomena in Ceramics [M]. New York: Plenum, 1975.
[9] R. H. Doremus. Glass Science [M]. New York: Wiley, 1973.

(原载于 Journal of Applied physics 100, 104904 (2006))

# Influence of Additives on Kinetic Behavior of SiO$_2$ - C - N$_2$ System

## Zhuang Youqing   Li Wenchao

(Department of Physical Chemistry, University of Science and Technology Beijing, Beijing 100083)

Silicon nitride ceramics have been under active development for heat engines, bearings, wear parts and other demanding high temperature applications. To obtain silicon nitride with improved mechanical properties, interest in Si$_3$N$_4$ powder preparation and its related properties is increasing among many researchers. Of all the methods of preparation of Si$_3$N$_4$ described in the literatures[1,2], the carbo-thermal reduction of silica followed by nitridation provides the most viable commercial route for Si$_3$N$_4$ production.

This letter attempts to assess the effects of several additives on the yield, $\alpha:\beta$ ratio and particle morphology of Si$_3$N$_4$ powders obtained by nitriding compacted mixtures of doped silica and carbon, which could lead to a more efficient and economical method of producing Si$_3$N$_4$.

Doped silica powders were obtained by TEOS hydrolysis and coprecipitation with 5 wt% additives. The H$_2$O/TESO ratio was 25, gelling temperature 30℃, pH 9, yielding surface areas of 105m$^2$g$^{-1}$ for undoped silica and 210-270m$^2$g$^{-1}$ for doped silica. For the reaction:

$$3SiO_2(s) + 6C(s) + 2N_2(g) = Si_3N_4(s) + 6CO(g)$$

The mixing molar ratio of C/SiO$_2$ was 5. Nitridation was performed at 1400℃ for 16h under nitrogen gas at a flow rate of 200cm·min$^{-1}$, then the excess carbon was burned off in air at 600℃ for 8h. Quantitative determinations of the $\alpha:\beta$ ration of silicon nitride were made by measuring the relative diffracted intensities of the (102)$_\alpha$ and (101)$_\beta$, suggested by Gazzara and Messier[3].

The influence of the type of additives on the Si$_3$N$_4$ formation and phase content is shown in Fig. 1. Note that $f$ representing the increment of Si$_3$N$_4$ content from doped SiO$_2$ compared with that from undoped SiO$_2$. The additives that are beneficial in accelerating the rate of nitride formation can be separated into two groups: (1) metal oxides in which the M-Si-O-N liquid formation temperature is lower than the nitriding temperature (1400℃) (high $\beta$ region, group II) and (2) metal oxides with which M-Si-O-N liquid cannot be formed at the nitriding temperature (low $\beta$ region, group I).

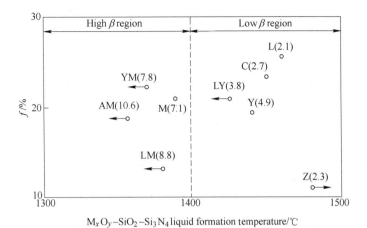

Fig. 1  Influence of the type of additive on $Si_3N_4$ formation and phase content

The figures in parentheses are relative $\beta$-modification contents in $Si_3N_4$ from doped $SiO_2$ compared with that from undoped $SiO_2$(L, $La_2O_3$; C, $Ce_2O_3$; Y, $Y_2O_3$; M, MgO; Z, $ZrO_2$; LY, $La_2O_3 + Y_2O_3$; YM, $Y_2O_3 + $MgO; AM, $Al_2O_3 + $MgO; LM, $La_2O_3 + $MgO)

With high $\beta$ region additives, because of the formation of M-Si-O-N liquid when the reaction precursor is subjected to the nitriding temperature (1400℃), liquid-phase reaction predominating over the synthesis process in this system would be proposed. Schaeffer[4] suggested that the process is a substitution reaction in which an Si-O bond is replaced by an Si-N bond in the presence of carbon. Fig. 2 shows the variations of nitrogen content and $\beta$ modification content with the change of nitriding temperature. There are more increments both in nitrogen content (1.7 wt%) and $\beta$ content (1.3 wt%) of MgO doped sample compared with the undoped one at 1350℃. The result corresponds entirely to the eutectic point of Mg-Si-O-N (1390℃) higher than the nitriding temperature, so no liquid which accelerates the rate of nitridation can be formed at this temperature. When the nitriding temperature is higher than 1390℃, say 1400℃, a drastic increase in both nitrogen content (15.8 wt%) and $\beta$ content (31.0 wt%) takes place. It is thought that the liquid formation would promote the nitridation reaction but favour the $\alpha \to \beta$ transformation (shown in the high $\beta$ region of Fig. 1.), so the resulting powder has high $\beta$ content. The investigation by

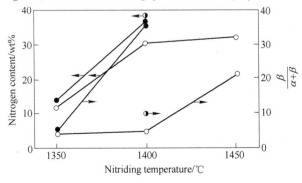

Fig. 2  Nitrogen content and $\beta$-modification content variations with change of nitriding temperature

○—Undoped; ●—Mg-doped; ◐—La-doped

Wusirika and Chyung[5] shows that the liquid which contain nitrogen also exhibits a very high ability to nucleate $Si_3N_4$. $\Delta G(\alpha) = -1167.3 + 0.596T$ (kJ·mol$^{-1}$), $\Delta G(\beta) = -925.2 + 0.450T$ (kJ·mol$^{-1}$), for $T > 1380℃$, $\Delta G(\alpha \rightarrow \beta) < 0$, so the $\beta$-modification nitride predominates over the precipitation process from the liquids[6,7].

In another aspect, the liquid formation can promote the growth of $\beta$-rich particles, hence coarser powder is obtained in this case. Table 1 shows the composition, resulting particle size and phase content of powders obtained from undoped and doped silica. Note the distinct difference in $\beta$ content and particle size between samples in groups Ⅰ and Ⅱ. This shows that the additives in group Ⅱ can effectively accelerate the rate of nitride formation, and moreover facilitate the $\alpha \rightarrow \beta$ transformation as well as particle growth, which is not the expected result.

**Table 1   Characteristics of the $Si_3N_4$ powder obtained**

| Group | Sample① | Nitrogen content /wt% | Oxygen content /wt% | $\beta/\alpha+\beta$ | Particle size /μm |
|---|---|---|---|---|---|
| Ⅰ | Undoped | 30.2 | 13.1 | 0.05 | 3.0 |
| | L | 37.8 | 3.0 | 0.10 | 0.7 |
| | Y | 35.9 | 5.4 | 0.25 | 0.5 |
| | LY | 36.5 | 4.7 | 0.19 | 1.2 |
| Ⅱ | M | 36.5 | 4.5 | 0.36 | 3.3 |
| | YM | 36.9 | 4.2 | 0.40 | 3.7 |
| | AM | 35.9 | 5.5 | 0.54 | 3.4 |

①L, $La_2O_3$; Y, $Y_2O_3$; LY, $La_2O_3 + Y_2O_3$; M, MgO; YM, $Y_2O_3 + $MgO; AM, $Al_2O_3 + $MgO.

With low $\beta$ region additives, because no liquid can be formed at the nitriding temperature, another accelerating mechanism of additives must be proposed. Metallic ions ($M^{n+}$: $La^{3+}$, $Y^{3+}$ etc.) can be introduced into the $SiO_2$ structure and the Si-O-M cluster can be formed during sol-gel processing. The electronegative difference between $Si^{4+}$ and $M^{n+}$ drives the bonding electrons to the oxygen atoms, which leads to a high potential of the $M^{n+}$ coordinatively bonding with the absorbed electron-pair donors, CO in this case. This signifies that $M^{n+}$ is a Lewis acid site, denoted as:

$$M(\leftarrow :\ddot{O}: --- \underset{|}{\overset{|}{S}}-)_n$$
Lewis acid site

Because $SiO_2(s) + CO(g) = SiO(g) + CO_2(g)$ predominates over the silica reduction in this system[8], the absorption of CO on M is very important for the process. CO can be absorbed on a Lewis acid site in the form of $CO_3^{2-}$, and then disabsorbed as SiO(g) and $CO_2(g)$, and the partial pressure of SiO is increased. Hendry and Jack[9] considered that a high partial pressure of SiO produces a nitride high in $\alpha$. Fig. 3 shows the difference in $Si_2N_4$ content for various preparation processes from $SiO_2$ precursors. In the case of Mg doping, $Si_3N_4$ content increases for both preparation processes compared with that from undoped $SiO_2$, which is attributed to the identical eutectic temperature (1390℃) of Mg-Si-O-N system for both processes, while in the case of La doping, the nitriding results are dramatically different for the two processes: the additives can dramatically accelerate the rate of $\alpha$-$Si_3N_4$ formation in which the doped $SiO_2$ precursor is derived from hydrolysis of TEOS, while the same additive mixed with $SiO_2$ by milling is without catalytic effect on the nitride formation, because no Si-O-M cluster formed. The results of other observators can support this hypothesis [10,11]. It can be concluded that the catalytic effect on nitride formation of low $\beta$ region additives is conditioned by the metallic ion being introduced into the Si-O network.

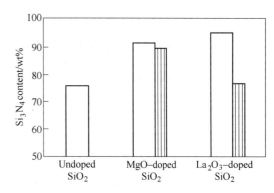

Fig. 3  Silicon nitride content of undoped. Mg-doped and La-doped silica prepared by (□), sol-gel process and (▥), milling

In the $SiO_2$-C-$N_2$ system, the additives that are beneficial in accelerating the rate of nitride formation can be separated into two groups. High $\beta$ region additives can promote the $\alpha \rightarrow \beta$ transformation for the M-Si-O-N liquid at the nitriding temperature, so the resulting powder has high $\beta$ modification (up to 54 wt %).

Low $\beta$ region additives participate in the Si-O network, and form Si-O-M cluster which is active in $SiO_2$ reduction. Conditions for a high yield of product with a high $\alpha$:$\beta$ ratio require a higher silicon monoxide partial pressure which, at present, is achieved by Si-O-M cluster formation.

The particle size of $Si_3N_4$ is principally determined by additives. La-adoped $SiO_2$ produced $\alpha$-$Si_3N_4$ powder 0.7 μm in diameter, while Mg-doped $SiO_2$ produced much coarser $Si_3N_4$ powder (3.3 μm) because of Mg-Si-O-N liquid formation at the nitriding temperature.

**References**

[1] I. B. Cutler, W. J. Croft. Powder Met. Int., 1974, 92 (6): 143.
[2] C. Schwiex. In " Progress in nitrogen ceramics". Edited by F. L. Riley. The Netherlands: Martinus NIjhoff, 1983.
[3] C. P. Gazzara, R. R. Messier. Amer. Ceram. Soc. Bull., 1977, 56: 777.
[4] H. A. Schaeffer. In "Progress in nitrogen ceramics". Edited by F. L. Riley. The Netherlands: Martinus NIjhoff, 1977.
[5] R. R. Wusirika, C. K. Chyung. J. Non-Cryst. Solids, 1980, 38: 39.
[6] T. G. Chart. High temp. -High Press [M]. 1970 (2): 461.
[7] A. Hendry. In"Progress in nitrogen ceramics". Edited by F. L. Riley. The Netherlands: Martinus NIjhoff, 1977.
[8] Y. Zhuang, W. Li, X. Zhong. In "Advanced structural materials". The Netherlands: Elsevier Science, 1999.
[9] A. Hendry, K. H. Jack. In " Special ceramics". Edited by P. Popper. UK: British Ceramic Research Association, Stoke-on-Trent, 1975.
[10] I. A. Rahaman, F. L. Riley. J. Eur. Ceram. Soc., 1989, 11 (5).
[11] W. L. Li., H. R. Zuang, X. R. Fu. In "Ceramic powders". Edited by P. Vincenzini. The Netherlands: Elsevier Science, 1983.

(原载于 *Journal of Materials Science Letters*, 1994, 13 (6): 410-412)

# Silica Photonic Crystals with Quasi-full Band Gap in the Visible Region Prepared in Ethanol

Zhang Hui[1,2]　Wang Xidong[1]　Zhao Xiaofeng[2]
Li Wenchao[1]　Tang Qing[2]

(1. University of Science and Technology Beijing, Beijing 100083;
2. Institute of Process Engineering, Chinese Academy of Sciences, Beijing 100080)

**Abstract:** Monodisperse silica spheres of 252nm with a standard deviation of 5.7% are prepared by Stöber method. By comparison of both of media, ethanol instead of water is used to assemble opal, and the artificial opal has been prepared by the sedimentation in ethanol of silica spheres. The structure of the opal prepared has been examined and discussed. The results show that the artificial opal has a structure similar to the face-centered cubic (fcc) type packed system with silica spheres. Transmission measurements of the artificial opal have been conducted, which shows that the artificial opal is quasi-full band gap silica photonic crystals in the visible region.
**Keywords:** photonic crystals; artificial opal; quasi-full band gap

Three dimensional (3D) photonic band gap materials or photonic crystals are 3D periodic dielectric structures, they receive increased attention of scientists mainly due to their potential applications in optoelectronic field[1-3]. 3D photonic crystals can be used to improve the properties of semiconductor lasers, to make new types of optoelectronic devices and to enhance the efficiency of fiber optical communication systems.

In view of the fact that 3D periodic solid structures are restricted to a few of layers through the conventional microfabrication techniques[4], the self-assembly of latex or silica spheres in water suspensions has been used to fabricate 3D periodic dielectric structures[5-7]. These colloidal crystals, with the 3D fcc ordering for silica colloidal crystals and the 3D body-centered cubic (bcc) or fcc ordering for polystyrene colloidal crystals, show Bragg diffraction effects in the visible region[8,9]. However, these systems are rather difficult to manage because they are in liquid suspensions.

Opal, amorphous silica spheres with 3D closely packed arrangements is a high quality natural photonic crystal as it diffracts white light into color strips. Recently, artificial opals were proposed as promising photonic band gap materials[10,11]. Considering the different properties of water and ethanol, using ethanol instead of water will probably make the fabrication of artificial opals of high quality easier.

---

Supported by the National Science Foundation of China (Grant No. 59974026).
张辉，1999~2003 年于北京科技大学师从李文超教授攻读材料学学位，获得北京科技大学优秀博士学位论文荣誉称号。目前在北京交通大学工作，副教授，发表论文30余篇，获奖3项。

In the present work, a high ordering opal crystal was prepared by means of sedimentation of silica spheres in ethanol and its structure and optical properties were examined.

# 1 Experimental procedure

## 1.1 Preparation of monodisperse silica spheres

Ethyl ester of orthosilicate (TEOS), ammonia water, distilled water and ethanol, all in analytically pure, were used as raw materials to prepare monodisperse silica spheres. Stöber method [12] was used and the reactions are as follows:

$$Si(OC_2H_5)_4 + 4H_2O = Si(OH)_4 + 4C_2H_5OH$$
$$Si(OH)_4 = SiO_2 \downarrow + 2H_2O$$
$$Si(OC_2H_5)_4 + 2H_2O = SiO_2 \downarrow + 4C_2H_5OH$$

The average diameter of silica spheres obtained was about 252nm with a standard deviation of 5.7%. The SEM morphology of monodisperse silica spheres with a diameter of 252nm is shown in Fig. 1.

## 1.2 Fabrication of the artificial opal

A glass tube of $\phi$18mm and a fine polished optical glass plate with a roughness of 30nm, sealed with cyanoacrylate adhesive, were used to prepare colloidal crystals. Colloidal crystal samples were obtained from the slow sedimentation of silica microspheres in colloidal suspension. The ethanol was extracted carefully when the sedimentation of

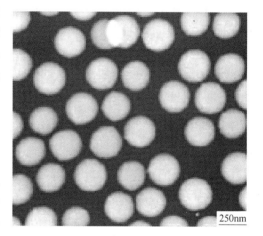

Fig. 1 SEM image of monodispersed silica spheres with a diameter of 252nm

particles was complete. The colloidal crystals were then dried at room temperature. On the average, it took about two weeks to prepare a 0.2mm thick opal. When the artificial opal was placed under an incandescent lamp, diffractive colors from green to red could be seen with the change of the observation angle.

## 1.3 Examination of the artificial opal

After the preparation, the morphology of the artificial opal sample made by sedimentation in ethanol of silica spheres was characterized with a scanning electron microscope (Cambridge Stereoscan 250MK 2).

In order to optically characterize the artificial opal, light transmission measurements of the sample's top surface were carried out with a spectrophotometer (CARY2390). In transmission spectra, the information of every Bragg diffraction plane was contained. Light had to cross hundreds of layers with the opal sample having an average thickness of 0.2mm. Band gap, caused by Bragg diffraction of the planes parallel to the top surface, could be observed.

## 2 Results and discussion

### 2.1 Comparison of water with ethanol as the sedimentation media

Water was often chosen as a sedimentation medium to prepare silica photonic crystals. In this work, ethanol, instead of water, was used as the medium. Comparing with water as medium, there are several advantages to use ethanol as a sedimentation medium, such as: (1) its surface tension is lower (about one third of that of water: $\sigma_{ethanol} = 22.1\rm{N} \cdot m^{-1}$ and $\sigma_{water} = 72.75\rm{N} \cdot m^{-1}$). Thus the dispersivity of the microparticles in ethanol medium is better; (2) in ethanol, it is easier for the silica spheres to be in the most stable position and the arrangement of silica spheres would be more regular after the sample is dried; (3) as ethanol volatilizes faster than water does, so the natural drying speed of ethanol colloidal crystal is faster than that of water, moreover, this kind of somewhat fast drying speed won't destroy the order of crystal. But the situation will probably be different if the other organic solvents are chosen as sedimentation media.

From the above-mentioned reasons, ethanol was used as a sedimentation medium in the present experimental work.

### 2.2 Observation and analysis of structure of the artificial opal

The SEM image from the top surface of the prepared sample is shown in Fig. 2. The triangular arrangement observed can be corresponded, in principle, to either a (111) surface of a fcc system or a (001) surface of a hexagonal closely packed (hcp) system. During the transition of spheres from colloid to solid phase, the volume-filling fraction is changed from around 40% for the colloidal phase to 76% for the solid phase as the solvent (medium) is gradually vaporized. Therefore, the floatage and Coulombian force become smaller and smaller. The crystallinity of silica spheres can become the stacking of hard spheres, if the vaporization lasts a rather long time[10]. Woodcock[13], using computer simulations, showed that the most stable stacking of hard spheres was fcc arrangement. Van Blaaderen et al.[6] have grown fcc sphere packings through a template directed crystallization. It was confirmed that artificial opals obtained from ethanol indeed belong to an analogue fcc structure through their microscopic and transmission characterizations[14].

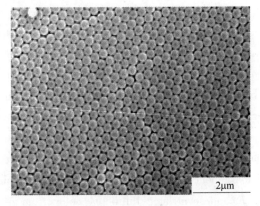

Fig. 2 SEM image of the top (111) surface of the artificial opal made from 252nm spheres

A typical SEM photograph taken from a cleaved edge of the artificial opal sample is shown in Fig. 3. Wide sections with triangular arrangements corresponding to {111} planes can be observed, whereas some square alignments corresponding to {100} planes can also be seen. This sort of lateral order can only be seen in a fcc structure. The three surfaces shown in Fig. 3. correspond to (100), ($1\bar{1}\bar{1}$) and ($1\bar{1}1$) crystalline planes of an fcc structure. A number of SEM images of cleaved edges of the artificial opal have been observed and results are the same as given in Fig. 3.

Fig. 3  SEM image of a lateral part of the artificial opal,
Hexagonal alignment corresponds to a {111}
surface and square alignment corresponds to a {100} surface

## 2.3  Band gap feature of the artificial opal

The spectrum for an opal sample of 0.2mm in thickness is shown in Fig. 4. The spectrum was measured in the normal incidence direction ($\theta = 0°$). A clear attenuation band in the optical transmission could be observed due to the Bragg diffraction of the (111) plane. Around the center of the band gap, whose wavelength (marked $\lambda_c$) was about 568nm, the transmitted intensity was reduced by more than one order of magnitude. Such a high drop also indicates a fine alignment of the (111) plane.

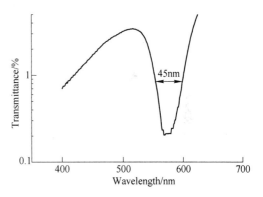

Fig. 4  Transmission spectrum of the top (111) surface of a 0.2mm thick artificial opal

The center wavelength can also be calculated with the Bragg law: $\lambda_c = 2n_{\text{eff.}} d_{\text{hkl}} \cos\theta$, where $n_{\text{eff}}$ is the effective refractive index of silica/air composite refraction ($n_{\text{eff.}}^2 = n_{\text{SiO}_2}^2 f + n_{\text{air}}^2 (1-f)$, $f$ is the filling ratio, $f = 0.74$ for a close-packed structure, then, $n_{\text{eff.}} \approx 1.347$), $d_{\text{hkl}}$ is the crystalline plane spacing distance (for a fcc structure, $d_{111} = 0.816\phi = 205.6$nm, $\phi = 252$nm), $\theta$ is the incidence angle. Then for normal incidence, the Bragg reflection maximum ($\lambda_c$) can be calculated and its value is 554nm, which matches the experimental value (568nm) quite well.

In order to estimate the peak broadening effect, a comparison of the experimental results with the analytical expression derived by Tarhan and Watson[15] has been carried out. In the model of Tarhan and Watson, $(\Delta\omega/\omega_c) = 0.054$ for the [111] stop band of an analogous fcc photonic crystal. For 0.2mm thick sample prepared in the present experiment, $(\Delta\omega/\omega_c) = 0.079$, where $\Delta\omega = 45$nm obtained from Fig. 4., the difference between the value of experimental and that of calculation is because that the opal sample prepared is not a single crystal structure.

Transmission measurements in the visible range were also performed at different angles $\theta$ with respect to the top surface. The results obtained for a sample of 0.2mm in thickness are shown in Fig. 5. At normal incidence direction ($\theta = 0°$), the dip of band gap is the deepest compared with other directions. It is suggested that the order of packing of microparticles be the best in crystal growing direction. As the angle of incidence increases, the dip (center wavelength) of band gap shifts to shorter wavelengths according to the Bragg law.

The relation between the angle and the center of wavelength is shown in Fig. 6, and results show that the opal synthesized in the present work is quasi-full band gap silica photonic crystals. The opal prepared, with a contrast of refractive index of about 1.45 and a filling factor of 74%, also suggests that the opal be a quasi-full band gap photonic crystal.

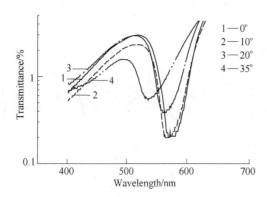

Fig. 5  Transmission spectra for different incidence angles with respect to the top (111) surface in a 0.2mm thick opal sample

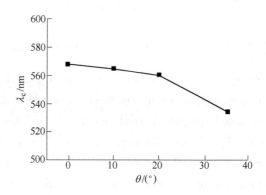

Fig. 6  The center wavelengths ($\lambda_c$) of the [111] bands plotted against the incidence angle

Thus, although the arrangement of packed layers of silica spheres could follow either fcc or hcp structure, both scanning electron microscopy (SEM) photographs and transmission measurements indicate that the cubic structure is favored. This tendency of the fcc arrangement has also been observed in solid samples made in dilute water suspensions[10].

## 3  Conclusions

The artificial opal was prepared by sedimentation in ethanol of silica spheres. The structure and optical properties of the opal have been measured, and results indicate that the packing of silica submicron spheres is 3D photonic crystals in the visible region. Microscopy and optical evidence of the 3D ordering strongly support an analogous fcc structure of the packing. In addition, a quasi-full band gap has also been observed.

## References

[1] Yablonovitch, E. Inhibited spontaneous emission in solid-state physics and electronics [J]. Phys. Rev. Lett., 1987, 58 (20): 2059.

[2] John, S. Strong localization of photons in certain disordered dielectric superlattices [J]. Phys. Rev. Lett., 1987, 58 (20): 2486.

[3] Gates, B., et al. Fabrication and characterization of chirped 3D photonic crystals [J]. Adv. Mater., 2000, 12 (18): 1329.

[4] Cheng, C. C., et al. Fabrication of photonic band-gap crystals [J]. J. Vac. Sci. Technol. B, 1995, 13 (6): 2696.

[5] Mei, D. B., et al. Visible and near-infrared silica colloid crystals and photonic gaps [J]. Phys. Rev. B, 1998, 58 (1): 35.

[6] Van Blaaderen, A., et al. Template-directed colloid crystallization [J]. Nature, 1997, 385 (23): 321.

[7] Sun, H. B., et al. Growth and property characterizations of photonic crystal structures consisting of colloidal microparticles [J]. J. Opt. Soc. Am. B, 2000, 17 (3): 476.

[8] Tsuyoshi Yoshiyama, et al. Kossel images as direct manifestations of the gap structure of the dispersion surface for colloidal crystals [J]. Phys. Rev. Lett., 1986, 56 (15): 1609.

[9] Willem, L., et al. Strong effects of photonic band structures on the diffration of colloidal crystals [J]. Phys. Rev. B, 1996, 53 (24): 16231.

[10] Cheng, B. Y., et al. More direct evidence of the fcc arrangement for artificial opal [J]. Optics Communications, 1999, 170: 41.

[11] Míguez, H., et al. Photonic crystal properties of packed submicrometric $SiO_2$ spheres [J]. Appl. Phys. Lett., 1997, 71 (9): 1148.

[12] Stöber, W., et al. Controlled growth of monodisperse silica spheres in the micron size range [J]. J. Colloid and Interface Science, 1968, 26: 62.

[13] Woodcock, L. V. Entropy difference between the face-centered cubic and hexagonal close-packed crystal structures [J]. Nature, 1997, 385 (9): 141.

[14] Zhang, H., et al. Preparation and characterization of silica photonic crystals made from ethanol [J]. Journal of University of Science and technology Beijing (in Chinese), 2002, 24 (2): 157.

[15] Tarhan, I. I., et al. Analytical expression for the optimized stop bands of fcc photonic crystals in the scalar-wave approximation [J]. Phys. Rev. B, 1996, 54 (11): 7593.

(原载于 *Progress in Natural Science*, 2003, 13(9): 79-82)

# Fe-C-$j$ ($j$ = Ti、V、Cr、Mn) 熔体的热力学性质规律

王海川[1]　王世俊[1]　乐可襄[1]　董元篪[1]　李文超[2]

(1. 安徽工业大学冶金与材料学院，马鞍山　243002；
2. 北京科技大学冶金学院，北京　100083)

**摘　要**：利用本文提出的热力学性质方法，根据 Fe-C-$j$ ($j$ = Ti、V、Cr、Mn) 熔体中的 C 溶解度与温度及第三组元 $j$ 之间的关系式，研究了 Fe-C-$j$ 熔体中第三组元 $j$ 对 C 溶解度的影响因子 $k_j$、$m_j$ 以及组元的活度相互作用系数与第三组元 $j$ 原子序数间的关系，探讨了 Fe-C-$j$ ($j$ = Ti、V、Cr、Mn) 熔体中不同组元热力学性质的变化规律。

**关键词**：Fe-C-$j$ ($j$ = Ti、V、Cr、Mn) 熔体；C 溶解度；活度相互作用系数；原子序数

## The Rules of Thermodynamic Properties in Fe-C-$j$ ($j$ = Ti, V, Cr, Mn) Melts

Wang Haichuan[1]　Wang Shijun[1]　Yue Kexiang[1]　Dong Yuanchi[1]　Li Wenchao[2]

(1. School of Metallurgy and Material, Anhui Polytechnical University, Maanshan 243002;
2. School of Metallurgy, University of Science and Technology Beijing, Beijing 100083)

**Abstract**: By a calculating method for thermodynamic property and based on the relationship expression of carbon solubility in Fe-C-$j$ ($j$ = Ti, V, Cr, Mn) melt and temperature and third component $j$ obtained in this work, it is studied that the relationship between not only the effect factor of the third component $j$ on carbon solubility $k_j$, $m_j$ but also activity interaction coefficient of component $j$ upon carbon and atomic number of third component $j$, and it is also probed that the objective rules among thermodynamic property of components in melt.

**Keywords**: Fe-C-$j$ ($j$ = Ti, V, Cr, Mn) melt; carbon solubility; activity interaction coefficient; atomic number

　　Ti、V、Cr、Mn 是元素周期表中与 Fe 近邻的重要金属元素，这类元素的铁合金是炼钢生产中重要的脱氧和合金化剂，因此，研究这类元素铁合金熔体的热力学性质对于冶金生产实践具有重要的实际意义。由于这类铁合金易于形成 C 饱和熔体，最常用的热力学研究方法是 C 溶解度法。然而，从已有的利用 C 溶解度研究三元 Fe-C-$j$ 金属熔体热力学性质的大量文献报道[1-9]来看，虽然不同研究者采用的数据处理方法、热力学性质计算方法不尽相同，但所有的研究报道均按温度条件的不同，进行实验数据处理和热力学性质计

---

王海川，1996～2000 年于北京科技大学师从李文超教授攻读冶金物理化学专业工学博士学位。目前在安徽工业大学工作，任冶金工程学院副院长。获得全国优秀教师、安徽省优秀教师、宝钢教育基金优秀教师奖、中国金属学会冶金先进青年科技工作者等荣誉称号，发表论文 120 余篇，获奖省部级科技奖励 6 项。

算,且由于实验误差等原因,不同研究者即使在相同温度下,研究相同体系所得的 C 溶解度关系式也不一致,同时计算过程中所引用的文献不同,选取的参考数据也不同,从而导致熔体组元的热力学性质参数的报道存在较大分歧[10],这种状况不利于深入研究熔体组元的热力学性质以及探讨热力学性质之间的规律。

本文根据已有大量文献报道[11]及本研究测定(表1)的 Fe-C-$j$($j$ = Ti、V、Cr、Mn)熔体中 C 溶解度的实验数据,利用文献 [11] 的数据处理方法,得到 Fe-C-$j$ 熔体中的 C 溶解度($x_C$)与温度($T$)和第三组元 $j$($x_j$)之间的关系式,利用此关系式计算 Fe-C-$j$ 熔体中组元的活度相互作用系数,研究组元 $j$ 对 C 溶解度的影响及组元的活度相互作用系数与组元 $j$ 的原子序数之间的关系,并揭示 Fe-C-$j$ 熔体中组元的热力学性质规律。

表1 1673K 温度下 Fe-C-V 熔体中的 C 溶解度的实验数据及处理结果
Table 1 Experimental data and treatment result of carbon solubility in Fe-C-V melt at 1673K

| No. | [%V]/mass% | [%C]/mass% | Δ[%C]/mass% | $x_V$/mol·mol$^{-1}$ | $x_C$/mol·mol$^{-1}$ | $\Delta x_C$/mol·mol$^{-1}$ |
| --- | --- | --- | --- | --- | --- | --- |
| 1 | 0.00 | 4.87 | 0.000 | 0.0000 | 0.1923 | 0.0000 |
| 2 | 1.96 | 5.13 | 0.260 | 0.0181 | 0.2006 | 0.0083 |
| 3 | 3.11 | 5.40 | 0.530 | 0.0284 | 0.2092 | 0.0169 |
| 4 | 3.90 | 5.43 | 0.560 | 0.0356 | 0.2101 | 0.0178 |
| 5 | 5.62 | 5.52 | 0.650 | 0.0511 | 0.2127 | 0.0204 |
| 6 | 5.90 | 5.66 | 0.790 | 0.0534 | 0.2171 | 0.0248 |
| 7 | 6.80 | 5.70 | 0.830 | 0.0614 | 0.2182 | 0.0259 |
| 8 | 8.20 | 5.91 | 1.040 | 0.0735 | 0.2246 | 0.0323 |

## 1 热力学性质计算方法

文献 [11] 研究了不同温度条件下 Fe-C-Mn 熔体中 C 的溶解度,其数据处理方法如式(1)所示:

$$\Delta x_C^{Mn} = x_C - x_C^b = 0.1 x_{Mn} \tag{1}$$

式中,$x_C$ 为 Fe-C-Mn 熔体中 C 的溶解度;$x_{Mn}$ 为 Fe-C-Mn 熔体中 Mn 的摩尔分数;$x_C^b$ 为纯 Fe 熔体中 C 的溶解度。该文还研究了 Fe 基熔体中 P、Si 对 C 溶解度的影响,但并未研究 C 溶解度关系式之间的关系,式(1)也未在热力学性质计算中应用。文献 [8] 采用类似的数据处理方法研究过 1350 ℃下 Mn-Ca-$j$ 熔体中 Cr、Al 对 Ca 的溶解度的影响。因此,按这种数据处理方法得到的 C 溶解度关系式在 Fe-C-$j$ 熔体组元的热力学研究中没有得到应用。

本文利用文献 [11] 的数据处理方法,将不同温度条件下 C 在三元 Fe-C-$j$($j$ = Ti、V、Cr、Mn)熔体中的溶解度分别按摩尔分数($x_C$)和质量百分数浓度([%C])进行处理,得到 C 的溶解度与纯 Fe 熔体中 C 的溶解度之差与第三组元 $j$ 之间的关系式:

$$x_C - x_C^b = k_j \cdot x_j \tag{2}$$

$$[\%C] - [\%C]_b = m_j \cdot [\%j] \tag{3}$$

式中,$x_C$、[%C] 分别为 Fe-C-$j$ 熔体中组元 $j$ 对 C 溶解度;$x_j$、[%$j$] 分别为 Fe-C-$j$ 熔体中组元 $j$ 的摩尔分数和质量百分数浓度;$k_j$、$m_j$ 为 Fe-C-$j$ 熔体中组元 $j$ 对 C 溶解度的影响因子，$k_j$、$m_j$ 仅与组元 $j$ 的浓度有关、与温度无关，$j$ 确定时 $k_j$、$m_j$ 为定值（见表2）；$x_C^b$ 和 $[\%C]_b$ 为 C 在纯 Fe 熔体中的溶解度，$x_C^b$ 和 $[\%C]_b$ 仅仅与温度有关，与组元 $j$ 无关。由于 $x_C^b$ 和 $[\%C]_b$ 为离散的实验值，为了利用式（2）、式（3）计算不同温度下熔体组元的热力学性质，对 Fe 基熔体，用纯 Fe 熔体中 C 的溶解度与温度的关系式（4）、式（5）中的 $x_C^{Fe}$、$[\%C]_{Fe}$ 分别代替式（2）、式（3）中的 $x_C^b$、$[\%C]_b$，$x_C^{Fe}$、$[\%C]_{Fe}$ 与温度的关系分别为：

$$x_C^{Fe} = 0.0520 + 0.0848 \times 10^{-3} T \text{ (K)} \tag{4}$$

$$[\%C]_{Fe} = 0.6021 + 2.5786 \times 10^{-3} T \text{ (K)} \tag{5}$$

将式（4）、式（5）分别代入式（2）、式（3），并重新整理后，得到：

$$x_C = (0.0520 + 0.0848 \times 10^{-3} T) + k_j \cdot x_j \tag{6}$$

$$[\%C] = (0.6021 + 2.5786 \times 10^{-3} T) + m_j \cdot [\%j] \tag{7}$$

按照式（6）、式（7），可将 Fe-C-$j$ 熔体中 C 的溶解度进行简单分离，分别得到温度（$T$）和第三组元 $j$ 对 C 溶解度的影响因子 $x_C^{Fe}$、$[\%C]_{Fe}$ 和 $k_j$、$m_j$，使 Fe-C-$j$ 熔体中的 C 溶解度的物理意义更加明确，对深入研究 Fe-C-$j$ 熔体中组元的热力学性质间的关系提供了可能。

根据等活度相互作用系数 $\dot\varepsilon_C^j$、$\dot\rho_C^j$、$\dot e_C^j$ 和 $\dot r_C^j$ 的定义[13]，在 C 等活度条件下，得到用影响因子 $k_j$、$m_j$ 和温度 $T$ 表示的组元 $j$ 对 C 的一阶和二阶等活度相互作用系数分别为：

$$\dot\varepsilon_C^j = \left(\frac{-\partial \ln x_C}{\partial x_j}\right)_{\substack{a_C \\ x_j \to 0}} = -\frac{k_j}{0.0520 + 0.0848 \times 10^{-3} T} \tag{8}$$

$$\dot\rho_C^j = \frac{1}{2}\left(\frac{-\partial^2 \ln x_C}{\partial x_j^2}\right)_{\substack{a_C \\ x_j \to 0}} = \frac{1}{2}\left(\frac{k_j}{0.0520 + 0.0848 \times 10^{-3} T}\right)^2 \tag{9}$$

$$\dot e_C^j = \left(\frac{-\partial \lg[\%C]}{\partial [\%C]}\right)_{\substack{\%C \\ [\%C] \to 0}} = -\frac{1}{2.303} \times \frac{m_j}{0.6021 + 2.5786 \times 10^{-3} T} \tag{10}$$

$$\dot r_C^j = \frac{1}{2}\left(\frac{-\partial^2 \lg[\%C]}{\partial [\%C]^2}\right)_{\substack{\%C \\ [\%C] \to 0}} = -\frac{1}{4.606} \times \left(\frac{m_j}{0.6021 + 2.5786 \times 10^{-3} T}\right)^2 \tag{11}$$

根据等活度相互作用系数与活度相互作用系数之间的关系[13]，如式（12）所示，可以将 Fe-C-$j$ 熔体组元的等活度相互作用系数转化为常用的活度相互作用系数，如式（12）所示：

$$\varepsilon_C^j = \dot\varepsilon_C^j \times (1 + \varepsilon_C^C \cdot x_C^{Fe} + 2\rho_C^C \cdot (x_C^{Fe})^2) \tag{12}$$

式中，$\varepsilon_C^C$、$\rho_C^C$ 分别为 C 非饱和条件下 C 对 C 的一阶和二阶活度相互作用系数，$\varepsilon_C^C$、$\rho_C^C$ 的值参见文献 [11]。将式（8）~式（12）应用到 Fe 基熔体组元的活度相互作用系数的计算中，能够使利用 C 溶解度研究 Fe-C-$j$ 熔体热力学性质的工作难度降低，计算过程和公式大大简化，热力学性质参数的获取变得非常容易，只需利用一些特殊温度条件下 C 的饱和溶解度数据，求得第三组元 $j$ 对 C 溶解度的影响项 $k_j$ 和 $m_j$，就很容易地计算出任意温度（熔体状态存在）下 Fe-C-$j$ 熔体中组元的热力学性质参数，进而还可以得到组元的活度相互作用系数与温度之间的关系式。

## 2 实验结果及计算分析

利用上述计算方法，本文首先对 Fe-C-Ti、Fe-C-V、Fe-C-Cr、Fe-C-Mn 熔体中的 C 溶解度进行处理,以 1673K 温度下 Fe-C-V 熔体中 C 的溶解度为例,实验数据及处理结果见表 1,1623~1923K 温度下 Fe-C-V 熔体中第三组元 V 对 C 的溶解度的影响如图 1 所示,其他 Fe-C-$j$ 熔体中 C 的溶解度实验数据及处理结果参见文献 [11]。

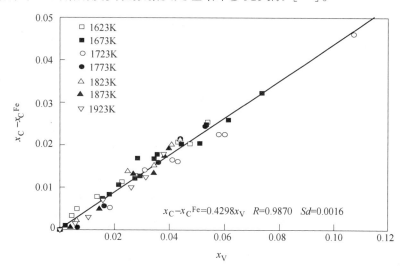

图 1　不同温度下 Fe-C-V 熔体中 V 对 C 饱和溶解度的影响

Fig. 1　Effect of on the carbon saturated solubility in Fe-C-V melt at different temperatures

## 2.1　第三组元 $j$ 对 C 溶解度的影响因子之间的关系

根据式(2)、式(3)得到的不同 Fe-C-$j$ 熔体中第三组元 $j$ 对 C 溶解度的影响因子 $k_j$、$m_j$ 见表 2,对于 Fe 基熔体,令 Fe 元素的 $k_j=0$ 和 $m_j=0$。将各体系的第三组元 $j$ 对 C 溶解度的影响因子 $k_j$ 和 $m_j$ 进行比较,可以发现 $k_j^*$、$m_j$ 与第三组元 $j$ 的原子序数有直接对应关系。

表 2　Fe-C-$j$ 熔体中第三组元 $j$ 对 C 溶解度的影响因子 $k_j$、$m_j$ 比较

Table 2　Effect factor $k_j$、$m_j$ of component $j$ on carbon solubility in Fe-C-$j$ melt

| No. | Element $j$ | Ti | V | Cr | Mn | Fe |
|---|---|---|---|---|---|---|
| 1 | atomic number | 22 | 23 | 24 | 25 | 26 |
| 2 | $k_j$ | 0.5500 | 0.4298 | 0.2544 | 0.1093 | 0.0000 |
| 3 | $m_j$ | 0.1721 | 0.1247 | 0.0762 | 0.0291 | 0.0000 |
| 4 | Range of Temp. /K | 1623~1773 | 1623~1973 | 1573~2273 | 1563~1963 | — |

图 2 为不同 Fe-C-$j$ 熔体中的第三组元 $j$ 对 C 溶解度的影响因子 $k_j$、$m_j$ 与组元 $j$ 的原子序数之间的关系,从图中可以看出,四种元素对 C 溶解度的影响因子 $k_j$、$m_j$ 均大于 0,加入上述元素均增大 Fe-C-$j$ 熔体中 C 的溶解度,而且 $k_j$、$m_j$ 与原子序数之间呈现非常好的线性关系,说明元素周期律影响 Fe-C-$j$ 熔体中 C 的溶解度。

在元素周期表中,Ti、V、Cr、Mn 为 ⅣB~ⅦB 族元素,原子序数从左至右依次递增,在周期表中均处于 Fe 元素的左侧,而 C 元素在元素周期表中处于 Fe 元素的右侧(ⅣA 族),Ti、V、Cr、Mn 元素与 C 元素之间的相互结合力比 Fe 与 C 之间的相互结合力要大,

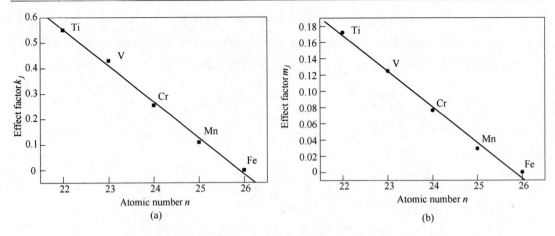

图 2 第三组元 $j$ 对 C 饱和溶解度的影响因子 $k_j$、$m_j$ 与元素 $j$ 的原子序数间的关系

Fig. 2 Relationship of the effect factor $k_j$、$m_j$ and atomic number of third component $j$

在 C 饱和状态下，Ti、V、Cr、Mn 能与 C 形成比 $Fe_3C$ 更加稳定的化合物，因此能增加 Fe-C-$j$ 熔体中的 C 溶解度。

同时 Ti、V、Cr、Mn 与 Fe 原子的半径相近，这类原子质点形成代位型熔体，C 原子半径与之相比很小，与这类原子质点形成间隙型熔体。Ti、V、Cr、Mn 元素随着原子序数的增大，原子半径依次减小，在 C 原子充填进入这类原子间隙的比例就增大，则 C 的溶解度也随之增大。

### 2.2 等活度相互作用系数之间的关系

利用式（8）~式（11）可以计算不同熔体中任意温度条件下组元的等活度相互作用系数 $\dot{\varepsilon}_C^j$、$\dot{\rho}_C^j$、$\dot{e}_C^j$ 和 $\dot{r}_C^j$，以 1823K 为例，计算得到的组元的等活度相互作用系数见表 3。对于 Fe 基熔体，令 Fe 元素的 $\dot{\varepsilon}_C^j=0$、$\dot{\rho}_C^j=0$、$\dot{e}_C^j=0$ 和 $\dot{r}_C^j=0$。从表中可以发现，第三组元 $j$ 对 C 的活度相互作用系数与组元 $j$ 的原子序数也存在定量关系。

表 3 Fe-C-$j$ 熔体中第三组元 $j$ 对 C 的等活度相互作用系数的影响（1823K）

Table 3 Effect of third component $j$ on activity interaction coefficient of C at constant activity in Fe-C-$j$ melt

| No. | element $j$ | Ti | V | Cr | Mn | Fe |
|---|---|---|---|---|---|---|
| 1 | atomic number | 22 | 23 | 24 | 25 | 26 |
| 2 | $\dot{\varepsilon}_C^j$ | -2.662 | -2.080 | -1.231 | -0.529 | 0.000 |
| 3 | $\dot{\rho}_C^j$ | 3.544 | 2.164 | 0.758 | 0.140 | 0.000 |
| 4 | $\dot{e}_C^j \times 10^2$ | -1.409 | -1.021 | -0.624 | -0.238 | 0.000 |
| 5 | $\dot{r}_C^j \times 10^4$ | 2.287 | 1.201 | 0.448 | 0.065 | 0.000 |

文献［14］报道了 1823K 温度条件下一阶等活度相互作用系数 $\dot{\varepsilon}_C^j$ 与组元 $j$ 的原子序数之间的周期关系，并从已知的 $\dot{\varepsilon}_C^j$ 估计尚未进行过实验研究的其他元素的 $\dot{\varepsilon}_C^j$ 值。本文对 Ti 元素 $\dot{\varepsilon}_C^{Ti}$ 的计算结果比该文的估计值要大，而且其他三种元素 $\dot{\varepsilon}_C^j$ 的计算结果也要比该文报道的大一些，但本文的 $\dot{\varepsilon}_C^j$ 根据大量的 C 溶解度的实验数据计算得出，因此本文的热力

学性质参数更为可靠。同时本文还给出了一阶活度相互作用系数 $\dot{e}_C^j$、二阶活度相互作用系数 $\dot{\rho}_C^j$ 和 $\dot{r}_C^j$,而且它们与第三组元 $j$ 的原子序数之间也存在比较好的关系。通过研究这些金属熔体中组元的热力学性质之间的关系,有助于更好地探索金属熔体的热力学规律和本质。

图 3 给出了 Fe-C-$j$ 熔体中组元的等活度相互作用系数 $\dot{\varepsilon}_C^j$、$\dot{\rho}_C^j$、$\dot{e}_C^j$ 和 $\dot{r}_C^j$ 与第三组元 $j$ 的原子序数之间的关系,不同组元的一阶等活度相互作用系数 $\dot{\varepsilon}_C^j$、$\dot{e}_C^j$ 与第三组元 $j$ 的原子序数之间为线性关系,二阶等活度相互作用系数 $\dot{\rho}_C^j$、$\dot{r}_C^j$ 与组元 $j$ 的原子序数之间为二阶非线性关系。

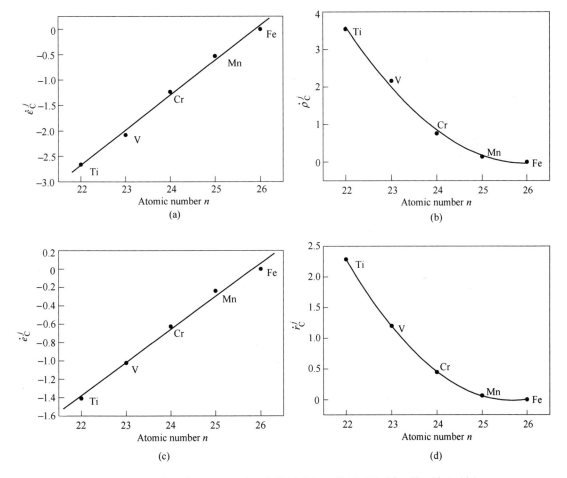

图 3 Fe-C-$j$ 熔体中第三组元 $j$ 对 C 的等活度相互作用系数 $\dot{\varepsilon}_C^j$、$\dot{\rho}_C^j$、$\dot{e}_C^j$ 和 $\dot{r}_C^j$ 与 $j$ 的原子序数间的关系(1823K)

Fig. 3 Relationship of activity interaction coefficient $\dot{\varepsilon}_C^j$ (a), $\dot{\rho}_C^j$ (b), $\dot{e}_C^j$ (c) and $\dot{r}_C^j$ (d) in Fe-C-$j$ melt and atomic number of third component $j$ (1823K)

## 2.3 活度相互作用系数与温度的关系

将不同的温度 $T$ 代入式(8)~式(11),可以得到不同温度下的 Fe-C-$j$ 熔体中的组元

$j$ 对 C 的一阶、二阶等活度相互作用系数,进而可以得出等活度相互作用系数与温度间的关系式。以 Fe-C-V 为例,一阶、二阶等活度相互作用系数与温度的关系式分别为:

$$\dot{\varepsilon}_C^V = -0.547 - 0.279 \times (10^4/T) \tag{13}$$

$$\dot{\rho}_C^V = -1.098 + 0.595 \times (10^4/T) \tag{14}$$

$$\dot{e}_C^V \times 100 = -0.126 - 0.163 \times (10^4/T) \tag{15}$$

$$\dot{r}_C^V \times 10^4 = -0.973 + 0.396 \times (10^4/T) \tag{16}$$

将 Ti、V、Cr、Mn 对 C 的一阶等活度相互作用系数分别代入式(12),就可以得到 Ti、V、Cr、Mn 对 C 的一阶活度相互作用系数与温度之间的关系式,分别如式(17)~式(20)所示:

$$\varepsilon_C^{Ti} = -5.374 - 0.807 \times (10^4/T) \tag{17}$$

$$\varepsilon_C^V = -5.067 - 5.313 \times (10^4/T) \tag{18}$$

$$\varepsilon_C^{Cr} = -0.467 - 1.426 \times (10^4/T) + 0.099 \times (10^4/T)^2 \tag{19}$$

$$\varepsilon_C^{Mn} = -1.240 - 0.521 \times (10^4/T) + 0.035 \times (10^4/T)^2 \tag{20}$$

得到这些 Fe-C-$j$ 熔体中组元的活度相互作用系数与温度之间的关系式后,在实际生产问题的热力学分析时就可以很方便地根据实际温度选择所需要的热力学性质参数。

## 3 结论

(1) Fe-C-$j$ ($j$ = Ti、V、Cr、Mn) 熔体中 C 的溶解度可分离为温度和第三组元 $j$ 的影响项,C 的溶解度可用式 $x_C = (0.0520 + 0.0848 \times 10^{-3}T) + k_j \cdot x_j$ 或 $[\%C] = (0.6021 + 2.5786 \times 10^{-3}T) + m_j \cdot [\%j]$ 表示,影响因子 $k_j$、$m_j$ 仅与第三组元 $j$ 有关,而与温度无关。

(2) Fe-C-$j$ ($j$ = Ti、V、Cr、Mn) 熔体中的 C 溶解度的影响因子 $k_j$、$m_j$ 与第三组元 $j$ 的原子序数之间存在较好的线性关系,$k_j$、$m_j$ 随着原子序数的增大而减小,影响 Fe-C-$j$ 熔体中 C 溶解度的因素为不同组元的原子间亲和力和第三组元 $j$ 的原子半径。

(3) 利用 Fe-C-$j$ 熔体中 C 的溶解度与温度和第三组元 $j$ 之间的关系式计算出的组元 $j$ 对 C 的等活度相互作用系数与第三组元 $j$ 的原子序数之间也存在一一对应关系。

(4) 利用 Fe-C-$j$ 熔体中组元 $j$ 对 C 的等活度相互作用系数可以计算出熔体中组元的活度相互作用系数,并可以得到活度相互作用系数与温度之间的关系式。

### 参 考 文 献

[1] Ni R. M., Ma Z. T., Wei S. K.. Steel Res., 1990, 61 (3): 397.

[2] Chen E. B., Dong Y. C., Guo S. X., Acta Metallurgica Sinica, 1997, 33 (8): 831.
陈二保,董元篪,郭上型. 金属学报,1997,33 (8): 831.

[3] Ni R. M., Ma Z. T., Wei S. K.. Acta Metallurgica Sinica, 1990, 26 (2): B87.
倪瑞明,马中庭,魏寿昆. 金属学报,1990,26 (2): B87.

[4] Che Y. C., Ji C. L., Qi G. J.. J. of Iron & Steel Res., 1988, 8 (3): 7.
车荫昌,冀春林,齐国均,钢铁研究学报,1988,8 (3): 7.

[5] Wei S. K.. Steel Res., 1992, 63 (4): 159.

[6] Ding X. Y.. Doctoral Thesis, University of Science & Technology Beijing, 1992, 08: 156.
丁学勇,博士学位论文,北京科技大学,1992,08: 156.

[7] Qi G. J., Ji C. L.. Proceeding of 5$^{th}$ Physical & Chemistry Annual Meeting, Xi'an, China, 1984: 227.
    齐国均, 冀春林. 第五届物理化学年会论文集, 西安, 1984: 227.
[8] Banya S., Suzuki M.. Tetsu-to-Hagane, 1975, 61: 2933.
    萬谷志郎, 鈴木幹雄. 鉄と鋼, 1975, 61: 2933.
[9] Ma Zh. T., Ni R. M., Cheng W.. Steel Res., 1991, 62 (11): 481.
[10] Wang H. C., Zhang Y. P., Dong Y. C., et al. Ferroalloy, 1999, 30 (4): 1.
     王海川, 张友平, 董元篪, 等. 铁合金, 1999, 30 (4): 1.
[11] Wang H. C.. Doctoral Thesis, University of Science & Technology Beijing, 2000: 121.
     王海川, 博士学位论文, 北京科技大学, 2000: 121.
[12] Turkdogan E. T., Leake L. E.. J. Iron & Steel Inst., 1955, 179: 39.
[13] Lupis C. H. P.. Acta Metallurgica, 1968, 16: 1365.
[14] Neuman F., Schenck. H.. Arch Eisenhuttenwes, 1959, 30: 477.

(原载于《金属学报》, 2001, 37 (9): 952-956)

# A Data Treatment Method of the Carbon Saturated Solubility in Fe-C-Cr Melt

Wang Haichuan[1,2]　Wang Shijun[1]　Yue Kexiang[1]　Dong Yuanchi[1]　Li Wenchao[2]

(1. Department of Metallurgical Engineering, ECUM, Maanshan 243002;
2. School of Metallurgy, USTB, Beijing 100083)

**Abstract:** Based on the current situation of studying the thermodynamic property of Fe-C-Cr melt using the carbon saturated solubility, a new experimental data treatment method of the carbon saturated solubility was put forward in this article. With this method a linear relationship expression of the carbon saturated solubility in Fe-C-Cr melt was obtained, which intercept is dependent of temperature and independent of third component [Cr] but which slope is dependent of third component [Cr] and independent of temperature. Through this expression activity interaction coefficients at different temperatures were calculated and the relationship between activity interaction coefficients and temperature is obtained also.

**Keywords:** data treatment method; intercept; slope; carbon saturated solubility; Fe-C-Cr melt

## 1 Introduction

It is the most popular data treatment method in metallurgical thermodynamic study that using the linear relationship expression of the carbon saturated solubility to calculate the thermodynamic properties of components in metallic melt. Though the carbon saturated solubility in different metallic melt has been reported in a large number literature, it is a pity that the lack of effective data treatment method leads that the different researchers obtain different linear relationship expressions of the carbon saturated solubility and calculate out different results of the thermodynamic properties under same temperature for same system. At the same time these thermodynamic property data are gained at some fixed temperature, so it is inconvenient and incorrect to analyse the practical problem using these thermodynamic parameters.

In this article a new data treatment method of the carbon saturated solubility in Fe-C-Cr melt has been put forward. Using this method we can obtain relationship among the carbon saturated solubility and temperature and third component [Cr], and relationship between activity interaction coefficient of components and temperature, it can be enhance the calculating precision of activity interaction coefficient of components in Fe-C-Cr melt.

---

王海川，1996～2000 年于北京科技大学师从李文超教授攻读冶金物理化学专业工学博士学位。目前在安徽工业大学工作，任冶金工程学院副院长。获得全国优秀教师、安徽省优秀教师、宝钢教育基金优秀教师奖、中国金属学会冶金先进青年科技工作者等荣誉称号，发表论文 120 余篇，获奖省部级科技奖励 6 项。

## 2  Experiment

This experiment studied the carbon saturated solubility in Fe-C-Cr ferroalloy melt. In this experiment 10g metal and graphite power which has been compounded in established proportion are put into every hole of the porous graphite crucible. The experimental temperature is measured with Pt-Rh30 and Pt-Rh6 thermocouple and controlled by DWK702 automatic controller. The experimental temperature is at 1723K.

After metal in the crucible all melt, to keep equilibrium for 5 hours to make carbon's dissolve reach to equilibrium, and take out the graphite crucible from the experimental furnace, after water-cooling to take out metal and break it into power to analyse carbon with burning method and Cr with potassium permanganate capacity method, and these analysing results are given in Table 1.

In order to deal with the data conveniently in the latter paragragh, the difference between the carbon saturated solubility in Fe-Cr-C melt and that in Fe-C melt are given in the Table 1 also, the calculating formula is:

$$\Delta x_C = x_C - x_C^{Fe} \tag{1}$$

$$\Delta[\%C] = [\%C] - [\%C]_{Fe} \tag{2}$$

where, $[\%C]$, $x_C$, $[\%C]_{Fe}$ and $x_C^{Fe}$ are the carbon saturated solubility in molar fraction and mass percentage in Fe-Cr-C and Fe-C melt under different temperatures respectively.

**Table 1  The Carbon Saturated Solubility in Fe-C-Cr Melt at 1723K**

| No. | [%Cr] | [%C] | Δ[%C] | $x_{Cr}$ | $x_C$ | $\Delta x_C$ |
|---|---|---|---|---|---|---|
| 1 | 0.00 | 5.04 | 0.00 | 0.0000 | 0.1980 | 0.0000 |
| 2 | 2.21 | 5.11 | 0.07 | 0.0200 | 0.2000 | 0.0020 |
| 3 | 5.38 | 5.46 | 0.42 | 0.0480 | 0.2110 | 0.0131 |
| 4 | 8.10 | 5.53 | 0.49 | 0.0720 | 0.2129 | 0.0150 |
| 5 | 10.81 | 5.82 | 0.78 | 0.0951 | 0.2218 | 0.0238 |
| 6 | 12.97 | 6.12 | 1.08 | 0.1130 | 0.2308 | 0.0329 |
| 7 | 13.31 | 6.10 | 1.06 | 0.1160 | 0.2302 | 0.0322 |
| 8 | 15.04 | 6.33 | 1.29 | 0.1300 | 0.2370 | 0.0390 |

## 3  Conventional data treatment method

The conventional data treatment method is:

(1) Drawing up these data of the carbon saturated solubility $x_C$ (or $[\%C]$) and the third component $x_j$ (or $[\%j]$) in ferroalloy melt at some fixed temperature into linear relationship expression as following:

$$x_C = x'_C + k'x_j \tag{3}$$

$$[\%C] = [\%C]' + m'[\%j] \tag{4}$$

$x'_C$ and $[\%C]'$ are the interception of the linear relationship expression about the carbon saturated solubility, $k'$ and $m'$ are the slope of the linear relationship about the carbon saturated solubility in molar fraction and mass percentage respectively, and the slope shows the effective extent of the third component on the carbon saturated solubility.

(2) According to the definition of activity interaction coefficient, using the interception and slope of the linear relationship expression to calculate the activity interaction coefficient of components in multi-component metallic melt at some known experimental temperature.

Because different researchers study different concentration ranges, with regarding to experimental error and analysis error, different researchers obtain different varying tendencies of the slope of the linear relationship expression of the carbon saturated solubility, even their results differ greatly. Some results are that the slope increase with increasing temperature, but some are that the slope decrease with the increasing temperature.

This research obtains the linear relationship expression of $x_C$ and $x_{Cr}$, [%C] and [%Cr].

$$x_C = 0.1953 + 0.3012 x_{Cr} \tag{5}$$

$$[\%C] = 4.9588 + 0.0861[\%Cr] \tag{6}$$

The slope of the relationship expression of the carbon saturated solubility in Fe-Cr melt ($x_C$) and $x_{Cr}$ with different researchers is shown Fig. 1 (these straight lines are drown according to the linear relationship expression and the concentration range from literature), three among six lines are mutual intersect. Moreover the line from literature [3] at higher temperature (1873k) is lower than that line from literature [2] at lower temperature (1823k).

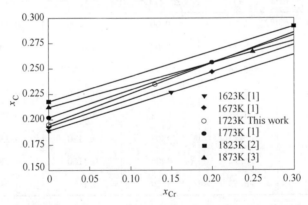

Fig. 1  The carbon saturated solubility of in Fe-C-Cr melts at different temperatures

These thermodynamics parameters of Fe-C-Cr melt have bigger divergence which are calculated by using $x_C'$, $k'$, $[\%C]'$, $m'$ based on the definition of activity interaction coefficient, these activity coefficients and activity interaction coefficients about Cr and C in Fe-C-Cr melt are shown in Table 2. Among which the $\varepsilon_C^{Cr}$ of the first-order differs more than 2 times and the second-order differs more greatly, and these activity interaction coefficients are studied at some fixed temperature and there are no explicit relationship among them. If we need some parameters at willfully temperature it will be very difficult to choice these parameters.

Table 2  Activity interaction coefficient of components in Fe-C-Cr melt from different literature

| No. | $\ln\gamma_{Cr}^0$ | $\varepsilon_C^{Cr}$ | $\rho_C^{Cr}$ | Temp. /K | Literature |
|---|---|---|---|---|---|
| 1 | -0.218 | -5.41 | 4.59 | 1673 | [1] |
| 2 | -0.321 | -4.93 | 4.26 | 1773 | [1] |
| 3 |  | -3.81 |  | 1823 | [2] |
| 4 | -0.35 | -7.6 | -1.0 | 1873 | [3] |
| 5 |  | -3.15 |  | 1873 | [4] |
| 6 |  | -3.60 | 19.35 | 1873 | [5] |
| 7 | -0.30 | -5.10 | -0.07 | 1873 | [6] |

## 4  Relationship of the carbon saturated solubility and [Cr]

By analyzing a great deal of experimental data of the carbon saturated solubility in Fe-Cr-C melts, which were measured by this work and reported in the literature, it is found that there is a very well linear relationship between the difference $\Delta x_C$ ( $= x_C - x_C^{Fe}$) and $x_{Cr}$, it is shown in Fig. 2.

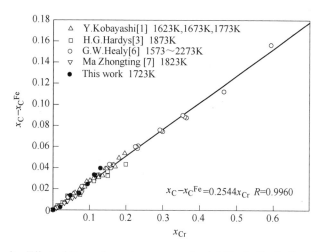

Fig. 2  Effect of Cr on the carbon saturated solubility in Fe-C-Cr melt at different temperatures

$$\Delta x_C = x_C - x_C^{Fe} = k \cdot x_{Cr} = 0.2544 x_{Cr} \tag{7}$$

while the unit of concentration is mass%, there is

$$\Delta[\%C] = [\%C] - [\%C]_{Fe} = m \cdot [\%Cr] = 0.0762[\%Cr] \tag{8}$$

where, $m_j$ is constant and independence of temperature. The relationship between the carbon saturated solubility in Fe-C melt $x_C^{Fe}$ or $[\%C]_{Fe}$ and temperature is shown as following[8]:

$$x_C^{Fe} = 0.0520 + 0.0848 \times 10^{-3} T \tag{9}$$

$$[\%C]_{Fe} = 0.6021 + 2.5786 \times 10^{-3} T \tag{10}$$

substituting equations (9) and (10) into equations (7) and (8) respectively, we obtain

$$x_C = 0.0520 + 0.0848 \times 10^{-3}T + 0.2544 x_{Cr} \tag{11}$$

$$[\%C] = 0.6021 + 2.5786 \times 10^{-3}T + 0.0762[\%Cr] \tag{12}$$

where, the intercept is dependent of temperature and independent of third component [Cr], the slope $k = 0.2544$ or $m = 0.0762$ is a fixed value and dependent of third component [Cr] and independent of temperature.

## 5 Calculation of activity interaction coefficient

Based on the definition of activity interaction coefficient at constant activity[9], in the case of carbon saturation, the first-order activity interaction coefficients of third component [Cr] upon [C] at constant activity are, respectively:

$$\dot{\varepsilon}_C^{Cr} = \left(\frac{-\partial \ln x_C}{\partial x_{Cr}}\right)_{\substack{a_C \\ x_{Cr} \to 0}} = -\frac{0.2544}{0.0520 + 0.0848 \times 10^{-3}T} \tag{13}$$

$$\dot{e}_C^{Cr} = \left(\frac{-\partial \lg[\%C]}{\partial[\%Cr]}\right)_{\substack{\%C \\ [\%Cr] \to 0}} = -\frac{1}{2.303} \times \frac{0.0762}{0.6021 + 2.5786 \times 10^{-3}T} \tag{14}$$

and the second-order activity interaction coefficients of third component [Cr] upon [C] at constant activity are, respectively:

$$\dot{\rho}_C^{Cr} = \frac{1}{2}\left(\frac{-\partial^2 \ln x_C}{\partial x_{Cr}^2}\right)_{\substack{a_C \\ x_{Cr} \to 0}} = \frac{1}{2}\left(\frac{0.2544}{0.0520 + 0.0848 \times 10^{-3}T}\right)^2 \tag{15}$$

$$\dot{r}_C^{Cr} = \frac{1}{2}\left(\frac{-\partial^2 \lg[\%C]}{\partial[\%Cr]^2}\right)_{\substack{\%C \\ [\%Cr] \to 0}} = \frac{1}{4.605}\left(\frac{0.0762}{0.6021 + 2.5786 \times 10^{-3}T}\right)^2 \tag{16}$$

while $T = 1573K, 1673K, \cdots, 2273K$, $\dot{\varepsilon}_C^{Cr}$, $\dot{\rho}_C^{Cr}$, $\dot{e}_C^{Cr}$ and $\dot{r}_C^{Cr}$ at different temperatures are shown in Table 3, the relationship between $\dot{\varepsilon}_C^{Cr}$, $\dot{\rho}_C^{Cr}$, $\dot{e}_C^{Cr}$, $\dot{r}_C^{Cr}$ and temperature are, respectively:

$$\dot{\varepsilon}_C^{Cr} = -0.296 - 0.170 \times (10^4/T) \quad R = 0.9996 \tag{17}$$

$$\dot{\rho}_C^{Cr} = 0.365 + 0.205 \times (10^4/T) \quad R = 0.9999 \tag{18}$$

$$\dot{e}_C^{Cr} \times 100 = -0.068 - 0.101 \times (10^4/T) \quad R = 0.9999 \tag{19}$$

$$\dot{r}_C^{Cr} \times 10^4 = -0.331 + 0.143 \times (10^4/T) \quad R = 0.9993 \tag{20}$$

Using the relationship between the activity interaction coefficient at constant activity and activity interaction coefficient at constant concentration[9], these calculated results are showed in Table 4 the relationship between $\varepsilon_C^{Cr}$, $\rho_C^{Cr}$, $\rho_C^{C,Cr}$, $\rho_{Cr}^C$ and temperature are, respectively:

$$\varepsilon_C^{Cr} = -0.467 - 4.426 \times (10^4/T) + 0.099 \times (10^4/T)^2 \tag{21}$$

$$\rho_C^{Cr} = 0.876 - 0.694 \times (10^4/T) + 0.061 \times (10^4/T)^2 \tag{22}$$

$$\rho_C^{C,Cr} = 1.407 + 0.571 \times (10^4/T) \quad R = 0.9984 \tag{23}$$

$$\rho_{Cr}^C = -0.568 - 1.119 \times (10^4/T) + 0.077 \times (10^4/T)^2 \tag{24}$$

Table 3 Activity interaction coefficient at constant activity $\dot{\varepsilon}_C^{Cr}, \dot{\rho}_C^{Cr}, \dot{e}_C^{Cr}$ and $\dot{r}_C^{Cr}$ at different temperatures

| Temp./K | 1573 | 1673 | 1773 | 1873 | 1973 | 2073 | 2173 | 2273 |
|---|---|---|---|---|---|---|---|---|
| $\dot{\varepsilon}_C^{Cr}$ | -1.372 | -1.312 | -1.257 | -1.207 | -1.160 | -1.117 | -1.077 | -1.039 |
| $\dot{\rho}_C^{Cr}$ | 0.942 | 0.861 | 0.790 | 0.728 | 0.673 | 0.624 | 0.580 | 0.540 |
| $\dot{e}_C^{Cr} \times 100$ | -0.710 | -0.673 | -0.639 | -0.609 | -0.582 | -0.556 | -0.533 | -0.512 |
| $\dot{r}_C^{Cr} \times 10^4$ | 0.581 | 0.522 | 0.471 | 0.427 | 0.389 | 0.356 | 0.327 | 0.302 |

Table 4 Activity interaction coefficient at constant concentration $\varepsilon_C^{Cr}, \rho_C^{Cr}, \rho_C^{C,Cr}$ and $\rho_{Cr}^C$ at different temperatures

| Temp./K | 1573 | 1673 | 1773 | 1873 | 1973 | 2073 | 2173 | 2273 |
|---|---|---|---|---|---|---|---|---|
| $\varepsilon_C^{Cr}$ | -5.548 | -5.464 | -5.369 | -5.268 | -5.162 | -5.052 | -4.941 | -4.827 |
| $\rho_C^{Cr}$ | -1.059 | -1.083 | -1.089 | -1.082 | -1.067 | -1.045 | -1.045 | -0.991 |
| $\rho_C^{C,Cr}$ | 5.073 | 4.823 | 4.613 | 4.435 | 4.284 | 4.155 | 4.044 | 3.951 |
| $\rho_{Cr}^C$ | -4.556 | -4.490 | -4.416 | -4.337 | -4.254 | -4.168 | -4.080 | -3.991 |

Based on these thermodynamic parameters, the thermodynamic analysis of de-carbonization with blowing oxygen is carried out while medium and low carbon ferrochrome are smelted with converter practice, the theoretic analysis result coincides with the data of practical product[10].

Using the data treatment method above mentioned to calculate the thermodynamic parameters of components in other metallic melts, it will be not difficult to study the thermodynamic property of metallic melt with carbon saturated solubility and very easy to obtain the thermodynamic parameters because calculating formula has been greatly simplified.

Only by determining the carbon saturated solubility at some fixed temperatures and obtaining the slope $k_j$ or $m_j$ of the expression of the carbon saturated solubility, it is also very easy to calculate activity interaction coefficient of components at constant activity at demanded temperature, and activity interaction coefficient of components at constant concentration based on the relationship between activity interaction coefficient at constant activity and that at constant concentration.

## 6 Conclusion

(1) Based on experimental data of this study and some literatures, using new data treatment method put forward in this work, the relationship expression of the carbon saturated solubility in Fe-Cr-C melt are obtained as follows:

$$x_C = 0.0520 + 0.0848 \times 10^{-3}T + 0.2544 x_{Cr}$$
$$[\%C] = 0.6021 + 2.5786 \times 10^{-3}T + 0.0762[\%Cr]$$

where, the intercept is dependent of temperature and independent of third component [Cr], the slope $k = 0.2544$ or $m = 0.0762$ is a fixed value and dependent of third component [Cr] and independent of temperature.

(2) Using the expression of carbon saturated solubility, activity interaction coefficient of components in metallic melt at demanded temperature can be calculated easily, and the relationship be-

tween the first-order or second-order activity interaction coefficient at constant activity or at constant concentration and temperature are, respectively:

$$\dot{\varepsilon}_C^{Cr} = -0.296 - 0.170 \times (10^4/T)$$

$$\dot{\rho}_C^{Cr} = 0.365 + 0.205 \times (10^4/T)$$

$$\dot{e}_C^{Cr} \times 100 = -0.068 - 0.101 \times (10^4/T)$$

$$\dot{r}_C^{Cr} \times 10^4 = -0.331 + 0.143 \times (10^4/T)$$

$$\varepsilon_C^{Cr} = -0.467 - 4.426 \times (10^4/T) + 0.099 \times (10^4/T)^2$$

$$\rho_C^{Cr} = 0.876 - 0.694 \times (10^4/T) + 0.061 \times (10^4/T)^2$$

$$\rho_C^{C,Cr} = 1.407 + 0.571 \times (10^4/T)$$

$$\rho_{Cr}^{C} = -0.568 - 1.119 \times (10^4/T) + 0.077 \times (10^4/T)^2$$

(3) We can calculate the thermodynamic parameters of components in many metallic melts with the data treatment method put forward in this work, it will be not difficult to study the thermodynamic property of metallic melt with carbon saturated solubility, and very easy to obtain the thermodynamic parameters because calculating formula has been greatly simplified.

## References

[1] Y. Kabayashi, K. Morita, N. Sano. Thermodynamics of molten Fe-Cr-C alloys [J]. ISIJ International, 1996, 36 (8): 1009.

[2] Guo Shangxing, Dong Yuanchi. Study on Fe-Cr-C-P melt Containing high Cr [J]. J. Iron & Steel Research (in Chinese), 1995, 7 (2): 15.

[3] H. G. Hardys, M. G. Froberg, J. F. Elliott. Activities in the liquid Fe-Cr-C, Fe-P-C and Fe-Cr-P systems at 1600℃ [J]. Met. Trans., 1970, 1 (7): 1867.

[4] Zhou Jinhua. FERROALLOY (in Chinese) [M]. Beijing: Metallurgical Engineering Press, 1992.

[5] M. G. Frohberg, J. F. Elliott, H. G. Hardys. Contribution to the study the thermodynamics of complex solution in Fe-Cr-P-C melts [J]. Arch Eisenhuttenwsen, 1968, 39 (8): 587.

[6] G. W. Heady. The thermodynamics of chromite ore smelting [J]. Trans. ISS, in Iron & Steel-maker, 1987, 9: 51.

[7] Ma Zhongting, J. Ohser, D. Janke. Thermodynamic description of multi-component systems by interaction parameters [J]. Acta Metallurgica Sinica (in Chinese), 1998, 34 (7): 753.

[8] Wang Haichuan, Zhang Youping, Li Wenchao, et al. The saturated solubility of the carbon in Fe-C and Mn-C and thermodynamic property of Mn-Fe-C melt [J]. Ferroalloys (in Chinese), 2000, 31 (2): 1.

[9] C. H. P. Lupis. On the use of polynomial for the thermodynamics of dilute metallic solutions [J]. Acta Metallurgica, 1968, 16 (11): 1365.

[10] Wang Haichuan, Dong Yuanchi, Li Wenchao, et al. The thermodynamic analysis on smelting medium carbon and low carbon ferrochrome with converter practice [J]. Ferroalloys (in Chinese), 2000, 4, 31 (3): 5.

(原载于 *Journal of University of Science and Technology Beijing*, 2002, 9 (1): 16-19)

# 低碳 FeMnSiAl 系 TWIP 钢冶炼技术研究

刘建华[1]　庄昌凌[1]　李世琪[1]　张庆雷[2]　雷　冲[2]　张凯亮[2]

(1. 北京科技大学冶金工程研究院，北京　100083；
2. 中原特钢股份有限公司，济源　454685)

**摘　要**：探索研究了氩氧精炼—模铸—电渣重熔工艺流程生产大尺寸 TWIP 钢坯料的工艺技术。研究表明低碳 Fe-Mn-Si-Al 系 TWIP 钢锰高碳低，氩氧精炼中完全依靠脱碳保锰技术措施较为困难，采用精炼前期脱碳保锰与精炼后期金属锰合金化相结合技术较为合理；TWIP 钢模铸坯凝固组织主要为粗长的柱状晶与中心杂乱取向的枝晶，组织不均匀；通过电渣重熔精炼，TWIP 钢凝固组织得到明显改善，夹杂物得到良好控制。

**关键词**：TWIP 钢；氩氧精炼；模铸；电渣重熔；凝固组织；夹杂物；脱碳保锰

## Technologies for Low Carbon FeMnSiAl TWIP Steel Smelting

Liu Jianhua[1]　Zhuang Changling[1]　Li Shiqi[1]　Zhang Qinglei[2]
Lei Chong[2]　Zhang Kailiang[2]

(1. Engineering Research Institute, University of Science and Technology Beijing, Beijing 100083;
2. Zhongyuan Special Steel Co., Ltd., Jiyuan 454685)

**Abstract**: Technologies for large scale TWIP blank production with argon oxygen refining (AOD) -ingot casting-electroslag remelting (ESR) process were studied. Results show that it is difficult to refine the low carbon FeMnSiAl TWIP steel with high Mn content hot iron in AOD refining by decarburization and conserving chromium technology, while it is practicable by assembling of decarburization and conserving chromium with lower Mn content hot iron in the early stage of AOD refining and alloying with pure manganese in the late stage; solidification structure of TWIP ingot is mainly comprised with rough and long columnar and central random oriental dendrite, and the structure is uneven; the solidification structure is improved obviously and the inclusions is well controlled though ESR.

**Keywords**: TWIP steel; AOD; ingot casting; ESR; solidification structure; inclusion; decarburization and conserving chromium

## 1　引言

TWIP 钢也称为孪晶诱导塑性钢（Twinning induced plasticity）[1,2]，是一种集高强度、

---

刘建华，1988~1991 年于北京科技大学师从李文超教授攻读硕士学位。目前在北京科技大学冶金工程研究院工作，教授。发表论文 150 余篇，获省部级奖 2 项。

高塑性和高应变硬化性能于一体的理想汽车用安全、减重、节能结构材料。目前国外对TWIP钢做了较多的研究工作,处于商业化量产TWIP钢的前夕;国内对TWIP钢研究的还处于实验室研究阶段,在基本理论与结构优化、组织控制方面进行了广泛研究,而对冶炼工艺技术的研究还未见涉及[3,4]。本课题组通过感应炉和AOD冶炼、模铸、电渣重熔精炼工艺流程,探索研究了冶炼大尺寸低碳Fe-Mn-Si-Al系TWIP钢坯料的工艺技术。

## 2 TWIP钢成分特点和冶金难点

TWIP钢按成分体系可分为Fe-Mn-Al-Si系、Fe-Mn-C系和Fe-Mn-Al-C系,表1列举了几种典型TWIP的成分和机械性能[5],表明Fe-25Mn-3Al-3Si的延伸率高,强塑积达到60000MPa·%以上,具有良好的高强度、高塑性特点,本研究重点研究该钢种冶炼工艺技术,其成分控制范围见表2。

表1 典型TWIP钢的成分与机械性能
Table 1 Composition and mechanic properties of classic TWIP steel

| 合金成分 | 室温机械性能 | | | |
|---|---|---|---|---|
| | 屈服强度/MPa | 抗拉强度/MPa | 延伸率/% | 断面收缩率/% |
| Fe-20Mn-3Al-3Si | 300 | 850 | 82 | — |
| Fe-25Mn-3Al-3Si | 270 | 640 | 95 | — |
| Fe-23Mn-0.5C | 480 | 1140 | 57 | — |
| Fe-17Mn-0.8C | — | 1074 | 70 | — |
| Fe-18Mn-0.6C-0Al | 130 | 880 | — | 38 |
| Fe-18Mn-0.6C-1Al | 330 | 890 | — | 41 |
| Fe-18Mn-0.6C-2Al | 430 | 860 | — | 50 |

表2 Fe-25Mn-3Al-3Si TWIP钢成分控制范围
Table 2 Composition range of Fe-25Mn-3Al-3Si TWIP steel　　　　　(wt%)

| 元素 | C | Si | Mn | P | S | Al |
|---|---|---|---|---|---|---|
| 化学成分 | ≤0.05 | 2.70~3.20 | 24.00~26.00 | ≤0.025 | ≤0.005 | 2.80~3.50 |

钢中Al含量较高,易形成$Al_2O_3$夹杂,该夹杂与钢水润湿性差,容易黏附在耐材上;同时,钢水Al含量高,钢水流动性也较差。因此,如采用连铸工艺生产Fe-25Mn-3Al-3Si TWIP钢,保证连铸顺行将是比较难以克服的困难。

高锰钢热导率较低,如国内经常生产的Mn13钢热导率仅为中碳钢的1/4~1/2[6]。导热率低,钢液凝固缓慢,连铸时如果速度过快,凝壳强度低,容易出现拉漏的现象。同时,高锰钢自由线收缩值(2.4%~3.0%)比碳素钢的线收缩值大得多;再加上导热率低,铸坯凝固组织粗大,柱状晶异常发达,易产生热裂纹[7]。因此,国内Mn13生产大都采用模铸生产。Fe-25Mn-3Al-3Si TWIP钢锰含量远大于Mn13钢,因此连铸生产的困难要远大于Mn13钢。

另一方面,Fe-25Mn-3Al-3Si TWIP钢的碳含量较低,要求低于0.05%,而锰含量较高,这也为该钢生产制造了困难。如采用价格较低的中高锰铁作为锰合金原料生产,将必须采用保锰降碳工艺;如采用低碳锰铁,由于锰含量较高,也很困难引起碳超标。

同时，Fe-25Mn-3Al-3Si TWIP 钢合金含量达到 31%，为高合金钢，生产过程中温度控制也是难点之一。

## 3 TWIP 钢冶炼实践

为了促进 TWIP 钢的量化生产，本研究尝试通过工业化生产制备一重量达 10 吨的大尺寸坯料，坯料宽度达到 1.0m，制备的坯料经过轧制、退火等处理后送汽车厂试用。同时，在产业化实践中探索 TWIP 的冶炼特性与生产工艺技术。

鉴于 Fe-25Mn-3Al-3Si TWIP 钢铝、锰及硅含量较高，连铸较困难的情况，本实验选择模铸生产工艺。但由于模铸保护浇铸难以达到连铸保护浇铸的水平，模铸过程钢水的过氧化可能会较为严重，钢中夹杂物会较多[8]；另一方面，模铸的偏析及组织控制也比较难，直接用模铸锭轧制 TWIP 钢存在一定风险。

为了解决模铸生产可能出现的夹杂物超标、偏析程度严重等问题，本实验采用电渣重熔工艺对模铸锭继续电渣重熔处理。

### 3.1 AOD 精炼

为了保证 Fe-25Mn-3Al-3Si TWIP 钢碳成分小于 0.05%，本实验采用 AOD 对铁水进行精炼。

为了节约成本，本课题希望采用中高碳锰铁进行精炼，通过脱碳保锰达到低成本冶炼高锰低碳 Fe-25Mn-3Al-3Si TWIP 钢。

$$MnO(s) + [C] = \{CO\}(g) + [Mn] \quad \Delta G^{\ominus} = 290160 - 173.14T \quad (1)$$

通过热力学计算得到各种条件下 AOD 精炼碳锰选择性氧化转化温度。表 3 结果表明转化温度随 C 含量的降低而升高，随 Mn 含量的升高而升高。直接采用 25.0% Mn 的铁水在 AOD 中冶炼，终点温度及氩氧比必须控制在较高水平，才能达到脱碳保锰的目的，控制难度非常大。但可以在冶炼初期采用一定锰含量的铁水，进行冶炼，通过脱碳保锰，减少精炼后期锰合金化的压力，并显著降低生产成本。

表 3　TWIP 钢冶炼转化温度
Table 3　Transition temperature for TWIP steel smelting

| 分类 | C/% | Mn/% | P/% | S/% | $P_{CO}$ | $T_{转化}$/℃ |
|---|---|---|---|---|---|---|
| 1 | 0.5 | 20 | 0.03 | 0.03 | 1 | 1801 |
| 2 | 1.0 | 20 | 0.03 | 0.03 | 1 | 1692 |
| 3 | 1.5 | 20 | 0.03 | 0.03 | 1 | 1623 |
| 4 | 2.0 | 20 | 0.03 | 0.03 | 1 | 1575 |
| 5 | 2.5 | 20 | 0.03 | 0.03 | 1 | 1534 |
| 6 | 1.5 | 20 | 0.03 | 0.03 | 1 | 1623 |
| 7 | 1.5 | 20 | 0.03 | 0.03 | 0.5 | 1553 |
| 8 | 1.5 | 20 | 0.03 | 0.03 | 0.2 | 1475 |
| 9 | 1.5 | 20 | 0.03 | 0.03 | 0.1 | 1414 |
| 10 | 1.5 | 22 | 0.03 | 0.03 | 1 | 1638 |

续表3

| 分类 | C/% | Mn/% | P/% | S/% | $P_{CO}$ | $T_{转化}$/℃ |
|---|---|---|---|---|---|---|
| 11 | 1.5 | 20 | 0.03 | 0.03 | 1 | 1623 |
| 12 | 1.5 | 15 | 0.03 | 0.03 | 1 | 1599 |
| 13 | 1.5 | 10 | 0.03 | 0.03 | 1 | 1507 |
| 14 | 0.04 | 20 | 0.03 | 0.03 | 1 | 2207 |
| 15 | 0.04 | 12.41 | 0.03 | 0.03 | 0.01 | 1527 |
| 16 | 0.04 | 17.98 | 0.03 | 0.03 | 0.01 | 1577 |
| 17 | 0.05 | 22 | 0.03 | 0.03 | 0.2 | 1941 |
| 18 | 0.05 | 22 | 0.03 | 0.03 | 0.1 | 1847 |
| 19 | 0.05 | 20 | 0.03 | 0.03 | 0.2 | 1920 |
| 20 | 0.05 | 20 | 0.03 | 0.03 | 0.1 | 1828 |

本实验在中原特钢通过 8t AOD 炉成功冶炼了 2 炉共 15t TWIP 钢,钢水成分见表4。表4表明 AOD 冶炼的 TWIP 钢成分基本接近目标成分,仅第二炉硅含量比目标成分稍低,但在随后的电渣重熔精炼中有增硅发生,电渣精炼后成分可控制在目标范围内。

表 4  AOD 冶炼终点 TWIP 钢组成
Table 4  Composition of TWIP steel at the end of AOD refining  (wt%)

| 炉次 | C | Si | Mn | P | S | Al |
|---|---|---|---|---|---|---|
| 1 | 0.05 | 2.75 | 25.78 | 0.019 | 0.009 | 3.00 |
| 2 | 0.05 | 2.52 | 25.27 | 0.016 | 0.006 | 3.27 |

## 3.2 模铸

本研究首先采用 DTA 技术对一系列 TWIP 钢的固液相线温度进行了确定,并回归总结得出 TWIP 钢的液相线温度 $T_L$ 计算公式:

Si < 0.23% 时    $T_L = 1537 - 55.3(\%C) - 18.7(\%Si) + (-3.76 - 0.86\%C)(\%Mn)$    (2)

Si ≥ 0.23% 时    $T_L = 1537 - 55.3(\%C) - 16.9(\%Si) - 7.92(\%Si)^2 + 2.17(\%Si)^3 + (-3.76 - 0.86\%C)(\%Mn)$    (3)

采用式(2)和式(3)可在试制过程中随时对 AOD 生产的 TWIP 钢凝固温度进行计算。本次试制的两炉 TWIP 钢的液相线温度分别为 1375.0℃ 和 1380.0℃。

由于 TWIP 钢的流动性较差,模铸时采用高过热度浇铸,分别浇铸生产出 2 根 7.5 吨左右直径为 600mm 的自耗电极。

图 1 为自耗电极的低倍组织照片。图 1 表明 TWIP 电极坯在表层存在 2~3mm 厚急冷层;急冷层内侧为 50mm 左右厚度的柱状晶组织,相对于一般的碳钢和微合金钢,柱状晶显著粗长;铸坯内部其他区域为杂乱取向的枝晶组织,枝晶长度一般为 10mm 左右,也较一般碳钢和微合金钢铸锭中心的等轴晶粗大。一般铸坯的凝固组织为急冷层、柱状晶和中心等轴晶组成,而 TWIP 钢铸坯内部未发现明显的中心等轴晶。这可能与 TWIP 合金含量

图 1　TWIP 钢自耗电极锭凝固组织照片

Fig. 1　Solidification structure of TWIP ingot

较高、固液相线温度差较大有关。相对于等轴晶而言,内部杂乱的枝晶组织容易引起裂纹、疏松、偏析等缺陷。

## 3.3　电渣重熔

为了控制钢中夹杂物、凝固组织及成分偏析,本研究通过电渣重熔精炼对模铸的自耗电极锭进行精炼。

电渣精炼时采用直径为 1000mm 的结晶器,采用两根电极交互重熔的方法,制备 14t 左右直径为 1000mm 左右的电渣锭。

TWIP 钢铝含量较高,在电渣过程中存在铝的烧损。为了防止铝的烧损,本研究不仅在电渣过程中采用氩气保护,而且在渣中配置了一部分铝,补充电渣过程中铝的烧损。

同时,TWIP 钢中锰含量较高,电渣时熔池中温度较高,可能存在锰的挥发损失。为此,实验中采用向熔池中添加电解锰片的方法,以弥补锰的损失。

另一方面,为了更好地控制凝固组织、偏析及中心疏松,本研究采用低电流密度、低熔速的工艺措施,平均熔化率控制在 750~950kg/h,目标熔化率为 800kg/h。

图 2 为电渣锭中段凝固组织照片。图 2 表明电渣锭表层同样存在很薄的一层急冷层,厚度只有 1~3mm;急冷层内侧为柱状晶组织,厚度为 200mm 左右;中心部位为等轴晶组织。凝固组织与电极锭存在明显差异,具体表现为:(1)柱状晶区较电极宽,但柱状晶短小,一般长度只有 5~10mm;(2)中心出现等轴晶,且等轴晶较为细小,一般为 2~5mm;(3)组织较为致密、均匀。相对而言,电渣锭凝固组织较电极锭有明显改善。

图 2　TWIP 钢电渣重熔锭凝固组织照片

Fig. 2　Solidification structure of TWIP ESR ingot

采用扫描电镜和能谱分析对 TWIP 钢电渣锭中夹杂物进行了分析,结果表明电渣锭中夹杂物主要为细小的 AlN 夹杂和 $Al_2O_3$ 夹杂，尺寸一般为 2~5μm，图 3 为钢中该两类夹杂的典型形貌。在电渣锭中未见有簇状 AlN 夹杂和 $Al_2O_3$ 夹杂。

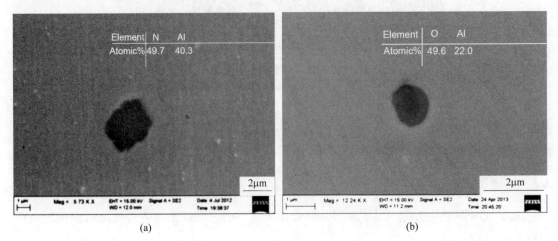

图 3　TWIP 电渣锭中夹杂物（(a) AlN 夹杂；(b) $Al_2O_3$ 夹杂）

Fig. 3　Inclusions in TWIP ESR ingot（(a) AlN inclusion；(b) $Al_2O_3$ inclusion）

将电渣锭加热，通过锻造成型及切割处理制备得 3000mm×1060mm×340mm 的 TWIP 钢坯料，按密度为 7300kg/$m^3$ 计算，坯料重为 7900kg，图 4 为本研究试制的 TWIP 钢坯料照片。

图 4　TWIP 钢坯

Fig. 4　TWIP blank

## 4　结论

（1）采用 AOD—模铸—电渣重熔工艺成功制备得 3000mm×1060mm×340mm 的 TWIP 钢坯料；

(2) 在 AOD 精炼中完全采用脱碳保锰工艺冶炼低碳高锰 TWIP 钢难度较大，但结合脱碳保锰工艺及后期金属锰合金化冶炼低碳高锰 TWIP 可行；

(3) 采用模铸生产 TWIP 钢铸锭凝固组织主要为粗长柱状晶组成的柱状晶和中心区域杂乱取向的枝晶组织，组织不均匀；

(4) 通过电渣重熔显著改善了 TWIP 钢的凝固组织，同时钢中夹杂物也得到较好控制，电渣锭中夹杂物主要为尺寸细小的 AlN 夹杂和 $Al_2O_3$ 夹杂。

## 参 考 文 献

[1] Zhuang C, Liu J, Jiang H, et al. Non-Metallic Inclusion in TWIP Steel [J]. Steel Research, 2013, 85 (4).

[2] Grässel O, Krüger L, Frommeyer G, et al. High strength Fe-Mn-(Al, Si) TRIP TWIP steels development properties application [J]. International Journal of Plasticity, 2000, 16: 1391-1409.

[3] 张贵杰, 宋卓霞. 汽车用高强度高塑性 TWIP 钢的开发研究 [J]. 中国科技产业, 2009 (3): 119-120.

[4] 黎倩, 李麟, 熊荣刚, 等. Fe-Mn-Si-Al 系 TWIP 钢显微组织与力学性能的研究 [J]. 上海金属, 2008, 30 (1): 8-11.

[5] 马凤仓. TWIP 钢的研究现状 [J]. 宝钢技术, 2008 (6): 62-66.

[6] 张增志. 耐磨高锰钢 [M]. 北京: 冶金工业出版社, 2002.

[7] 傅排先, 李殿中, 李依依. 厚断面高锰钢铸件凝固组织模拟与控制 [J]. 铸造, 2009, 58 (10): 1030-1037.

[8] Liu J, Wang G, Bao Y, et al. Inclusion variations of hot working die steel H13 in refining process [J]. Journal of iron and steel research, International, 2012, 19 (11): 01-07.

(原载于《第十八届全国炼钢学术会议论文集》——S08：品种开发与质量控制，2014 年 5 月：2-7)

# Influence of Vanadium on Microstructure and Properties of Medium-chromium White Cast Iron

Liu Keming[1]  Wang Fuming[1]  Li Changrong[2]  Su Liuyan[3]

(1. School of Metallurgical and Ecological Engineering, University of Science and Technology Beijing, Beijing 100083;
2. School of Materials Science and Engineering, University of Science and Technology Beijing, Beijing 100083;
3. College of basic medical sciences, China medical University, Shenyang 110001)

**Abstract:** White cast iron containing 5wt% chromium and different vanadium contents (up to 10 wt% V) were studied. The microstructures of the samples were analyzed by use of SEM and EDS. The impact energy, Rockwell hardness and wear resistance of the samples were determined. The results indicate that with an increase in vanadium content, the microstructure of medium chromium white cast iron become finer, the impact energy and wear resistance are improved.

**Keywords:** vanadium; medium chromium white cast iron; microstructure; wear resistance

## 1 Introduction

High chromium cast irons have sufficient wear resistance (in particular, under oxidizing and corrosive conditions), as their microstructures have $M_7C_3$ carbide in matrix. The discontinuous distribution of rod-like $M_7C_3$ carbide in matrix makes their toughness higher than that of low alloy white cast irons, which contain continuous brittle cementite. Because of this superior properties high chromium castings have been widely applied to wearing parts of machines in mineral engineering, steelmaking plants, and so forth. However, the demand for higher quality wear resistant materials that can length service life under even more sever conditions becomes greater in many fields. The desire to eliminate frequent shutdown of equipment for replacement of worn or broken castings, plus recognition of the resulting costs in terms of lost productivity, have encouraged engineers to evaluate candidate alloys on a cost/performance basis and, furthermore, to develop and specify new alloys which provide superior abrasive wear resistance along with adequate toughness[1,2]. It has been reported that the addition of strong carbide-forming elements, such as vanadium, tungsten, niobium and titanium, improves the mechanical properties of high chromium white irons[3]. Vanadium can form vanadium carbide, the Vickers hardness of VC is 2800, which is much harder than that of $M_7C_3$ (HV 1200~1800) in high chromium cast iron. The round morphology of VC can reduce splitting to matrix, which may be useful to get superior toughness. Alloying high chromium iron with

---

刘克明,1998~2001 年于北京科技大学师从李文超教授攻读硕士学位。目前在中国钢研科技集团有限公司工作。

vanadium makes the structure finer. Vanadium is soluble in eutectic $M_7C_3$ carbides as well as in austenite and influences the transformation of austenite in high chromium iron. Where the vanadium content is higher than 4%, precipitation of dispersive secondary carbides of VC type in austenite are observed, which is favourable for martensitic transformation[4-6]. In this paper, the influence of vanadium on microstructure and properties of medium-chromium white cast iron is examined.

## 2  Experimental procedures

The raw materials for melting vanadium containing medium-chromium white cast irons are pig iron, ferrovanadium and high carbon ferrochrome. The melting of cast irons was carried out in 25kg non-vacuum induction furnace, the melting temperature was about 1500℃, then cast by sand mould, the size of the mould was 80mm × 80mm × 200mm. The final chemical compositions of the samples are shown in Table 1.

Table 1  Chemical compositions of the samples (wt%)

| Sample No. | C | Cr | V |
|---|---|---|---|
| 1 | 2.25 | 4.95 | 2.03 |
| 2 | 2.38 | 5.11 | 4.16 |
| 3 | 2.33 | 5.21 | 6.10 |
| 4 | 2.24 | 5.30 | 8.00 |
| 5 | 2.35 | 5.06 | 10.20 |

The non-notch samples with a size of 10mm × 10mm × 55mm were made from the cast irons by wire cutting and grinding. The impact energy was measured using a Charpy's tester. The samples for hardness measurement were selected from the bars after impact test and measured by a Rockwell tester.

Samples were heat treated in an electric furnace with vacuum atmosphere at 980℃ for 2 hours, followed by oil-quenching to room temperature. The tempering temperature was 250℃. The technological process of heat treatment is shown as Fig. 1.

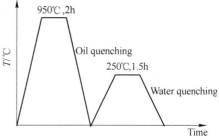

Fig. 1  The technological process of heat treatment

Samples for structural analysis were selected from the bars after impact test. After the samples were roughly and finely ground, polished and etched by use of 4% nital, metallographic observation and analysis were made by a scanning electron microscope, JSM-6301F, using an accelerating voltage of 25kV.

Abrasion wear experiments were carried out in an ML-10 tester. The cylindrical samples with a diameter of 5mm were made by wire cutting. Silicon carbide waterproof abrasive papers with a mesh of 120 were used in the abrasive wear test. The load was 2 kg. The weight loss was determined by a balance with a sensitivity of 0.1mg. The relative wear resistance coefficient $\varepsilon$ was used to judge wear resistance of the cast iron, the reference specimen was high chromium white cast iron. The following equation was used to calculate the value of $\varepsilon$.

$$\varepsilon = \frac{\text{weight loss of reference specimen}}{\text{weightloss of test specimen}}$$

## 3 Experimental results and discussion

### 3.1 Influence of vanadium on microstructure

Fig. 2 (a) – (e) shows the back scattering micrographs of the samples with different vanadium contents. The vanadium content is 2.03%, 4.16%, 6.10%, 8.00% and 10.20% respectively. Samples

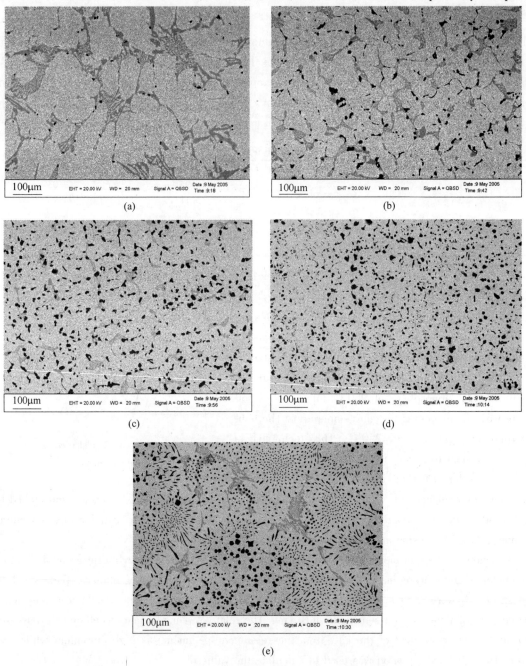

Fig. 2 Back scattering micrographs of the samples with different V contents: 2.03% (a), 4.16% (b), 6.10% (c), 8.00% (d) and 10.20% (e) in mass fraction

1-4 are hypoeutectic alloys and sample 5 is a eutectic alloy. It can be seen from Fig. 2 that the metallographic microstructure of the samples is composed of white metal matrix, gray eutectic carbides and black granular or rod-like eutectic carbides. Analyzed by EDS and X-ray diffraction patterns, when the vanadium content is lower, the gray eutectic carbides were identified as cementite ($M_3C$); when the vanadium content is higher, the gray eutectic carbides were identified as (Cr, Fe, V)$_7C_3$ type of chromium carbide ($M_7C_3$); the black granular or rod-like eutectic carbides were identified as (V, Fe, Cr) C type of vanadium carbides (MC). The EDS results for the three types of carbides are shown in Fig. 3 (a) – (c). For the hypoeutectic alloys, it is found from Fig. 2 (a) and (b) that samples 1 and 2 exhibit apparent dendrites morphology, with an increase in vanadium content from 2% to 4%, the microstructure becomes finer, the vanadium carbides distribution is along grain boundary. When the content of vanadium exceeds 6%, it is found from Fig. 2 (c) and (d) that the dendrite morphology is not apparent, and a great amount of granular MC type carbides occur, which prevents the formation of the dendrite structure, the vanadium carbides distribution is uniform. For eutectic alloys, the vanadium carbides take on chrysanthemum distribution (in Fig. 4 (a)), during solidification, a group of rod-like MC seemed to grow radially from a nucleus and formed a spherical eutectic cell together with austenite as schematically shown in Fig. 4 (b)[7,1]. By increasing the content of vanadium, the amount of MC type carbides increases.

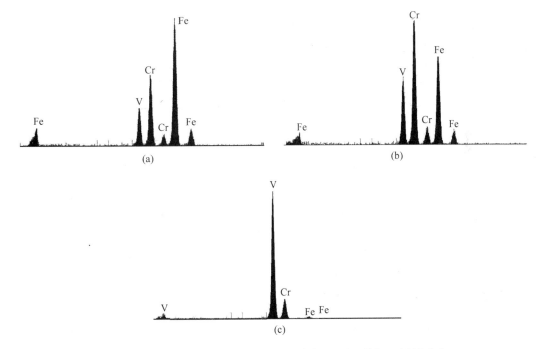

Fig. 3  EDS of three type carbides: $M_3C$ (a), $M_7C_3$ (b) and MC (c)

## 3.2 Influence of vanadium on impact toughness and hardness

Table 2 shows the experimental results of impact test, hardness test and abrasion wear test. The effects of vanadium on the impact toughness and hardness are shown in Fig. 5. With an increase in

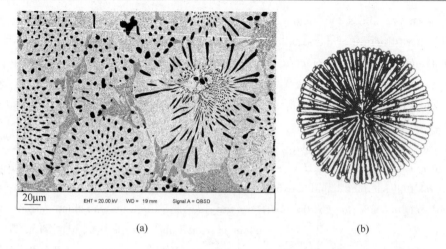

Fig. 4  Back scattering micrograph of sample 5 and schematic illustration of eutectic MC

vanadium content, the impact toughness increases, while hardness decreases. Vanadium changed the microstructure parameters of phases existing in microstructure of the alloys examined, including the size of primary austenite dendrites, the morphology and distribution of vanadium carbides and the degree of martensitic transformation, therefore vanadium affected the properties of the alloys examined. Finer primary austenite dendrites and round morphology of MC type carbides, which can reduce the splitting to matrix, result in the increase of the impact toughness. While more alloy content causes the stabilization of austenite, which leads to more retained austenite in the matrix after heat treatment, therefore resulting in the decrease in the hardness.

Table 2  Experimental results of impact test, hardness test and abrasion wear test

| Sample No. | Vanadium content /wt. % | Impact toughness /J·cm$^{-2}$ | Hardness (HRC) | Relative wear resistance, $\varepsilon$ |
|---|---|---|---|---|
| 1 | 2.03 | 2.47 | 62.8 | 1.21 |
| 2 | 4.16 | 2.7 | 62.5 | 1.36 |
| 3 | 6.10 | 3.15 | 62.5 | 1.68 |
| 4 | 8.00 | 4.32 | 55.6 | 2.01 |
| 5 | 10.20 | 4.42 | 52.1 | 2.13 |

## 3.3  Influence of vanadium on wear

The influence of vanadium on abrasion resistance is presented in Fig. 6. Relative wear resistance improved as the content of vanadium increased.

The wear mechanism of the matrix changes depending on the magnitude of the impact energy. Under a small impact energy the wear is mainly cutting[8]. In abrasion wear test there were no impact, the matrix was preferentially worn by a cutting action. Some investigations show that when the wear is mainly cutting, if the ratio of the hardness of the material worn to the hardness of the abrasive material is more than 0.8, the abrasion loss will decrease greatly[9]. Table 3 shows the

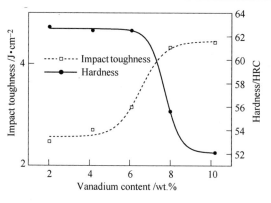
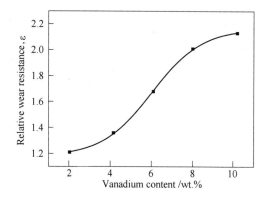

Fig. 5  Impact toughness and hardness as a function of vanadium in the alloy

Fig. 6  Relative wear resistance as a function of vanadium content in the alloy

ratios of the microhardness of the phases in cast irons examined and the microhardness of the abrasive material SiC. Since the hardness of the MC type carbides (HV2800) is greater than the hardness of the abrasive material SiC (HV2600), the MC type carbides can obstruct the cutting action of abrasive particles and protect the matrix from direct attack of abrasive particles. Therefore, with an increase of vanadium content, the amount of MC type carbides increases (Fig. 2), then the abrasion resistance improved.

Table 3  Ratios of the microhardness of the phases in the cast irons examined and microhardness of SiC

| Phases | MC (HV2800) | $M_7C_3$ (HV1200 – 1800) | $M_3C$ (HV840 – 1100) | Martensite (HV500 – 1000) | Austenite (HV300 – 600) |
|---|---|---|---|---|---|
| SiC (HV2600) | 1.08 | 0.46 – 0.69 | 0.32 – 0.42 | 0.19 – 0.38 | 0.12 – 0.24 |

In addition to the volume fraction of MC type of carbides, the size of phases present in the structure is another microstructure variable which affects the abrasion resistance of the alloys examined. The smaller size of primary austenite dendrites and the mean free path, i. e. the average distance between carbides particles caused by increasing the content of vanadium in the alloy, better protect the matrix from direct attack by abrasive particles.

It is important to note that wear resistance under low-stress abrasion conditions also depend on the matrix microstructure. In addition to the fact that the matrix help control the penetration depth of abrasive particles, it also plays an important role in preventing bodily removal of smaller carbides and cracking of massive ones. Martensitic matrix more adequately reinforces MC carbides than austenitic matrix. With an increase in vanadium content, there is more retained austenite in the matrix, the hardness of the alloys examined decreases, while the abrasion resistance increases, it seems contradictory. The reason is that the contribution to improvement of the abrasion resistance is mainly due to the amount increase of MC type carbides. This also suggests that hardness was not a sufficient indicator of a material's wear resistance.

## 4 Conclusions

(1) Vanadium influenced the microstructure characteristics of medium chromium white cast iron. By increasing the vanadium content, the structure became finer and the amount of MC type carbides increased.

(2) With an increase in vanadium content, the impact toughness and wear resistance were improved. Compared with high chromium white cast iron, the vanadium containing medium chromium white cast iron investigated has better wear resistance.

**Acknowledgements:** This research was supported financially by the Vanadium International Technical Committee and supervised by the Chinese Society for Metals.

### Reference

[1] A. Sawamoto, K. Ogi, K. Matsuda. Solidification Structures of Fe-C-Cr-(V-Nb-W) Alloys [J]. AFS Transactions, 1986, 72: 403-416.

[2] J. Dodd, J. L. Parks. Factors Affecting the Production and Performance of Thick-Section High Chromium-Molybdenum Alloy Iron Castings [J]. Metals Forum, 1980, 3 (1): 3-12.

[3] M. Radulovic, M. Fiset, K. Peev, M. Tomovic. The Influence of Vanadium on Fracture Toughness and Abrasion Resistance in High Chromium White Cast Irons [J]. Journal of Materials Science, 1994, 29: 5085-5094.

[4] Wang Yuwei, Shi Wen. The Effect of V in High Cr Cast Iron [J]. Fundry, 1989, 5: 9-12.

[5] Su Junyi, Guang Xiangzhong, Wang Enze. Study on High Chromium White Cast Iron Containing Vanadium with Martensite Matrix in the As-cast Condition [J]. Journal of Xi'an Jiaotong University, 1984, 18 (5): 23-28.

[6] Ye Yifu, Fan Tongxiang. Spheroidizing of Carbides of White Cast Iron [J]. Modern Cast Iron, 1995, 3: 28-32.

[7] Xu Liujie, Wei Shizhong, Long Rui, et al. Research on Morphology and Distribution of Vanadium Carbide in High Vanadium High Speed Steel [J]. Foundry, 2003, 52 (11): 1069-1073.

[8] Juntong Xi, Qingde Zhou. Influence of Retained Austenite on the Wear Resistance of High Chromium Cast Iron under Various Impact Loads [J]. Wear, 1993, 162-164: 83-88.

[9] Hao Shijian. High Chromium Wear Resistance Cast Iron [M]. Beijing: China Coal Industry Publishing House, 1993.

(原载于 *Iron and Steel*, 2005, 40 (S): 202)

# Phase Relationships and Thermodynamic Properties in the Mn-Ni-C System

Teng Lidong[1]   Ragnhild Aune[1]   Li Wenchao[2]   Seshadri Seetharaman[1]

(1. Dept. of Materials Science and Engineering, Royal Institute of Technology,
Stockholm SE 10044, Sweden;
2. Dept. of Physical Chemistry, University of Science and Technology
Beijing, Beijing 100083, China)

**Abstract:** In the present work, phase relationships in selected phase regions of the Mn-Ni-C system have been investigated at 1073K and 1223K by use of an equilibration technique. Alloys of Mn-Ni-C were prepared from pure Mn, Ni and C powders by powder metallurgy method. The phase identification of the heat treated samples was carried out by scanning electron microscope (SEM) and transmission electron microscope (TEM). The main phase compositions of the alloys have been analyzed by X-ray diffraction (XRD). The experimental results show that the site fraction of Ni in the metallic sublattice of the carbides $M_{23}C_6$, $M_7C_3$ and $M_5C_2$ is quite low and the value is around 2~3 percent. The thermodynamic activities of manganese in 16-composition Mn-Ni-C alloys have been studied by solid-state galvanic cell technique with $CaF_2$ as the solid electrolyte in the temperature range 940K to 1165K. The results are discussed in the light of the existing thermochemical information.

**Keywords:** phase equilibrium; Mn-Ni-C system; thermodynamic activity; galvanic cell technique

## 1 Introduction

The knowledge of the phase relationships in the Mn-Ni-C ternary system is of great importance for an understanding of the properties of alloy steels. The thermodynamics of the three binary systems in the system, i.e. Mn-C[1-3], Ni-C[4,5] and Mn-Ni[6,7], are relatively well established, but to the knowledge of the present authors, there is no experimental information available on the solid phase relationships of the ternary system Mn-Ni-C[8,9]. Because of the lack of the experimental data, there is no phase equilibrium assessment available for the Mn-Ni-C ternary system[10,11]. In a previous paper[12], some of the thermodynamic activities of Mn in the Mn-Ni-C ternary system have been reported. As a continuation of the earlier study, the present work is aimed at more experimental results of the activities of manganese in the system by the galvanic cell method as well as phase relationships by use of X-ray diffraction (XRD), scanning electron microscopy (SEM) and transmission electron microscopy (TEM).

---

滕立东,1998~2002年于北京科技大学师从李文超教授攻读博士学位。目前在ABB工作,任首席高级工程师。发表论文60余篇,获奖5项。

## 2 Experimental

The nominal compositions and sample numbers are identical to those described in a previous work[12] and listed in Table 1. Fig. 1 illustrates the locations of the alloy compositions and corresponding sample numbers in the isothermal section of Mn-Ni-C system at 1100K on the basis of the estimated Therml-Calc calculation[13]. Alloys of Mn-N-C were prepared from pure Mn, Ni and C powders by mixing the components in required proportions, compacting into pellets and sintering them in a closed alumina crucible at 1273K for two weeks under a stream of purified argon. Details of the argon gas cleaning apparatus and the oxygen meter used in this study have been reported elsewhere[14]. The sintered samples were placed inside quartz capsules and sealed under vacuum (0.1 Pa). The capsules were then placed in the uniform temperature zone (±0.5K) of a vertical alumina reaction tube and heat-treated. The duration of the heat-treatment was 336 h at 1073K and 288 h at 1223K. After heat treatment, the samples were quenched by dropping the capsules into liquid nitrogen and crushing them immediately. The main phase compositions of the alloys have been analyzed by X-ray diffraction (XRD). The phase identification of the heat treated samples were carried out by scanning electron microscope (SEM, JEOL 8900R) and transmission electron microscope (TEM, H-800). The chemical analysis of the various phases present was obtained by the electron microprobe and electron dispersion spectroscopy (EDS) detector attached to the SEM and TEM units. The analysis volume to identify each phase is typically a few cubic microns. The composition of the different phases was measured by a selection of three analyzing points from each phase. In the previous work[12], the results of EMF measurements of 4 alloys (MNC-21-24) have been reported. In the present work, the thermodynamic activities of Mn in all the studied alloys have been given. The cell arrangement and EMF measuring procedure are the same as those described in the previous work[12].

The galvanic cell used to determine the electromotive force can be represented as:

$$(-)Pt, Mn + MnF_2 + CaF_2 \mid CaF_2 \mid CaF_2 + MnF_2 + Mn-alloy, Pt(+) \quad (1)$$

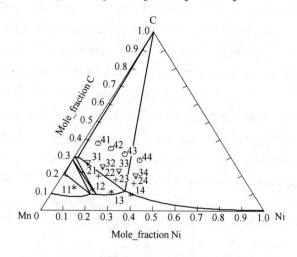

Fig. 1  The locations of various alloys in the Mn-Ni-C isthermal diagram at 1100K

The individual electrode reactions can be written as follows:

$$Mn + 2F^- = MnF_2 + 2e \quad (\text{left}) \tag{2}$$

$$MnF_2 + 2e = Mn-\text{alloy} + 2F^- \quad (\text{right}) \tag{3}$$

$$Mn = Mn-\text{alloy} \quad (\text{overall}) \tag{4}$$

## 3 Results and discussions

### 3.1 Phase composition and microstructure

The main phase compositions of all the alloys quenched at 1073K have been identified by XRD at room temperature. The XRD patterns are shown in In order to identify the chemical compositions of the equilibrium phases, some of the alloys were analyzed by use of the SEM and electron microprobe JEOL 8900R. Fig. 2 (a)-(f) present micrographs from SEM analyses of MNC-12, MNC-14, MNC-21, MNC-24, MNC-33 and MNC-41 alloys equilibrated at 1073K or 1223K. In Fig. 2 (a) the grey phase is $M_7C_3$ and the light phase is $\gamma$ solid solution. In Fig. 2 (b), the main phase is $\gamma$ (light phase) and the dark grey phase is $M_7C_3$. In Fig. 2 (c), three phases have been observed and the light phase is $\gamma$, the grey phase is $M_5C_2$, and the dark phase is $M_7C_3$. In Fig. 2(d), (e) and (f), $M_7C_3$ and $\gamma$ phases have been detected in the various alloys. The chemical compositions determined by electron microprobe are presented in Table 1. Chemical composition is given in site fractions of

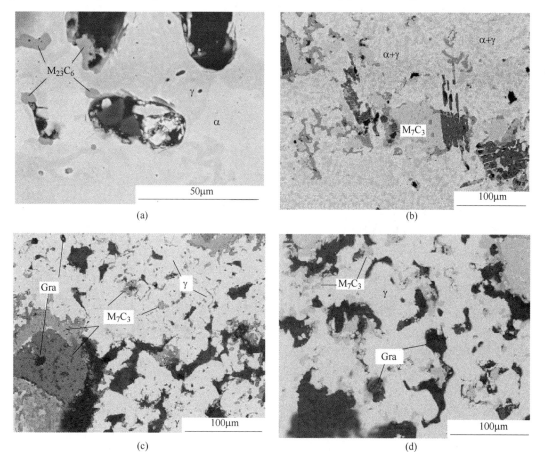

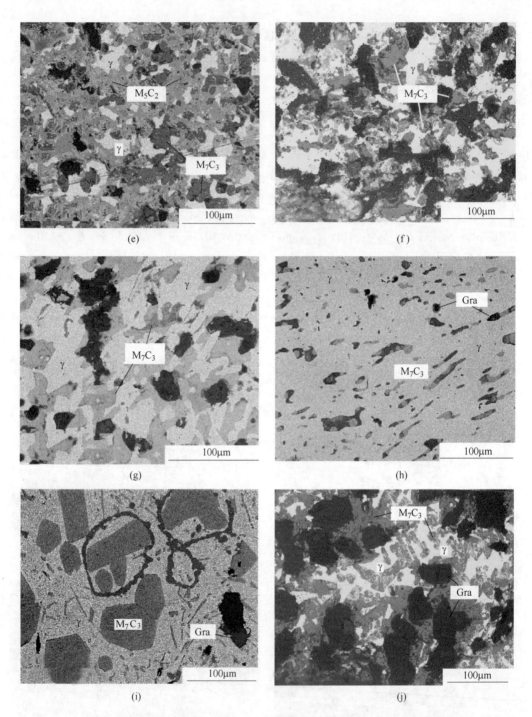

Fig. 2  SEM photographs of the selected alloys quenched at 1073K or 1223K
(a) MNC-11, 1223K; (b) MNC-12, 1223K; (c) MNC-14, 1073K; (d) MNC-14, 1223K;
(e) MNC-21, 1073K; (f) MNC-21, 1223K; (g) MNC-22, 1073K; (h) MNC-24, 1223K;
(i) MNC-33, 1073K; (j) MNC-41, 1223K

metal components $y_{Mn}$ and $y_{Ni}$ ($y_i = x_i/(x_{Mn} + x_{Ni})$; $x_i$ molar fraction of element, $i$ denotes Mn or Ni). As can be seen from Table 1, the site fractions of Mn and Ni in $M_{23}C_6$ or $M_7C_3$ and even $M_5C_2$ in the case of MNC-21 are nearly constant in the different alloys. Further, the solubility of Ni in the carbides is quite low and the value is around 2 percent. The solubility of Ni in the $M_5C_2$ is less than 3 percent. On the other hand, Mn and Ni can form continuous series of solid solutions in the $\gamma$ alloys.

**Table 1  Experimentally determined phase compositions of the alloys**

| Specimens | Phases | $y_{Mn}$ | $y_{Ni}$ |
| --- | --- | --- | --- |
| MNC-11 1223K | $M_{23}C_6$ | 97.82 | 2.18 |
|  | $\gamma$ | 84.35 | 15.65 |
| MNC-12 1223K | $M_7C_3$ | 98.23 | 1.77 |
|  | $\gamma$ | 64.81 | 35.19 |
| MNC-14 1073K | $M_7C_3$ | 97.94 | 2.06 |
|  | $\gamma$ | 51.48 | 48.52 |
| MNC-14 1223K | $M_7C_3$ | 97.77 | 2.23 |
|  | $\gamma$ | 54.43 | 45.57 |
| MNC-21 1073K | $M_5C_2$ | 97.41 | 2.59 |
|  | $\gamma$ | 54.12 | 45.87 |
|  | $M_7C_3$ | 98.87 | 1.13 |
| MNC-21 1223K | $M_7C_3$ | 98.73 | 1.27 |
|  | $\gamma$ | 66.33 | 33.67 |
| MNC-22 1073K | $M_7C_3$ | 98.71 | 1.29 |
|  | $\gamma$ | 66.84 | 33.16 |
| MNC-24 1223K | $M_7C_3$ | 97.89 | 2.11 |
|  | $\gamma$ | 53.84 | 46.16 |
| MNC-33 1073K | $M_7C_3$ | 97.57 | 2.43 |
|  | $\gamma$ | 51.72 | 48.28 |
| MNC-41 1223K | $M_7C_3$ | 98.09 | 1.91 |
|  | $\gamma$ | 53.17 | 46.83 |

Note: Analyzed by electron microprobe JEOL 8900R. Chemical composition is given in site fractions of the metal components $y_{Mn}$ and $y_{Ni}$ ($y_i = x_i/(x_{Cr} + x_{Fe})$; $x_i$ is the molar fraction of element Mn or Ni).

## 3.2  The EMF results and Mn activities

The experimental EMF values measured in the temperature range from 940K to 1165K were found to be reproducible during heating and cooling cycles. The experimentally measured EMF values for the galvanic cell represented by Eq. (1) at different temperatures for all the alloys are presented in Table 2. The values in this table are the average values of a number of measurements both during heating as well as cooling cycles.

Table 2  The experimental results of the EMF values and activities of Mn measured in the temperature range from 940K to 1165K for the Mn-Ni-C alloys

| \multicolumn{3}{c|}{MNC-11} | \multicolumn{3}{c|}{MNC-12} | \multicolumn{3}{c|}{MNC-13} | \multicolumn{3}{c}{MNC-14} |

| MNC-11 ||| MNC-12 ||| MNC-13 ||| MNC-14 |||
|---|---|---|---|---|---|---|---|---|---|---|---|
| $T$/K | EMF/mV | $a_{Cr}$ | $T$/K | EMF/mV | $a_{Cr}$ | $T$/K | EMF/mV | $a_{Cr}$ | $T$/K | EMF/mV | $a_{Cr}$ |
| 984 | 21.07 | 0.608 | 991 | 37.78 | 0.413 | 944 | 53.04 | 0.271 | 998 | 69.61 | 0.198 |
| 1015 | 20.14 | 0.631 | 1016 | 38.86 | 0.412 | 971 | 52.55 | 0.285 | 1023 | 70.17 | 0.203 |
| 1046 | 20.00 | 0.642 | 1048 | 39.15 | 0.420 | 994 | 51.44 | 0.301 | 1055 | 70.66 | 0.211 |
| 1079 | 20.51 | 0.643 | 1076 | 40.22 | 0.420 | 1026 | 50.75 | 0.317 | 1083 | 71.19 | 0.217 |
| 1108 | 22.05 | 0.630 | 1098 | 41.23 | 0.418 | 1048 | 50.52 | 0.327 | 1112 | 71.99 | 0.222 |
| 1137 | 23.93 | 0.614 | 1103 | 41.57 | 0.417 | 1080 | 51.12 | 0.333 | 1140 | 69.72 | 0.242 |
| 1165 | 25.24 | 0.605 | 1133 | 41.65 | 0.426 | 1111 | 52.89 | 0.331 | 1154 | 69.72 | 0.244 |
|  |  |  | 1158 | 41.13 | 0.438 | 1157 | 53.53 | 0.342 |  |  |  |

| MNC-21 ||| MNC-22 ||| MNC-23 ||| MNC-24 |||
|---|---|---|---|---|---|---|---|---|---|---|---|
| $T$/K | EMF/mV | $a_{Cr}$ | $T$/K | EMF/mV | $a_{Cr}$ | $T$/K | EMF/mV | $a_{Cr}$ | $T$/K | EMF/mV | $a_{Cr}$ |
| 952 | 54.69 | 0.264 | 953 | 52.06 | 0.281 | 970 | 58.09 | 0.249 | 944 | 68.09 | 0.187 |
| 1030 | 50.85 | 0.318 | 982 | 50.50 | 0.303 | 1010 | 61.62 | 0.243 | 972 | 69.03 | 0.192 |
| 1068 | 48.65 | 0.347 | 1001 | 48.97 | 0.321 | 1052 | 64.13 | 0.243 | 1011 | 69.63 | 0.202 |
| 1104 | 47.24 | 0.370 | 1039 | 47.07 | 0.349 | 1089 | 66.00 | 0.245 | 1027 | 69.43 | 0.208 |
| 1123 | 46.20 | 0.385 | 1066 | 44.92 | 0.376 | 1128 | 68.65 | 0.243 | 1059 | 69.28 | 0.219 |
| 1124 | 46.47 | 0.383 | 1093 | 43.96 | 0.393 | 1130 | 68.36 | 0.246 | 1084 | 67.42 | 0.236 |
| 1159 | 42.33 | 0.428 | 1122 | 44.17 | 0.401 |  |  |  | 1085 | 67.43 | 0.236 |
|  |  |  | 1123 | 43.89 | 0.404 |  |  |  | 1115 | 64.61 | 0.261 |
|  |  |  | 1149 | 45.31 | 0.400 |  |  |  | 1144 | 65.11 | 0.267 |

| MNC-31 ||| MNC-32 ||| CCF-33 ||| MNC-34 |||
|---|---|---|---|---|---|---|---|---|---|---|---|
| $T$/K | EMF/mV | $a_{Cr}$ | $T$/K | EMF/mV | $a_{Cr}$ | $T$/K | EMF/mV | $a_{Cr}$ | $T$/K | EMF/mV | $a_{Cr}$ |
| 947 | 71.88 | 0.172 | 1003 | 70.72 | 0.195 | 973 | 70.43 | 0.186 | 981 | 72.34 | 0.181 |
| 977 | 71.64 | 0.182 | 1028 | 70.66 | 0.203 | 998 | 70.67 | 0.193 | 1009 | 72.37 | 0.189 |
| 1010 | 71.55 | 0.193 | 1061 | 70.68 | 0.213 | 1027 | 71.00 | 0.201 | 1037 | 72.42 | 0.198 |
| 1040 | 71.41 | 0.203 | 1091 | 70.68 | 0.222 | 1060 | 71.05 | 0.211 | 1061 | 72.52 | 0.205 |
| 1067 | 71.26 | 0.212 | 1113 | 70.82 | 0.228 | 1085 | 71.13 | 0.218 | 1094 | 72.63 | 0.214 |
| 1094 | 71.22 | 0.221 | 1118 | 69.98 | 0.234 | 1086 | 70.68 | 0.221 | 1123 | 72.09 | 0.225 |
| 1124 | 70.89 | 0.231 | 1141 | 70.12 | 0.240 | 1118 | 71.23 | 0.228 | 1150 | 72.01 | 0.234 |
| 1150 | 70.23 | 0.242 |  |  |  | 1144 | 71.01 | 0.237 |  |  |  |
|  |  |  |  |  |  | 1145 | 71.05 | 0.237 |  |  |  |

| MNC-41 ||| MNC-42 ||| MNC-43 ||| MNC-44 |||
|---|---|---|---|---|---|---|---|---|---|---|---|
| $T$/K | EMF/mV | $a_{Cr}$ | $T$/K | EMF/mV | $a_{Cr}$ | $T$/K | EMF/mV | $a_{Cr}$ | $T$/K | EMF/mV | $a_{Cr}$ |
| 945 | 71.03 | 0.175 | 946 | 70.74 | 0.176 | 947 | 69.97 | 0.180 | 954 | 71.05 | 0.177 |
| 971 | 71.65 | 0.180 | 976 | 70.80 | 0.186 | 975 | 70.13 | 0.188 | 983 | 70.99 | 0.187 |

Continued Table 2

| MNC-41 | | | MNC-42 | | | MNC-43 | | | MNC-44 | | |
| --- | --- | --- | --- | --- | --- | --- | --- | --- | --- | --- | --- |
| $T/K$ | EMF/mV | $a_{Cr}$ | $T/K$ | EMF/mV | $a_{Cr}$ | $T/K$ | EMF/mV | $a_{Cr}$ | $T/K$ | EMF/mV | $a_{Cr}$ |
| 1002 | 71.77 | 0.190 | 1008 | 70.66 | 0.19648 | 1004 | 70.25 | 0.197 | 1010 | 70.95 | 0.196 |
| 1032 | 71.71 | 0.199 | 1032 | 70.48 | 0.205 | 1030 | 70.38 | 0.205 | 1038 | 70.85 | 0.205 |
| 1061 | 71.64 | 0.209 | 1061 | 70.23 | 0.215 | 1060 | 70.56 | 0.213 | 1068 | 70.72 | 0.215 |
| 1084 | 71.55 | 0.216 | 1065 | 70.22 | 0.216 | 1089 | 70.63 | 0.222 | 1094 | 70.64 | 0.223 |
| 1117 | 71.73 | 0.225 | 1093 | 70.05 | 0.226 | 1118 | 68.32 | 0.242 | 1123 | 70.78 | 0.232 |
| 1145 | 72.02 | 0.232 | 1120 | 69.95 | 0.235 | 1146 | 68.17 | 0.251 | 1147 | 70.6 | 0.238 |
| | | | 1134 | 69.95 | 0.239 | | | | | | |
| | | | 1147 | 69.59 | 0.245 | | | | | | |

The activity of Mn in the Mn-alloy can be calculated by equation (5):

$$\Delta_r G = -2EF = RT \ln a_{Mn-alloy} \qquad (5)$$

The standard state for the activity of Mn is pure β-Mn. The calculated activities of Mn by equation (5) for all the alloys are plotted as functions of temperature in Fig. 3 (a)-(d). The activities of Mn for alloys with 10 wt.% nickel content are shown in Fig. 3 (a). As can be seen from Fig. 3 (a), the activities of Mn for MNC-11 alloy is nearly constant ($a_{Mn} = 0.61 - 0.64$) in the temperature range 984 - 1165K. This demonstrates that the solubility of Mn in the solid solution γ-phase which is in equilibrium with $M_{23}C_6$ phase changes only very slightly in this temperature range. For MNC-21, the activity of Mn increase with the increase of temperature. There is a noticeable increase of Mn activity between 1124K and 1159K. A rough calculation by Thermo-Calc[13] shows that the $M_5C_2$ phase will transform to $M_7C_3$ at 1157K. According to the SEM results, there are 3 equilibrium phases ($M_5C_2$, γ, and $M_7C_3$) at 1073K and only 2 equilibrium phases ($M_7C_3$ and γ) at 1223K in MNC-21 alloy. Accordingly, the activity change between 1124K and 1159K is likely related with the disappearing of the $M_5C_2$. For MNC-31 and MNC-41 alloys, the activity of Mn increase slightly with the increase of temperature and the temperature coefficient of activity is constant in the temperature range 952-1159K; moreover, the activities of Mn for alloy MNC-31 are nearly the same as those for alloy MNC-41, as can be seen in Fig. 3 (a). This indicates that these two alloys are in the same 3-phase region ($M_7C_3$, γ, and graphite). Further, the activities of components are independent of the alloy composition in this region of the Mn-Ni-C ternary system. This result is in conformity with the XRD phase composition analysis at 1073K. XRD patterns of the MNC-31 and MNC-41 alloys equilibrated at 1073K and quenched in liquid nitrogen are shown in Fig. 4. The equilibrium phase compositions are $M_7C_3$, γ and graphite for the two alloys, though the graphite content in MNC-31 is very low. The activities of Mn for alloys with 20 wt.% Ni content are shown in Fig. 3 (b). The activities of Mn for MNC-12 alloy is nearly constant ($a_{Mn} = 0.41 - 0.44$) in the temperature range 991 - 1158K. The activity of Mn for MNC-22 alloy increases with the increase of temperature. According to the XRD analysis results, the equilibrium phase of MNC-22 at 1073K is γ-Mn and $M_7C_3$. The increase in $a_{Mn}$ with temperature may be affiliated to increase in the solubility of Mn in the solid solution γ-phase which is in equilibrium with $M_7C_3$ phase increases with temperature in this temperature range. The activities of Mn for alloy MNC-32 fall very close to those for alloy MNC-42, as can be seen in Fig. 3 (b). No phase transformation could be detected for these 4 alloys by EMF measurements in the experimental temperature ranges. The activities of Mn of alloys with 30 wt. % Ni are shown in Fig. 3 (c). The activities of Mn for MNC-13 alloy exhibit an apparent increase with increasing temperature at 944 – 1048K and afterwards the activities are nearly constant at $a_{Mn} = 0.33$. It is not certain if the effect would be attributed to any possible phase transformation around 1048K. For the MNC-23 alloy, the activities of Mn are independent of temperature at 970 – 1130K; further, the activities of Mn at 1128K and 1130K are quite reproducible and they are very close to the values of MNC-33 which is in the 3-phase-region ($M_7C_3$, γ, and graphite), as shown in Fig. 3 (c). The activities of Mn for alloy MNC-33 fall very close to those for alloy MNC-43. These two alloys are also in the same 3-phase-region ($M_7C_3$, γ, and graphite). The activities of Mn for alloys with 40 wt. % nickel are shown in Fig. 3 (d). As can be seen from the figure, the activities of Mn for all the four alloys are almost the same. For alloy MNC-24, the activities are slightly higher than the others by 10 percent when the temperature is above 1085K. To check if there is a possible phase transformation, the samples of MNC-24 alloy were quenched at 1073K and 1223K. XRD analysis results show that the equilibrium phases are the same

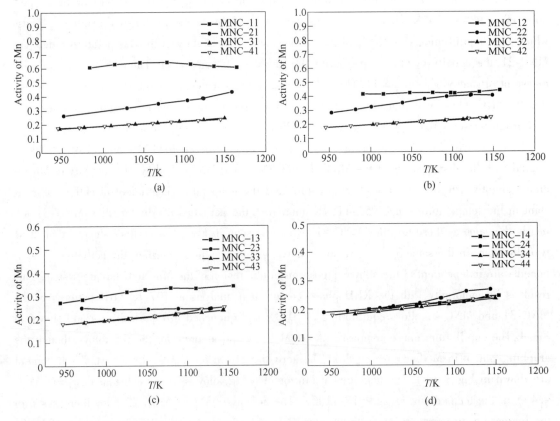

Fig. 3 (a)-(d) The plots of manganese activities vs temperatures for all the alloys
(a) Alloys with 10 wt. % Ni; (b) Alloys with 20 wt. % Ni; (c) Alloys with 30 wt. % Ni; (d) Alloys with 40 wt. % Ni

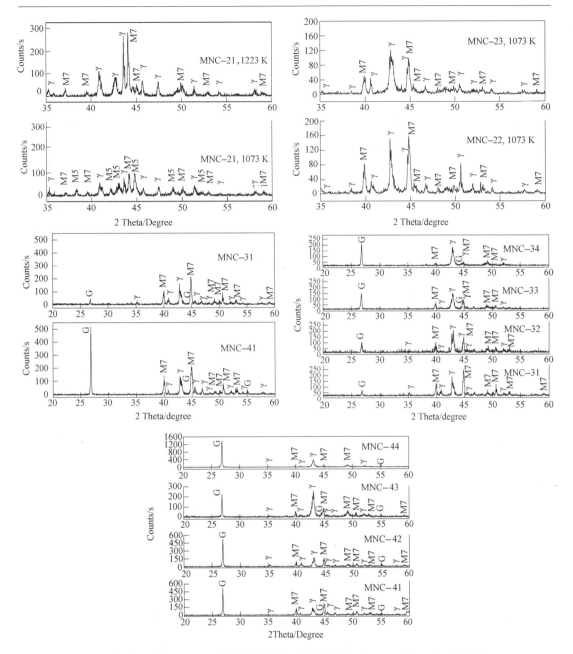

Fig. 4 The XRD patterns of the MNC-31 and MNC-41 alloys equilibrated at 1073K

and they are $M_7C_3$, $\gamma$ and graphite, as shown in Fig. 5. It can be concluded that these four alloys are located in the same 3-phase-region ($M_7C_3$, $\gamma$ and graphite) in the ternary diagram. Although no graphite phase could be detected in MNC-14 by XRD (shown in Fig. 6), the EMF results demonstrate that a small amount of graphite should be present in the alloy. To verify the existence of graphite in the alloy, the samples of MNC-14 alloy, quenched in liquid nitrogen from 1073K, were carefully analyzed by transmission electron microscopy (TEM, Japan H-800) equipped with energy dispersive spectroscopy (EDS). The graphite phase was detected by the TEM analysis. Fig. 7 and Fig. 8 shows the transmission electron micrographs of different phase in MNC-14 quenched at 1073K. The bright-field image of the $M_7C_3$ crystal particles is shown in Fig. 7 (a) and the $[12\bar{9}]$ SAD pattern of $M_7C_3$ is shown in Fig. 7 (b). The interfacial structure between graphite and γ phase was shown in Fig. 8 (a). Fig. 8 (b) shows the multiple crystal diffraction of graphite phase. The solid solution phase near the grain boundary of graphite has a lattice-deformation feature which is induced by the interfacial stress. The SAD pattern of γ phase presents a lattice-deformation feature, as shown in Fig. 8 (c). Graphite/γ interfacial stress resulted from the mismatch of the thermal expansion coefficients of the two phases.

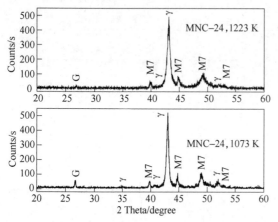

Fig. 5 The XRD patterns of MNC-24 alloy equilibrated at 1073K and 1223K

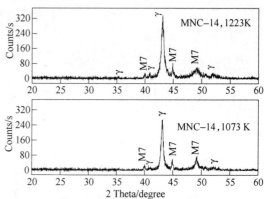

Fig. 6 The XRD patterns of the MNC-14 equilibrated at 1073K and 1223K

Fig. 7 TEM bright-field image and selected area diffraction (SAD) patterns from MNC-14 alloy quenched at 1073K

(a) the morphology of the $M_7C_3$ crystal particles; (b) the $[12\bar{9}]$ diffraction pattern of $M_7C_3$

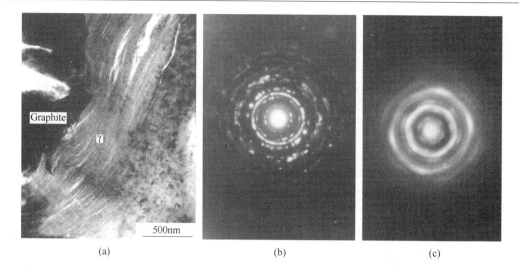

Fig. 8  TEM bright-field image and SAD patterns from MNC-14 alloy quenched at 1073K
(a) the morphology of the interface between graphite and γ phase;
(b) the multiple crystal diffraction of graphite;  (c) the SAD pattern of γ phase which showed a
lattice-deformation feature induced by interfacial stress

By combining the experimental results of XRD, SEM, TEM and EMF, the equilibrium phases in the Mn-Ni-C system at different temperatures are presented in Table 3. The solubility of carbon in γ phase is about 3 at.%. Fig. 9 presents an isothermal section of the Mn-Ni-C system at 1073K on the basis of the estimated Therml-Calc calculation. The experimental results show that the solubility of Ni in the γ phase which is in equilibrium with $M_7C_3$ and graphite is higher than those indicated by Thermo-Calc in preliminary calculations. The three-phase region of γ/$M_7C_3$/graphite predicted by the experimental results is shown as dashed lines in Fig. 9.

Table 3  The sample numbers, nominal chemical compositions (in wt.%)
and equilibrium phases of the Mn-Ni-C alloys

| Sample No. | Ni | Mn | C | Temp./K | Main phases by XRD |
|---|---|---|---|---|---|
| MNC-11 | 10 | 86.8 | 3.2 | 1223 | γ-Mn, $M_{23}C_6$ |
| MNC-12 | 20 | 77.2 | 2.8 | 1223 | γ-Mn, $M_{23}C_6$ |
| MNC-13 | 30 | 67.6 | 2.4 | 944-1048 | γ-Mn, $M_7C_3$ |
| MNC-14 | 40 | 57.9 | 2.1 | 998-1154 | γ-Mn, $M_7C_3$, Graphite |
| MNC-21 | 10 | 84.4 | 5.6 | 1073<br>1159-1223 | γ-Mn, $M_5C_2$, $M_7C_3$<br>γ-Mn, $M_7C_3$ |
| MNC-22 | 20 | 75.1 | 4.9 | 953-1149 | γ-Mn, $M_7C_3$ |
| MNC-23 | 30 | 65.7 | 4.3 | 1028-1030 | γ-Mn, $M_7C_3$, Graphite |
| MNC-24 | 40 | 56.3 | 3.7 | 944-1144 | γ-Mn, $M_7C_3$, Graphite |
| MNC-31 | 10 | 82.5 | 7.5 | 947-1150 | γ-Mn, $M_7C_3$ |
| MNC-32 | 20 | 73.4 | 6.6 | 1003-1141 | γ-Mn, $M_7C_3$, Graphite |

Continued Table 3

| Sample No. | Ni | Mn | C | Temp./K | Main phases by XRD |
|---|---|---|---|---|---|
| MNC-33 | 30 | 64.2 | 5.8 | 973-1145 | γ-Mn, $M_7C_3$, Graphite |
| MNC-34 | 40 | 55.0 | 5.0 | 981-1150 | γ-Mn, $M_7C_3$, Graphite |
| MNC-41 | 10 | 78.4 | 11.6 | 945-1145 | γ-Mn, $M_7C_3$, Graphite |
| MNC-42 | 20 | 69.7 | 10.3 | 946-1147 | γ-Mn, $M_7C_3$, Graphite |
| MNC-43 | 30 | 61.0 | 9.0 | 947-1146 | γ-Mn, $M_7C_3$, Graphite |
| MNC-44 | 40 | 52.3 | 7.7 | 954-1147 | γ-Mn, $M_7C_3$, Graphite |

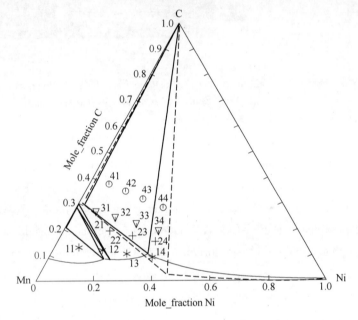

Fig. 9 Isothermal section of the Mn-Ni-C system at 1073K. The solid lines show the results calculated by Thermo-Calc. The dashed lines show the three-phase region (γ-Mn, $M_7C_3$, C) obtained by the present experiment

## 4 Summary

Phase relationships in selected phase regions of Mn-Ni-C system have been investigated at 1073K and 1223K by use of an equilibration technique. The site fraction of Ni in the metallic sublattice of the carbides $M_{23}C_6$, $M_7C_3$ and $M_5C_2$ is around 2-3 percent. The thermodynamic activities of manganese in 16 Mn-Ni-C alloys have been studied by solid-state galvanic cell technique with $CaF_2$ as the solid electrolyte in the temperature range 940K to 1165K. The three-phase region of γ/$M_7C_3$/graphite has been constructed at 1073K.

**Acknowledgements:** The authors wish to thank Professor Nobuo Sano for his valuable discussions during the preparation of this paper. The authors also thank Mr. Zuotai Zhang for the help on the

TEM analysis. This work is financially supported by the Swedish Research Council (VR).

## References

[1] Du S, Seetharaman S, Staffansson L-I. Metallurgical Transactions B, 1989; 20: 747.
[2] W. Huang, Scand. J. Metall., 1990, 19: 26-32.
[3] R. Benz, J. F. Elliot, J. Chinpman. Metallurgical Transactions, 1973, (4): 1449-1452.
[4] J. Natesan, T. F. Kassner. Metallurgical Transactions, 1973, (4): 2557-2566.
[5] T. B. Massalski. Binary Phase Diagrams ($2^{nd}$ edition)[M]. Ohio: the ASM international, Materials Park, 1990.
[6] B. R. Coles, W. Hume-Rothery. J. Insti. Met. (London), 1951-52, 80: 85-92.
[7] L. Pal, E. Kren, G. Kadar, et al. J. Appl. Phys., 1968, 39: 538-544.
[8] E. A. Brandes, R. F. Flint. Manganese Phase Diagram [M]. Paris: the manganese center, 1980.
[9] R. E. Aune, S. C. Du, S. Seetharaman. The Japan Society of Calorimetry and Thermal Analysis-Netsu Sokutei, 1997.
[10] W. Huang. Metall. Trans., 1990, 21A: 2115-2123.
[11] A. Gabriel, P. Gustafson, I. Ansara. Calphad, 1987, 11: 203-218.
[12] L. D. Teng, R. Aune, S. Seetharaman. Intermetallics, 2003, 11: 1229-1235.
[13] B. Sundman, B. Jansson, J. O. Andersson. Calphad, 1985, 9 (2): 153-190.
[14] L. D. Teng, X. G. Lu, R. E. Aune, et al. Thermodynamic investigations of $Cr_3C_2$ and reassessment of the Cr-C system [J]. Metallurgical and Materials Transactions (A), 2004.

(原载于 *Metallurgical and Materials Transactions A*, 2005, 36A: 2585-2593)

# Experimental Investigation and Modeling of Cooling Processes of High Temperature Slags

Sun Yongqi  Shen Hongwei  Wang Hao  Wang Xidong  Zhang Zuotai

(Department of Energy and Resources Engineering and Beijing Key Laboratory for Solid Waste Utilization and Management, College of Engineering, Peking University, Beijing 100871)

**Abstract**: This paper investigated the cooling processes of high temperature slags using SHTT (single hot thermocouple technique) and CFD (computational fluid dynamics) method for the purpose of recovering the waste heat. A series of slags with the $CaO/SiO_2$ ratio of 1.1 and different $Al_2O_3$ contents were designed. The Continuous Cooling Transformation diagrams were established, aiming to bridge the gap between slag properties and practical operations. The variation of slag properties during cooling processes, including crystallization ability and glass forming ability, was investigated by SHTT and the results indicated that the variation of $Al_2O_3$ content greatly changed the slag properties, which had a great effect on heat recovery of slags. Slags with $Al_2O_3$ content 15 wt% had the smallest critical cooling rate, which was suitable for heat extraction. A 3-D model was developed to simulate the natural heat transfer between a slag droplet and the ambient air in this study using CFD package Fluent software. The results indicated that the temperature differences could be more than 100℃ between core and surface of slag droplets with 3mm in diameter. These large temperature differences could cause crystallization inside the slag droplets, which should be considered in the processes of heat recovery.

**Keywords**: slag; heat recovery; cooling process; glass forming ability; cooling rate

## 1 Introduction

Metallurgical industry is one of most energy-intensive industries, accounting for around 15% of total energy consumption in China. Although metallurgical industry has made a significant progress in its energy efficiency during past decades through implementing numerous new technologies, the results of waste energy recovery are still unsatisfactory. Several studies [1-3] have investigated the energy intensity in metallurgical industry and reported that the waste heat from BFG (blast furnace gas) has been essentially recovered. However, the waste heat recovery from high temperature slags still faces strong challenges and has not been effectively achieved. According to the previous estimation [4,5], metallurgical slags, carrying a substantial amount of high quality thermal energy, represent the largest undeveloped energy sources in metallurgical industry. In China, the output of pig iron was more than 650 million tons and 710 million tons of crude steel in 2012 [6], which produced around 200 million tons BF (Blast Furnace Slags) and 70 million tons steel slags, respectively. The total waste heat of

---

张作泰，2000~2006年于北京科技大学师从李文超教授攻读博士学位。目前在北京大学工作，研究员/博士生导师。发表论文80余篇，专利20余项，获省部级二等奖3项。

these slags is more than $4.80 \times 10^{19}$ J, corresponding to around 16 million tons of standard coal. Cai et al. [7] have estimated that the waste-heat recovery rate of metallurgical slags is less than 2% in China. Furthermore, Barati [8] discussed the energy intensity and the $CO_2$ emission in metallurgical industry and Milford et al. [9] pointed out that the waste heat recovery from slags was effective to reduce $CO_2$ emission in metallurgical industry. It is well known that the ferrous slags contribute over 95% of the waste heat existing in the slags, and BF slags alone make up more than 70% energy of the ferrous slags [10]. Therefore, the heat recovery from BF slags is especially significant and the previous investigations were generally concentrated on BF slags in this field [11,12].

Conventionally, BF slags are gradually cooled by air or rapidly cooled by water for different subsequent utilizations. Air-gradually-cooled slags have limited utilizations, while water quenched slags are increasingly used as aggregates in construction or raw materials in cement manufacture nowadays because water quenched slags consist of high content of glassy phases to avoid subsequent swelling of the concrete [11]. However, wet granulation is faced with the series of problems [12,13]: (1) Too much water is consumed; (2) The alkaline leached from the slags pollutes the water; (3) $SO_2$ and $H_2S$ are emitted into the air; (4) The solid slags should be dried before used as cementious material; (5) The thermal energy is wasted without recovery. To solve these problems and realize recovery of both waste heat and resources of slags, a series of dry granulation methods have been proposed and studied in the past several decades [10,13]. It has been expected that slags could be cooled fast enough using dry granulation methods instead of water quenched method and the heated medium (air) could be used for heat recovery [13]. However, none of the dry granulation methods has realized industrial application because of the fundamental constraint that the thermal conductivity of slags is too low to extract heat efficiently during cooling processes. The thermal conductivity of slag is generally around $1-2$ W/(m·K) for glassy phases and $0.1-0.3$ W/(m·K) for liquid slags [14]. This needs large quantities of coolant and the liquid slags should be broken up into small droplets (3 – 6mm in diameter) with large specific surface areas during the operations of dry granulation. Therefore, it is significant to know how the high cooling rate influences the phase transformation regularities of high temperature slags, which is necessitated to establish the operation procedures of dry granulation technique. However, to the knowledge of present authors, few works have been carried out using experimental methods so far. The present paper was therefore motivated. Furthermore, In view of the large quantities of BF slags, the present study was focused on the investigation of BF slags. Meanwhile, iron ore has been degraded and the gangue contents such as $Al_2O_3$ content is increasing [15,16], which shows great influence on slag properties and subsequently affects waste heat recovery from slags. The $Al_2O_3$ content in general slags was less than 10%, while that in high content slags can exceed 20%. Therefore a series of slags with $Al_2O_3$ content of $0-25\%$ were designed in this study.

This investigation utilizes a SHTT (single hot thermocouple technique) for visualizing phase transformation in the BF slags. SHTT combines the advantages of optical observation and the low inertial of the system, which makes it possible to achieve a high heating and cooling rate with a maximum rate of 50K/s and to observe crystallization phenomena at elevated temperatures. Furthermore, the CFD (computational fluid dynamics) package Fluent software was used to predict the cooling process of high temperature slags [17]. These investigations may provide the clues of tailoring the procedure of dry granulation technique.

## 2 Materials and Methods

### 2.1 Sample preparation

In this study, 5 samples were prepared using AR (analytically regent) pure CaO, MgO, SiO$_2$, Al$_2$O$_3$ with the range of Al$_2$O$_3$ content 0–25 wt%. These oxides were mixed and melted in a Pt crucible ($\phi$40 × 45 × $H$40mm) under argon atmosphere at 1500℃ for 2h to homogenize the chemical compositions. Then the molten slags were water quenched to obtain glassy slags and subsequently dried at 120℃ for 12h, crushed and grinded to 300 meshes for SHTT tests. Table 1 presents the XRF (X-Ray fluoroscopy) results of samples and it can be seen that the XRF values showed a small deviation compared with designed compositions. Basicity represents the mass ratio of CaO to SiO$_2$. To confirm the glassy phase of the quenched slags, XRD (X-Ray Diffraction) analysis was carried out, as shown in Fig. 1. The XRD test was performed in a $2\theta$ degree range of 10°–80° with a speed of 4°/min, where $\theta$ is the angle between the incident ray and the scattered plane and intensity of the reflected ray is shown in $Y$-axis.

Table 1  Chemical compositions (wt%) of samples

| Samples | | Basicity($m_{CaO}/m_{SiO_2}$) | CaO | SiO$_2$ | MgO | Al$_2$O$_3$ |
|---|---|---|---|---|---|---|
| A1 | Designed | 1.1 | 46 | 41 | 8 | 5 |
|    | XRF | 1.05 | 43.5 | 41.4 | 9.2 | 6.0 |
| A2 | Designed | 1.1 | 43 | 39 | 8 | 10 |
|    | XRF | 1.05 | 40.7 | 38.9 | 9.2 | 11.3 |
| A3 | Designed | 1.1 | 40 | 37 | 8 | 15 |
|    | XRF | 1.02 | 37.5 | 36.7 | 9.2 | 16.6 |
| A4 | Designed | 1.1 | 39 | 35 | 8 | 18 |
|    | XRF | 1.02 | 35.8 | 35.1 | 9.1 | 20.0 |
| A5 | Designed | 1.1 | 37 | 34 | 8 | 21 |
|    | XRF | 1.02 | 34.3 | 33.5 | 9.0 | 23.1 |

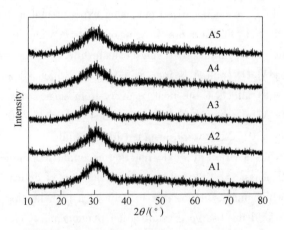

Fig. 1  XRD results of the pre-melted slags

## 2.2 Experimental procedures

Experiments were carried out using SHTT, and the principle has been described in detail elsewhere [18] and is briefly outlined here. As schematically shown in Fig. 2, a Pt-Rh thermocouple was used to heat sample and measure the temperature simultaneously, and the temperature was controlled by a computer program. A microscope equipped with a video camera was used to capture images. About 10mg sample was mounted on the top of the thermocouple, heated to 1500℃, and held for 120s to eliminate the bubbles and homogenize the chemical compositions. Then liquid slags were continuously cooled at a constant cooling rate. It should be pointed out that the temperature of the sample was accurately detected and the heat needed to confirm the constant cooling rate was compensated by the electric current. The sample images were captured by the video camera, as shown in Fig. 3, based on which CCT (continuous cooling transformation) diagrams were obtained. Pure $K_2SO_4$ with the constant melting point of 1067℃ was used to calibrate the temperature. The crystalline phases precipitated in the melts were identified by XRD technique.

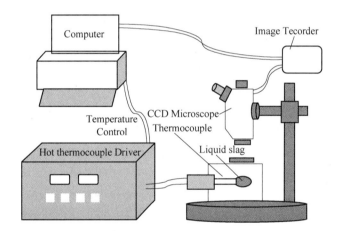

Fig. 2　Schematic of SHTT apparatus

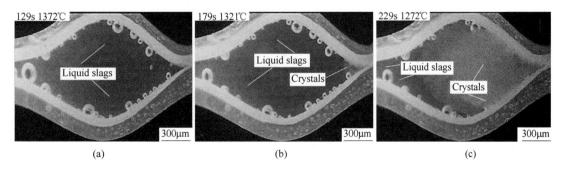

Fig. 3　SHTT images of sample A3

## 2.3 Simulation methods

Using CFD software, a 3-D model was developed to simulate the heat transfer between a slag droplet and the ambient air, shown in Fig. 4 as a schematic diagram. The simulation domain consisted of a slag droplet with $r_0$ in radius and the ambient air ($10r_0$). Through numerous attempts, the present simulation domain, i.e., $10r_0$, can effectively cover the heat transfer region, taking fully account of the calculation speed. To simplify the calculation, mass transfer was not considered and heat transfer was computed in this model. The energy conservation equation, including conduction and convection was founded inside and outside the droplet as below:

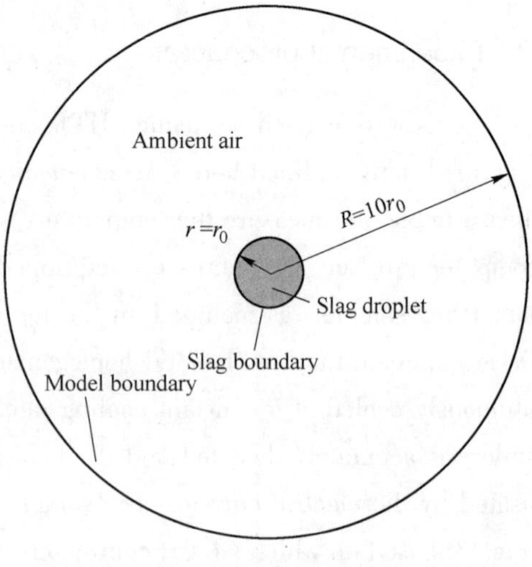

$$\rho c_p \frac{\partial T}{\partial t} + \rho c_p \vec{v} \nabla T = K\Delta T \quad (1)$$

Fig. 4  Schematic diagram of simulation model

where $\rho$ is density, $c_p$ is heat capacity, $\vec{v}$ is flow velocity, and $K$ is thermal conductivity. $c_p$ is calculated by Factsage [19] and average value was used in this study. Thermal conductivity $K$ was obtained referred to several studies [20-23] and interpolation method was used to approximate deduce the thermal conductivity. Radiative heat inside the slag droplet was expressed by RTE (radiative heat transfer) equation for an absorbing, emitting, and scattering medium [17], shown in Eq. (2):

$$\frac{dI(\vec{r},\vec{s})}{ds} + (a + \sigma_s)I(\vec{r},\vec{s}) = an^2\frac{\sigma T^4}{\pi} + \frac{\sigma_s}{4\pi}\int_0^{4\pi} I(\vec{r},\vec{s})\Phi(\vec{s},\vec{s}')d\Omega' \quad (2)$$

where $\vec{r}$ is position vector, $\vec{s}$ is direction vector, $\vec{s}'$ is scattering direction vector, $s$ is path length, $a$ is absorption coefficient, $n$ is refractive index, $\sigma_s$ is scattering coefficient, $\sigma$ is Stefan-Boltzmann constant ($5.669 \times 10^{-8}$ W/(m$^2 \cdot$ K$^4$)), $I$ is radiation intensity, which depends on position $\vec{r}$ and direction $\vec{s}$, $T$ is local temperature, $\Phi$ is phase function, and $\Omega'$ is solid angle. To simplify calculation, $a$, $n$ and $\sigma_s$ kept constant (50/m, 1.5, 0) [24,25]. The simulation parameters used in this model are summarized and listed in Table 2.

Table 2  Parameters used by simulation model

| Sample | | A1 | A2 | A3 | A4 | A5 |
| --- | --- | --- | --- | --- | --- | --- |
| $M$/g $\cdot$ mol$^{-1}$ | | 58.66 | 60.88 | 63.10 | 64.40 | 65.74 |
| $\rho$/g $\cdot$ cm$^{-3}$ | glassy | 2.82 | 2.84 | 2.89 | 2.84 | 2.78 |
| | liquid | 2.73 | 2.75 | 2.80 | 2.75 | 2.70 |
| $c_p$/J $\cdot$ K$^{-1}$ $\cdot$ mol$^{-1}$ | | 64.71 | 69.99 | 75.27 | 78.32 | 81.51 |
| $D$/mm | | 3.00 | 3.00 | 3.00 | 3.00 | 3.00 |
| Adsorption coefficient/m$^{-1}$ | | 60 | 60 | 60 | 60 | 60 |
| Scattering coefficient/m$^{-1}$ | | 0 | 0 | 0 | 0 | 0 |
| Refractive index | | 1.5 | 1.5 | 1.5 | 1.5 | 1.5 |
| Thermal conductivity(slag) /W $\cdot$ m$^{-1}$ $\cdot$ K$^{-1}$ | glassy | 0.90 | 0.90 | 0.95 | 0.95 | 0.95 |
| | liquid | 0.15 | 0.15 | 0.20 | 0.20 | 0.25 |

Continued Table 2

| Sample | A1 | A2 | A3 | A4 | A5 |
|---|---|---|---|---|---|
| Thermal conductivity (air) /W·m$^{-1}$·K$^{-1}$ | 0.06 | 0.06 | 0.06 | 0.06 | 0.06 |
| Glassy forming temperature/℃ | 940 | 930 | 945 | 962 | 975 |

Constant pressure boundary condition was used in this model, shown as below:

$$P = P_0 \qquad (3)$$

where $P_0$ is atmospheric pressure. Initially the temperature of the slag droplet was 1500℃ and that of the ambient air was 25℃. The governing equations were solved using CFD software ANSYS Fluent[17] with the Gambit preprocessor. The calculated domain was divided into 86,160 tetrahedral and hexahedral cells and 201,640 faces and time step size was 0.03s. Through numerous attempts, the present cells, faces, and time step can effectively confirm the calculation accuracy, considering the calculation speed.

## 3 Results and discussions

### 3.1 Temperature control

To recover waste heat and obtain glassy slags simultaneously, the cooling path should be determined in advance, and the variation of the slag properties during the cooling process can be therefore figured out. CCT diagrams describe the crystallization phenomena of slags during cooling process, which expresses the relationship between cooling rate and crystallization temperature. When slags were cooled at different cooling rates, crystals were precipitated at certain temperatures and CCT diagrams illustrated these changes. The significance of CCT diagrams was to provide important clues to control heat exchange between slags and working medium.

The CCT diagrams of samples presented apparent tendencies, as shown in Fig. 5. The crystallization time of CCT diagrams was measured several times and the average value was used. Firstly, the crystallization temperature decreased with increasing cooling rate for each sample. For example, crystallization temperature of sample A1 was 1345℃ at a cooling rate of 1K/s, whereas it decreased to 1200℃ at 10K/s. It is well known that undercooling was a necessary condition for crystal precipitation [26]. The higher the undercooling, the larger is the driving force of crystallization. However, the viscosity of slag melts increased with decreasing temperature, which indicates that a larger undercooling is required to crystalize. Thus, the crystallization temperature decreased with the increase of cooling rate. Secondly, it can be seen that the CCT curves moved to right with increasing $Al_2O_3$ content for samples A1, A2, and A3, as shown in Fig. 5 (a) whereas the CCT curves moved to left with further increasing $Al_2O_3$ content for samples A3, A4, and A5, as shown in Fig. 5 (b), which indicated that the influence of $Al_2O_3$ on crystallization properties of slag melts changed at around 15 wt% content. This suggested that the initial increasing $Al_2O_3$ content in the BF slag may be beneficial for waste heat recovery, and further increasing $Al_2O_3$ content increase the difficulties of waste heat recovery.

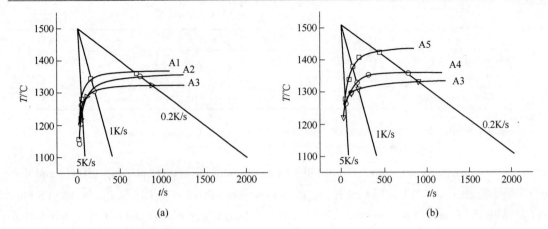

Fig. 5 CCT diagrams of the samples (a) A1, A2, and A3 (b) A3, A4, and A5

To further clarify the cooling properties of slags, critical cooling rate is acquired. Critical cooling rate is a cooling rate larger than which there is no crystals forming in the melts. In other words, it is a significant criterion to depict the glass forming ability of slags. Critical cooling rate was directly measured by CCT tests in this study. Fig. 6 illustrated the variation of the critical cooling rate with error bars of samples. It can be seen that the critical cooling rate first considerably decreased with increasing $Al_2O_3$ content (less than 15%), and then increased with further increasing $Al_2O_3$ content. The increasing critical cooling rate indicated that glass forming ability was suppressed by $Al_2O_3$ addition, which suggested that there was limited heat exchange time and slags must be rapidly cooled. According to the studies of Uhlmann et al.[26] and P. Rocabois et al.[27], glass forming ability is mainly determined by liquidus temperature and viscosities of slags. Uhlmann et al.[26] have established the following expression for critical cooling rate:

$$R_c = \frac{AT_m^2}{\eta(0.77T_m)}\exp(-0.212B)\left[1 - \exp\left(-\frac{0.3\Delta H_m}{RT_m}\right)\right]^{3/4} \quad (4)$$

where $A = 40000 \text{J} \cdot \text{m}^{-3} \cdot \text{K}^{-1}$, $T_m$ is liquidus temperature, $\eta$ is viscosity, $B$ is a kinetic barrier to crystallization, and $\Delta H_m$ is melting enthalpy of the oxides. As expected from Eq. (4), a lower liquidus temperature and a larger viscosity value were beneficial for glass formation. Liquidus temperatures of samples were calculated by Factsage[19], which were 1400℃, 1388℃, 1410℃, 1437℃, 1455℃ for samples A1-A5, respectively. The liquidus temperatures first slightly decreased and then increased with increasing $Al_2O_3$ content (varied at 15 wt%), which agreed with the variation regularity of critical cooling rate. The influence of $Al_2O_3$ on viscosities of $CaO$-$SiO_2$-$MgO$-$Al_2O_3$ system have been studied in some previous papers[28,29] and Kim J R et al.[29] found that the viscosities decrease with increasing $Al_2O_3$ content at high temperatures (1500℃ and 1450℃) within the $Al_2O_3$ content in the range of 10-18 wt% when the ratio, $(CaO + MgO)/SiO_2$ equals 1.45, which was similar to the present study. The decreasing viscosity and increasing liquidus temperature with increasing $Al_2O_3$ content enhanced crystallization and suppressed glass formation of slags, i.e., a larger critical cooling rate was determined.

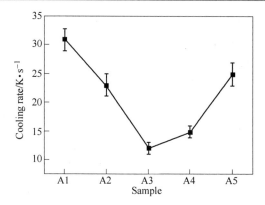

Fig. 6 Critical cooling rate of samples

## 3.2 Crystalline phase in slag melts

To identify the crystalline phase precipitated in the melts, the slags were cooled from 1500℃ at a rate of 5 K/s to a certain temperature to hold 30min, and then quenched and examined using XRD technique. Fig. 7 showed the XRD results of samples A1 and A4 held at 1380℃ and 1120℃, respectively. The primary crystals were Melilite ($Ca_2(Al, Mg)[(Si, Al)SiO_7]$) for these samples, and $Ca_2SiO_4$ was formed when temperature decreased to 1120℃.

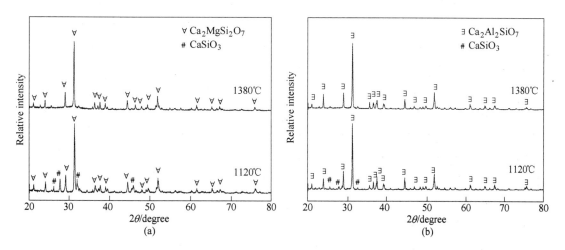

Fig. 7 XRD patterns of the quenched samples ((a) A1, (b) A4)

The primary phase could be changed from Akermanite ($Ca_2MgSi_2O_7$) to Gehlenite ($Ca_2Al_2SiO_7$) with increasing $Al_2O_3$ content. Based on the XRD results, the variation of slag properties, including CCT diagrams and critical cooling rate could be explained. In this study, the lowest liquidus temperature appeared and the variation trend of CCT diagrams changed at around 15 wt% $Al_2O_3$ content, which indicated that the effect of $Al_2O_3$ on slag properties correspondingly changed. It is generally accepted that $Al_2O_3$ is a typical amphoteric oxide. $Al_2O_3$ decomposes into $Al^{3+}$ and $O^{2-}$ in acid melts, whereas $[AlO_4]^{5-}$ ion forms in basic melts [20]. The slags in this study were near-neutral and both $Al^{3+}$ and $[AlO_4]^{5-}$ ions formed in the slag melts. When $Al_2O_3$ content increased less than 15 wt%, the increasing $Al^{3+}$ and $[AlO_4]^{5-}$ ions restricted the linkage between $Ca^{2-}$, $Mg^{2-}$ and $[SiO_4]^{4-}$ to form $Ca_2MgSi_2O_7$, i.e., crystallization was suppressed, whereas $Al^{3+}$ ion replaced $Mg^{2-}$ and $[AlO_4]^{5-}$ ion partly replaced $[SiO_4]^{4-}$ to form $Ca_2Al_2SiO_7$, because of the substantial increase of $Al^{3+}$ and $[AlO_4]^{5-}$ ions when $Al_2O_3$ content further increased, i.e., crystallization was enhanced.

## 3.3 Simulation analysis

Using the present model, temperature evolutions with time and temperature distributions inside slag droplets were acquired. Considering the symmetry of temperature distribution inside the slag droplet, the following function was used to express the temperatures of the slag droplet: $T = T(r, t)$, where $r$ is the radius and $t$ is time. When $r$ keeps constant and $t$ varies, the temperature evolution of a position can be obtained. For example, the temperature of slag droplet core can be expressed by Ref. $T = T(0, t)$ when $r = 0$, and the temperature of slag droplet surface can be expressed by Ref. $T = T(r_0, t)$ when $r = r_0$. And when $t$ keeps constant and $r$ varies, the temperature distribution inside the slag droplet at a given time can be attained.

Fig. 8 presents the core and surface temperature evolution for samples A1 and A3. It can be observed that the temperature decreased with time and the cooling rate became smaller because of the radiation reduction with falling temperature. The temperature evolution for these samples showed similar tendencies. When temperature started to decrease, the surface temperature decreased faster than the core temperature, causing larger temperature difference between center and surface with time. Then the cooling rate of core became larger than that of surface because of higher temperatures in core and the temperature difference became smaller. During this process, the largest temperature difference appeared, as presented in Fig. 8. It can be seen that the temperature difference first increased and then decreased with time and the maximum value was 112℃, which appeared at about 4s for sample A1. The temperature difference continually exceeded 50℃ in the time range of 0.36 ~ 11.5s, which may promote the crystal formation in the core of slag droplets. Table 3 presents the maximum values of temperature difference (denoted as $TD_{max}$) and its corresponding time (denoted as $t_{max}$) for different samples. It can be observed that $TD_{max}$ was more than 85℃ for all samples, which could become larger if the size of granulated slag droplets was larger. Crystal precipitation had great effect on heat transfer and subsequent utilization of slags, which must be avoided in practical cooling processes. In a previous study [30], the temperature distribution inside a slag droplet with 5mm in diameter was measured, and it was found that the temperature difference could exceed 200℃, which agreed with the present simulation results considering the diameter difference. Additionally, Meng Y et al. [31] have computerized the temperature gradients of slags using numerical simulations and they have also found the large temperature difference of 90-200℃ with varying time.

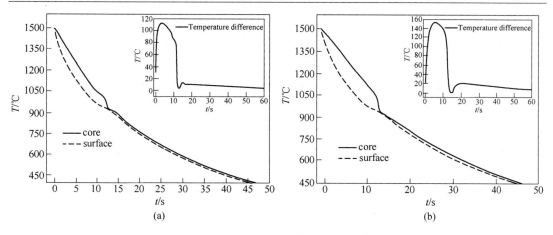

Fig. 8  Cooling process and difference between core and surface temperature of samples (a) A1, (b) A3 by simulation model

Table 3  Calculated results by simulation model

| Sample | A1 | A2 | A3 | A4 | A5 |
|---|---|---|---|---|---|
| $TD_{max}/℃$ | 112.26 | 111.57 | 153.43 | 88.7 | 129.18 |
| $t_{max}/s$ | 3.48 | 3.72 | 5.4 | 3.24 | 4.68 |
| $R_1/℃·s^{-1}$ | 12.77 | 12.42 | 11.58 | 11.77 | 11.51 |
| $R_2/℃·s^{-1}$ | 45.18 | 41.29 | 41.34 | 40.63 | 42.68 |
| $R_n/℃·s^{-1}$ | 42.35 | 44.61 | 44.39 | 42.70 | 43.45 |

Temperature distribution in radial direction is a significant property, which suggests crystallization ability and glass forming ability in different positions. Fig. 9 presents the temperature distribution in radial direction at 4s, 8s and 12s, respectively for sample A1. It can be seen that the temperature decreased from center to surface because of the low thermal conductivity. It is interesting noted that the temperature of different positions inside slag droplets dropped in different cooling paths, and the temperature gradient in radial direction first

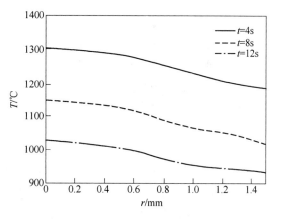

Fig. 9  Temperature distribution inside slag droplet at the time of 4s, 8s and 12s for sample A1

increased and then decreased, which indicated that the maximum value appeared in a certain region of slag droplets. In other words, there was a region where temperature varied dramatically, which might have great influence on crystallization and glass formation.

To further analyze the cooling process, average cooling rate $R_1$ was defined as the average temperature variation rate before the temperature decreased to 100℃. Meanwhile, average cooling rate at high temperatures $R_2$ was defined as the average temperature variation rate before core and surface temperature first became less than 5 ℃. Based on the cooling curves by Fluent simulation, $R_1$ and $R_2$ were calculated as listed in Table 3. It can be seen that $R_1$ slightly changed for different samples because of the similar thermal conductivity and heat capacity, while $R_2$ is considerably larger than $R_1$ because of the strong radiation at high temperatures. To validate the simulation results, the natural cooling rates of slag particles, defined as $R_n$, were measured by SHTT, as listed in Table 3. It can be seen that the value of $R_n$ is very close to that of $R_2$, and the difference between $R_2$ and $R_n$ was in the range of 1.77% -7.44%. This also proved the reasonability of the simulation results. However, several factors could lower the cooling rate in a practical process, such as crystallization during granulation processes, non-uniform of droplet size and interaction effect of different slag droplets, which caused a smaller cooling rate in a practical process and should be incorporated in the future work. As aforementioned, crystallization ability and glass forming ability greatly changed with $Al_2O_3$ content. However, the average cooling rate slightly varied under the condition of natural heat transfer between slag droplet and air, which could not satisfy the requirements of heat extraction. In other words, with the variation of $Al_2O_3$ content, there was great difference between the substantial change of $R_c$ by experimental measurements and the approximately constant $R_1$ ($R_2$ and $R_n$) by Fluent simulation, i.e., it should adjust the operating conditions with varying slag compositions during the waste heat recovery process. Considering the varying crystallization properties with $Al_2O_3$ content, especially critical cooling rate, different operations were required to treat slags, such as a smaller droplet size, a new working medium or a faster flow rate of working medium.

## 4  Conclusions

This study provides an analysis of cooling processes of high temperature slags for heat recovery. A series of slags containing different levels of $Al_2O_3$ were designed and investigated using SHTT experiments and CFD simulation. The CCT diagrams of different samples were obtained. The results indicated that the crystallization ability during cooling processes was greatly influenced by $Al_2O_3$, which subsequently affect the waste heat recovery and further utilization of slags. Critical cooling rate was measured and it increased with increasing $Al_2O_3$ content (more than 15 wt.%), suggesting a larger cooling rate required to confirm the glass formation during cooling processes. The simulation results clearly showed that a large temperature difference, more than 100 ℃, appeared during the cooling processes between core and surface of slag droplets in this study. Crystals could form inside the slag droplet because of the relative low cooling rate, and suppressing crystallization during cooling processes is quite important for further utilization. Considering the similar cooling processes of natural heat transfer and the substantial variation of slag properties during cooling processes, quite different operations should be performed to heat recovery with varying $Al_2O_3$ content.

**Acknowledgements:** The authors gratefully acknowledge financial support by the Common Development Fund of Beijing and the National Natural Science Foundation of China (51074009, and 51172001). Supports by the National High Technology Research and Development Program of

China (863 Program, 2012AA06A114) and Key Projects in the National Science & Technology Pillar Program (2011BAB03B02) are also acknowledged.

## References

[1] Hasanbeigi A, Morrow W, Sathaye J, et al. A bottom-up model to estimate the energy efficiency improvement and $CO_2$ emission reduction potentials in the Chinese iron and steel industry [J]. Energy, 2013, 50: 315-25.

[2] Pardo N, Moya J A. Prospective scenarios on energy efficiency and $CO_2$ emissions in the European iron & steel industry [J]. Energy, 2013, 54: 113-28.

[3] Meng F K, Chen L G, Sun F R, et al. Thermoelectric power generation driven by blast furnace slag flushing water [J]. Energy, 2014, 66: 965-72.

[4] Fruehan R J, Fortini O, Paxton H W, et al. Theoretical minimum energies to produce steel for selected conditions [R]. Washington DC, U. S. A: U. S. Department of Energy: 2000.

[5] Bisio G. Energy recovery from molten slag and exploitation of the recovered energy [J]. Energy, 1997, 22 (5): 501-9.

[6] World steel Association Media center Press Releases, 2013 press releases: 2012 statistics table. See also: < www.worldsteel.org/media-centre/press-releases/2013/12-2012-crude-steel.html >.

[7] Cai J J, Wang J J, Chen C X, et al. Recovery of residual heat integrated steelworks [J]. Iron Steel, 2007, 42: 1-6.

[8] Barati M. Energy intensity and greenhouse gases footprint of metallurgical processes: a continuous steelmaking case study [J]. Energy, 2010, 35 (9): 3731-7.

[9] Milford RL, Pauliuk S, Allwood JM, et al The roles of energy and material efficiency in meeting steel industry $CO_2$ targets [J]. Environ Sci Technol, 2013, 47 (7): 3455-62.

[10] Barati M, Esfahani S. Utigard TA. Energy recovery from high temperature slags [J]. Energy, 2011, 36 (9): 5440-9.

[11] Liu J X, Yu Q B, Dou C X, et al. Experimental study on heat transfer characteristics of apparatus for recovering the waste heat of blast furnace slag [J]. Adv Mater Res, 2010, 97: 2343-6.

[12] Mizuochi T, Akiyama T, Shimada T, et al. Feasibility of rotary cup atomizer for slag granulation [J]. ISIJ Int, 2001, 41 (12): 1423-8.

[13] Zhang H, Wang H, Zhu X, et al. A review of waste heat recovery technologies towards molten slag in steel industry [J]. Appl Energ, 2013, 112: 956-66.

[14] Goto K S, Linder K H. Thermal conductivities of blast furnace slags and continuous casting powders in the temperature range 100 to 1550 degree C [J]. Stahl U Eisen, 1985, 105: 1387.

[15] Yajima K, Matsuura H, Tsukihashi F. Effect of simultaneous addition of $Al_2O_3$ and MgO on the liquidus of the $CaO$-$SiO_2$-$FeO_x$ system with various oxygen partial pressures at 1573 K [J]. ISIJ Int, 2010, 50 (2): 191-4.

[16] Sunahara K, Nakano K, Hoshi M, et al. Effect of high $Al_2O_3$ slag on the blast furnace operations [J]. ISIJ Int, 2008, 48 (4): 420-9.

[17] ANSYS, FLUENT 13.0 theory guide. Pittsburgh, U. S. A: ANSYS, Inc., 2010.

[18] Kashiwaya Y, Cicutti CE, Cramb AW, et al. Development of double and single hot thermocouple technique for in site observation and measurement of mold slag crystallization [J]. ISIJ Int, 1998, 38 (4): 357-65.

[19] Bale C W, Chartrand P, Degterov S A, et al. FactSage thermochemical software and databases [J]. Calphad, 2002, 26 (2): 189-228.

[20] Kang Y, Morita K. Thermal conductivity of the CaO-Al$_2$O$_3$-SiO$_2$ system [J]. ISIJ Int, 2006, 46 (3): 420-6.

[21] Nishioka K, Maeda T, Shimizu M. Application of square-wave pulse heat method to thermal properties measurement of CaO-SiO$_2$-Al$_2$O$_3$ system fluxes [J]. ISIJ Int, 2006, 46 (3): 427-33.

[22] Mcdavid R M, Thomas B G. Flow and thermal behavior of the top surface flux/powder layers in continuous casting molds [J]. Metall Mater Trans B, 1996, 27: 672-85.

[23] Mills K C, Susa M. Slag altas. 2nd ed [M]. Dusseldorf, Germany: Verlag Stahleisen GmbH, 1995.

[24] Chen ZS. Measurement of the absorption coefficient of optical glass [J]. J Chin Ceram Soc, 1982, 10 (2): 231-6.

[25] Boling N L, Glass A J, Owyoung A. Empirical relationships for predicting nonlinear refractive index changes in optical solids [J]. IEEE J Quantum Electron 1978, 14 (8): 601-8.

[26] Uhlmann DR, Yinnon H. In: Uhlmann DR, Kreidl. Editors, Glass. New York: science and technology, vol 1, Academic Press, 1983.

[27] Rocabois P, Pontoire J N, Lehmann J, et al Crystallization kinetics of Al$_2$O$_3$-CaO-SiO$_2$ based oxide inclusions [J]. J Non-Cryst Solids, 2001, 282: 98-109.

[28] Mills KC. Slag altas, 2nd ed [M]. Dusseldorf. Germany: Verlag Stahleisen GmbH, 1995.

[29] Kim J R, Lee Y S, Min DJ, et al. Influence of MgO and Al$_2$O$_3$ contents on viscosity of blast furnace type slags containing FeO [J]. ISIJ Int, 2004, 44 (8): 1291-7.

[30] Yoshinaga M, Fujii K, Shigematsu T, et al. Dry granulation and solidification of molten blast furnace slag [J]. Trans Iron Steel Inst Jpn, 1982, 22 (11): 823-9.

[31] Meng Y, Thomas B G. Heat-transfer and solidification model of continuous slab casting: CONID [J]. Metall Mater Tran B, 2003, 34 (5): 685-705.

(原载于 *Energy*, 2014, 76: 761-767)

# Thermodynamic Investigation of Synthesizing Metastable β-Sialon-Alon Composite Ceramic

Huang Xiangdong　Li Wenchao　Wang Fuming　Shao Yu

(University of Science and Technology Beijing, Beijing 100083)

**Abstract:** Based on its thermodynamic analysis, β-Sialon-Alon metastable composite ceramic has been prepared by hot pressing sintering. XRD results indicate that the product of hot pressing sintering is indeed Sialon-Alon metastable composite ceramic, which is in accordance with thermodynamic analysis.

## 1　Introduction

Refractory industry is challenged by the development of new metallurgical technologies, such as reverse solidification and super steel. Impurities in super steel, such as oxygen and carbon, must be about several ppms or less, the carbon-bearing refractories can not meet these demands. A new carbon-free refractory material must be developed. Therefore, recently more and more attention has been paid to the development of high technology ceramic.

Sialon material, a kind of high performance ceramic materials, has been developed in last twenty years. Its application in refractory just begin, usually as bonded phase. Some European countries[1,2] used Sialon-bonded SiC brick in blast furnace. Compared to $Si_3N_4$-bonded SiC brick, Sialon-bonded SiC brick has better alkali resistance, which has wide applications.

Because of Alon's excellent oxidation and erosion resistance, it is important to refractory industry. T. Hos ka, M. Kato[3] reported that introducing Alon in castable for trough of blast furnace, the erosion resistance was improved greatly and thermal shock resistance also enhanced with the increase of Alon, while conditions of construction was not affected. Kazumasa and Akira Iwasaki[4] added Alon in order to improve durable performance of $Al_2O_3$-C slide nozzle. The results indicated that compared with the nozzle without Alon the enlarging rate of aperture was 20 percent lower and useful span also increased. Addition of Alon to $Al_2O_3$-C submerged nozzle, K. Taketa and T. Hsaka[5] made the erosion resistance and striping resistance improve. It has not been reported whether β-Sialon-Alon composites can be synthesized and how the composites obtained. In present work, the possibility and process of synthesizing β-Sialon-Alon composites have been analyzed by use of thermodynamic method and confirmed through experiment.

---

Supported by the National Science Funds of China 59872002.

黄向东，1997~2001年于北京科技大学师从李文超教授攻读博士学位。目前定居英国。

## 2 Thermodynamic analysis about stability of β-Sialon-Alon composites

### 2.1 The stable phase diagrams of Al-O-N and Si-O-N

The synthesis process is discussed using stable phase diagrams of Al-O-N and Si-O-N. It is adopted that some thermodynamic data about β-Sialon and Sialon polytype are evaluated by our research group[7] using quasiparaboloidal rule.

Table 1 and Table 2 list the equilibrium of phases and the relation of partial pressures in Al-O-N and Si-O-N.

**Table 1  Equilibra of phases and the relation of partial pressures in Si-O-N (at 1800K)**

| | |
|---|---|
| (1) $SiO_{2(s)} = Si_{(s)} + O_{2(g)}$ | $\lg(P_{O_2}/P^\ominus) = -16.81$ |
| (2) $\beta\text{-}Si_3N_{4(s)} = 3Si_{(s)} + 2N_{2(g)}$ | $\lg(P_{N_2}/P^\ominus) = -1.68$ |
| (3) $2SiO_{2(s)} + N_{2(g)} = Si_2N_2O_{(s)} + 3/2O_{2(g)}$ | $\lg(P_{N_2}/P^\ominus) = 3/2\lg(P_{O_2}/P^\ominus) + 21.18$ |
| (4) $2\beta\text{-}Si_3N_{4(s)} + 3/2O_{2(g)} = 3Si_2N_2O_{(s)} + N_{2(g)}$ | $\lg(P_{N_2}/P^\ominus) = 3/2\lg(P_{O_2}/P^\ominus) + 30.61$ |
| (5) $Si_2N_2O_{(s)} = 2Si_{(s)} + N_{2(g)} + 1/2O_{2(g)}$ | $\lg(P_{N_2}/P^\ominus) = 1/2\lg(P_{O_2}/P^\ominus) - 12.44$ |

**Table 2  Equilibra of phases and the relation of partial pressures in Al-O-N (at 1800K)**

| | |
|---|---|
| (1) $2Al_{(s)} + 3/2O_{2(g)} = Al_2O_{3(s)}$ | $\lg(P_{O_2}/P^\ominus) = -21.19$ |
| (2) $Al_{(s)} + 1/2N_{2(g)} = AlN_{(s)}$ | $\lg(P_{O_2}/P^\ominus) = -6.94$ |
| (3) $2Al_7O_9N_{(s)} + 3/2O_{2(g)} = 7Al_2O_{3(s)} + N_2$ | $\lg(P_{O_2}/P^\ominus) = 3/2\lg(P_{O_2}/P^\ominus) + 23.72$ |
| (4) $2Al_7O_9N_{(s)} = 14Al_{(s)} + 9O_{2(g)} + N_{2(g)}$ | $\lg(P_{N_2}/P^\ominus) = -9\lg(P_{O_2}/P^\ominus) - 199.3$ |

In order to determine coexisting condition of β-Sialon and Alon, superimposed stable phase diagram of β-Sialon and Alon are plotted based on Table 1 and Table 2. It is shown in Fig. 1.

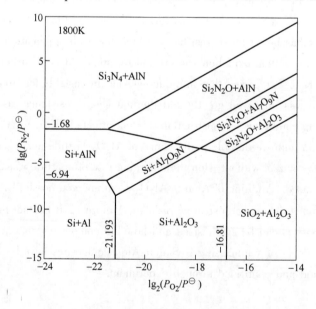

Fig.1  Superimposed stable phase diagram of Si-O-N and Al-O-N system at 1800K

## 2.2 The thermodynamics analysis of synthesis for β-Sialon-Alon composites

In the process of synthesis β-Sialon-Alon composites, the following reactions could be involved. According to thermodynamic data of Si-Al-O-N system from literature[6,7], $\Delta_r G$ of the above reactions can be determined.

$$14\ Si_2N_2O_{(s)} + 2\ Al_7O_9N_{(s)} + 6\ N_2 = 7\ Si_4Al_2O_2N_{6(s)} + 9\ O_2$$
$$\Delta_r G_1^\ominus = 5865.2 - 0.1372\ T \tag{1}$$

$$42\ Si_4Al_2O_2N_6 + 24\ Al_7O_9N_{(s)} + 123\ O_2 = 14\ Si_{12}Al_{18}O_{39}N_{8(s)} + 82\ N_2$$
$$\Delta_r G_2^\ominus = -76277.97 + 2.6614\ T \tag{2}$$

$$Si_4Al_2O_2N_6 + 2\ Al_7O_9N_{(s)} + 4\ N_2 = 4\ SiAl_4O_2N_{4(s)} + 6\ O_2$$
$$\Delta_r G_3^\ominus = 6592.72 - 0.9184\ T \tag{3}$$

$$7\ Si_4Al_2O_2N_6 + 6\ Al_7O_9N_{(s)} + 4\ N_2 = 14\ Si_2Al_4O_4N_4 + 6\ O_2$$
$$\Delta_r G_4^\ominus = 4513.3 - 0.6286\ T \tag{4}$$

$$10\ N_2 + 2\ Si_{12}Al_{18}O_{39}N_{8(s)} + 6\ SiAl_4O_2N_{4(s)} = 15\ Si_2Al_4O_4N_{4(s)} + 15\ O_2$$
$$\Delta_r G_5^\ominus = 5843.451 + 0.3239\ T \tag{5}$$

$$9\ Si_5AlON_{7(s)} + 12\ SiAl_4O_2N_{4(s)} + 12\ O_2 = 19\ Si_3Al_3O_3N_{5(s)} + 8\ N_2$$
$$\Delta_r G_6^\ominus = -16153.59 + 2.2305\ T \tag{6}$$

(1) In $N_2$ atmosphere $\quad P_{O_2} = 1.025 \times 10^{-19}\ MPa$

$$P_{N_2} = 1.025 \times 1\ MPa$$

At 1600℃:

$\Delta_r G_1 = -201.4474\ kJ \quad \Delta_r G_2 = 8106.682\ kJ$

$\Delta_r G_3 = 999.3803\ kJ \quad \Delta_r G_4 = -537.2439\ kJ$

$\Delta_r G_5 = -3232.733\ kJ \quad \Delta_r G_6 = -4229.509\ kJ$

At 1720℃:

$\Delta_r G_1 = -590.1274\ kJ \quad \Delta_r G_2 = 13513.07\ kJ$

$\Delta_r G_3 = 641.0244\ kJ \quad \Delta_r G_4 = -860.8239\ kJ$

$\Delta_r G_5 = -3814.229\ kJ \quad \Delta_r G_6 = -3465.553\ kJ$

The above calculated results indicate that $Si_2N_2O$ reacts with Alon to form β-Sialon either at 1600℃ or 1700℃ in $N_2$ atmosphere; at the same time β-Sialon reacts with Alon to make Z of β-Sialon increase and 15R reacts with Alon to form β-Sialon with a higher Z.

(2) In Ar atmosphere $\quad P_{O_2} = 1.025 \times 10^{-19}\ MPa$

$$P_{N_2} = 1.025 \times 10^{-6}\ MPa$$

At 1600℃:

$\Delta_r G_1 = 874.4037\ kJ \quad \Delta_r G_2 = -6597.116\ kJ$

$\Delta_r G_3 = 1716.739\ kJ \quad \Delta_r G_4 = 180.1149\ kJ$

$\Delta_r G_5 = -1439.711\ kJ \quad \Delta_r G_6 = -5664.04\ kJ$

At 1720℃:

$\Delta_r G_1 = 554.6516 \text{kJ} \quad \Delta_r G_2 = -2132.779 \text{kJ}$

$\Delta_r G_3 = 1404.343 \text{kJ} \quad \Delta_r G_4 = -97.50507 \text{kJ}$

$\Delta_r G_5 = -1906.331 \text{kJ} \quad \Delta_r G_6 = -4991.992 \text{kJ}$

The calculated results indicate that β-Sialon reacts with Alon to form $X$ phase; and then $X$ phase reacts with 15R to form β-Sialon; at the same time β-Sialon with a lower $Z$ reacts with Alon to form β-Sialon with a higher $Z$; in contrast to that in $N_2$ atmosphere, 15R reacts with Alon. to form β-Sialon with a higher $Z$ only at 1720℃.

## 3  Experimental

Based on the above analysis, β-Sialon and Alon powder in certain proportion were attrition-milled in ethyl alcohol with agate balls for 2 hours and after being dried and sieved, the prepared powder was sintered under 250MPa hot pressing at 1720℃, respectively, in $N_2$ or Ar atmosphere.

## 4  Results and discussion

The XRD spectra of sintered specimens and raw materials are shown in Figs. 2-5.

From results of the XRD analysis. it can be seen that the main phase of β-Sialon material is β-Sialon besides definite amount diffraction peaks of 15R. Based on the relationship of $Z$ and lattice parameter of β-Sialon, it can be determined that $Z$ of β-Sialon phase in β-Sialon material is 0.96; it is 3.5 for hot pressing sintered body in $N_2$ atmosphere; 3.0 for hot pressing composite in Ar atmosphere. Only a small amount of 15R apparently less than that of β-Sialon material was found in $N_2$ or Ar atmosphere and no $X$ phase.

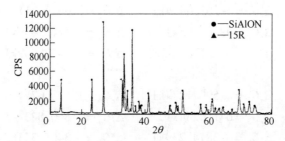

Fig. 2  XRD spectrum of β-Sialon powder

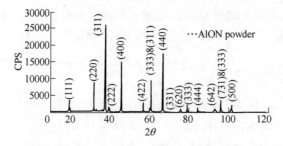

Fig. 3  XRD spectrum of Alon powder

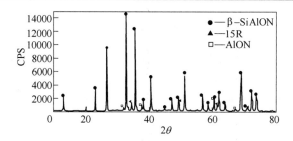

Fig. 4  XRD spectrum of synthesized β-Sialon-Alon metastable composite in $N_2$ atmosphere at 1720℃

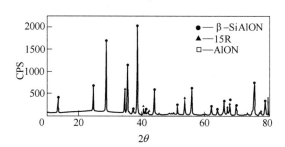

Fig. 5  XRD spectrum of synthesized β-Sialon-Alon metastable composite in Ar atmosphere at 1720℃

Based on reaction 5, β-Sialon reacts with Alon to form β-Sialon with a higher $Z$, $N_2$ is propitious for reaction 5 to proceed. Therefore, $Z$ of the obtained β-Sialon in $N_2$ atmosphere is higher than that in Ar and the amount of Alon reduced. At the same time, since β-Sialon reacts with 15R to form β-Sialon with a higher $Z$, the content of 15R in sintered body is less than that in β-Sialon material in $N_2$ or Ar atmosphere. According to thermodynamic analysis results, although β-Sialon reacts with Alon to form $X$ phase in Ar atmosphere, there is no $X$ phase in final sythesized material since the formed $X$ phase further reacts with 15R to form β-Sialon.

## 5  Conclusions

The thermodynamic analysis about synthesis process of β-Sialon-Alon metastable composite indicate:

(1) β-Sialon reacts with Alon to form β-Sialon with a higher $Z$.
$N_2$ is propitious for that procedure to proceed.

(2) β-Sialon reacts with Alon to form $X$ phase in Ar atmosphere.
$X$ phase further reacts with 15R to form β-Sialon.

(3) Whether in $N_2$ or Ar atmosphere, β-Sialon reacts with 15R to form β-Sialon with a higher $Z$.

(4) In certain condition β-Sialon-Alon metastable composite can be synthesized.

The results of thermodynamic analysis are in accordance with those experimental results.

## References

[1] J. M. Bauer. Development trend about European refractory used as blast funace lining [J]. Foreign Iron And Steel, 1992, (9): 6-15.

[2] D. A. Gunn. A theoretical evaluation of sialon-bonded silicon carbide in the blast furnace environment [J]. J. Eur. Ceram. Soc., 1993, 11: 35-41.

[3] T. Hosaka, M. Kato. A study of compositional modification of trough mixture by using aluminum oxynitride [J]. Taikabutsu, 1985, 37 (10): 22-26.

[4] K. Murakami, A. Iwasaki, Y. Akatsuka, et al. One result of sliding nozzle refractories using aluminum oxynitride [J] Taikabutsu, 1986, 38 (1): 18-20.

[5] K. Takeda, T. Hosaka. Characteristics of new raw materials "Alon" for refractory [J]. Interceram, 1989, 38 (1): 18-20.

[6] H. X. Willems, et al. Thermodynimics of alon I: Stability at lower temperatures [J]. J. Eur. Ceram. Soc., 1992, 10, 327-37.

[7] Wen Hongjie. Foundational investigation on sythesized β-Sialon using andalusite [D]. Beijing: University of Science and Technology Beijing, 1998.

(原载于 *China's Refractories*, 1999, 8 (4): 8-11)

## 高炉渣合成 Ca-α-Sialon-SiC 粉的热力学分析及工艺优化

刘克明　王福明　李文超　谢丽君　郭宇艳

（北京科技大学物理化学系，北京　100083）

**摘　要**：在热力学分析的基础上，以高炉渣为原料，引入适当的添加剂，利用碳热还原—氮化的方法制备了 Ca-α-Sialon-SiC 粉，得到的 Ca-α-Sialon 含量最高可以达到81%；利用统计模式识别结合人工神经元网络优化了工艺。

**关键词**：Ca-α-Sialon；合成制备；碳热还原—氮化；显微结构

## Synthesis of Ca-α-Sialon-SiC Powders from Blast Furnace Slag and Optimization of Its Synthesis Process

Liu Keming　Wang Fuming　Li Wenchao　Xie Lijun　Guo Yuyan

(Department of Physical Chemistry, UST Beijing, Beijing 100083)

**Abstract**: Blast furnace slag is a kind of metallurgical solid wastes. In order to make better use of it and enhance its additional value, Ca-α-Sialon-SiC powders were synthesized from blast furnace slag and some additives by carbothermal reduction and nitridition (CRN) based on thermodynamic analysis. The content of Ca-α-Sialon in the sintered samples was up to 81%. The synthesis process of CRN was optimized by Statistical Pattern Recognition (SPR) and Artificial Neural Network (ANN).

**Keywords**: Ca-α-Sialon; synthesis; CRN; microstructure

冶金炉渣目前利用率较低，且多数作为筑路或一般建筑材料。1992年起，国内外开始研究如何提高高炉渣的附加值，降低污染。高炉渣是常见的一种冶金废渣，它的主要成分 $CaO$、$SiO_2$ 和 $Al_2O_3$ 等是合成 Ca-α-Sialon 的原料。本研究提出了使用高炉渣碳热还原氮化合成 Ca-α-Sialon-SiC 复合材料。该材料可以作为新一代耐火材料用于诸如反向凝固炉底狭缝材料等领域。使用纯原料合成 Ca-α-Sialon 价格比较昂贵，而利用天然原料合成 α-Sialon 的报道却比较少，Thommy Ekstrom 等人利用黏土为原料，采用碳热还原氮化的方法两步合成 Y-α-Sialon 陶瓷[1]。

本文以冶金固体废弃物高炉渣为原料，并加入合适的添加剂，利用碳热还原氮化的方法合成 Ca-α-Sialon-SiC 粉；采用统计模式识别获得目标优化区，用模式识别逆映照和人工神经元网络验证模式识别的可靠性。

---

国家自然科学基金资助课题（No.59934090，No.59872002。）

刘克明，1998~2001年于北京科技大学师从李文超教授攻读硕士学位。目前在中国钢研科技集团有限公司工作。

# 1 热力学分析

## 1.1 Ca-α-Sialon 合成温度评估

Ca-α-Sialon 表达式为 $Ca_{m/2}Si_{12-(m+n)}Al_{m+n}O_nN_{16-n}$，通常用 $m$ 和 $n$ 表征 Ca-α-Sialon，如 $Ca_{0.75}Si_{9.5}Al_{2.5}O_1N_{15}$ 表示为 Ca (1510)。文献 [2] 给出了 Ca-Si-Al-O-N 体系 Ca-α-Sialon 平面的相关系图，但该图只是示意图，无法利用它准确选择所需成分，据文献 [2-4] 提供的单相 Ca-α-Sialon 的 $m$ 和 $n$ 值，绘制 Ca-α-Sialon 平面单相区和两相区，见图1。

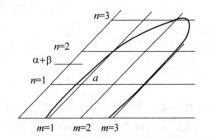

图 1  Ca-α-Sialon 平面单相区和两相区
Fig. 1  Single phase and two phases region in Ca-α-Sialon plane

选择成分 (1510)，$m = 1.5$，$n = 1.0$。选择该成分的原因是，它恰好处于单相区，可能获得单相 Ca-α-Sialon 材料。整体反应式如下：

$$3CaO(s) + 38SiO_2(s) + 5Al_2O_3(s) + 30N_2(g) + 90C(s) =$$
$$4Ca_{0.75}Si_{9.5}Al_{2.5}ON_{15}(s) + 90CO(g) \tag{1}$$

利用拟抛物线规则评估和预报 Ca-α-Sialon 的标准 Gibbs 自由能[5]（式（2）），计算了式（1）的反应标准 Gibbs 自由能（式（3））：

$$\Delta_f G^{\ominus}_{Ca-\alpha-Sialon} = -4009.358 + 1.551T(kJ \cdot mol^{-1}) \tag{2}$$

$$\Delta_r G^{\ominus} = 19791.961 - 11.184T(kJ \cdot mol^{-1}) \tag{3}$$

由该反应的标准 Gibbs 自由能计算出反应的开始发生温度为 1769.7K，只有当温度高于 1496℃，式（1）反应才开始有 Ca-α-Sialon 生成。

## 1.2 碳热还原氮化合成 Ca-α-Sialon 气氛选择

为确定合成材料的气氛条件，利用文献 [5] 的热力学数据，计算了 1873K 下 Si-O-N、Al-O-N 和 Ca-O-N 3 个体系的凝聚相平衡分压，由此绘制了这 3 个体系叠加的热力学参数状态图（见图2）。

图 2 中阴影部分满足合成 Ca-α-Sialon 所需的气氛条件。本工作使用高纯氮气（氧分压 $p_{O_2}/p^{\ominus} = 10^{-6}$）常压合成 Ca-α-Sialon，$p_{N_2} = 0.1MPa$。由图 1 可以看出，当 $p_{N_2} = 0.1MPa$ 时，氧分压需要处于 $10^{-19} \sim 10^{-23}$ 范围内，本实验利用 B-B 反应控制氧分压使其落在所需的范围内。

## 1.3 碳热还原氮化合成 Ca-α-Sialon 过程中杂质的走向

由于高炉渣等合成原料中除含有 Ca、Si、Al 和 O 等合成 Ca-α-Sialon 的元素外，还含有 Na、K、Fe 和 Mg 等杂质元素的氧化物，在烧结过程中杂质的去向是人们普遍关注的一个问题，因为这些元素的存在将对最终材料的力学性能及其高温性质有一定的影响。本文利用有关热力学数据计算分析了这几种杂质元素的走向。

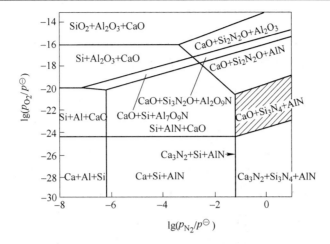

图 2　1873K 下 Si-O-N，Al-O-N 和 Ca-O-N 体系叠加热力学参数状态图

Fig. 2　Overlapped phase stability diagram as a function of partial pressure of nitrogen and oxygen at 1873K

计算了 Mg-O-N 体系 1873K 下的凝聚相平衡分压（见表 1），并由此绘制了该体系的热力学参数状态图（见图 3）。

表 1　1873K 下 Mg-O-N 体系凝聚相及其平衡分压

Table 1　Condensed phases and their equilibrium partial pressures for Mg-O-N system at 1873K

| 反 应 式 | 平 衡 分 压 |
| --- | --- |
| $3Mg(g) + N_2(g) = Mg_3N_2(s)$ | $\lg(p_{N_2}/p^{\ominus}) = -1.74$ |
| $Mg(g) + \frac{1}{2}O_2(g) = MgO(s)$ | $\lg(p_{O_2}/p^{\ominus}) = -21.86$ |
| $\frac{1}{3}Mg_3N_2(s) + \frac{1}{2}O_2(g) = \frac{1}{3}N_2(g) + MgO(s)$ | $\lg(p_{O_2}/p^{\ominus}) = -23.02 + \lg(p_{N_2}/p^{\ominus})$ |

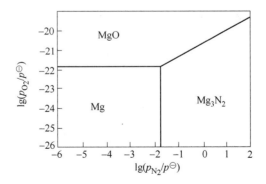

图 3　1873K 下 Mg-O-N 体系热力学参数状态图

Fig. 3　Phase stability diagram of Mg-O-N system at 1873K

由于氮气分压为 0.1MPa，从图 3 可看出在该实验条件下，Mg 元素以 MgO 或 $Mg_3N_2$ 的形式存在，在合成反应过程中，Mg 元素固溶进入 Ca-α-Sialon 的晶格，形成 Ca(Mg)-α-

Sialon。

在合成过程末期，碳粉被反应完全损耗，体系的氧分压约为 $10^{-6}$（高纯氮气中的氧分压约为 $1\times10^{-6}$），体系中的 Fe 元素主要存在以下两个反应：

$$2Fe(l) + O_2(g) = 2FeO(l)$$

$$\Delta_r G^{\ominus} = -459400 + 87.4T(J\cdot mol^{-1}) \tag{4}$$

$$6FeO(l) + O_2(g) = 2Fe_3O_4(s)$$

$$\Delta_r G^{\ominus} = -978200 + 459T(J\cdot mol^{-1}) \tag{5}$$

通过计算得出，1873K 下两反应的平衡氧分压 $\lg(p_{O_2}/p^{\ominus})$ 分别为 $-8.25$ 和 $-3.30$，所以此时在该体系中液相 FeO 的形式是稳定的。烧结过程中 FeO 与原料中未反应的 $SiO_2$ 或 $Al_2O_3$ 结合进入到玻璃相中。

杂质元素 Na 和 K 在碳热还原过程的前期中，主要存在以下的反应：

$$4Na(g) + O_2(g) = 2Na_2O(l)$$

$$\Delta_r G^{\ominus} = -518800 + 2340.7T(J\cdot mol^{-1}) \tag{6}$$

$$4K(g) + O_2(g) = 2K_2O(s)$$

$$\Delta_r G^{\ominus} = -487700 + 252.95T(J\cdot mol^{-1}) \tag{7}$$

由于本实验条件下，Na、K 的平衡分压未知，但肯定小于 1 个标准压力，令 $\lg(p_M/p^{\ominus})=0$（$M$ 代表 Na 或 K 的单质气体），通过计算得出，1873K 下两反应的吉布斯自由能的值皆大于 0，实际条件下 $\lg(p_M/p^{\ominus})<0$，则反应吉布斯自由能的值一定为正值，因此反应是逆向进行的，即杂质元素 Na 和 K 以单质气体的形式存在。Na 和 K 的单质气体随流动的氮气流排出反应器外，再氧化生成各自稳定的氧化物。

## 2 实验验证

本实验采用上海宝钢高炉水淬渣作为原料，通过添加硅砂和蓝晶石来调整成分使之落于 Ca-α-Sialon 单相区，各种原料的成分见表 2。

表 2 实验原料成分（质量分数）
Table 2 Compositions of the starting material powders (%)

| 化合物 | 高炉渣 | 蓝晶石 | 硅砂 |
|---|---|---|---|
| $SiO_2$ | 33.80 | 31.75 | 99.50 |
| $Al_2O_3$ | 14.50 | 59.76 | 0.20 |
| CaO | 40.52 | — | — |
| MgO | 8.50 | — | — |
| $K_2O$，$Na_2O$ | — | 0.98 | — |
| $Fe_2O_3$ | — | 0.25 | 0.10 |
| 杂质 | 0.68 | — | — |

各种固体原料按设计成分进行配料，6h 球磨混料，然后干燥，加入黏结剂，并在 150MPa 下机压成型。将成型的样品放入刚玉坩埚内，埋粉，将坩埚放于立式钼丝炉中，底部通入高纯 $N_2$，常压下烧结。利用 X 射线衍射分析确定了烧结样品的相组成，并用

TEM 进行显微形貌观察。

## 2.1 热力学分析的可靠性

表 3 给出了烧结保温时间为 4h 的实验条件和实验结果，并着重考察 CaO 含量与温度两个因素对合成 Ca-α-Sialon 的影响。由于在烧结过程中，产生大量气体，因此样品很疏松。

表 3 中的实验 1、4 和 7 使用的是同一配料的样品，只是烧结温度不同，实验 2、5、8 和实验 3、6、9 分别使用两种配料的样品。从实验 1、2 和 3 可以看出，同一温度下（1550℃），实验 1 没有生成 Ca-α-Sialon，而 CaO 含量为化学计量的实验 2 和 CaO 含量过量 10% 的实验 3 生成了 Ca-α-Sialon，由此可见 CaO 的过量加入可以降低 Ca-α-Sialon 的合成温度。同种配料的样品不同温度下烧结的结果表明，随着温度的升高 Ca-α-Sialon 含量增加，SiC 含量减少，说明较高的温度可以促进 Ca-α-Sialon 的合成。其中实验 8 得到了约 81%（质量分数）的 Ca-α-Sialon。图 4 是实验 8 烧结样品的 X 射线衍射分析图谱。

表 3 实验条件与结果
Table 3 Experimental conditions and results

| No. | $t/℃$ | CaO | Phases | Ca-α-Sialon/SiC |
|---|---|---|---|---|
| 1 | 1550 | Lower | SiC，α-$Si_3N_4$，α-$Si_3N_4$ | 0 |
| 2 | 1550 | Proper | Ca-α-Sialon，SiC | 0.48 |
| 3 | 1550 | Higher | Ca-α-Sialon，SiC | 0.90 |
| 4 | 1600 | Lower | Ca-α-Sialon，SiC | 0.86 |
| 5 | 1600 | Proper | Ca-α-Sialon，SiC | 0.79 |
| 6 | 1600 | Higher | Ca-α-Sialon，SiC | 1.15 |
| 7 | 1650 | Lower | Ca-α-Sialon，SiC | 2.38 |
| 8 | 1650 | Proper | Ca-α-Sialon，SiC | 4.63 |
| 9 | 1650 | Higher | Ca-α-Sialon，SiC | 1.72 |

实验结果表明：在 1650℃，氮分压为 0.1MPa，氧分压在 10～23MPa 的条件下，CaO 含量为化学计量有利于合成 Ca-α-Sialon。

## 2.2 TEM 显微形貌观察及选区电子衍射结果

实验 8 烧结样品在玛瑙研钵中细磨成粉，过 300 目筛，在无水乙醇中超声波振荡进行均匀分散，滴于有醋酸显微微栅膜的铜网上，喷碳后进行透射电镜显微形貌观察，图 5 是实验 8 烧结样品粉末样的 TEM 形貌图，分别对图中块状的颗粒 A 和长柱状的颗粒 B 做电子选区衍射（见图 5），通过标定证明 A 和 B 均为 Ca-α-Sialon。

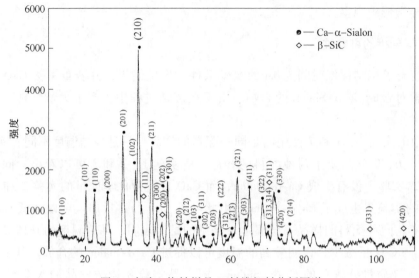

图4 实验8烧结样品X射线衍射分析图谱

Fig. 4 XRD pattern of sintering sample in No. 8

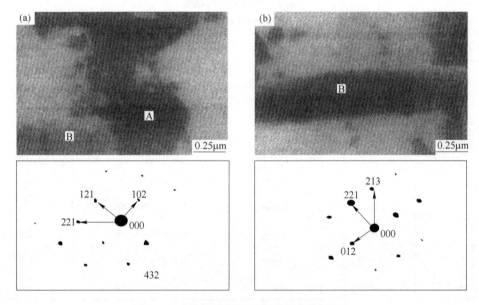

图5 A，B颗粒形貌及选区衍射花样标定

（a）A颗粒；（b）B颗粒

Fig. 5 Microstructure and diffraction pattern of A and B

这种长柱状晶粒的存在可以使材料在受外界载荷的作用下，内部发生裂纹偏转、裂纹桥连及晶粒拔出等造成能量耗散增大的变化，从而优化材料的机械性能。Ca-α-Sialon 中这种长柱状和块状晶粒的同时存在，说明 Ca-α-Sialon 有潜在的自增韧作用[6]。

## 3 合成 Ca-α-Sialon-SiC 工艺优化

用高炉渣碳热还原氮化制备 Ca-α-Sialon-SiC 粉的影响因素很多，因此材料的研制工作量很大，而采用计算机统计模式识别对材料合成工艺进行优化，可以大大提高工作效率。本文选择了烧结温度、保温时间、配碳量和高炉渣加入量 4 个影响因素，以是否合成了 Ca-α-Sialon 为判据，对 19 组实验组成的原始样本集（表 4）进行了统计模式识别处理，得到目标优化区（见图 6）。

表 4  实验样本点
Table 4  Experimental samples

| No. | $t/℃$ | $\tau/h$ | C[①] | $w_{渣}/\%$ | Ca-α-Sialon[②] |
|---|---|---|---|---|---|
| 1 | 1300 | 3 | 1.0 | 11.72 | 0 |
| 2 | 1300 | 3 | 1.5 | 10.76 | 0 |
| 3 | 1300 | 3 | 2.0 | 10.02 | 0 |
| 4 | 1350 | 3 | 1.0 | 11.95 | 0 |
| 5 | 1350 | 3 | 1.5 | 11.34 | 0 |
| 6 | 1350 | 3 | 2.0 | 9.280 | 0 |
| 7 | 1500 | 3 | 1.0 | 12.60 | 0 |
| 8 | 1500 | 3 | 1.5 | 10.19 | 0 |
| 9 | 1500 | 3 | 2.0 | 9.790 | 0 |
| 10 | 1550 | 4 | 1.5 | 6.300 | 1 |
| 11 | 1550 | 4 | 1.5 | 8.690 | 1 |
| 12 | 1550 | 4 | 1.5 | 9.500 | 1 |
| 13 | 1600 | 4 | 1.5 | 6.300 | 1 |
| 14 | 1600 | 4 | 1.5 | 8.690 | 1 |
| 15 | 1600 | 4 | 1.5 | 9.500 | 1 |
| 16 | 1650 | 4 | 1.5 | 6.300 | 1 |
| 17 | 1650 | 4 | 1.5 | 8.690 | 1 |
| 18 | 1650 | 4 | 1.5 | 9.500 | 1 |
| 19 | 1600 | 3 | 1.5 | 6.370 | 1 |

① 为理论配碳量的倍数。
② 0 代表无 Ca-α-Sialon 生成，1 代表有 Ca-α-Sialon 生成。

由图 6 可以看出，好点与坏点明显分布在两个不同的区域中。为验证本模式识别的结果，在图 6 的好区和坏区中取 6 个点 A、B、M、N、P 和 Q，通过模式识别逆映照得到这 6 个样本点的工艺参数（见表 6），对于样品 A 和 B 根据其参数制备合成了试样，并对试样进行 X 射线衍射分析物相组成，分析结果是样品 A 中有约 50% 的 Ca-α-Sialon，而样品 B 中没有发现有 Ca-α-Sialon 存在，实验结果与模式识别的分类相一致，证明了本模式识别的可靠性。

表 5 中的数据经归一化处理后，转换为人工神经元网络（ANN）识别的数值，其中 4 个因素作为输入值，Ca-α-Sialon 的有无（－1 表示无 Ca-α-Sialon 生成，1 表示有 Ca-α-

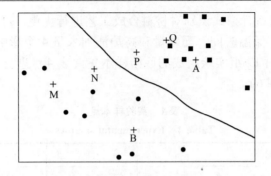

图 6 模式识别分类图

Fig. 6 Classifying diagram of statistic pattern recognition

■ Good point ● Bad point + Predicted point

Sialon 生成）作为输出结果的目标值。用这些处理后的数据对 ANN 进行训练，把表 6 中 M、N、P 和 Q 的参数值归一化后，代入训练后的 ANN 中得到以下输出结果（见表6）。

表 5 选取样本点的参数

Table 5 Parameters of selected samples

| 样本点 | 轴值 | $t/℃$ | $\tau/h$ | $C^{①}$ | $w_{渣}/\%$ |
|---|---|---|---|---|---|
| A | $x = 0.927692$<br>$y = 0.61003$ | 1627 | 4 | 1.51 | 9.25 |
| B | $x = -0.36963$<br>$y = -1.74531$ | 1420 | 3 | 1.32 | 12.07 |
| M | $x = -1.98851$<br>$y = -0.17508$ | 1372 | 3 | 1.32 | 10.42 |
| N | $x = -1.15689$<br>$y = 0.329632$ | 1485 | 3 | 1.87 | 8.85 |
| P | $x = -0.29201$<br>$y = 0.871735$ | 1539 | 3 | 1.72 | 10.16 |
| Q | $x = 0.306752$<br>$y = 1.264290$ | 1536 | 3 | 1.54 | 8.30 |

① 为理论配碳量的倍数。

表 6 选取样本点的 ANN 输出值

Table 6 ANN outputs of selected samples

| 样本点 | M | N | P | Q |
|---|---|---|---|---|
| ANN 输出 | -1 | -1 | 1 | 1 |

该结果与模式识别的分类结果相一致，再次证明了本模式识别的可靠性。

## 4 结论

（1）在热力学分析的基础上，利用高炉渣碳热还原氮化合成了 Ca-α-Sialon-SiC 粉，并用 XRD 分析和 TEM 观察、电子选区衍射标定进行了验证,证实了热力学分析的可靠性。

(2) 利用统计模式识别优化了合成 Ca-α-Sialon-SiC 粉工艺,并利用模式识别逆映照和人工神经元网络验证了统计模式识别分类的可靠性。

## 参 考 文 献

[1] Thommy Ekstrom, Zhijian J Shen, Kenneth J D MacKenzie, et al. α-Sialon Ceramics Synthesised from a Clay Precursor by Carbothermal Reduction and Nitridation [J]. J. Mater. Chem., 1998, 8 (4): 977.

[2] Catherine L Hewett, Cheng Yibing, Barry C Muddle. Phase Relationship and Related Microstructural Observations in the Ca-Si-Al-O-N system [J]. J. Am. Ceram. Soc., 1998, 81 (7): 1781.

[3] Wang P L, Zhang C, Sun W Y, et al. Characteristics of Ca-α-Sialon-Phase Formation, Microstructure and Mechanical properties [J]. J. Eur. Ceram. Soc., 1999, 19: 553.

[4] Hewett C L, Cheng Y B, Muddle B C, et al. Thermal Stability of Calcium α-Sialon Ceramics [J]. J. Eur. Ceram. Soc., 1998, 18: 417.

[5] 甄强. O'-Sialon-BN 复合材料的研究及相关热力学性质预报 [D]. 北京:北京科技大学, 2000, 36.

[6] Chris A Wood, Zhao Hong, Cheng Yibing. Microstructural Development of Calcium α-Sialon Ceramics with Elongated Grains [J]. J. Am. Ceram. Soc., 1999, 82 (2): 421.

(原载于《北京科技大学学报》, 2001, 23 (5): 404-408)

# 第二部分

## 新型高温结构陶瓷
### New Type High Temperature Structural Ceramics Materials

从热力学上讲，温度可以无限提高，而高温材料的发展限制了温度使用范围。随着航天航空（火箭导弹、超音速飞机）、原子能工业，以及冶金新工艺、新技术等科学技术的发展，材料的使用温度从钢铁冶金的1600℃到化学火箭的4000℃，再到宇航的17000℃，甚至更高，因此对高温陶瓷材料提出了新的要求，耐高温、耐腐蚀、耐侵蚀、耐烧蚀等性能，具有高强度、高韧性，以及一些特殊功能。

本文集应用物理化学的原理和方法，对一些高温陶瓷材料进行化学设计并合成新型材料，在此基础上利用计算机进行优化和寻优，简化了实验研究的过程，为材料科学研究者提供一些可参考的研究方法和途径。

# Kinetic Studies of Oxidation of MgAlON and a Comparison of the Oxidation Behaviour of AlON, MgAlON, O'-SiAlON-ZrO$_2$ and BN-ZCM Ceramics

Wang Xidong[1]  Li Wenchao[2]  Seshadri Seetharaman[1]

(1. Division of Metallurgy, Royal Institute of Technology, Stockholm SE 10044, Sweden;
2. Department of Physical Chemistry, University of Science
and Technology Beijing, Beijing 100083, China)

**Abstract:** The kinetics and morphology of the oxidation process of magnesium-aluminium oxynitride (MgAlON), aluminium oxynitride (AlON), O'-SiAlON-ZrO$_2$ and BN-ZCM have been studied in the temperature range 1373-1773K. Oxidation experiments with powder and plate samples of the above materials have been carried out in air. MgAlON shows the best resistance to oxidation at lower temperatures (<1473K), whereas at higher temperatures (~1773K), AlON shows the best resistance. O'-SiAlON-ZrO$_2$ shows very good oxidation resistance up to 1673K. But its oxidation rate increases greatly above 1673K, presumably due to the formation of liquid phase. BN-ZCM has the poorest oxidation resistance due to the evaporation of B$_2$O$_3$.

The activation energies for the chemical oxidation reaction of AlON, MgAlON and O'-SiAlON-ZrO$_2$ are 214, 330 and 260kJ/mol, respectively. The overall diffusion activation energies for AlON, MgAlON, O' SiAlON-ZrO$_2$ and BN-ZCM are 227, 573, 367 and 289kJ/mol, respectively.

**Keywords:** oxidation; MgAlON; kinetics; morphology

## 1 Introduction

Magnesium aluminum oxynitride (MgAlON) was first synthesized by Jack[1] in 1976. Since then, the phase relations and synthesis of MgAlON ceramics have been investigated intensively [2-6]. It has almost the same structure and properties as aluminium oxynitride (AlON), but it is thermodynamically stabe over a wider temperature range. This provides an edge for MgAlON as a superior ceramic material at high temperatures. In addition, its semitransparent property makes it a promising high-temperature window material.

Like other nitrogen ceramics[7], MgAlON can be oxidised at high temperatures when it is exposed to oxygen. An understanding of its oxidation resistance is very important, in order to examine the areas of potential applications. In view of the lack of information on the susceptibility MgAlON to oxidation, a study of the oxidation kinetics of MgAlON was undertaken in the present work. The measurements were carried out in the temperature range 1373-1773K. The oxidation resistance of this material is also compared with the similar property of AlON, O'-SiAlON-ZrO$_2$ and BN-ZCM ceramics in order to understand the relative merits of application of these materials.

## 2 Experimental

### 2.1 Materials

$Al_2O_3$, AlN and MgO were used as raw materials to prepare MgAlON ceramics by sintering under high pressure. The preparation steps are as follows:

(1) Fine powders of $Al_2O_3$, AlN and MgO were weighed in the molar ratio of 0.60, 0.30 and 0.10, respectively. The powders were then mixed by ball milling in ethanol medium for 24 hours.

(2) The mixture, contained in a graphite mould covered with boron nitride, was sintered at 2073K and at 25MPa in a $N_2$ gas flow for 3h.

(3) MgAlON synthesized had the composition $MgO \cdot 3AlN \cdot 6Al_2O_3$ and its X-ray diffraction pattern is shown in Fig. 1.

(4) The MgAlON ceramics, as well as AlON, O'-SiAlON-$ZrO_2$ and BN-ZCM ceramics, were cut, ground and polished into pieces (1mm × 6mm × 20mm) or ground into powders (5-30μm) before thermogravimetric analysis.

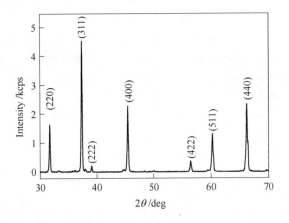

Fig. 1 XRD pattern of MgAlON. kcps = $10^3$ counts per second

The synthesis of AlON was discussed in ref. [8]. The composition of the synthesized material was $3.5AlN \cdot 6.5Al_2O_3$. O'-SiAlON-$ZrO_2$ was also prepared from a mixture of 70% O'-SiAlON and 30% $ZrO_2$ by a similar sintering technique. BN-ZCM, prepared in the same way, has the composition of 50% BN and 50% ZCM (15% $ZrO_2$, 26% $Al_2O_3$ and 9% $3Al_2O_3 \cdot 2SiO_2$) [% = wt. %].

Argon gas (<5ppm $O_2$) and air (21% $O_2$) were provided by AGA GAS Stockholm, Sweden. The argon was purified by passing through silica gel and dehydrite [$Mg(ClO_4)_2$] to remove water, through ascarite to remove carbon dioxide, and through tube furnaces containing Cu and Mg, respectively, at 773K to remove residual oxygen.

## 2.2 Apparatus and procedure

The experiments on the oxidation of MgAlON, AlON, O'-SiAlON-ZrO$_2$ and BN-ZCM were carried out in a SETARAM TGA92 system. The balance of the unit had a detection limit of 1μg. The system was fully controlled by an IBM 386PC through a CS92 controller. Carefully weighed amounts of both powder as well as plate samples were held in suitable Pt containers and suspended in the transducers.

In order to choose the oxidation temperature range, preliminary experiments were conducted under non-isothermal conditions with MgAlON, AlON, O'-SiAlON-ZrO$_2$ and BN-ZrO$_2$ powders (5-30μm). The sample was held in Pt crucibles and placed in the transducers. Air was let into the reaction tube at a fixed flow (800mL/min). The furnace was then heated from room temperature to a maximum temperature of 1773K at fixed heating rates (20K/min). The results are shown in Fig. 2.

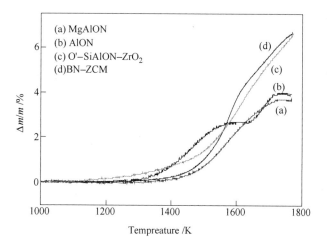

Fig. 2 Oxidation curves for the materials studied

Isothermal oxidation was carried out at intervals of 100K in the temperature range 1273-1773K. The reaction tube was initially evacuated for 15min and then flushed with purified Ar gas. The furnace was then heated at a fixed rate (25K/min) to the required temperature. After maintaining at the required temperature for 5min, air was led into the reaction tube at a flow rate of ~800mL/min. Preliminary experiments were carried out at different flow rates to ensure that this gas flow was above the starvation rate. It was found that the gas flow had no effect on the reaction rate above 600mL/min. At the end of the experimental run, the reaction was arrested by replacing the air flow by Ar and the furnace was allowed to cool at the rate of 99K/min to room temperature.

The powders after oxidation were taken to XRD analysis to determine the oxidation products. In the case of plate samples, after the oxidation, the samples were cut, ground and polished for microscopic examination.

## 3 Results

### 3.1 Oxidation process

Isothermal oxidation experiments in the case of MgAlON, AlON, O'-SiAlON-$ZrO_2$ and BN-ZCM powders (5-30μm) were carried out in the temperature range 1273-1773K (Fig. 3 and Fig. 4). As the temperature increased, the oxidation rate showed an increase. But the reaction was found to slow down at longer time intervals. In the case of BN-ZCM, a decrease in the weight of the sample was observed. This is most probably due to the evaporation of the boron oxide at higher temperatures.

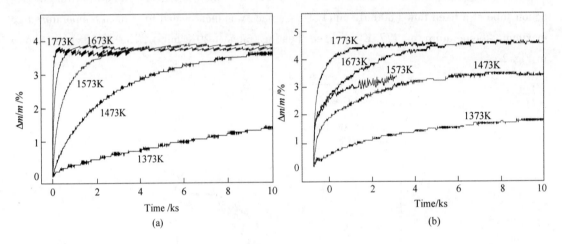

Fig. 3  Oxidation of MgAlON (a) and AlON (b) powders

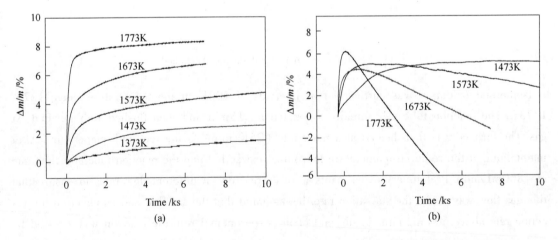

Fig. 4  Oxidation of O'-SiAlON-$ZrO_2$ (a) and BN-ZCM (b) powders

In the case of the experiments with plate samples, MgAlON, AlON, O'-SiAlON-$ZrO_2$ and BN-ZCM plates were cut, ground and polished. The oxidation was carried out from 1373K to 1773K, and the results are shown in Fig. 5 and Fig. 6.

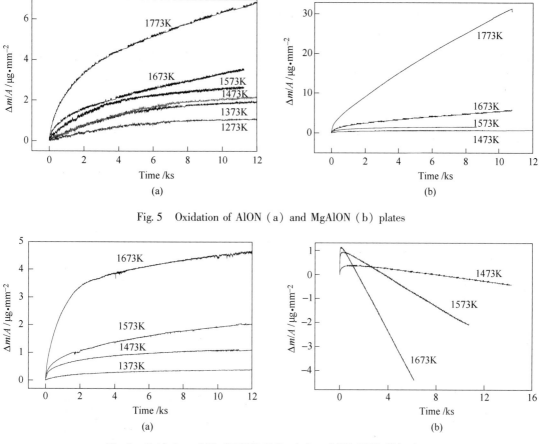

Fig. 5  Oxidation of AlON (a) and MgAlON (b) plates

Fig. 6  Oxidation of O'-SiAlON-ZrO$_2$ (a) and BN-ZCM (b) plates

## 3.2 Oxidation products

In order to determine the products of oxidation, X-ray diffraction analyses of the samples after oxidation were carried out and some of the results are shown in Fig. 7 and Fig. 8. The oxidation product of AlON at high temperatures is -Al$_2$O$_3$, while for MgAlON the products are -Al$_2$O$_3$ and MgAl$_2$O$_4$.

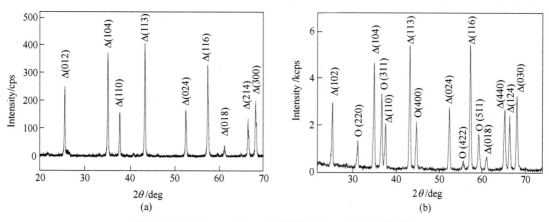

Fig. 7  Oxidation products of AlON (a) and MgAlON (b)

Δ—α-Al$_2$O$_3$; O—MgAl$_2$O$_4$

The main products of oxidised O'-SiAlON-ZrO$_2$ are SiO$_2$ and ZrSiO$_4$ and for BN-ZCM they are -Al$_2$O$_3$ and ZrO$_2$.

After oxidation, the surface of the oxidation layer was measured by AFM (atomic force microscopy) and the results are shown in Figs. 9-12. The plates were cut and the sections were prepared for analysis by scanning electron microscopy (SEM) (Fig. 13 and Fig. 14).

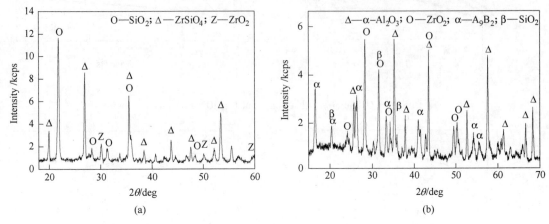

Fig. 8  Oxidation products of O'-SiAlON-ZrO$_2$ (a) and BN-ZCM (b)

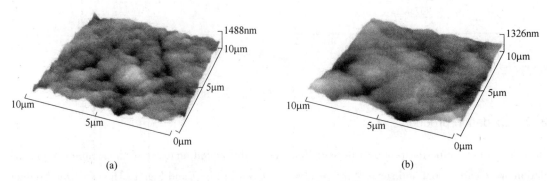

Fig. 9  AFM of AlON at 1473K (a) and 1673K (b)

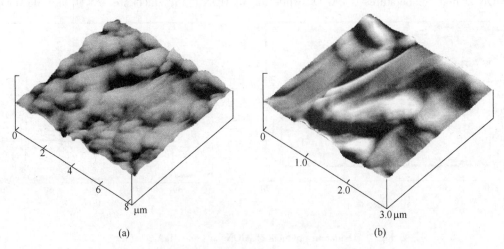

Fig. 10  AFM of MgAlON at 1473K (a) and 1673K (b)

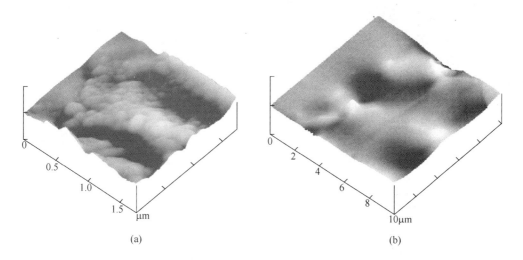

Fig. 11　AFM of O'-SiAlON-ZrO$_2$ at 1473K (a) and 1673K (b)

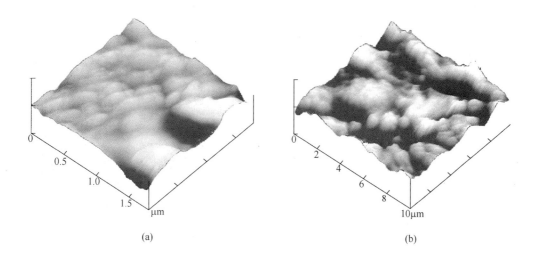

Fig. 12　AFM of BN-ZCM at 1473K (a) and 1673K (b)

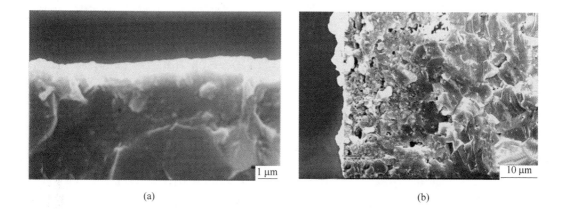

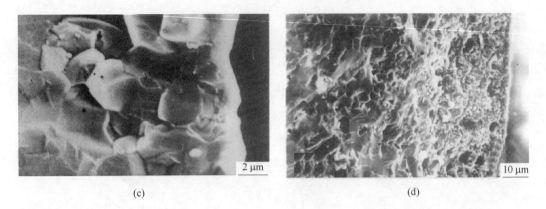

Fig. 13 SEM micrographs of AlON (a, b) and MgAlON (c, d) oxidation sections at 1373K (a, c) and 1673K (b, d)

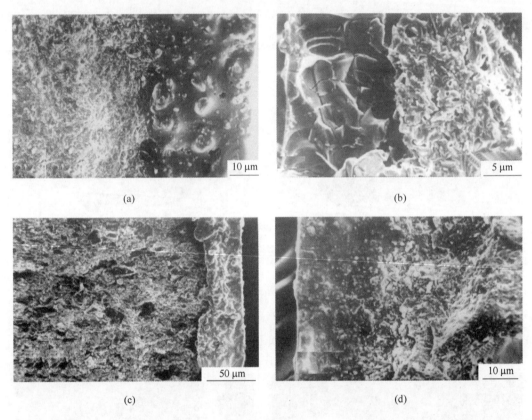

Fig. 14 SEM micrographs of O'-SiAlON-ZrO$_2$ (a, b) and BN-ZCM (c, d) at 1473K (a, c) and 1673K (b, d)

## 4 Discussion

The oxidation reactions can described as follows:

$$MgAl_{15}O_{19}N_3 + 2.25O_2 = 6.5Al_2O_3 + MgAl_2O_4 + 1.5N_2 \quad (1)$$

$$Al_{33}O_{39}N_7 + 5.25O_2 = 16.5Al_2O_3 + 3.5N_2 \quad (2)$$

As some $Al_2O_3$ can be dissolved into $MgAl_2O_4$, the ratio of $MgAl_2O_4$ and $Al_2O_3$ is higher than the result of Eq (1). This can be seen from Fig. 7.

During the oxidation of O'-SiAlON-$ZrO_2$ composite ceramics, $ZrO_2$ will remain unaffected. Hence, the oxidation reaction, which is restricted to $Si_4AlO_{3.5}N_4$, can be written as

$$Si_4AlO_{3.5}N_4 + 3O_2 = 4SiO_2 + 0.5Al_2O_3 + 2N_2 \tag{3}$$

During the oxidation process, $SiO_2$ and $ZrO_2$ will react and some $ZrSiO_4$ is likely to be produced. This is evidenced in the X-ray diffraction pattern of the product formed.

In the case of BN-ZCM oxidation, BN will be oxidised as:

$$2BN + 1.5O_2 = B_2O_3 + N_2 \tag{4}$$

Some $B_2O_3$ will combine with $Al_2O_3$ and produce a new compound ($9Al_2O_2 \cdot 2B_2O_3$) as can be seen in Fig. 8. However, most of $B_2O_3$ will be evaporated. ZCM [$ZrO_2$, $Al_2O_3$ and $3Al_2O_3 \cdot 2SiO_2$ (mullite)] will not be oxidised. Mullite, on the otherhand, will decompose into $Al_2O_3$ and $SiO_2$.

## 4.1 Kinetics study of oxidation

As discussed above, the oxidation mechanism of MgAlON, AlON, O'-SiAlON-$ZrO_2$ and BN-ZCM are different. In the case of O'-SiAlON-$ZrO_2$ and BN-ZCM composite ceramics, only parts of the material, viz. O'-SiAlON and BN, are susceptible to oxidation. For BN-ZCM, the oxidation product, $B_2O_3$, is likely to be lost in the vapour phase due to its relatively high vapour pressure at the experimental temperatures.

### 4.1.1 MgAlON

The oxidation process can be divided into the following steps:

(1) Oxygen transfer from the bulk of gas flow to the surface of MgAlON.

(2) Oxygen transfer from the surface to the interface between the oxidised and non-oxidised layers.

(3) Oxygen reaction with MgAlON to produce $MgAl_2O_4$ or $Al_2O_3$ and nitrogen.

(4) Nitrogen transfer from the interface to the surface.

(5) Nitrogen transfers from the surface of MgAlON to the gas flow.

As mentioned earlier, the experiments were conducted at flow rates higher than the starvation rate for the reaction. Hence, under the present experimental conditions, the reaction step (1) and (5), viz. oxygen and nitrogen transfer in the bulk gas flow, may be fast and is not likely to influence the reaction rate.

Initially, the oxidation of MgAlON powder takes place at the surface and the resistance of diffusion is relatively small. Thus, the oxidation rate is only controlled by the interfacial chemical reaction. Let

$$\alpha = \frac{m - m_0}{m_\infty - m_0} = \frac{\Delta m}{\Delta m_{max}} \tag{5}$$

where $m$ is the weight of the oxidation specimen at time $t$, $m_0$ is the weight at the beginning and $m_\infty$ is the weight after completely oxidised. Assuming that the sample powder is spherical, the following equation can be deduced.

$$(1-\alpha)^{1/3} = 1 - k_r t \tag{6}$$

Plots of $(1-\alpha)^{1/3}$ as a function of t are shown in Fig. 15. The constant $k_r$ can be obtained at each temperature from the slopes of the curves at the initial linear stage. In accordance with the Arrhenius equation

$$k_r = k_{r(0)} \exp(-E_r/RT) \tag{7}$$

the activation energy ($E_r$) for the chemical reaction can be obtained from a plot of $\ln k_r$ as a function of $1/T$ (Fig. 16): $E_r = 330$ kJ/mol.

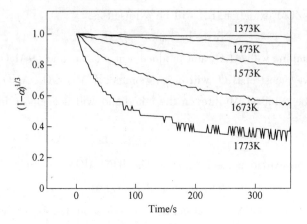

Fig. 15 $(1-\alpha)^{1/3}$ as a function of time for MgAlON at different temperatures

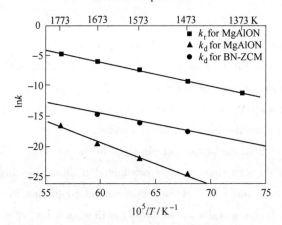

Fig. 16 $\ln k$ as a function of temperature for MgAlON and BN-ZCM

Initially, the oxidation of AlON plate takes place at the surface. In this case, the chemical reaction is controlling the rate of oxidation and the following equation can be deduced:

$$\Delta m / A = k_r t \tag{8}$$

where $A$ is the area of the interface between the oxidised and non-oxidised layers. As the oxidation proceeds, the oxidised layer becomes thicker and, consequently, the diffusion path increases for the reactant gas. Hence, the resistance offered by the diffusion step increases. After some certain time interval, the diffusion step determines the rate of oxidation. The chemical reaction reaches equilibrium at the interface. Assuming that the area of interface is constant under the experimental conditions (the product layer was relatively thin, <50μm), the following equation can be deduced:

$$\left(\frac{\Delta m}{A}\right)^2 = k_d t \tag{9}$$

The relations between $(\Delta m/A)^2$ and $t$ is linear during longer time intervals of the oxidation process. The constant $k_d$ can be calculated from the slopes of the lines. Plots of $\ln k_d$ as a function of $1/T$ are shown in Fig. 16. the activation energy, $E_d = 573 \text{kJ/mol}$.

#### 4.1.2 BN-ZCM

The oxidation of BN-ZCM is a complicated process. At high temperatures, BN will be oxidised into $B_2O_3$. Simultaneously, $B_2O_3$ will evaporate into air because of its high vapour pressure. During the oxidation process, as BN oxidises into $B_2O_3$, the specimen will have a weight gain. On the other hand, as $B_2O_3$ evaporates into air, the specimen has a weight loss. For the oxidation of BN-ZCM plates, the rate of evaporation of $B_2O_3$ can be reasonably assumed constant. Assuming that the volume of the plate keeps almost the same during the process, the weight gain during the oxidation of a BN-ZCM plate would be

$$\frac{d(\Delta w)}{dt} = A_0(\theta' \rho_{B_2O_3} - \theta \rho_{BN}) \frac{dx}{dt} \tag{10}$$

where $\Delta w$ is the weight gain of oxidation process (without considering the evaporation of $B_2O_3$), $A_0$ is the surface area of the sample, $\theta'$ is the volume content of $B_2O_3$ in the oxidised layer, $\theta$ is the content of BN in the non-oxidised body, $x$ is the thickness of oxidation layer, and $l_{B_2O_3}$ and $l_{BN}$ are the densities of $B_2O_3$ and BN, respectively.

For the evaporation of $B_2O_3$, the weight loss can be expressed as

$$\frac{\Delta w'}{A_0} = \theta' x_2 \rho_{B_2O_3} = k_e t \tag{11}$$

where $\Delta w'$ is the weight loss of $B_2O_3$ evaporation during the oxidation process, $x_2$ the thickness of the evaporated layer and $k_e$ the rate constant of evaporation.

Initially, the oxidation occurs at the surface of BN-ZCM (the product layer is very thin) and the chemical reaction is the rate-controlling step. The rate equation can then be written as

$$r_c = \frac{d(\Delta w)}{dt} = A_0 k_r c \tag{12}$$

where $k_r$ is the rate constant of chemical reaction and $c$ is the content of oxygen at surface. Integration of Eq (12) yields

$$\frac{\Delta w}{A_0} = k_r c t = k_o t \tag{13}$$

where $k_o$ is the rate constant of oxidation. It is now possible to incorporate the weight loss due to $B_2O_3$ evaporation. Let the total weight change be expressed as $w_t$, then

$$\frac{\Delta w_t}{A_0} = \frac{\Delta w - \Delta w'}{A_0} = (k_o - k_e)t \tag{14}$$

As the oxidation proceeds, the thickness of the product layer increases, and the diffusion through the product layer becomes the rate-controlling step. The product layer (with the thickness of $x$) can be divided into two sub-layers: one inner layer containing $B_2O_3$ and ZCM with the thickness of $x_1$ and an outer layer containing only ZCM without $B_2O_3$, which has the thickness of $x_2$, so that $x = x_1 + x_2$. Due to the high porosity, the resistance of diffusion through the outside layer is relatively very small. In this case, the rate of the diffusion-controlled reaction will be

$$r_d = A_0 \Delta c \frac{D}{x_1} = A_0 (\theta' \rho_{B_2O_3} - \theta \rho_{BN}) \frac{dx}{dt} \tag{15}$$

where $\Delta c$ ($c - c_i$) is the difference in the oxygen content, during diffusion controlling step, $c_i$ equals to the equilibrium oxygen content, it is relatively very small, then $\Delta c = c$, D is the diffusion coefficient.

$$x_1 = x - x_2 = x - \frac{k_e t}{\theta' \rho_{B_2O_3}} \tag{16}$$

$$\frac{dx_1}{dt} = \frac{dx}{dt} - \frac{k_e}{\theta' \rho_{B_2O_3}} \tag{17}$$

Combination of Eqs (17) and (15) gets

$$(\theta' \rho_{B_2O_3} - \theta \rho_{BN}) \left( \frac{dx_1}{dt} + \frac{k_e}{\theta' \rho_{B_2O_3}} \right) = \frac{cD}{x_1} \tag{18}$$

$$\frac{x_1 dx_1}{\frac{cD}{\theta' \rho_{B_2O_3} - \theta \rho_{BN}} - \frac{k_e}{\theta' \rho_{B_2O_3}} x_1} = dt \tag{19}$$

$$\frac{\theta' \rho_{B_2O_3}}{k_e} \left( \frac{1}{1 - \frac{k_e (\theta' \rho_{B_2O_3} - \theta \rho_{BN})}{cD \theta' \rho_{B_2O_3}} x_1} - 1 \right) dx_1 = dt \tag{20}$$

Integration of Eq (20) yields:

$$-x_1 - \frac{cD \theta' \rho_{B_2O_3}}{k_e (\theta' \rho_{B_2O_3} - \theta \rho_{BN})} \ln \left( 1 - \frac{k_e (\theta' \rho_{B_2O_3} - \theta \rho_{BN})}{cD \theta' \rho_{B_2O_3}} x_1 \right) = \frac{k_e}{\theta' \rho_{B_2O_3}} t \tag{21}$$

As can be seen from Eq (21), when time $t \to \infty$, then

$$\frac{k_e (\theta' \rho_{B_2O_3} - \theta \rho_{BN})}{cD \theta' \rho_{B_2O_3}} x_1 = 1 \tag{22}$$

$$k_o = \frac{cD}{x_1} = \frac{\theta' \rho_{B_2O_3} - \theta \rho_{BN}}{\theta' \rho_{B_2O_3}} k_e = \text{constant} \tag{23}$$

where $k_o$ is the oxidation rate constant during the diffusion-controlling step and $k_e$ is evaporation rate constant. The total weight change rate is

$$r_t = \frac{d(\Delta w_t / A_0)}{dt} = k_o - k_e = \frac{-\theta \rho_{BN}}{\theta' \rho_{B_2O_3}} k_e \tag{24}$$

If the oxidation duration is long enough, the oxidation rate, as well as the evaporation rate, will be constant. From the slope of the oxidation curves, $r_t$ can be attained at different temperatures and $k_e$ can be calculated by Eq (24). Thus the weight gain by oxidation (after correction for evaporation of $B_2O_3$) can also be calculated at different temperatures. The results are shown in

Fig. 17. Also the rate constant $k_d$ can be attained at each temperature, and the overall activation energy $E_d$ can be calculated from Fig. 16: $E_d = 289 kJ/mol$.

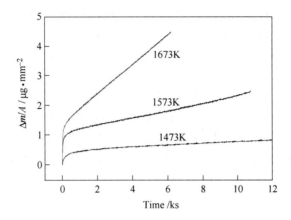

Fig. 17　Oxidation of BN-ZCM plate (corrected) at different temperatures

### 4.1.3　Comparison of oxidation results

A comparison of the oxidation rates of MgAlON, AlON, O'-SiAlON-$ZrO_2$ and BN-ZCM ceramics is significant to decide their application conditions. Therefore, the oxidation of O'-SiAlON-$ZrO_2$ and AlON were analysed as descripted in section 4.1.1. The results are shown in Fig. 18. The non-isothermal oxidation results can be seen in Fig. 2. As can be seen from Fig. 2, the oxidation of AlON exhibits two steps. The curve is also S-shaped due to the phase transformation of the oxidation product[8], while MgAlON, O'-SiAlON-$ZrO_2$ and BN-ZCM show only one step with a parabolic shape. As also can be seen, BN-ZCM has a low oxidation rate at low temperatures because a part of the surface is covered by ZCM which does not oxidise. However, at high temperatures, its oxidation rate increases greatly due to the evaporation of $B_2O_3$.

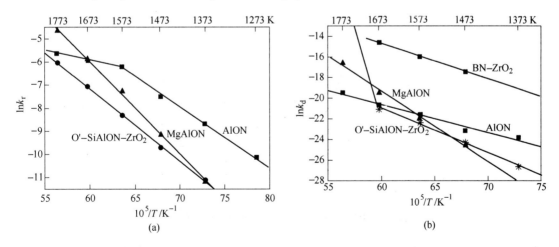

Fig. 18　Arrhenius plot of $k_r$ (a) and $k_d$ (b) for the different materials

As can be seen from Fig. 18, at lower temperatures, MgAlON shows the best resistance to oxidation. But at higher temperatures (1773K), AlON shows the best resistance to oxidation. O'-SiAlON-$ZrO_2$ shows a very good oxidation resistance up to 1673K. But for $T > 1673K$, the situation is complicated due to the formation of a liquid phase produced during the oxidation process. Bubble formation in the product layer will cause flaking-off of the layer and enhance the oxidation rate. BN-ZCM shows a good oxidation resistance during the interfacial chemical reaction-controlling step. But, due to the evaporation of $B_2O_3$, the resistance decreases greatly during the diffusion-controlling step.

The overall chemical reaction activation energies for the oxidation of AlON, MgAlON and O'-SiAlON-$ZrO_2$ are 214, 330 and 260kJ/mol, respectively. The overall diffusion activation energies at longer time intervals were found to be 227, 573, 367 and 289kJ/mol for AlON, MgAlON, O'-SiAlON-$ZrO_2$ and BN-ZCM, respectively.

## 4.2 Oxidation products

After oxidation, the samples were analysed by XRD. Part of the resultsis shown in Figs. 7 and 8. For AlON, the oxidation product, $\alpha$-$Al_2O_3$, is quite dense at the experimental temperatures. Hence, AlON exhibits the best oxidation resistance at high temperatures (1773K), while MgAlON has the highest oxidation resistance at low temperatures (1473K) due to the stability of $MgAl_2O_4$.

In the case of O'-SiAlON-$ZrO_2$ ceramics, the main oxidation product is $SiO_2$, which provides a very good oxidation resistance below 1673K. But the oxidation products (diffused together) will melt at higher temperatures (~1773K). Hence, its oxidation rate increases greatly (Fig. 19) as the oxidation temperature is higher than 1673K. Gas cavities and gas bubbles could be visually observed on the surface of the oxidation specimen at 1773K.

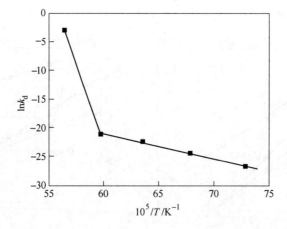

Fig. 19  Relation of $\ln k_d$ and $1/T$ of O'-SiAlON-$ZrO_2$

In the case of BN-ZCM, the main products of oxidation are $\alpha$-$Al_2O_3$ and $ZrO_2$. As a great deal of $B_2O_3$ vaporises, the oxidation layer with a high porosity allows further oxidation to go on. Therefore, it has the poorest oxidation resistance.

## 4.3 Structure and morphology of the oxidation layer

The structure and morphology of oxidation layers have been analysed by AFM (Figs. 9-12). The roughnesses of MgAlON, AlON, O'-SiAlON-$ZrO_2$ and BN-ZCM are 234, 174, 75 and 63nm, respectively, at 1473K, and 297, 284, 52 and 406nm, respectively, at 1673K.

The degree of roughness ($R_d$) is calculated by the following equation with 50 points ($N = 50$) in each photo:

$$R_d = [N^{-1} \sum (Z_i - Z_{av})^2]^{\frac{1}{2}} \quad (25)$$

where $R_d$ is the roughness degree, $N$ is the number of points, $Z_i$ is the height of point i and $Z_{av}$ is the average height of all points.

It can be seen from Figs. 9-12 that, as the temperature increases, the roughness degree of AlON and MgAlON increases slightly due to the growth of crystals. The roughness degree of BN-ZCM increases from 63nm to 406nm because of the evaporation of $B_2O_3$. However, the roughness of O'-SiAlON-$ZrO_2$ decreases from 75nm to 52nm as the temperature increases from 1473K to 1673K. The main reason is that the liquid phase (glass) produced during the oxidation process at high temperatures (1673K and 1773K).

The oxidation layers were also analysed by SEM, as can be seen from Figs. 13 and 14. At lower temperatures ( < 1473K), the oxidation layers of MgAlON, AlON and O'-SiAlON-$ZrO_2$ are dense. Therefore, they have a very good oxidation resistance. However, at high temperatures (1673K and 1773K), many cracks can be observed in the oxidation layers of MgAlON and AlON which no longer provide the protection from oxidation. For O'-SiAlON-$ZrO_2$, a glass phase can be observed due to the melting of some of the products of oxidation at high temperatures. Many cracks were also observed in the product layer, which might as well have been caused during cooling.

## 5 Conclusions

(1) Under non-isothermal oxidation conditions, the oxidation of AlON exhibits two steps with a S-shaped curve due to the phase transformation of the oxidation product. MgAlON, O'-SiAlON-$ZrO_2$ and BN-ZCM show only one step with parabolic curves. BN-ZCM has a low oxidation rate at low temperatures (chemical reaction controlling step) because possibly the sample surface is partly covered by ZCM that does not undergo oxidation. However, at high temperatures, its oxidation rate increases greatly due to the evaporation of $B_2O_3$.

(2) At lower temperatures, MgAlON shows the best resistance to oxidation. But at higher temperatures (1773K), AlON shows the best resistance to oxidation. O'-SiAlON-$ZrO_2$ shows a very good oxidation resistance in the low temperature range up to 1673K. But for $T > 1673$K, there is a liquid phase produced during the oxidation process. Gas bubbles are also formed in the product layer causing the flaking-off of some parts of the product layer. Therefore, its oxidation rate increases greatly as the temperature rises to 1673K. In the case of BN-ZCM ceramics, due to the evaporation of $B_2O_3$, the oxidation resistance seems to be poorest.

(3) The chemical reaction activation energies for the initial stage of oxidation of AlON, Mg-AlON and O'-SiAlON-ZrO$_2$ are 214, 330 and 260kJ/mol, respectively. The activation energies at the later stages are 227, 573, 367 and 289kJ/mol for AlON, MgAlON, O'-SiAlON-ZrO$_2$ and BN-ZCM, respectively.

(4) As the temperature increases, the degree of roughness of AlON and MgAlON surfaces increases slightly due to the crystal growth. The roughness of BN-ZCM increases greatly because of the evaporation of B$_2$O$_3$. However, the roughness of O' SiAlON-ZrO$_2$ decreases as the temperature increases from 1473K to 1673K. The main reason is that the liquid phase (glass) produced during the oxidation process at high temperatures (1673K and 1773K). The roughness of MgAlON, AlON, O'-SiAlON-ZrO$_2$ and BN-ZCM are 234, 174, 75 and 63 nm, respectively, at 1473K, and 297, 284, 52 and 406nm, respectively, at 1673K.

(5) For protective oxidation, such as MgAlON, AlON and O'-SiAlON-ZrO$_2$, the oxidation is determined by diffusion. While for the non-protective oxidation, such as BN-ZCM, the oxidation rate is determined by the evaporation process.

## References

[1] Jack, K. H. J. Mater. Sci, 1976, 11: 1135-1158.
[2] Weiss S., Greil P., Gauckler L J. J. Am. Ceram. Soc., 1982, 65: C68-C69.
[3] Sun W. Y., Ma L. T., Yan D. S. Chinese Sci. Bull, 1990, 35: 200-202.
[4] Willems H. X., De With G., Metsellar R. J. Eur. Ceram. Soc., 1993, 12: 43-49.
[5] Guyader A., Goeuriot F., Thevenot F, et al. J. Euro. Ceram. Soc., 1994, 13: 365-370.
[6] Granon A., Goeuriot P., Thevenot F. J. Euro. Ceram. Soc., 1995, 15: 249-254.
[7] Corbin N. D. J. Eur. Ceram. Soc., 1989, 5: 143-154.
[8] Wang X., Du S., Li W., Seetharaman S. Kinetic study of the oxidation of Aluminum Oxynitride [J]. Metallurgical and Materials Transaction B.

(原载于 *Zeitschrift für Metallkundek*, 2002, 93 (6): 545-553)

# Thermal Diffusivity/Conductivity of MgAlON-BN Composites

### Zhang Zuotai[1,2]　Li Wenchao[2]　S. Seetharaman[1]

(1. Department of Materials Science and Engineering, Royal Institute of Technology,
Stockholm SE 10044, Sweden;
2. Department of Physical Chemistry, University of Science and Technology
Beijing 100083, China)

**Abstract:** Thermal diffusivity and heat capacity of MgAlON and MgAlON-BN composites were measured in the temperature range of 25℃ to 1300℃ using a laser flash technique and a differential scanning calorimeter (DSC) technique, respectively. Based on these measurements, effective thermal conductivity of the composites was calculated using the values measured earlier in the same substance. The experimental effective thermal conductivity results of the composites containing different BN contents were found to show the similar trend, which decreased rapidly with increasing temperature below 900℃ followed by a slow decrease with further increasing temperature. This can be explained by the fact that thermal conduction in both components, MgAlON and BN, was dominated by phonons. The phonon mean free path decreased with increasing temperature, limited by the characteristic length between two neighboring atoms. The BN addition has significant influence on the effective thermal conductivity. The effective thermal conductivity of the composites containing BN exhibited a small degree of anisotropy with respect to preferred orientation of the BN phase. The degree of anisotropy of the composites increased with increasing BN content, which is particularly pronounced at the higher BN additions. An equation suitable for the present composites has been derived based on Luo's model. The model was slightly modified in the present article. The predicted values calculated by the model were in good agreement with experimental results.

## 1 Introduction

Thermophysical properties such as thermal diffusivity, thermal conductivity, and heat capacity have both theoretical and practical importance. In many engineering situations, a knowledge of their thermophysical properties is essential to analyzing heat-transfer data in order to calculate the ability of a structural component to conduct heat or to dissipate a large quantity of locally generated heat[1]. For example, thermal conductivities of refractories are important parameters that underlie the design of furnaces, metallurgical reactor, boilers, etc[2,3].

　　In recent years, there has been increased interest in the area of ceramic-ceramic composites. The primary reason is that ceramic-ceramic composites have better properties than monolithic ones[4-6].

---

　　张作泰，2000~2006年于北京科技大学师从李文超教授攻读博士学位。目前在北京大学工作，研究员/博士生导师。发表论文80余篇，专利20余项，获省部级二等奖3项。

One such system that has been shown to have improved high-temperature mechanical properties, thermal shock durability, and wetting characteristics is magnesium aluminate spinel-boron nitride (MgAlON-BN) composite[7-9]. It is expected that this composite could find applications as high performance refractory, which can be used, for example, in special refractory nozzles, tubes, and break rings for the continuous casting of steel. To the best knowledge of the present authors, no systematic study of the thermal diffusivity/conductivity of MgAlON and MgAlON-BN composites has been carried out so far. The present work is thus motivated.

Our goal in this work was to study the thermal diffusivity/conductivity of MgAlON-BN composites of variable composition, prepared by hot pressing technology and intended for operation under hostile environments.

## 2 Experimental Procedure

### 2.1 Sample preparation

Single-phase MgAlON samples were prepared by mixing $Al_2O_3$, AlN, and MgO powders. Based on previous works in the present laboratory[10], a mixture with a molar ratio of $Al_2O_3/AlN/MgO = 7/2/1$ was selected to synthesize MgAlON. Then, different amounts of BN were added to the mixture. The weighed powders were mixed in a ball mill using an ethanol medium for 12 hours and then dried. The mixtures were then pressed into cylindrical compacts. These cylindrical compacts, covered with boron nitride, were hot pressed at 1800 ℃ under 20MPa for 2 hours. X-ray diffraction analysis of the sample was carried out to confirm the formation of the MgAlON-BN. It should be pointed out that a small amount of MgO, AlN, and $Al_2O_3$ volatilized under the reaction conditions. Scanning electron microscopy (SEM) was used to examine the fracture samples.

Disks with about 12-mm diameter and 1- to 2-mm thickness were prepared for thermal diffusivity measurements in directions parallel and perpendicular to the hot-pressing direction. The surfaces were polished to obtain the parallelism between both faces. Disks with 5-mm diameter and 1-mm thickness were prepared for heat capacity measurements.

### 2.2 Measurement equipment

The measurements of the thermal diffusivity, $\alpha$, were performed with a laser flash device (model TC-7000H/MELT Ulvac-Riko, Yokohama, Japan). Fig. 1 gives the schematic diagram of the laser flash unit. In the process of measuring, the top surface of the disk sample is irradiated with laser beam, which provides an instantaneous energy pulse. The laser energy is absorbed on the top surface of a sample and gets converted into the heat energy. The heat energy travels through the sample. Immediately after the laser pulse, the temperature of the rear surface of the sample is monitored by collecting the radiation using a photovoltaic infrared detector. The computer uses the rear surface temperature vs time trace to obtain a value for half rise time, $t_{1/2}$, which is the time required for the rear face to reach half of its maximum value. The thermal diffusivity, $\alpha$, is then calculated from Eq. (1)[11,13]:

$$\alpha = \frac{1.37 \cdot L^2}{\pi^2 \cdot t_{1/2}} \tag{1}$$

where $L$ is the thickness of the sample. The thermal diffusivity measurements were carried out during the heating cycle at first and then during the cooling cycle.

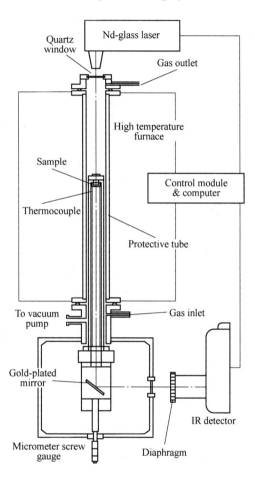

Fig. 1  A schematic diagram of the laser flash unit

Table 1  Densities and porosities of MgAlON-BN composites

| BN content/vol% | Measured density/g·cm$^{-3}$ | Theoretical density/g·cm$^{-3}$ | Porosity /% |
|---|---|---|---|
| 0 | 3.88 | 3.95 | 1.77 |
| 5 | 3.73 | 3.87[①] | 3.53 |
| 10 | 3.69 | 3.78[①] | 2.46 |
| 15 | 3.56 | 3.70[①] | 3.77 |
| 20 | 3.42 | 3.62[①] | 5.42 |
| 30 | 3.22 | 3.45[①] | 6.64 |

①Calculated using composite law.

The measurements of the heat capacity were performed with a differential scanning calorimeter (DSC) from Netzsch STA 449C Jupiter unit (Netzsch Instruments, Gerätebaw Gmbh, Germany). The experimental procedure has been described earlier in the present laboratory[12] and is briefly outlined here. The calorimeter covers a temperature range 25 ℃ to 1400 ℃. The measurements were performed in a platinum crucible with a platinum lid on the top. One sapphire disk was positioned between the measuring head and the platinum crucible. The crucible, sapphire disk, and lid were marked in order to ensure that these were always in the same position for each run. The same type of platinum disk and lid were used as the reference. A type S thermocouple was connected to the measuring head. The reaction chamber was evacuated three times before the measurements started. Argon gas ( <5ppm $O_2$) supplied by AGA Gas AB (Stockholm) was used to protect the sample and balance.

The argon gas was purified by passing through columns of silica gel and dehydrating to absorb the moisture, through ascarite to remove carbon dioxide, and through tube furnaces containing copper and magnesium at 600 ℃ and 500 ℃, respectively, to remove residual oxygen.

The DSC furnace was stabilized at 40 ℃ for 20 minutes before heating to 1300 ℃ at a heating rate of 10 ℃/min according to a predetermined temperature program. The standard run was conducted with a 1-mm-thick sapphire disc provided by Netzsch Instruments. The calibrationfles of the DSC apparatus including temperature file and sensitivity file were corrected by checking the melting points of five different metals and their enthalpies of transformation. The heat capacities of the composites were calculated by Netzsch Thermal Analysis software for MS windows. Parts of the experiments were repeated to confirm the reproducibility of the results.

## 3  Results

Table 1 lists the composition, density, and porosity of each composite. The amount of porosity was determined from the measured density and the theoretical density. As can be seen from Table 1, the porosity generally increases with increasing BN content.

Fig. 2 shows the thermal diffusivities as a function of temperature both parallel to and perpendicular to the hotpressing directions. The experimental results indicate that thermal diffusivities decreased quickly below 900 ℃, followed by a slow decrease with further increasing of temperature. Addition of the BN particles had a significant in fiuence on the thermal diffusivities and resulted in the anisotropy properties of the composites with respect to the hot-pressing direction. The addition of BN increased the thermal diffusivities in both hot-pressing directions.

The heat capacity measurements are shown in Fig. 3. As can be seen, there was a similar behavior for all composites. The heat capacities increased with increasing temperature below 900 ℃, and the slope for heat capacity curves is small with further increase in temperature. Addition of BN particles has a small effect on the heat capacity.

Effective thermal conductivities were calculated from the present experimental results of thermal diffusivities and heat capacity and density from Table 1 by the following equation[11]:

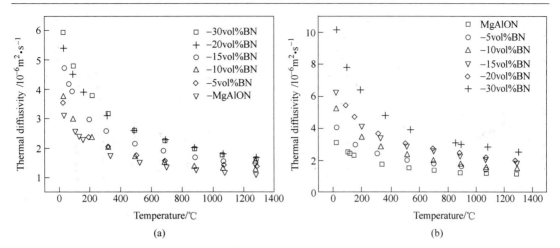

Fig. 2  Temperature dependence of thermal diffusivity of MgAlON with different BN contents:
(a) parallel to hot-pressing directions and (b) perpendicular to hot-pressing directions

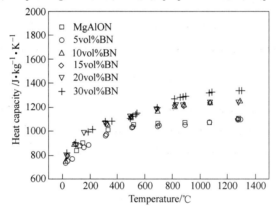

Fig. 3  Heat capacities as a function of temperatures for MgAlON and with different BN contents

$$\lambda = C_p \rho \alpha \qquad (2)$$

where $C_p$ is the specific heat capacity and $\rho$ is the density. It should be pointed out that the density of the composites at room temperature was used in the calculation of thermal conductivity because the decrease of density is less than 0.5 pct from room temperature up to 1300 ℃. Fig. 4 presents the computed effective thermal conductivities. As can be seen, the effective thermal conductivities decreased quickly with increasing temperature followed by a slow decrease at higher temperatures, the trend being the same as the thermal diffusivities of composites. The addition of BN has an obvious effect on the effective thermal conductivity. The composites, however, exhibited a small degree of anisotropy with respect to the hot-pressing direction.

## 4  Discussion

### 4.1  Temperature-dependent thermal diffusivity/conductivity

The effective thermal conductivity decreased with increasing temperature, as can be seen from Fig. 4. These can be explained by the change of mean free path. Generally, thermal conductivity is

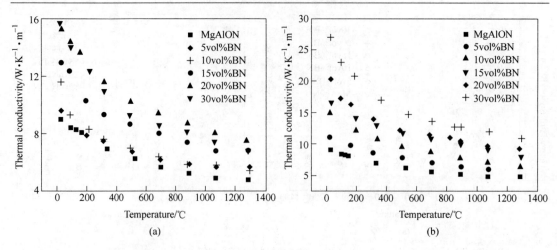

Fig. 4 Temperature dependence of effective thermal conductivities for MgAlON-BN composites with different BN contents: (a) parallel to hot-pressing directions and (b) perpendicular to hot-pressing directions

determined by lattice vibration in insulators. Debye[20] proposed that the thermal conductivity of insulator $\lambda$ can be described as the following equation using the kinetic theory of gases[11,13]:

$$\lambda = \frac{1}{3} C_v v l \tag{3}$$

where $C_v$ is the heat capacity at constant volume in J/(m$^3$ · K), $v$ the velocity of sound, and $l$ the phonon mean free path. Eq. (3) can be combined with Eq. (2). If the thermal expansion can be neglected for the solid state, then the relationship $C_v = \rho C_p$ can be used to obtain the following equation[13]:

$$\alpha = \frac{1}{3} v l \tag{4}$$

If it is assumed that $v$ is constant irrespective of temperature, the thermal diffusivity is proportional to the phonon mean free path. The phonon mean free path $l$ is limited by collisions with other phonons and with lattice defects of various kinds. These mechanisms can be assumed as additive[11]:

$$\frac{1}{l_{tot(q,s)}} = \frac{1}{l_{ph-ph(q,s)}} + \sum \frac{1}{l_{ph-def(q,s)}} \tag{5}$$

where the first term on the right-hand side refers to phononphonon collisions and the next terms to phonon scattering by faults such as point defects, dislocations, grain and phase boundaries, and the finite size of the sample. Generally, above room temperature, phonon collisions that limit the thermal conductivity must involve three or more phonons, and this process led to the phonon mean free path $l$ being proportional to $\frac{1}{T}$ (mainly for three phonon processes) or $1 \sim \frac{1}{T^2}$ for four phonon processes. The value of $l_{ph\text{-}def(q,s)}$ is independent of temperature. Thus, the thermal diffusivities may be expected to decrease with increasing temperature. At very high temperatures, the mean free path cannot be shorter than a characteristic distance between two neighboring atoms; this led to the very slow thermal diffusivity with increasing temperature at very high temperatures.

## 4.2 Effect of BN on the effective thermal conductivity

Boron nitride was always present as the hexagonal structure similar to that of graphite, which is an extremely good electrical insulator. Consequently, it has a fairly thermal conductivity up to very high temperature in its hot-pressing form. Hence, it is a useful refractory material. No evidence of reaction or formation of solid solution between the MgAlON and BN was observed in the present experiment[7]. At the same time, hexagonal-BN exhibits considerable anisotropy of its thermal conductivity/diffusivity, which results in the anisotropy for thermal diffusivity/conductivity of MgAlON-BN composites. The low Young's modulus, density properties, and nonreactive nature would lead to the porosity and defect of the composites. The thermal conductivity of BN along the $c$-axis ($k_{//}$) is much smaller than that along the $a$-axis ($k_{\perp}$). The term $k_{\perp}$ decreases with increasing of the randomness of orientation of BN particles. The term $k_{//}$, however, increases with increasing of the randomness. The BN particles are preferentially oriented with the basal plane perpendicular to the hot-pressing direction[14]. Since the thermal conductivity along the basal plane of BN is greater than that perpendicular to the basal plane, the effective thermal conductivity of MgAlON-BN composites should exhibit a similar trend, as can be seen from Fig. 5. This behavior has been demonstrated by the thermal diffusivity of hot-pressing composites of $Si_3N_4$-BN composites[15] and the SiC-BN composite[14]. The degree of anisotropy for the composites can be defined as $k$ (larger value)/$k$ (smaller value)[16]. The degree of anisotropy values containing different BN contents were calculated and are listed in Table 2. As can be seen, the value of degree of anisotropy is 1.15 to 1.71 at room temperature and increases with increasing BN content. Generally, the degree of anisotropy decreases with decreasing BN[26]. The BN particles distributed in composites are about hundreds of nanometers in thickness as well as several micrometers in length, as can be seen from Fig. 7. This results in a small degree of anisotropy.

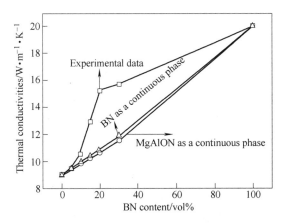

Fig. 5 Relationship between BN content and the effective thermal conductivities of MgAlON/BN composites parallel to the hot-pressing direction determined by Eucken's equation

Table 2  Degree of anisotropy for thermal conductivity for MgAlON-BN composites containing different BN content

| Temperature/℃ | BN content/vol% | | | | | |
|---|---|---|---|---|---|---|
| | 0 | 5 | 10 | 15 | 20 | 30 |
| 27 | 1 | 1.15 | 1.28 | 1.28 | 1.32 | 1.71 |
| 1300 | 1 | 1.06 | 1.16 | 1.24 | 1.26 | 1.59 |

The hexagonal BN particle also has anisotropy regarding the thermal expansion coefficient, which shows that the $\alpha$ along the $c$-axis is $7.51 \times 10^{-6} K^{-1}$, which is much larger than that along the $a$-axis, $1 \times 10^{-6} K^{-1}$. The thermal expansion coefficient of MgAlON is $5.31 \times 10^{-6} K^{-1}$. The mismatch of thermal expansion coefficients between BN and MgAlON would generate thermal stresses and consequently form some microcracks and dislocations, as can be seen from Fig. 8[7]. The thermal conductivity of a material is closely related to the microstructure of the materials. The phonons can strongly interact with point defects, line defects, planar defects, dislocations, grain and phase boundaries, and finite size of the sample, which were included in the second term on the right-hand side in Eq. (5). Therefore, the presence of the BN phase, porosity, together with the nature of the interface of BN phase will have a significant influence on the thermal conductivity of the composites[17].

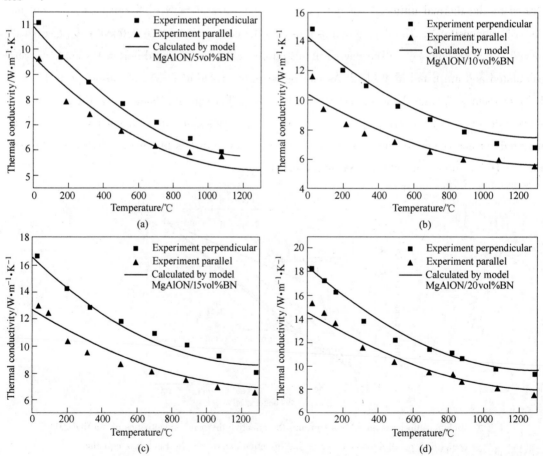

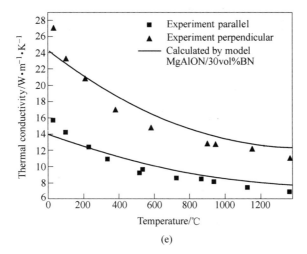

(e)

Fig. 6  Comparison of the effective thermal conductivities of MgAlON/BN composites between the experimental results and predicted lines determined by the present model:
(a) 5 vol pct BN composite, (b) 10 vol pct BN composite, (c) 15 vol pct BN composite,
(d) 20 vol pct BN composite, and (e) 30 vol pct BN composite

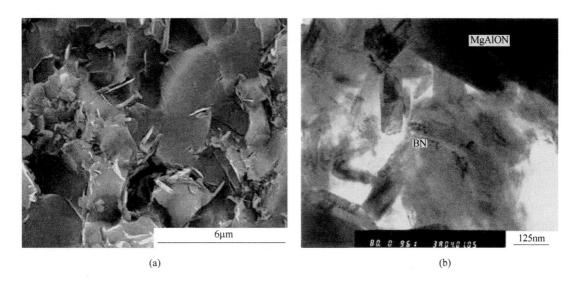

Fig. 7  SEM (a) and TEM (b) micrographs of the composites containing 15 vol pct BN particles

## 4.3 Modeling

The thermal conductivity of composites has been widely investigated theoretically over the past decades[17-22]. In most of these investigations, the inclusions in particulate composites were generally considered as being spherical[22]. Eucken[18,29] gave one equation to describe how a minor amount of a discontinuous second phase disperses uniformly in a continuous major phase. Fig. 5 shows the comparison of the experimental data with the thermal conductivities calculated by Eucken's equation. It can be seen that there is a large discrepancy between the experimental and calculated val-

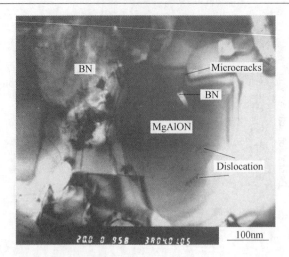

Fig. 8 TEM micrographs of microcracks and grain boundary dislocation in composites containing 15 vol pct h-BN

ues especially when the BN addition is beyond 10 vol pct. This may be due to the large amount of discontinuous second phase resulting in the defects or dislocation in composites, as can be seen in Fig. 8. Consequently, the composite ceramics always do not normally satisfy the isothermal interface condition[17]. Interfacial imperfections such as residual stresses, defects at the interface, and interfacial debonding caused by the mismatch of thermal expansion coefficient or impurities reduce the thermal conductivity. These have been extensively confirmed for the case of ceramic-matrix composites[17,23-25]. In composite ceramics, another important factor that cannot be ignored is the effect of porosity on the thermal conductivity. Dunn[26] proposed a model to calculate the thermal conductivity of multiphase ellipsoid inclusion composites on the basis of Eshelby's model[22]. The pores in composite ceramics were evaluated by treating the porosity containing materials as a special case of the inclusion composites, with a zero thermal conductivity for the inclusions. Luo considered the special case, that of a particulate composite containing ellipsoidal pores, and derived the equation that was suitable for SiC particle-reinforced MgO composite ceramics[17]. It should be pointed out that the inclusions were distributed homogenously in composites in Luo's equation. However, in the present composites, BN was distributed on the grain boundaries of MgAlON, and part of them agglomerated, as can be seen from Fig. 7. Therefore, we introduce one constant $f(F)$ to describe the BN distribution in the composites. The equation based on Luo's equation can be described as follows:

$$\frac{f(F)}{K_{CP}} = \frac{K_M}{K_C K_P} + \frac{(K_M - K_P)(K_C - K_M)}{3 K_M K_C K_P} \quad (6)$$

where $F$, $K_C$, $K_M$, $K_P$, and $K_{CP}$ are the volume fraction of BN, the thermal conductivities of the fully dense composites, the matrix, the matrix with some porosity, and the composites containing pores, respectively. The constant $f(F)$ is proportional to the volume of BN, $f(F) = 0.961 + 0.019F$ for the composites parallel to the hot-pressing direction, and $f(F) = 1.023 + 0.0363F$ for the composites perpendicular to hot-pressing direction. The term $K_C$ can be calculated by Eq. (7), which took into account the effect of interfacial thermal conditions[27]:

$$K_C = \frac{2\lambda(1-F) + \beta[3+(\lambda-1)(1+2F)]}{\lambda(2+F) + \beta[3+(\lambda-1)(1-F)]} K_M \quad (7)$$

where the parameter $\beta$ represents the interfacial thermal condition of the composites, and $\lambda = K_I / K_M$. The term $F$ is the volume fraction of the inclusions, and $K_P$ in Eq. (6) can be expressed as

$$K_P = \frac{1-F_P}{1+(\eta_2-1)F_P} K_M \quad (8)$$

where $F_P$ is the volume fraction of the porosity and $\eta_2$ is a parameter that is related to the shape of porosity. The minimum value of $\eta_2$ is 2, for ellipsoidal pores[17,28].

In order to elucidate the applicability of this model to the present work, the parameter $\beta$ value was needed. The parameter values $\beta = 3$ to 10 were used by Luo and Stevens[17] to calculate the SiC particle-reinforced MgO composites, which were also used in the present prediction. Fig. 6 shows the comparison of the experimental data with the thermal conductivity calculated by Eq. (6). It can be seen that there is little discrepancy between the experimental results and the calculated values, especially the composites containing 30 vol pct BN. The discrepancy can be attributed to many factors, as mentioned earlier. The addition of BN induced thermal stresses, dislocation, impurities, and interfacial conditions.

## 5 Conclusions

Thermal diffusivity and heat capacity of MgAlON-BN composites were measured in the present article. The thermal diffusivity decreased rapidly with increasing temperature below 900℃, while it decreased slowly with further temperature increase; this can be explained by the fact that the change of phonon mean free path gets a fixed value at very high temperatures. Effective thermal conductivity was calculated based on the thermal diffusivity and heat capacity, which exhibits a trend similar to thermal diffusivity. The addition of BN has a significant effect on thermal diffusivity/conductivity of the composites. The anisotropy of BN results in the anisotropy of composites. The degree of anisotropy is small compared with that of BN. This may be due to the agglomeration of BN at the grain boundaries.

An equation based on Luo's model was derived and was used to predict the effective thermal conductivity of the present composites. The predicted lines were in agreement with the experimental results. The small discrepancy between experimental results and predicted lines probably comes from thermal stresses, impurities, dislocation, and interfacial conditions.

**Acknowledgements:** The present authors thank Drs. Ragnhild E. Aune and Lidong Teng for their kind help during the measurement of heat capacity. Special thanks is also extended to Mr. Riad Abdul Abas and Dr. Matsushita Taishi for their constructive discussions and technical help. The authors also thank Miss Pengli Dong for her help in preparing the samples. Financial support for sample preparation from the National Nature Science Foundation of China (Grant Nos. 50332010 and 50425415) is acknowledged.

## References

[1] J. Ormerod, R. Tylor, J. Edward. Met. Technol., 1978, 4: 109-113.

[2] E. Y. Litovsky, M. Shapiro. J. Am. Ceram. Soc., 1992, 75 (12): 3425-3439.

[3] E. Y. Litovsky, M. Shapiro, A. Shavit. J. Am. Ceram. Soc., 1996, 75 (5): 1366-1376.

[4] W. Kobayashi, I. Terasaki. Appl. Phys. Lett., 2005, 87 (3): 032902-032904.

[5] T. Nobuyuki, T. Takahiro, I. Kondoh, et al. J. Ceram. Soc. Jpn., 1991, 99 (1): 87-91.

[6] L. M. Russell, L. F. Johnson, D. P. H. Hasselman, et al. J. Am. Ceram. Soc., 1987, 70: C226-C229.

[7] Z. T. Zhang, X. D. Wang, W. C. Li, et al J. Eur. Ceram. Soc., 2006.

[8] Z. T. Zhang, X. D. Wang, W. C. Li, et al. unpublished research.

[9] Z. T. Zhang, T. Matsushita, W. C. Li, et al. Metall. Mater. Trans. B, 2006, 37B.

[10] X. D. Wang, W. C. Li, S. Seetharaman. Z. Metallkd., 2002, 93 (6): 540-544.

[11] G. Grimvall. Thermophysical Properties of Materials—Enlarged and Revised Edition [M]. Amsterdam: Elsevier Science B. V., 1999.

[12] T. Sterneland, R. E. Aune, S. Seetharaman. Scand. J. Metall., 2003, 32: 233-240.

[13] M. Hayashi, A. A. Riad, S. Seetharaman. ISIJ Inter., 2004, 44 (4): 691-697.

[14] R. Rubert. J. Am. Ceram. Soc., 1984, 5: C83-C85.

[15] K. Niihara, L. D. Bentsen, D. P. H. Hasselman, et al. J. Am. Ceram. Soc., 1981, 64 (9): C117-C118.

[16] T. O. Kanai, K. Tanemoto. Jan. J. Appl. Phys., 1993, C132: 3544-3548.

[17] J. Luo, R. Stevens. J. Am. Ceram. Soc., 1997, 80 (3): 699-704.

[18] A. Eucken. Forsch. Geb. Ingenieurwes, 1932, 353: 1-16.

[19] D. W. Richarson. Modern Ceramic Engineering, 2nd ed [M]. New York: Marcel Dekker Inc., 1992.

[20] N. Nitani, T. Yamashita, T. Matsuda, et al. J. Nucl. Mater., 1999, 274, 15-22.

[21] Y. Benveniste. J. Appl. Phys., 1987, 61, 2840-2843.

[22] H. Hatta, M. Taya. Int. J. Eng. Sci., 1986, 24, 1159-1172.

[23] A. J. Reeves, R. Taylor, T. W. Clyne. Mater. Sci. Eng., A, 1991, 141: 129-138.

[24] L. M. Russell, L. F. Johnson, D. P. H. Hasselman, et al. J. Am. Ceram. Soc., 1987, 70: C226-C229.

[25] S. P. Turner, R. Taylor, F. H. Gordon, et al. J. Mater. Sci., 1993, 28: 3969-3976.

[26] M. L. Dunn, M. Taya, H. Hatta. J. Compos. Mater., 1993, 27: 1473-1519.

[27] D. P. H. Hasselman, L. F. Johnson. J. Compos. Mater., 1987, 21: 508-515.

[28] J. Luo, R. Stevens. J. Appl. Phys., 1996, 79: 9057-9063.

[29] K. Hayashi, T. M. Kyaw, Y. Okamoto. High Temp.-High Press., 1998, 30: 283-290.

(原载于 *Metallurgical and Materials Transactions B*, 2005, 37 (B): 615-621)

# 热压合成 AlON-VN 复相陶瓷的研究

赛音巴特尔  张作泰  李文超

(北京科技大学理化系,北京 100083)

**摘 要**：本文在对 V-Al-O-N 体系热力学分析的基础上热压合成了阿隆—氮化钒(AlON-VN)复相陶瓷,通过 XRD 分析证实合成了 AlON-VN 复相陶瓷。力学性能分析的结果是 AlON-VN 复相陶瓷的抗弯强度和断裂韧性都高于纯 AlON 陶瓷,并且随烧结温度的增加,AlON-VN 复相陶瓷的抗弯强度和断裂韧性先增加后减小,试样断口 SEM 分析结果证实断裂方式以沿晶断裂为主,并伴有少量的穿晶或解理断裂。

**关键词**：热力学分析；复相陶瓷；力学性能；断裂方式

## A Study on Synthesis of AlON-VN by Hot-press Sintering

Sianbaatar  Zhang Zuotai  Li Wenchao

(Department of Physical Chemistry of Metallurgy, University of Science and Technology Beijing, Beijing 100083)

**Abstract**: The AlON-VN composite ceramics are synthesized by hot-pressing technique in accordance with the thermodynamic analysis of the V-Al-O-N system. XRD results verify the credibility of the thermodynamic analysis. Mechanical properties are also measured, the results show that the strength and fracture toughness of AlON-VN composites enhance from AlON matrix. The mechanical properties increase, then degrade with a increasing sintering temperature. The fracture section morphology are examined using scanning electron microscope (SEM), the results showed that the fracture mechanism is the intercrystalline fracture accompanied some transcrystalline and cleaver fracture.

**Keywords**: thermodynamic analysis; composite; mechanical properties; fracture mechanism

氮氧化铝尖晶石(AlON)是 AlN 和 $Al_2O_3$ 的固熔体,它具有优良的光学、力学和化学性能,因而近二十多年来引起了人们广泛的关注[1-4]。在 AlON 中加入其他组分可改善其力学性能[5-7],但通过添加高硬度、高熔点的氮化物 VN 形成 AlON-VN 复相陶瓷以提高 AlON 的力学性能,尚未见报道。

本文在热力学计算的基础之上,采用一步法热压合成了 AlON-VN 复相陶瓷；用 XRD、SEM 对其进行了相分析和断口形貌观察、并对其力学性能进行了测定。

---

赛音巴特尔,2001~2005 年于北京科技大学师从李文超教授攻读博士学位。目前在首钢技术研究院工作,教授级高工,首钢技术专家。申请和获得国家专利20余项,发表论文70余篇,主编、参编专著4部,获省部级、首钢级等奖6项。

## 1 AlON-VN 复相陶瓷的热力学初步分析

AlON-VN 复相陶瓷合成工艺条件的确定必须以 AlON 和 VN 的热力学性质为依据。1964 年 Lejus[8]绘制了 AlN-$Al_2O_3$二元系的第一张相关系图。此后很多研究工作者对 AlN-$Al_2O_3$二元系进行了热力学研究和相图评估[9-14]。尽管对 Al-O-N 系的相关系进行了很多研究，但对 AlON 的热力学稳定性没有一致的认识。

图 1 是 Willems 等人的实验相关系图。它是根据在阿隆固溶区内，不同温度下，阿隆的晶格常数与其组成具有线性关系的原理由实验得出的。经王习东的计算与评估认为图 1 是目前较为合理的确定阿隆固溶区的 AlN-$Al_2O_3$二元系相图。由图 1 可以看出在 1896K 以下，AlON 不稳定将分解为，文献报道 AlON 在 650℃ 开始分解，本实验研究表明添加 VN 可以抑制 AlON 的分解；当 VN 加入量超过 16% 时，在实验研究的温度下，未见 AlON 的分解，因此本文合成 AlON-VN 复合材料，以期成为化学性能稳定的高温结构材料。

以图 1 的热力学稳定区为依据，利用拟抛物线规则和一些计算结果，可以估算不同温度下组成在 EA 和 EB 线上 AlON 的 Gibbs 生成能。利用估算的 AlON 的热力学数据及 JANAF 热力学数据，可绘制不同温度下的 Al-O-N 体系的热力学优势区图[15]。同样利用 JANAF 热力学数据，可绘制不同温度下的 V-O-N 体系的热力学优势区图[15]。图 2 是在 2073K 下 Al-O-N 和 V-O-N 二个体系叠加的热力学优势区图。

从图 2 可以看出：在 2073K 下的温度下，AlON、VN 两个化合物稳定存在的气氛条件为：$N_2$ 为 0.1MPa 和氧分压为：$10^{-13.7} \sim 10^{-14.3}$Pa 范围内。所以，在合成 AlON-VN 复合陶瓷时，必须有高纯氮气保护，并在适当的埋粉环境下进行反应烧结。

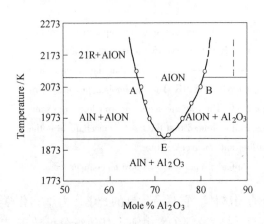

图 1 在 $N_2$ 气氛下 AlN-$Al_2O_3$二元系中 AlON 单相区图

Fig. 1 The homogeneity region of AlON in the AlN-$Al_2O_3$ system (In $N_2$ gas)

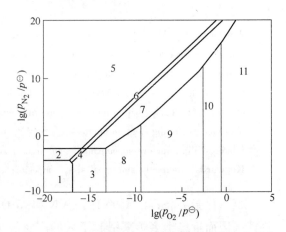

图 2 V-O-N 系及 Al-O-N 系热力学优势区叠加图（$T$ = 2073K）

Fig. 2 The overlapped potential phase diagram of V-O-N and Al-O-N system

1—Al + V；2—V + AlN；3—V + $Al_2O_3$；4—V + AlON；
5—VN + AlN；6—AlON + VN；7—VN + $Al_2O_3$；
8—VO + $Al_2O_3$；9—$V_2O_3$ + $Al_2O_3$；
10—$VO_2$ + $Al_2O_3$；11—$V_2O_5$ + $Al_2O_3$

## 2 实验部分

### 2.1 原料

实验选用的原料为 $Al_2O_3$（北京化学试剂公司生产，分析纯，粒度小于 $3\mu m$），AlN（北京钢铁研究总院生产，纯度大于 99%，粒度小于 $0.5\mu m$），$V_2O_3$（通过 $V_2O_5$ 还原制备，平均粒度 $3.9\mu m$）。

### 2.2 热压烧结合成 AlON-VN 复相陶瓷

AlON-VN 复相陶瓷合成的工艺流程为：原料配料→湿法球磨混合→干燥→预压成型→热压烧结。考虑一步合成 AlON-VN 时原料中 AlN 和 $V_2O_3$ 会发生化学反应生成 $Al_2O_3$ 和 VN，故原料的摩尔配比分别为：$Al_2O_3$ 61.5%；AlN 38.5%；$V_2O_3$ 5%。热压烧结在日本富士电波株式会社生产的 HIGH-MULTI5000 多功能热压炉中进行。烧结气氛为 $N_2$，烧结压力为 25kPa，合成温度分别为 1750℃、1800℃、1850℃、1900℃，保温时间为 3 小时。

### 2.3 性能测试

在 SJ-TA 型三轴剪力仪上采用三点弯曲法测定试样的抗折强度，加载速率为 0.5mm/min。跨距 $L$ = 30mm，试样尺寸：3mm×4mm×35mm。

采用单边开槽三点弯曲法，即 SENB 法来测定断裂韧性，切口深度为试样高度的 0.4~0.6 倍，加载速率为 0.05mm/min，跨距 $L$ = 30mm，试样尺寸：4mm×6mm×35mm。

### 2.4 相分析与断口形貌观测

在 MAC Science Co. Ltd 生产的 X 射线衍射仪上对合成的 AlON-VN 复相材料物相组成进行分析；在日本公司生产的 JSM-6460LV 型扫描电子显微镜下观测 AlON-VN 复相材料的断口形貌。

## 3 实验结果

### 3.1 XRD 分析结果

XRD 分析结果如图 3 所示，热压合成的 AlON-VN 复相材料的主晶相为 AlON，次晶相为 VN，微量相有待于进一步分析。

### 3.2 烧成温度对 AlON-VN 复合材料的室温抗弯强度的影响

烧成温度对 AlON-VN 复合材料的室温抗弯强度的影响如图 4 所示：AlON-VN 复相陶瓷的抗折强度要高于纯 AlON 陶瓷，纯 AlON 陶瓷的常温抗折强度为 248MPa[16]；随着烧成温度的增高，AlON-VN 复相陶瓷的室温抗折强度先增大后减小。这是由于升温使基体密实，强度增加；而温度过高，出现晶粒长大使强度下降。

### 3.3 烧成温度对 AlON-VN 复合材料断裂韧性的影响

烧成温度对 AlON-VN 复合材料断裂韧性的影响如图 5 所示：AlON-VN 复相陶瓷的断

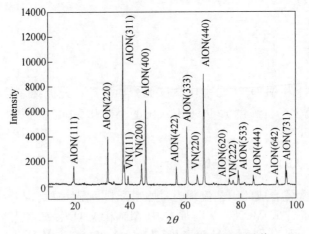

图 3　1850℃ 下 AlON-VN 试样的 XRD 图谱
Fig. 3　X-Ray diffraction of AlON-VN composites

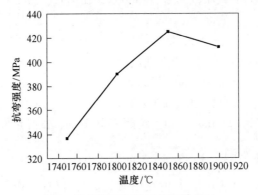

图 4　抗折强度与烧成温度的关系
Fig. 4　Correlation between strength and sintering temperature

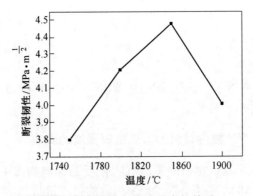

图 5　断裂韧性与烧成温度的关系
Fig. 5　Correlation between fracture toughness and sintering temperature

裂韧性也高于纯 AlON 陶瓷，这是由于加入了高弹性模量、高硬度、高强度的 VN，增加了复合陶瓷的断裂表面能的缘故。本实验条件下测得的纯 AlON 陶瓷的断裂韧性为 3.75MPa·m$^{1/2}$，随着烧成温度的增高，AlON-VN 复相材料的显为微结构发生变化，其断裂韧性也出现了先增大后又减小的变化规律。

### 3.4 AlON-VN 复相材料的断口形貌

AlON-VN 复相材料的断口形貌如图 6 所示：AlON-TiN 复合材料在不同烧结温度（(a) 1800℃、(b) 1850℃、(c) 1900℃）的断口形貌图。从中可以看出，其断裂方式均为混合断裂，即以沿晶断裂为主，并伴有少量的穿晶或解理断裂；随着温度的增加，晶粒经过了逐渐长大的过程，当温度达到 1850℃时对应强度最大值出现穿晶断裂，而当温度达到 1900℃时，晶粒尺寸长大，并伴有解理断裂，强度下降。

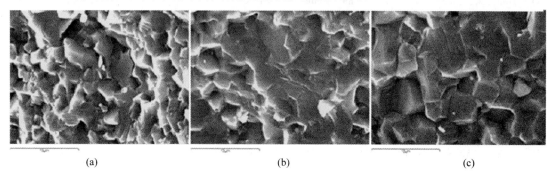

图 6 AlON-VN 复相陶瓷试样的断口形貌
Fig. 6 Fracture microstructure of AlON-VN composites
(a) 1800℃; (b) 1850℃; (c) 1900℃

## 4 结论

在热力学分析的基础上，确定了 AlON-VN 复相陶瓷合成的工艺条件，热压合成了 AlON-VN 复相陶瓷；XRD 的分析表明，其主晶相为 AlON，次晶相为 VN。对 AlON-VN 复相陶瓷力学性能的分析表明，AlON-VN 复相陶瓷的抗弯强度断裂韧性都高于纯 AlON 陶瓷，随着烧成温度的增高，AlON-VN 复相陶瓷的室温抗折强度和断裂韧性出现先增大而后又减小的变化规律，1850℃时达到最大值。通过 SEM 对 AlON-TiN 复合陶瓷在不同烧结温度的断口形貌图观察，其断裂方式与材料的力学性能相一致。

### 参 考 文 献

[1] Corbin N D. Aluminium oxynitride spinel: A Review [J]. J Eur Ceram Soc, 1989, (5): 143-154.
[2] James W, McCauley, Corbin N D. Phase relations and reaction sintering of transparent cubic aluminium oxynitrede spinel [J]. J Amer Ceram Soc, 1979, 62: 476-479.
[3] Earl K, Graham, Cmunly W. Elastic Properties of Polycrystalline Aluminum Oxynitride Spinel and Their Dependence on Pressure, Temperature, and Composion [J]. J of Amer Ceram Soc, 1988, 71 (10): 807-812.
[4] Bertil Forslund, Jie Zheng. On the thermal and oxidation behaviour of aluminium oxynitride powder [J]. Chinese J Mater research, 1998, 12 (5): 597-606.
[5] Djenkal D, Goeuriot D. SiC-reiforcement of an $Al_2O_3$-AlON composite [J]. Journal of the European Ceramic Society, 2000, 20 (12): 2585-2590.

[6] Shimpo, Akihiro. AlON and its composite ceramics [J]. Journal of the Ceramic society of Japan, 1992, 100 (4): 504-508.

[7] 张作泰. AlON-TiN 复合材料的性能与结构 [D]. 北京：北京科技大学, 2003.

[8] Lejus A M. Sur la formation a hautc temperature de spinelles non stoetcchiometriques et de phase derives [J]. Rev. Hautes tempr. et Refrac., 1964, 1: 53-95.

[9] McCauley J W, Corbin N D. NATO ASI Sci., Ser. E, 65 [Prog. Nitrogen Ceram.] 1983: 111-118.

[10] Kaufman L. Calculation of quasibinary and quasiternary oxtnitride systems-3 [J]. Calphad, 1979, 3 (4): 275-291.

[11] Hillert M, Jonsson S. Thermodynamic calculation of Al-O-N system [J]. Z. Metallkd, 1992, 83 (10): 714-719.

[12] Willems H X. Thermodynamic of Alon 2: Phase relations [J]. J. Eur. Ceram. Soc., 1992, 10: 339-346.

[13] Dumitrescu L, Sundman B. A thermodynamic reassessment of the Si-Al-O-N system [J]. J. Am. Ceram. Soc. 1995, 78 (15): 239-247.

[14] Qiu C, Metselaar R. Phase relations in aluminum carbide- aluminum nitride- aluminum oxide system [J]. J. Am. Ceram. Soc., 1997, 80 (8): 2013-2020.

[15] 王习东, 李文超. 热压合成 AlON 陶瓷的研究 [J]. 硅酸盐学报, 2001, 29 (1): 31-34.

[16] 王习东. AlON 及 MeAlON 陶瓷的性能与结构 [D]. 北京：北京科技大学, 2001.

（原载于《技术科学与工程》, 2004, 4 (3): 218-221）

# Synthesis of TiN/AlON Composite Ceramics

Wang Xidong  Gao Lichun  Li Guobao  Li Wenchao

(Department of Physical Chemistry, University of Science and
Technology Beijing, Beijing 100083)

**Abstract:** The synthesis process of TiN/AlON composite ceramics was studied, the thermodynamics, mechanical properties and microstructures of TiN/AlON have also been investigated. The TiN/AlON composite ceramics has been synthesized by both hot-pressing and pressureless sintering. The characterizations of the material synthesized were analyzed with XRD (X-ray diffraction) and TEM (transmission electronic microscope). The density and toughness strength of TiN/AlON are $3.57g/cm^3$ and $4.74 MPa \cdot m^{1/2}$, respectively. The bending strength was measured at both room temperature and high temperatures and the results are 399MPa (room temperature), 406MPa (1073K), 417MPa (1273K) and 323MPa (1573K). Pattern Recognition (PR) and Artificial Neural Network (ANN) were used to optimize the parameters and to predict the expected values. A proper parameter for pressureless sintering of TiN/AlON has been obtained and testified, the parameters are temperature (1978K), AlN/ (AlN + $Al_2O_3$) ratio (0.22), MgO (4.7%) and $TiO_2$ (7.2%).

**Keywords:** synthesis; TiN; AlON; composite ceramics

## 1 Introduction

AlON, a solid solution of AlN and $Al_2O_3$, has drawn the attentions of scientists during the past 40 years because of its excellent optical, mechanical and chemical properties[1-3]. However, AlON is not thermodynamically stable below some certain temperature (about 1913K). Therefore, investigations of introducing fourth element, such as Si, Mg, Ti and R ($R$ = Ce, Pr, Nd, Sm or Eu), into Al-O-N system have been done [4-16]. A series of solid solutions and compounds, such as O'-SiAlON, β'-SiAlON, MgAlON, TiAlON and RAlON have been found.

Ti-Al-O-N system makes it a very interesting field because it contains several important refractory and ceramics such as $Al_2O_3$, AlN, AlON, $TiO_2$, TiN and $Al_2TiO_5$. D. S. Persser et al reported the preliminary Ti-Al-O-N Quaternary phase diagram shown as reference[12]. The syntheses of TiAlON and TiAlON/Spinel composite ceramics have been published by Mocellin[13], Perera[14], Heystek[15] and Hoyer[16]. But, the thermodynamic analysis has not been reported on this system. The processing parameters of TiAlON and TiN/AlON synthesis have not been investigated.

This paper is going to investigate the synthesis process of TiN/AlON composite ceramics. The thermodynamics, mechanical properties and microstructures of TiN/AlON have also been investigated.

---

This work was financially supported by the National Natural Science Foundation of China (No. 50074004).

## 2 Raw materials and experimental procedures

$TiO_2$, AlN and $Al_2O_3$, which are all analytically pure and provided by Fangda Ceramics Company of Tsinghua University, were selected as raw materials to prepare TiN/AlON ceramics. Both pressureless sintering and hot-pressing sintering was carried out. The experimental procedures are as follows.

(1) Appropriate amounts of the fine powders of $Al_2O_3$, AlN and $TiO_2$ were mixed by ball milling in ethanol medium for 24h.

(2) For hot-pressing sinter, the mixture thus obtained, was pre-pressed into a compact cylinder. The cylinder was placed in a graphite mould covered with boron nitride and was sintered at 2073K, at 25MPa, in $N_2$ gas flow for 3h. The mole fraction of $Al_2O_3$, AlN and $TiO_2$ are listed in Table 1.

**Table 1  Raw material compositions (mole fraction) for the synthesis of TiN/AlON (%)**

| No. | $Al_2O_3$ | AlN | $TiO_2$ |
|---|---|---|---|
| 1 | 69.3 | 29.7 | 1.0 |
| 2 | 67.9 | 29.1 | 3.0 |
| 3 | 65.8 | 28.2 | 6.0 |
| 4 | 63.0 | 27.0 | 10.0 |

(3) For pressureless sintering, MgO was added as sintering agent, the mixtures were pressed into a rectangle bar and buried at AlN powder at $N_2$ gas flow at different temperature and time. The compositions of starting materials are listed in Table 2.

(4) XRD of the synthesized material was carried out.

(5) The properties, crystalline characteristics and microstructures of the synthetic TiN/AlON materials were determined.

**Table 2  Compositions (mole fraction) and strength of TiN/AlON**

| No. | $T$/K | AlN/(AlN + $Al_2O_3$) | Composition/% | | $\sigma_b$/MPa |
| | | | MgO | $TiO_2$ | |
|---|---|---|---|---|---|
| 1 | 1923 | 0.20 | 0 | 3 | 249 |
| 2 | 1973 | 0.20 | 5 | 8 | 304 |
| 3 | 2023 | 0.20 | 10 | 20 | 115 |
| 4 | 1923 | 0.25 | 5 | 20 | 196 |
| 5 | 1973 | 0.25 | 10 | 3 | 216 |
| 6 | 2023 | 0.25 | 0 | 8 | 84 |
| 7 | 1923 | 0.30 | 10 | 8 | 143 |
| 8 | 1973 | 0.30 | 0 | 20 | 273 |
| 9 | 2023 | 0.30 | 5 | 3 | 195 |
| 10 | 1973 | 0.25 | 20 | 0 | 191 |
| 11 | 1973 | 0.25 | 20 | 8 | 173 |
| 12 | 2023 | 0.25 | 20 | 8 | 203 |

Continued Table 2

| No. | T/K | AlN/(AlN+Al$_2$O$_3$) | Composition/% MgO | Composition/% TiO$_2$ | $\sigma_b$/MPa |
|---|---|---|---|---|---|
| 13 | 2023 | 0.25 | 0 | 3 | 138 |
| 14 | 1973 | 0.25 | 0 | 8 | 163 |
| 15 | 1973 | 0.25 | 0 | 13 | 296 |
| 16 | 2023 | 0.40 | 0 | 0 | 164 |
| 17 | 1973 | 0.30 | 0 | 0 | 146 |
| 18 | 2023 | 0.25 | 0 | 0 | 121 |
| 19 | 2023 | 0.15 | 0 | 0 | 257 |
| 20 | 1973 | 0.40 | 0 | 0 | 143 |
| 21 | 1973 | 0.25 | 0 | 0 | 105 |
| 22 | 1973 | 0.15 | 0 | 0 | 296 |

## 3 Experimental results

The TiN/AlON composite ceramics has been synthesized by both hot-pressing and pressureless sintering. The materials synthesized were introduced to X-ray diffraction analysis and parts of the results are shown in Figs. 1-3.

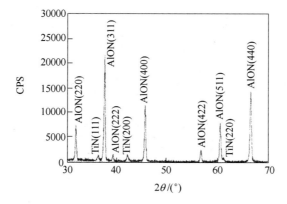

Fig. 1   X-ray diffraction pattern for No. 1 sample

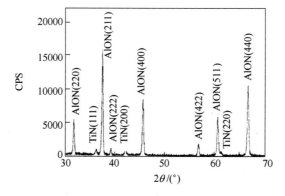

Fig. 2   X-ray diffraction pattern for No. 2 sample

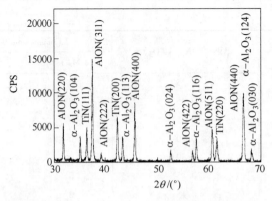

Fig. 3  X-ray diffraction pattern for No. 4 sample

Physical properties of synthesized materials by hot pressing sintering were measured. The density of TiN/AlON, with the composition of No. 2, is $3.57 g/cm^3$. The toughness strength is $4.74 MPa \cdot m^{1/2}$. The bending strength was measured at both room temperature and high temperatures, and the results are 399MPa (room temperature), 406MPa (1073K), 417MPa (1273K) and 323MPa (1573K).

The microstructures of TiN/AlON were examined by TEM (transmission electronic microscope). The TEM photo and diffraction patterns are shown in Fig. 4. As can be seen from Fig. 4, TiN and AlON grains interlaced each other. AlON particles are equiaxed grains, while TiN grains are colum-

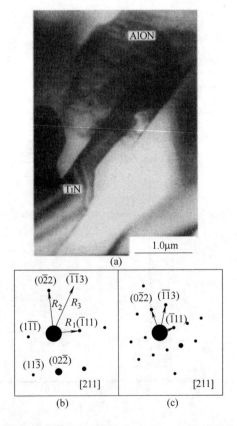

Fig. 4  TEM Photos (a) and diffraction pattern of TiN (b)/AlON (c)

nar and tabular crystals. The two interlaced phases could make the composite material strengthened and toughened.

TiN/AlON composite ceramics has been synthesized by pressureless sintering at different conditions. Threepoint strength of the materials synthesized by pressureless sintering has been measured (by preparing the samples in 3mm × 4mm × 40mm cuboid bars). The results are listed in Table 2.

## 4 Discussion

### 4.1 Thermodynamic analysis

Perera et al has investigated the ohase relations in Ti-Al-O-N system. As can be seen in Fig. 5[12], a spinel phase of AlON which dissolved with TiN were found. However the thermodynamic properties of the spinel phase have not been reported, and the synthesis parameters are not determined.

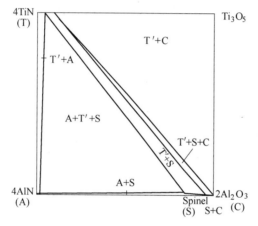

Fig. 5 Phase diagram of Ti-Al-O-N system at 2073K

In $TiO_2$-$Al_2O_3$-AlN system, the following reactions usually occurred and produced $Al_2TiO_5$. The reaction along with the standard Gibbs free energy are shown as follows:

$$TiO_2(s) + Al_2O_3(s) = Al_2TiO_5(s) \qquad (1)$$
$$\Delta_r G^\ominus = -6047 - 11.65T \ (J)$$

In equation (1), $TiO_2$ and $Al_2O_3$ can react and produce $Al_2TiO_5$. However, at the experimental conditions (at high $O_2$ pressure), the equilibrium product is TiN. In nitrogen gas, if oxygen partial pressure is very low, the following reaction happens,

$$4Al_2O_5(s) + 4AlN(s) = 4TiN(s) + 6Al_2O_3(s) + O_2(g) \qquad (2)$$
$$\Delta_r G^\ominus = 331585 + 160.73T \ (J/mol)$$

Figs. 1-3 indicate that TiN phase existed in all the TiN/AlON samples, hence, the following reactions are also reasonbly to be considered.

$$TiO_2(s) + 0.5N_2(g) = TiN(s) + O_2(g) \qquad (3)$$
$$\Delta_r G^\ominus = 578035 + 56.2T \ (J/mol)$$
$$6TiO_2(s) + 8AlN(s) = 6TiN(s) + 4Al_2O_3 + N_2(1) \qquad (4)$$

$$\Delta_r G^\ominus = -541270 + 115.8T \; (\text{J/mol})$$

$$4TiO_2(s) + 4AlN(s) = 4TiN(s) + 2Al_2O_3(s) + O_2(g) \tag{5}$$

$$\Delta_r G^\ominus = 307400 + 114.1T \; (\text{J/mol})$$

In the above reactions, $O_2$ will react with carbon and produce CO as in the existance of carbon, thus $O_2$ pressure is actually very low. As the Gibbs free energy of reaction (3) is very high, this reaction was hardly occurred. However, at very low partial pressure of oxygen, reaction (4) and (5) were possibly happened. This was confirmed by Figs. 1-3. As reaction (4) or (5) proceeded, $Al_2O_3$ produced, and excess $\alpha$-$Al_2O_3$ was detected by XRD (as shown in Fig. 4).

In Fig. 1, there is a small amount of TiN phase detected even the content of $TiO_2$ is very low. This indicates that the stability region of TiAlON in Fig. 5 is very narrow. However, in the present of TiN, the lattice paremeter of AlON was changed a little, that might suggest that a very small amount of TiN be dissolved into AlON. The reason is that the lattice parameters of a solid solution often shifted when it's composition changed or when it dissolved other elements.

## 4.2 Optimization of synthesis technology

In order to reduce the cost of synthesis of TiN/AlON spinel composite ceramics, pressureless sintering is required. With $TiO_2$, AlN and $Al_2O_3$ as raw materials and MgO as sintering agent, pressureless sintering at different conditions have been done, three-point bending strength of the samples have been measured, the results are shown in Table 2.

To determine a proper sintered parameters is very important. In the present work, Pattern Recognition (PR) and Artificial Neural Network (ANN) were used to optimize the parameters and to predict the expected values.

The three-point bending strengths are chosen to be the criterion target (those are defined "good" when their strength are higher than 200MPa, or defined "bed" when their strength are less than 200MPa). The sinter temperature, $TiO_2$ mole content, MgO mole content and the ratio of AlN and ($Al_2O_3$ + AlN) are chosen as 4 features (characteristic parameters). The analysis results are shown in Fig. 6.

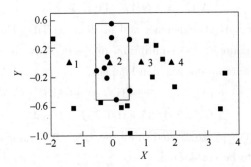

Fig. 6  Pattern recognition of TiN/AlON synthesis
●—good point; ■—bad point; ▲—inverse point

The circle marks in Fig. 6 present the "good points" (strength higher than 200MPa), the square marks present the "bad points" (strength lower than 200MPa). As can be seen from Fig. 6, "good" region and "bad" region was obviously separated. For any giving sintered parameters, the expected strength can be predicted by PR analysis. In addition, the proper parameter can also be determined by a reverse projection. The corresponding parameters of an inverse projection of rtiangle marks 1 in Fig. 6 are shown in Table 3.

Table 3  Inversed parameters and their targets by ANN

| No. | $T$/K | AlN/(AlN+Al$_2$O$_3$) | $x_{MgO}$/% | $x_{TiO_2}$/% | $\sigma_b$(ANN)/MPa |
| --- | --- | --- | --- | --- | --- |
| 1 | 1940 | 0.20 | 12.8 | 10.8 | 167 |
| 2 | 1978 | 0.22 | 4.7 | 7.2 | 294 |
| 3 | 1999 | 0.24 | 0.4 | 4.6 | 193 |
| 4 | 2021 | 0.26 | 0 | 2.4 | 134 |

In order to verify the results of PR, ANN was also employed to analyze the experimental results in Table 2. The training time is 300000, the error is 0.3%, and the results are shown in Fig. 7. As can be seen, the predicted value is in accordance with the experimental data.

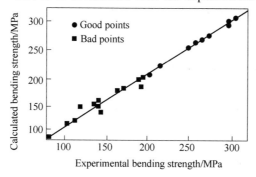

Fig. 7  The results of ANN for TiN/AlON synthesis

ANN was also used to predict the value of the parameters (by inverse projection) in Table 3, and the results are shown in Table 3. As can be seen, the results of PR and ANN are in agreement. Among the 4 parameters, the expected value of the mark 2, which is the only one located in the "good" region, was predicted (by ANN) to be 294MPa (higher than 200MPa), the predicted values of the other three (located in "bad region") are all less than 200MPa. As also can be seen, a proper parameter for pressureless sinter of TiN/AlON is temperature (1978K), AlN and (AlN and Al$_2$O$_3$) ratio (0.22), MgO (4.7%) and TiO$_2$ (7.2%).

With the parameters optimized above, TiN/AlON ceramics was synthesized. The three-point bending strength of the above material was measured to be 292MPa, only 12MPa lower than the best value. Therefore, the optimized parameters are quite reasonable.

# 5  Conclusion

The TiN/AlON composite ceramice has been synthesized by both hot-pressing and pressureless sintering. The characterizations of the material synthesized were analyzed with XRD and TEM. The

TEM photo indicated that TiN and AlON grains interlaced each other. The two interlaced phases could make the composite material strengthened and toughened.

Physical properties of synthesized materials by hot-press sintering were measured. The density and toughness strength of TiN/AlON are 3.57g/cm$^3$ and 4.74MPa·m$^{1/2}$ respectively. The bending strength was measured at both room temperature and high temperatures, and the results are 399MPa (room temperature), 406MPa (1073K), 417MPa (1273K) and 323MPa (1573K).

Pattern Recognition (PR) and Artificial Neural Network (ANN) were used to optimize the parameters and to predict the expected values. For any giving sintered parameters, the expected strength can be predicted by PR of ANN analysis. In addition, the proper parameter can also be determined by a reverse projection. A proper parameter for pressureless sintering of TiN/AlON is obtained and testified, the parameters are temperature (1978K), AlN and (AlN and $Al_2O_3$) ratio (0.22), MgO (4.7%) and $TiO_2$ (7.2%).

## References

[1] N. D. Corbin. Aluminium oxynitride spinel: a review [J]. J. Eur. Ceram. Soc., 1989, 5: 143.

[2] H. X. Willems, M. M. R. Hendrix, G. D. With, et al. Thermodynamic of alon 2: phase relations [J]. J. Eur. Ceram. Soc., 1992, 10: 339.

[3] L. Kaufman. Calculation of quasibinary and quasiternary oxynitride system-3 [J]. CALPHAD, 1979, 3: 275.

[4] Lucia Dumitrescu, Bo Sundman. A thermodynamic reassessment of the Si-Al-O-N system [J]. J. Am. Ceram. Soc., 1995, 78: 239.

[5] Mats Hillert, Stefan Jonsson. Thermodynamic calculation of Si-Al-O-N system [J]. Z. Metallkd., 1992, 183: 720.

[6] K. H. Jack. Review: sialons and related nitrogen ceramics [J]. J. Mater. Sci., 1976, 11: 1135.

[7] S. Weiss, P. Greil, L. J. Gauckler. The system Al-Mg-O-N [J]. Commun. Amer. Ceram. Soc., 1982: C68.

[8] H. X. Willems, R. Metsellar. Thermodynamics of $AlON_3$: stabilization of AlON with MgO [J]. J. Eur. Ceram. Soc., 1993, 12: 43.

[9] A. Granon, P. Goeuriot, F. Thevenot, et al. Reactivity in $Al_2O_3$-AlN-MgO system-the MgAlON spinel phase [J]. J. Euro. Ceram. Soc., 1994, 13: 365.

[10] A. Granon, P. Goeuriot, F. Thevenot. Aluminum Magnesium Oxynitride: a new transparent spinel ceramic [J]. J. Euro. Ceram. Soc., 1995, 15: 249.

[11] W. Y. Sun, J. Chen, Y. Jia, et al. Phase relations of $R_2O_3$-AlN-$Al_2O_3$ (R = Ce, Pr, Nd and Sm) system [J]. Chinese Sci. (A) (in Chinese), 1990, 9: 990.

[12] D. S. Perera. Phase relationships in the Ti-Al-O-N system [J]. Br. Ceram. Trans. J., 1990, 89: 57

[13] A. Mocellin, G. Bayer. Chemical and microstructural investigations of high-temperature interactions between AlN and $Al_2O_3$ [J]. J. Mat. Sci., 1985, 20: 3687.

[14] D. S. Perera, M. E. Boweden. Nitrogen-containing aluminum titanate [J]. J. Mat. Sci., 1991, 26: 1585.

[15] H. Heystek. Ceramic research at the U. S. Bureau of mines [J]. Am. Ceram. Bull, 1988, 67: 1345.

[16] J. L. Hoyer, J. P. Bennett, K. J. Liles. Properties of TiAlON/spinel ceramic composites [J]. Ceram. Eng. Sci. Proc., 1990, 11: 1423.

(原载于 *Journal of University of science and Technology Beijing*, 2003, 10(1): 49-53)

# The Effect of $Al_2O_3$ (Mul.) on Phase Compositions of O'-Sialon Ceramics

## Zhong Weibin　Li Wenchao　Zhong Xiangchong

(University of Science and Technology Beijing, Beijing 100083)

**Abstract:** O'-Sialon ceramics are formed by reaction sintering of silicon nitride and silica with $Al_2O_3$ ($Y_2O_3$) additive. The phase compositions of O'-Sialon ceramics are highly affected by addition of $Al_2O_3$ (mul.) and are measured by X-ray diffraction technique. The range of additions which could impair the structure of O'-Sialon ceramics is given in this article.

## 1 Introduction

Silicon oxynitride ceramics are formed by reaction-sintering of silicon nitride and silica with $Al_2O_3$ ($Y_2O_3$) to promote densification[1]. O'-Sialon can be represented by the formula $Si_{2-x}Al_xO_{1+x}N_{2-x}$. Hot pressed $Si_2N_2O$ exhibits a very good oxidation resistance up to 1600℃[2]. The addition of 50 mol% $Al_2O_3$ into $Si_2N_2O$ under hot. Pressing (23MPa) and 1700℃ for 2h in 0.1MPa nitrogen atmosphere can achieve a better strength value (RT, 492MPa)[3]. But according to the reaction,

$$SiO_2 + Si_3N_4 \Longleftrightarrow 2Si_2N_2O$$
$$(2-x)Si_2N_2O + xAl_2O_3 \Longleftrightarrow 2Si_{2-x}Al_xO_{1-x}N_{2-x} \quad [4]$$

the limit at 1800℃ is 10 mol% $Al_2O_3$ in the O'-sialon general formula of equation, $x \leq 0.2$. Thus the material has a lot of $Al_2O_3$ out of the range of solid solubility with silicon oxynitride. If equilibrium is reached, since the line between $Al_2O_3$ and O'-Sialon is divided by χ-phase (1700℃) (see Fig. 1), it is impossible for $Al_2O_3$ and O'-Sialon to occur simultaneously. So the structure of O'-Sialon ceramics will be impaired. In order to investigate the reaction sintering of various ratio of $Al_2O_3$ (Mul.) to O'-Sialon, the full research work is placed on the phase compositions.

## 2 Experimental procedure

The starting materials are: (1) silicon nitride (with major impurities 2.08%); (2) silica (99.8% purity); (3) alumina (mullite) (99.5% purity); and yttria (99.9% purity). The composition of starting mixes lies on the line of O'-Sialon and $Al_2O_3$ (mul.), see Table 1 and Fig. 1. Mixed powders of $Y_2O_3$, $SiO_2$, $Si_3N_4$ and $Al_2O_3$ were ball-mined. The resulting slurries were then dried at 115℃. Powder mixes were pressed at 200MPa before firing in nitrogen atmosphere for 1h at 1650℃.

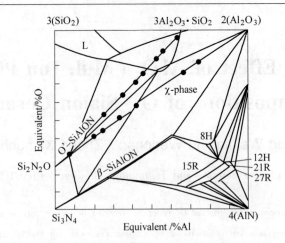

Fig. 1  Behavior diagram of subsystem, illustrated as a reciprocal salt system at 1700-1730℃

Note: (1) The points on the line of O'-Sialon and mullite represent the compositions of starting mixes, M0, M15 ($Y_3M_{15}$), M20 ($Y_3M_{20}$), M30, M45, M55, M65, M75, M80, M91;

(2) The points on the line of O'-Sialon and $Al_2O_3$ represent $Y_3A_{15}$ ($Y_8A_{15}$), $Y_3A_{20}$ ($Y_8A_{20}$), $Y_3A_{30}$ ($Y_8A_{30}$), $Y_8A_{65}$, $Y_8A_{75}$.

**Table 1  The composition of starting mixes**

| wt% | $Y_2O_3$ | $SiO_2$ | "$Si_3N_4$" | Mul. | $Al_2O_3$ | YS/O' |
|---|---|---|---|---|---|---|
| M0 | 7.8 | 26.1 | 56.5 | 0 | 9.5 | 0.15 |
| M15 | 6.63 | 22.19 | 48.03 | 15 | 8.08 | 0.15 |
| M20 | 6.24 | 20.88 | 45.2 | 20 | 7.6 | 0.15 |
| M30 | 5.54 | 18.68 | 40.39 | 28.6 | 6.79 | 0.15 |
| M45 | 4.35 | 14.54 | 31.39 | 44.44 | 5.28 | 0.15 |
| M55 | 3.51 | 11.75 | 25.43 | 55 | 4.28 | 0.15 |
| M65 | 2.73 | 9.14 | 19.78 | 65 | 3.33 | 0.15 |
| M75 | 1.95 | 6.53 | 14.13 | 75 | 2.38 | 0.15 |
| M80 | 1.56 | 5.22 | 11.3 | 80 | 1.90 | 0.15 |
| M91 | 0.17 | 2.37 | 5.14 | 90.22 | 0.86 | 0.15 |
| Y3M15 | 2.38 | 21.51 | 52.28 | 15 | 8.84 | 0.05 |
| Y3M20 | 2.24 | 20.24 | 49.20 | 20 | 8.32 | 0.05 |
| Y3A15 | 2.38 | 21.51 | 52.28 | 0 | 8.84 + 15 | 0.05 |
| Y3A20 | 2.24 | 20.24 | 49.2 | 0 | 8.32 + 20 | 0.05 |
| Y3A30 | 1.96 | 17.71 | 43.05 | 0 | 7.28 + 30 | 0.05 |
| Y8A15 | 6.63 | 22.19 | 48.03 | 0 | 8.08 + 15 | 0.15 |
| Y8A20 | 6.24 | 20.88 | 45.2 | 0 | 7.60 + 20 | 0.15 |
| Y8A30 | 5.54 | 18.68 | 40.39 | 0 | 6.79 + 28.60 | 0.15 |
| Y8A65 | 2.73 | 9.14 | 19.78 | 0 | 3.33 + 65 | 0.15 |
| Y8A75 | 1.95 | 6.53 | 14.13 | 0 | 2.38 + 75 | 0.15 |

Notes: (1) YS/O' represents the weight ratio $Y_2Si_2O_7$/O'-Sialon.

(2) The extent of solid solubility of alumina in $Si_2N_2O$ was assumed to be $x = 0.2$.

(3) In order to investigate the effect of liquid on the final compositions, the $Y_2O_3$ were added in different amount.

## 3 Results and discussion

### 3.1 The effect of different amount of mullite additions on the XRD results

From the appearance of the sintered samples, in the range of mullite addition: 0-15%, 75%-90.90%, the samples have no bubbles. But out of the range, there are obvious bubbles and the structure are degraded. The amount of $Y_2O_3$ has no obvious influence on the appearance of samples, whereas the amount of $Y_2O_3$ could not be added too much.

In the synthesis of sialon materials, the process of densification is believed to be liquid sintering[5]. $\chi$-phase can become liquid above the temperature of 1600℃, and the liquid can promote densification, but afterwards crystallize β'-$Si_3N_4$[6]. The influence of $\chi$-phase can be interpreted from the transient sintering. The reactions are:

$$5Si_3N_4 + 21SiO_2 + 14Al_2O_3 = 4Si_9Al_2O_2N_5$$
$$(\chi\text{- phase})$$
$$Si_3N_4 + \chi\text{- phase} = β'\text{- }Si_3N_4$$

But from solid-liquid equilibrium phase diagram of $Si_3N_4$-AlN-$SiO_2$-$Al_2O_3$ system, we can estimate that melting point of $\chi$-phase is higher than 1730℃. $\chi$-phase can not turn into liquid above the temperature of 1600℃. Fig. 2 shows that with only a small amount of $Y_2O_3$ content, there is a larger liquid forming region, this may be the real source of liquid pahse. So the previous work might not consider the influence of $Y_2O_3$, and the source of liquid was attributed to the $\chi$-phase.

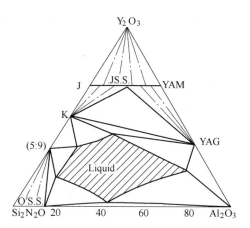

Fig. 2  Isothermal section of the system
$Si_2N_2O$ - $Al_2O_3$ - $Y_2O_3$ at 1550℃

Quantitative analysis of the O'-phase% was done from measurements of the peak intensities of the $Si_2N_2O$ (020) and (110) and β'-$Si_3N_4$ (101) and (200), using the following relation:

$$O'\text{- phase\%} = \frac{O'(020) + O'(110)}{O'(020) + O'(110) + β'(101) + β'(200)} \times 100\%$$

From Fig. 3, it can be concluded that:

(1) Before the range of bubbles, O'-phase% is 67.40% ~ 80.36%; after the range of bub-

bles, O'-phase% is <41.38%, thus χ-phase may react with $Si_3N_4$ to form $β'$-$Si_3N_4$, leading to the loss of O'-phase.

(2) Although the addition of M75 is with 75% mullite, the XRD result shows only 30% mullite, half of which lost. The addition of M15 with 15% mullite, the XRD result shows nomullite.

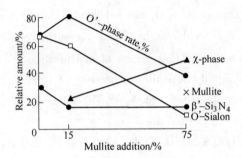

Fig. 3  X-ray diffraction analysis result with varied mullite addition ( YS/O' =0.15 )

## 3.2  The effect of different amount of $Al_2O_3$ additions on the XRD results

The appearance of the sintered samples demonstrates that in the range of $Al_2O_3$ addition: <15%, >75%, the samples have no bubbles. Out of the range of that, there are obvious bubbles. The amount of $Y_2O_3$ has no obvious influence on the appearance of samples.

From the Fig. 4, it can be concluded that:

(1) Before the range of bubbles, O'-phase has a bigger value 59.46%-67.6%, but has no χ-phase; After this range, samples have no O'-phase, but a lot of corundum phase.

(2) In the region of less $Al_2O_3$ addition, χ-phase has a slightly increase; after that, χ-phase has a highly decrease which is in accordance with phase diagram. The whole range of composition can not occur simultaneously for O'-Sialon and $Al_2O_3$.

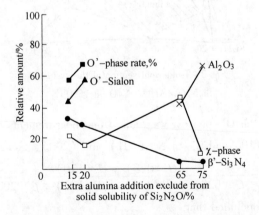

Fig. 4  X-ray diffraction analysis result with varied alumina addition

## 4 Conclusions

(1) In the range of mullite additions: 0-15%, 75%-90.90%; $Al_2O_3$ addition: <15%, >75%, the samples have no bubbles. Out of the range, there are obvious bubbles. There is a larger liquid forming region with only a small amount of $Y_2O_3$ content in the isothermal section of the system $Si_2N_2O$-$Al_2O_3$-$Y_2O_3$, and this may be the reason for bubbles.

(2) The whole range of composition can not occur simultaneously for O'-Sialon and $Al_2O_3$ (not include the solid solubility with $Si_2N_2O$).

(3) $\chi$-phase can degrade the structure of O'-Sialon ceramics, but a little of $\chi$-phase can promote densification.

### References

[1] J. Sjöberg et al. J. Eur. Ceram. Soc., 1992, 10: 41-50.
[2] J. Persson et al. J. Am. Ceram. Soc., 1992, 12: 3377-3384.
[3] M. Ohashi et al. J. Mat. Sci., 1991, 26: 2608-2614.
[4] M. B. Trigg. et al. J. Mat. Sci., 1988, 23: 481-487.
[5] Z. K. Huang et al. Ceramics International, 1984, 10: 14.
[6] S. P. Li. Technique of Special Ceramics (in Chinese). Wuhan Industry University, 1990.

(原载于《中国耐火材料（英文版）》, 1995, (1): 26-29)

# 氮化硼对锆刚玉莫来石材料力学性能及显微结构的影响

赵海雷　李文超　钟香崇　王　俭

（北京科技大学理化系，北京　100083）

**摘　要**：研究了锆刚玉莫来石—氮化硼复合材料的显微结构及力学性能，结果表明：在锆刚玉莫来石基质中引入氮化硼，降低材料的抗折强度，但可提高断裂韧性。这是氮化硼的微裂纹增韧作用所致。氮化硼的编织状结构可阻碍晶界的滑移，降低材料高温强度的衰减率。材料内生成的针状 $9Al_2O_3 \cdot 2B_2O_3$，在断裂过程中产生拔出效应，有利于力学性能的提高。

**关键词**：氮化硼；锆刚玉莫来石；复合材料；显微结构；力学性能

# Effect of BN Additive on Mechanical Properties and Microstructure of Corundum-mullite-zirconia Materials

Zhao Hailei　Li Wenchao　Zhong Xiangchong　Wang Jian

(Department of Physicochemiscry, University of Science and Technology Beijing, Beijing 100083)

**Abstract:** The micostructure and mechanical properties of corundum-mullite-zirconia/boron nitride composites have been studied. The results indicate that addition of boron nitride to corundum-mullite-zirconia material leads to decrease in modulus of rupture and increase in fracture toughness, which may be mainly attributed to the presence of microcracks. BN interwoven microstructure will prevent grain boundary slip and reduce the attenuation rate of high temperature strength. Needle shaped $9Al_2O_3 \cdot 2B_2O_3$ crystals formed in the material will create a pull-out effect when the material approaches to fracture, which may be beneficial to the enhancement of mechanical properties.

**Keywords:** boron nitride; corundum-mullite-zirconia; composite materials; microstructure; mechanical property

# 1　引言

为了适应钢铁冶炼和其他高温技术发展的多层次要求，近年来对耐火材料的质量、性能也要求向高新发展。综合考虑材料的高温使用性能，如高温强度、抗蠕变、抗热震、抗

---

赵海雷，1989～1993年于北京科技大学师从李文超教授攻博士学位。目前在北京科技大学工作，教授，博士生导师。获得教育部新世纪人才（2007）、北京市师德先进个人（2012）、宝钢优秀教师奖（2013）等荣誉称号，发表论文220余篇。

侵蚀、抗氧化等，氧化物与非氧化物的复合材料可能是一类很有发展前途的耐火材料[1,2]。在这方面锆刚玉莫来石与SiC复合的研究工作已经开展并取得较好的结果[3,4]，正在继续进行。锆刚玉莫来石是一种综合性能较好的材料，国内外已进行较详细的研究[5,6]，但它的抗热震性不很理想。BN具有良好的抗热震性、耐腐蚀性和高温下的化学稳定性，但它的力学性能不甚佳，抗氧化性也较差。为了改善锆刚玉莫来石的抗热震性，提高BN的力学性能和抗氧化性，我们拟对两者进行复合，通过综合韧化与强化，以期使材料的各方面性能得以互补。因此开展了对锆刚玉莫来石—氮化硼系复合材料的显微结构和高温性能的系统研究，整个工作分为两个部分进行：(1) 以氧化物为基料，引入10%~30%（以质量计，下同）BN进行复合；(2) 以BN为基料，引入10%~30%锆刚玉莫来石进行复合。对两组试样都进行了制备工艺、显微结构和高温性能（力学性能、抗热震性、抗热疲劳、抗氧化性等）的研究。本文仅对锆刚玉莫来石—氮化硼材料的显微结构及力学性能的研究结果进行报道。

## 2 试样的制备与试验方法

### 2.1 配方的选择

根据以往的研究结果，刚玉/莫来石比取3:1，$ZrO_2$质量含量为15%时，锆刚玉莫来石材料的性能比较优越[5]。在这种比例的氧化物基料中引入BN（质量含量）10%、20%、30%，观察不同BN引入量对锆刚玉莫来石（以下简称ZCM）材料性能的影响。同时外加5%（以质量计，下同）$CaCO_3$和$TiO_2$助烧剂以促进烧结。为了进行性能的对比，制备了ZCM试样。各试样的配方组成及密度如表1所示。

表1 试样的化学组成与密度
Table1 Chemical composition and density of samples

| Sample | Chemical composition w/% | | | | | | Bulk density $\rho_v$/g·cm$^{-3}$ | Relative density d/% |
| --- | --- | --- | --- | --- | --- | --- | --- | --- |
| | $ZrO_2$ | $Al_2O_3$ | Mullite | BN | $CaCO_3$ | $TiO_2$ | | |
| ZCM(C) | 15 | 63.75 | 21.25 | — | 5 | — | 3.87 | 98.0 |
| ZCM-BN$_1$(C) | 13.5 | 57.38 | 19.12 | 10 | 5 | — | 3.48 | 94.5 |
| ZCM-BN$_2$(C) | 12 | 51 | 17 | 20 | 5 | — | 3.01 | 87.6 |
| ZCM-BN$_3$(C) | 10.5 | 44.63 | 14.87 | 30 | 5 | — | 2.51 | 77.6 |
| ZCM(T) | 15 | 63.75 | 21.25 | — | — | 5 | 3.86 | 97.6 |
| ZCM-BN$_1$(T) | 13.5 | 57.38 | 19.12 | 10 | — | 5 | 3.43 | 93.2 |
| ZCM-BN$_2$(T) | 12 | 51 | 17 | 20 | — | 5 | 2.89 | 84.0 |
| ZCM-BN$_3$(T) | 10.5 | 44.63 | 14.87 | 30 | — | 5 | 2.53 | 72.8 |

### 2.2 试样的制备

原料用工业纯$Al_2O_3$、合成莫来石、m-$ZrO_2$及六方BN粉，以乙醇为介质湿磨至粉料粒度小于5μm，以聚乙二醇为结合剂，在100MPa下干压成型，然后在300MPa下等静压成型，试条尺寸为6mm×8mm×40mm，于钼丝炉中氮气氛下烧成，1750°C保温2h。

## 2.3 测试方法

采用三点弯曲法测定试样的抗折强度,试样的尺寸为 3mm×4mm×36mm,加载速率为 0.4mm/min,高温测试加载速率为 0.05mm/min。采用单边开槽三点弯曲法测定断裂韧性,试样尺寸 4mm×6mm×36mm,加载速率为 0.06mm/min。用 XRD 结合 EDAX 及 TEM 衍射斑点确定物相的成分,用 SEM 及 TEM 观察材料的显微结构特征及各晶相间的结合方式。

# 3 材料的力学性能

## 3.1 材料的室温力学性能

由图 1、图 2 可见,随着 BN 含量的增加,试样的抗折强度均明显下降。任何 BN 含量的试样,以 $TiO_2$ 为助烧剂时的抗折强度均低于以 $CaCO_3$ 为助烧剂时的抗折强度。各试样的断裂韧性随 BN 含量而变化,在 BN 为 10% 处出现峰值,此后随 BN 含量的增加,材料的韧性明显下降。

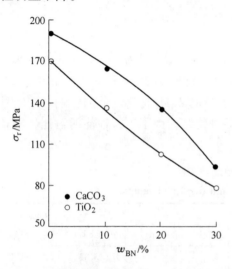 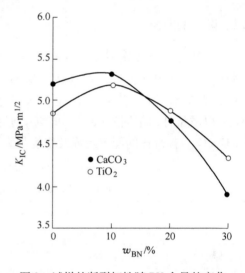

图 1 试样的抗折强度随 BN 含量的变化  图 2 试样的断裂韧性随 BN 含量的变化
Fig. 1 Modulus of rupture vs BN content for samples  Fig. 2 Fracture toughness vs BN content for samples

## 3.2 材料的高温力学性能

空气中不同温度下材料的抗折强度如图 3 所示。由图中可以看出:ZCM-BN 复合材料的抗折强度随温度的变化具有同一规律,即随温度的上升,抗折强度基本保持不变;当温度达到 900°C 后,抗折强度显著下降。未加 BN 的 ZCM(C)试样随温度变化,强度出现了峰值,然后迅速降低,属于文献[7]提出的 I 类曲线特征。$N_2$ 气中 ZCM-BN 材料的强度—温度曲线是典型的 I 类曲线(图 4),在 1000°C 下均出现了强度峰值。以上现象估计与 BN 的易氧化性有关。据文献[8]报道,BN 烧结体于 700°C 时已可观察到一定的氧化失重。BN 在加热过程中被氧化为 $B_2O_3$,$B_2O_3$ 熔点极低(450°C),高温下成为液相,分布于各颗粒间,有损于材料的高温抗折强度。在抗氧化研究中发现:ZCM-BN 材料开始氧化的温度为900℃左右，正是如此，影响了ZCM-BN材料空气中的强度-温度曲线特征。由图5可以看出：BN的引入，有利于材料高温强度保持率的提高。

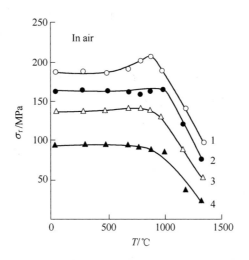

图3　空气中ZCM-BN材料的强度-温度曲线

Fig. 3　Strength-temperature curves for ZCM-BN samples in air

1—ZCM（C）；2—ZCM-$BN_1$（C）；3—ZCM-$BN_2$（C）；4—ZCM-$BN_3$（C）

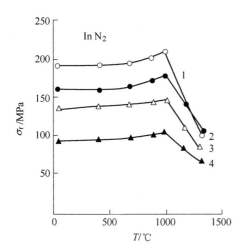

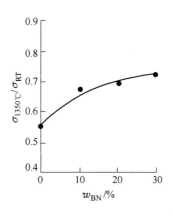

图4　$N_2$气中ZCM-BN材料的强度-温度曲线

Fig. 4　Strength-temperature curves for ZCM-BN samples in $N_2$

1—ZCM（C）；2—ZCM-$BN_1$（C）；3—ZCM-$BN_2$（C）；
4—ZCM-$BN_3$（C）

图5　试样的高温强度保持率

Fig. 5　Retention of strength at high-temperature vs BN content for ZCM-$BN_5$ samples

## 4　材料的显微结构特征

根据扫描电镜观察结果，锆刚玉莫来石—氮化硼复合材料的一般显微结构特征是：刚玉以多边大颗粒存在，大小为10~15μm；莫来石为10μm左右的短柱状，分布于刚玉的颗粒间，构成基质相；$ZrO_2$ 以圆颗粒均匀分布于基体中，大小为 $5\sim 8\mu m$; BN 易聚集联生为花束状分布于基体中（图6）。烧结过程中材料内生成的 $9Al_2O_3\cdot 2B_2O_3$ 则以类似短纤维的细条状分布于整个材料基体间（图7）。$TiO_2$ 在烧结过程中与 $Al_2O_3$ 生成的 $Al_2O_3\cdot TiO_2$ 存在于 $Al_2O_3$ 颗粒附近，形成晶间裂纹。

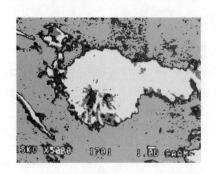 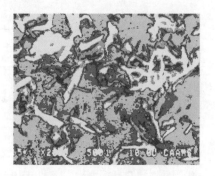

图6 ZCM-$BN_1$（C）试样的扫描电镜形貌  
Fig. 6 SEM photograph for sample ZCM-$BN_1$ (C) (5000×)

图7 ZCM-$BN_3$（C）试样中的 $9Al_2O_3\cdot 2B_2O_3$ 扫描电镜形貌  
Fig. 7 SEM photograph showing $9Al_2O_3\cdot 2B_2O_3$ in sample ZCM-$BN_3$ (C) (2000×)

通过透射电镜观察各颗粒间的接触情况，未发现 BN 的花束状聚集体，而是片状 BN 颗粒的编织状结构（图8）。扫描电镜中观察到的花束状结构，经透射电镜衍射斑点分析，确定为 BN。同时还可以观察到：各氧化物颗粒间大多为直接结合，大颗粒的 $Al_2O_3$ 构成骨架，莫来石与 $ZrO_2$ 充填于其中。在 BN 颗粒与氧化物颗粒周围有非晶质薄膜（$CaO\cdot Al_2O_3\cdot 2SiO_2$）的存在（图9），也有一部分晶界处的非晶质相已经微晶化。这是由 $CaCO_3$ 高温下与莫来石和 $Al_2O_3$ 作用所致。

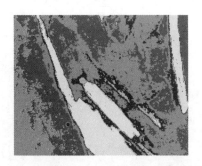 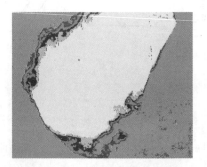

图8 ZCM-$BN_3$（T）试样中 BN 的编织状结构（TEM）  
Fig. 8 TEM photograph showing the interwoven microstructure of BN in sample ZCM-$BN_3$ (T) (60000×)

图9 ZCM-$BN_2$（C）试样中的 BN 与氧化物间的非晶质薄膜（TEM）  
Fig. 9 TEM photograph showing amorphous film between BN and oxide grains in sample ZCM-$BN_2$ (C) (30000×)

通过透射电镜还可以观察到：BN 颗粒周围有裂纹的偏转、弯曲及分叉效应（图10），氧化物颗粒间还存在一定的晶界位错（图11）。

图 10 ZCM-BN$_1$（C）试样中 BN 引起的
微裂纹（TEM）

Fig. 10  TEM photograph showing microcrack induced by BN in sample ZCM-BN$_1$（C）（35000×）

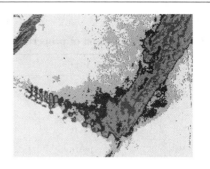

图 11 ZCM-BN$_3$（C）试样中的晶界
位错（TEM）

Fig. 11  TEM photograph showing grain boundary dislocation in sample ZCM-BN$_3$（C）（30000×）

在对材料的断口进行观察时，发现有 $9Al_2O_3 \cdot 2B_2O_3$ 的拔出痕迹（图12左上角）。

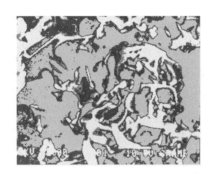

图 12 ZCM-BN$_2$（C）试样中 $9Al_2O_3 \cdot 2B_2O_3$ 的拔出效应（SEM）

Fig. 12  SEM photograph showing the pull-out effect of $9Al_2O_3 \cdot 2B_2O_3$ in sample ZCM-BN$_2$（C）（2000×）

## 5 显微结构与力学性能的关系

由于片状 BN 本身热膨胀系数的各向异性，一个方向的热膨胀系数比大多数氧化物高，而另一个方向则低，在材料内易产生较多的环形及放射状微裂纹，这些微裂纹将会影响到材料的力学性能。ZCM-BN 材料随 BN 引入量的增加，抗折强度下降。这主要是由于 BN 本身的弹性模量比较低（热压 BN 的弹性模量为 18GPa 左右），随 BN 引入量的增加，试样的弹性模量也逐渐下降（表2），从而导致强度的下降。同时随 BN 引入量增加，试样内微裂纹的增多和气孔率的上升，也是强度下降的一个原因。以 $TiO_2$ 为添加剂的试样，由于生成了易产生裂纹的 $Al_2O_3 \cdot TiO_2$，从而使其强度低于以 $CaCO_3$ 为添加剂的试样。

另外，由 BN 引起的微裂纹，在材料断裂过程中，将使主裂纹偏转、弯曲以及分叉（图10），消耗扩展功，阻止裂纹的前进，起到微裂纹增韧的作用，因而随 BN 的引入，材料的韧性值上升（图2）。但 BN 超过 10% 以后，材料的相对密度降低，材料内存在较多的空隙，此时微裂纹增韧的作用也就有限了。

表 2 ZCM-BN（C）系列试样的性能对比

Table 2 Comparison of properties for ZCM-BN (C) samples with high relative density

| Property | ZCM（C）<br>1750°C, 2h | ZCM-BN$_1$（C）<br>1750°C, 2h | ZCM-BN$_2$（C）<br>1800°C, 2h | ZCM-BN$_3$（C）<br>1850°C, 2h |
|---|---|---|---|---|
| $\rho_v/g \cdot cm^{-3}$ | 3.87 | 3.48 | 3.27 | 3.04 |
| $d/\%$ | 98.0 | 94.5 | 95.0 | 93.2 |
| $\sigma_f/MPa$ | 189.0 | 164.0 | 153.7 | 124.9 |
| $K_{IC}/MPa \cdot m^{1/2}$ | 5.20 | 5.80 | 6.48 | 6.38 |
| $E/GPa$ | 72.31 | 66.90 | 35.21 | 24.20 |

为了证实 BN 的微裂纹增韧作用，增加各试样间的可比性，在 $N_2$ 气氛下，提高 ZCM-BN$_2$(C) 和 ZCM-BN$_3$(C) 试样的烧成温度分别达 1800℃（保温 2h）和 1850℃（保温 2h），使各试样的相对密度都达到 93%~95%，然后再测其强度和韧性，其结果如表 2 所示。由表可见，随 BN 含量的增加，其断裂韧性也明显提高。

回顾前人的工作，将 60%（$ZrO_2$ + mullite）与 40% BN 进行复合，通过热压烧结后，其抗折强度也仅为 22.84MPa[9]。锆刚玉莫来石与 SiC 进行复合时，常压烧结后，其抗折强度也小于 120MPa[4]。而我们研究的锆刚玉莫来石与 BN 的复合材料，在常压下烧结后，其强度高于以往材料的主要原因可能是，材料内生成的细条状 $9Al_2O_3 \cdot 2B_2O_3$，使结构强化，在断裂时产生拔出效应所致。

随 BN 的引入，试样高温强度保持率的提高应归功于 BN 的编织状结构的存在（图 8）。当材料在高温下受到应力的作用时，这种编织状结构可以阻碍晶界的滑移，从而提高材料的高温抗折强度。因而随 BN 引入量增加，编织状结构增多，增强效果显著。

在升温过程中，材料内产生的位错会活化晶格，促进烧结的进行，但在冷却以后，这种位错的出现是晶界的薄弱环节，当受到应力作用时，其能量很易释放，成为材料的破坏源。因而必须控制位错的产生，保证材料的力学性能。

# 6 结论

（1）在锆刚玉莫来石—氮化硼材料中，氧化物颗粒构成骨架，片状 BN 构成编织状结构，呈花束状分布于基体中。氧化物颗粒间为直接结合。BN 与氧化物颗粒通过晶界非晶质薄膜结合。

（2）在锆刚玉莫来石基料中引入 BN，降低材料的室温及高温抗折强度，但对材料的韧性有一定程度的改善，这是 BN 所引起的微裂纹作用的结果。

（3）随 BN 引入量增加，材料的高温强度衰减率下降，其主要原因之一是 BN 的编织状结构可以阻碍晶界的滑移。

（4）材料中生成的 $9Al_2O_3 \cdot 2B_2O_3$，呈细条状分布于基体中，在材料断裂时有拔出效应，从而改善材料的力学性能。

## 参 考 文 献

[1] 钟香崇. 我国耐火材料基础研究的一些进展 [J]. 硅酸盐学报，1990，18 (5)：450.

[2] 钟香崇. 展望九十年代我国耐火材料的发展 [J]. 耐火材料, 1993, 27 (2): 63.
[3] 李庭寿, 钟香崇. 锆刚玉莫来石-碳化硅复合材料的显微结构 [J]. 硅酸盐学报, 1991, 19 (3): 241.
[4] 李庭寿. 锆刚玉莫来石-碳化硅复合材料的高温力学性能和抗热震性的研究 [D]. 北京: 北京科技大学, 1990.
[5] 刘新彧, 孙庚辰, 钟香崇. 反应烧结锆刚玉莫来石系熟料的高温性能及其应用 [J]. 硅酸盐通报, 1992, 11 (4): 4.
[6] Orange G, Fantozzi G, Cambier F, et al. High temperature mechanical properties of reaction-sincered mullite/zirconia and mullite/alumina/zirconia composites [J]. J Mater Sci., 1985, 20: 2533.
[7] 钟香崇. 碱性耐火材料热机械性质 [M]. 北京: 重工业出版社, 1957.
[8] 素木洋一, 刘达权, 陈世兴合译. 硅酸盐手册 (中译本) [M]. 北京: 轻工业出版社, 1982.
[9] Shaffer G W. Zirconia mullite/boron nitride composites [P]. US Patent: No. 4904626, 1990-2-27.

(原载于《硅酸盐学报》, 1994, 22 (5): 509-515)

# $ZrO_2$-CaO-BN 复合材料的研制

黄绵亮　李文超　钟香崇

（北京科技大学理化系，北京　100083）

**摘　要**：对目前使用的浸入式水口进行了分析。分析了 $ZrO_2$-CaO-BN 复合材料的热力学可行性。用热压烧结合成该复合材料。用 XRD 对其作了相分析，结果与热力学分析一致。

**关键词**：浸入式水口；$ZrO_2$-CaO-BN；复合材料；热力学分析

## Development of $ZrO_2$-CaO-BN Composite

Huang Mianliang　Li Wenchao　Zhong Xiangchong

(Physics and Chemistry Department, University of Science & Technology Beijing, Beijing 100083)

**Abstract**: The submerged nozzles used presently were investigated. The feasibility of $ZrO_2$-CaO-BN composite was discussed with thermodynamic method. The phase composition of the composite prepared in multifunctional furnace was identified by XRD method. The result was consistent with the thermodynamic calculation.

**Keywords**: submerged nozzle; $ZrO_2$-CaO-BN; composite; thermodynamic analysis

## 1　引言

在连铸用耐火材料中，浸入式水口是重要的功能耐火材料，它应具有好的耐热冲击性、良好的强度、耐钢水侵蚀性和对保护渣的耐蚀性[1]。随着冶金技术的发展，其材质和结构也发生了一系列的变化，材质上从熔融石英质到铝碳质、铝锆碳质，直到目前广泛使用的锆钙碳质浸入式水口，使用性能得以提高，使用寿命得以延长。但由于含碳而限制了其在低碳钢和超纯净钢中的使用，又由于添加了 $SiO_2$，不可避免会导致结瘤与堵塞。

## 2　浸入式水口的研究状况

为了防止水口的堵塞，在结构方面，主要采用复合式、阶梯式和狭缝式水口。这能起到一定的作用，但也存在缺点，如结构复杂，制造比较困难。另外，对于含二氧化硅和碳的狭缝式浸入式水口，在使用过程中，由于二氧化硅的还原和碳的氧化，气孔率会增加，造成吹气不稳的现象。

因此，要较好地解决堵塞问题，必须从材质方面入手。主要采用氮化物、氧化锆、氧化钙和其他非氧化物制作的含石墨质浸入式水口,诸如 BN-AlN-C、$ZrO_2$-C-$ZrB_2$、$ZrO_2$-SiC-C、CaO-$ZrO_2$-C 以及 Sialon-C 等[2]。这些材质的防止堵塞机理主要是靠本身或与钢中的夹杂物形成易被钢水冲刷的低熔相,从而达到减少 $Al_2O_3$ 黏附的目的。

而在含有 $SiO_2$ 的铝碳质浸入式水口中,在 1550℃ 时由于发生氧化还原反应,孔内的 $SiO_2$ 随着铸造的进行会与水口中的碳发生氧化还原反应,生成 SiO(g)、CO(g),其与钢中的 [Al] 反应,生成 $Al_2O_3$(s) 而沉积在水口内壁,造成堵塞。

碳会与水口中的氧化物发生氧化还原反应,使气孔率增加。同时,由水口内部向内壁扩散的 $Al_2O$(g)、SiO(g)、CO(g) 等气体与钢中的 [Al] 反应,生成 $Al_2O_3$(s) 而沉积在水口内壁,造成堵塞。

氮化硼有两种晶体结构:六方氮化硼(HBN)和立方氮化硼(CBN)。六方氮化硼系白色松散粉末,其晶体结构与石墨相似,具有良好的润滑性、导热性,并有"白石墨"之称。它的化学稳定性好,不受熔融的玻璃、硅、硼、非氧化物炉渣、熔盐和金属熔体腐蚀[3]。为了提高浸入式水口的抗热震性,拟引入氮化硼来代替 C 和 $SiO_2$,制成 $ZrO_2$-CaO-BN 复合材料,以期改善锆钙质水口的相关性能。

## 3 新材质浸入式水口的热力学基础

对于拟研制的 $ZrO_2$-CaO-BN 质浸入式水口,从性能结构上是可行的,但尚且不知该复相材料的化学稳定性,为此对这个体系进行了热力学计算。绘制了 Ca-O-N、B-O-N、Zr-O-N 热力学参数状态图,并对三者进行叠加,如图 1 所示。

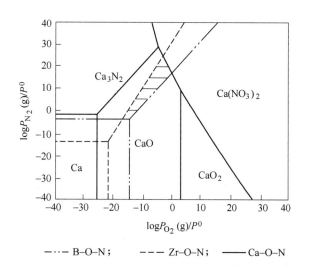

—·—·— B-O-N; ——— Zr-O-N; ——— Ca-O-N

图 1 Ca-O-N、B-O-N、Zr-O-N 的热力学参数状态叠加图

Fig. 1 Combined status diagram of thermal dynamic parameters of Ca-O-N、B-O-N and Zr-O-N systems

如图 1 所示,划横线区域为 $ZrO_2$-CaO-BN 三者在 1550℃ 的共存区,说明这三者能在此温度下共存。为了确定在 1550℃ 时这三者的反应,由化合物的标准生成 Gibbs 自由能可

算得相关反应的自由能值[4]：

$$CaO + ZrO_2 = CaZrO_3 \tag{1}$$

$$\Delta_f G^\ominus = -38.52 \text{kJ} \cdot \text{mol}^{-1}$$

$$3/2ZrO_2(s) + 5BN(s) = 3/2ZrB_2(s) + B_2O_3(l) + 5/2N_2(g) \tag{2}$$

$$\Delta_r G^\ominus = 1322.68 \text{kJ} \cdot \text{mol}^{-1}$$

$$4CaO(s) + 2BN(s) = Ca_3N_2(s) + CaB_2O_4 \tag{3}$$

$$\Delta_r G^\ominus = 494.56 \text{kJ} \cdot \text{mol}^{-1}$$

$$CaZrO_3(s) + 23/6BN(s) = ZrB_2(s) + 2/3B_2O_3(l) + 1/4Ca_3N_2(s) + 1/4CaB_2O_4(l) + 5/3N_2(g) \tag{4}$$

$$\Delta_r G^\ominus = 1043.95 \text{kJ} \cdot \text{mol}^{-1}$$

由式（1）~式（4）中的计算可知：在实际条件下 BN 不参与反应；BN 和 CaO、$ZrO_2$、$CaZrO_3$ 能稳定存在。

## 4 $ZrO_2$-CaO-BN 复合材料的制备

由 $ZrO_2$-CaO 系二元相图（图2）可知，当 CaO 的摩尔分数大于50%时，二者能合成 $CaZrO_3$，当 CaO 的摩尔分数大于50%时，则有部分游离的 CaO 存在，这时会由于 CaO 的水化而使材料劣化。因此，选择 CaO 的摩尔分数略小于50%，使其生成较为稳定的 $CaZrO_3$。将 $ZrO_2$、CaO、BN 三者共磨后于热压炉上烧结，制成复合材料。对烧制后的复合材料进行 XRD 相成分分析，结果如图3所示。从图中可以看出，主晶相为锆酸钙，还有四方氧化锆和氮化硼，这与理论计算相符。

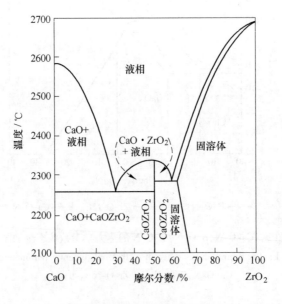

图2 $ZrO_2$-CaO 系二元相图

Fig. 2 Phase diagram of $ZrO_2$-CaO system

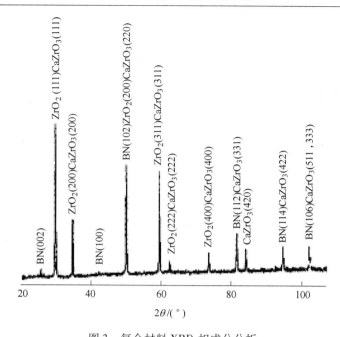

图 3 复合材料 XRD 相成分分析

Fig. 3　XRD pattern of the compound material

## 5　结论

（1）碳或二氧化硅的存在能引发浸入式水口中 $Al_2O_3$ 的结瘤。

（2）绘制了 Ca-O-N、B-O-N、Zr-O-N 三个三元系叠加的热力学参数状态图，并根据化学反应等温方程式计算了实际条件下相关反应的自由能值，证实了合成 $ZrO_2$-CaO-BN 复合材料热力学是可行的。

（3）热压烧结制备了该复合材料，并用 XRD 对其作了相分析，确定其主晶相为锆酸钙，还有四方氧化锆和氮化硼，这与理论分析一致。

### 参 考 文 献

[1] 张利华，译. 防止吹气式浸入水口的裂纹 [J]. 国外耐火材料，1996，20（2）：15-20.
[2] 仲维斌. O'-Sialon-$ZrO_2$ 高温力学性能和显微结构的研究 [D]. 北京：北京科技大学，1995.
[3] 赵海雷. 锆刚玉莫来石-BN 复合材料的高温力学性能和显微结构的研究 [D]. 北京：北京科技大学，1993.
[4] 梁英教，车荫昌. 无机物热力学数据手册 [M]. 沈阳：东北大学出版社，1996.

（原载于《耐火材料》，1999，33（5）：257-258，264）

# Synthesis Mechanism of Silicon Nitride Obtained from Silica Reduction

Zhuang Youqing (庄又青)[1]　Wang Jian (王俭)[1]　Li Wenchao (李文超)[1]
Zhong Xiangchong (钟香崇)[2]

(1. Department of Physical Chemistry, University of Science and Technology Beijing, Beijing 100083;
2. Luoyang Institute of Refractories Research, Luoyang 471039)

**Abstract:** The reduction mechanism of silica was studied by means of thermodynamic calculations and kinetic consideration for the system $SiO_2$-C-$N_2$. $Si_3N_4$ formation is independent of $SiO_2$-C contact condition, but dependent on CO partial pressure. It was considered that silica reduction would be divided into direct and indirect reduction. Silica reacts directly with solid carbon to form silicon monoxide only at initial stage of silica reduction, $SiO_2(s) + C(s) \rightarrow SiO(g) + CO(g)$. While silica can be reduced by carbon monoxide in the indirect reduction stage, $SiO_2(s) + CO(g) \rightarrow SiO(g) + CO_2(g)$, which predominates over the silica reduction in this system. The reaction $C(s) + CO_2(g) \rightarrow 2CO(g)$ can provide a thermodynamic driving force and replenish carbon monoxide required for the reaction $SiO_2(s) + CO(g) \rightarrow SiO(g) + CO_2(g)$ in the presence of carbon.

## 1 Introduction

The carbothermal reduction and nitridation of silica is a commercial route for preparing high purity ultrafine $Si_3N_4$ powders. Extensive investigations have been undertaken on its synthesis mechanism and technique by many ceramists[1-4]. The synthesis mechanism has generally been considered that $SiO_2$ can be directly reduced by solid carbon to form an intermediate SiO, followed by nitridation to obtain $Si_3N_4$. The reduction-nitridation is through the following two-step reactions:

$$SiO_2(s) + C(s) = SiO(g) + CO(g) \qquad (1)$$

$$3SiO(g) + 3C(s) + 2N_2(g) = Si_3N_4(s) + 3CO(g) \qquad (2)$$

Some publications, however, have considered $SiO_2$ would be reduced by carbon monoxide[5,6]:

$$SiO_2(s) + CO(g) = SiO(g) + CO_2(g) \qquad (3)$$

but the details of the reduction process have not been investigated up to now.

In the present paper, the mechanism for silica reduction will be discussed in more detail.

## 2 Thermodynamic considerations

Supposing reaction (3) is a main step for reducing $SiO_2$ in the system $SiO_2$-C-$N_2$, the equilibrium

---

庄又青，1987~1991 年于北京科技大学师从李文超教授攻读博士学位。目前在北京仪尊时代科技有限公司工作，任总经理。

$$CO_2(g) + C(s) \Longrightarrow 2CO(g) \tag{4}$$

would coexist in the presence of carbon. The standard Gibbs free energies of the reactions above are listed in Table 1[7].

**Table 1  Standard free energies of the reaction concerned**

($\Delta G^{\ominus}(T) = a + bT\log T + cT$ (J/mol))

| Reaction | $a$ | $b \times 10^3$ | $c \times 10^3$ |
|---|---|---|---|
| (1) | 681573.6 | 26986.8 | -434508.4 |
| (2) | -824666.4 | -105060.2 | 659607.6 |
| (3) | 535970.4 | 26986.8 | -260035.6 |
| (4) | 145603.2 | — | -174472.8 |

The free energies are given for real condition

$$\Delta G_1(T) = \Delta G_1^{\ominus}(T) + RTP_{CO} \cdot P_{SiO} \tag{5}$$

$$\Delta G_3(T) = \Delta G_3^{\ominus}(T) + RT\frac{P_{CO_2} \cdot P_{SiO}}{P_{CO}} \tag{6}$$

$$\Delta G_4(T) = \Delta G_4^{\ominus}(T) + RT\frac{P_{CO}^2}{P_{CO_2}} \tag{7}$$

Substituting (7) into (6) gives $\Delta G_1(T) = \Delta G_3(T)$, it signifies that the thermodynamic conditions are the same for reaction (1) and (3) when reaction (4) is in the equilibrium state.

On the other hand, if reaction (3) and (4) are in the equilibrium state simultaneously, thus

$$\Delta G_3(T) = 0, \Delta G_4(T) = 0$$

The $CO_2$ partial pressure in the system can be represented by a function of the form

$$\ln P_{CO_2} = -\frac{1}{RT}(1217544 + 53.977\log T - 694.54T + 2RT\ln P_{SiO}) \tag{8}$$

Substituting the equilibrium SiO vapor pressure, $\log P_{SiO} = -16750/T + 1.75\log T + 190$ kPa[8] into Eq. (8) thus gives the solid line in Fig. 1. The thermodynamic valid region for reaction (3) is above the line in the diagram, and the region below is not valid. However, reaction (4) is much faster than reaction (3) in the real system, hence it is impossible that the two reactions are in equilibrium state simultaneously, i.e. reaction (4) is in equilibrium state, but (3) is not. The atmosphere of the system is controlled by reaction (4).

Solving the equations $\Delta G_3(T) = 0$ and $\Delta G_4(T) = 0$ separately gives the relevance of temperature to equilibrium $CO_2$ partial pressure (shown in Fig. 1). The regions surround by dotted lines are the thermodynamic valid portions when $P_{CO} = 5$ kPa, $P_{CO} < 5$ kh respectively. It is apparent from Fig. 1 that the $CO_2$ partial pressure would be $P_2$ if reaction (3) was in equilibrium state, but the real $CO_2$ partial pressure is $P_1$ in the system, therefore the thermodynamic driving force for reaction (3) is

$$\Delta G_{DC}(T) = RT\int_{P_2}^{P_1} d\ln P_{CO_2} = RT(\ln P_1 - \ln P_2) \tag{9}$$

where $P_1$ denotes $CO_2$ pressure at point D, $P_2$ denotes $CO_2$ pressure at point C. At the present experimental condition ($T = 1350°C$), usually $P_{CO} < 5$ kPa, thus

$$\Delta G_{DC}(1623) < \Delta G_{BA}(1623)$$
$$= RT(\ln P_3 - \ln P_4)$$
$$= -10517 \text{J/mol}$$

where $P_3$ denotes $CO_2$ pressure at point B, $P_4$ denotes $CO_2$ pressure at point A. The existence of the thermodynamic driving $\Delta G_{DC}(T)$ enables reaction (3) to keep toward right.

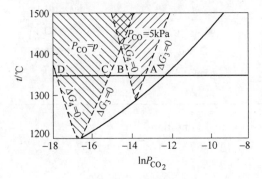

Fig. 1  Phase equilibrium diagram as a function of partial pressure and temperature

## 3  Kinetic aspect

Because reaction (1) is a solid-state reaction, the reaction rate would be related to $SiO_2$-C contact condition. Table 2 summarizes the relation between compact pressure of the precursors and $Si_3N_4$ formed from the system $SiO_2$-C-$N_2$. It can be seen that nitriding results are not dependent on compact pressure, which infers that the amount of $Si_3N_4$ formation is independent of $SiO_2$-C contact condition. So the reduction of silica would mainly be through some other route rather than reaction (1).

Table 2  Correlation of compact pressure with $Si_3N_4$ formation
($C/SiO_2 = 5$, 1350℃ for 12h)

| Compact pressure/MPa | 0 | 20 | 40 |
|---|---|---|---|
| $Si_3N_4$ content/wt% | 25.9 | 26.7 | 24.5 |

In the initial reaction stage, there is small amount of $SiO_2$ reduced directly by carbon to form CO and SiO because there is no carbon monoxide as reactant in the system. However, reaction (1) hardly proceeds at the later stage for kinetic reason, the silica reduction is mainly through reaction (3).

Assuming that the rate of reaction (1) is controlled by diffusion and chemical process simultaneously, the reaction rate can be expressed as

$$r_1 = k_1 S a_C^L a_S^L$$

where $k_1$ is a constant, the term $S$ is proportional to the concentration of reaction sites, related with $SiO_2$-C contact condition, and as are the average activities of carbon and silica in the $SiO_2$-C diffusion layer, respectively. Chemical reactivity can be considered as rate-limiting factor for reactions

(2) and (3)

$$r_2 = k_2 A_c P_{SiO}^3 P_{N_2}^2$$

$$r_3 = k_3 A_s P_{CO}$$

where $k_2$ and $k_3$ are specific reaction-rate constants for reaction (2) and (3) respectively, and $A_c$, $A_s$ are specific area of carbon and silica at time $t$ respectively.

Because of the existence of the intermediate SiO, reaction (1) and (3) can be considered as rate-controlling steps, so that $dP_{SiO}/dt = 0$. For the former mechanism (reaction (1) and (2)), then

$$\frac{dP_{SiO}}{dt} = k_1 S a_C^L a_S^L - 3k_2 A_c P_{SiO}^3 P_{N_2}^2 = 0 \tag{10}$$

The overall silicon nitride formation rate is given by

$$\frac{d[Si_3N_4]}{dt} = k_2 A_c P_{SiO}^3 P_{N_2}^2 \tag{11}$$

Substituting Eq. (10) into Eq. (11) and integration can be obtained

$$[Si_3N_4]_1 = 2\int_0^t \frac{1}{3} k_1 S a_C^L a_S^L dt \tag{12}$$

Now that reaction (3) predominates for $SiO_2$ reduction in the system, thus

$$\frac{dP_{SiO}}{dt} = k_1 S a_C^L a_S^L + k_3 A_s P_{CO} - 3k_2 A_c P_{SiO}^3 P_{N_2}^2 = 0 \tag{13}$$

Upon substituting this equation into Eq. (11), the integral expression can be obtained

$$[Si_3N_4]_2 = [Si_3N_4]_1 + \int_0^t \frac{1}{3} k_3 A_s P_{CO} dt \tag{14}$$

For a certain reaction time, Eq. (12) and Eq. (14) can be reduced as

$$[Si_3N_4]_1 = K \tag{15}$$

$$[Si_3N_4]_2 = K + K' P_{CO} \tag{16}$$

where $K$ and $K'$ are constants. Comparing (15) with (16) reveals that silicon nitride content is $P_{CO}$-independent if the $SiO_2$ reduction is through reaction (1), while it indicates a direct relationship between $P_{CO}$ and the amount of $Si_3N_4$ formation if reaction (3) is a main $SiO_2$ reduction step. The nitriding results of $C/SiO_2 = 5$, 1350℃ for 12h samples at various $P_{CO}$ are shown in Fig. 2. It shows that $Si_3N_4$ formation is significantly increased with increase in CO partial pressure up to 10kPa, this result is identical with Eq. (16) but contradicts with Eq. (15). Whereas there was a little $Si_3N_4$ formed when CO partial pressure increased to 15kPa. It is probably because CO partial pressure has exceeded the critical value.

Through thermodynamic calculations on the basis of Table 1, a $Si_3N_4$ stability diagram as a function of CO partial pressure and temperature can be obtained (Fig. 3). It indicates that critical CO partial pressure is 10.9kPa at the present experimental condition (1350℃). $P_{CO} = 15$kPa is out of the $Si_3N_4$ synthesis region, which does not meet thermodynamic conditions. So no obvious amount of silicon nitride is formed under this condition.

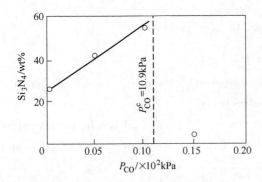 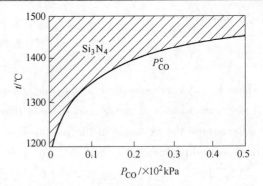

Fig. 2  $P_{CO}$ dependence of $Si_3N_4$ formation from $C/SiO_2 = 5$ sample, holding for 12h at 1350℃

Fig. 3  Thermodynamic stability relations for $Si_3N_4$ compared to $SiO_2$

## 4  Conclusions

The reduction mechanism of $SiO_2$ can be divided into direct reduction and indirect reduction for the system $SiO_2$-C-$N_2$. Carbon directly reacts with silica to form silicon monoxide in the initial stage; silica can be reduced by carbon monoxide in the indirect reduction stage, which is the main process for $SiO_2$ reduction. Reaction (4) can provide a thermodynamic driving force $\Delta G_{DC}$ and replenish carbon monoxide required for reaction (3) in the presence of carbon.

### References

[1] Komeya K., Inoue H. J. Mat. Sci., 1975, 10 (7): 1243.

[2] Hendry A., Jack K. H. Special Ceramics, Edited by P. Popper. British Ceramic Research Association. Stoke-on-Trent, England, 1975, 6: 199.

[3] Lee J. G., Cutler I. B. Nitrogen Ceramics. Edited by P. Popper. Noordhoff, Leyden, The Netherlands, 1977: 175.

[4] Klinger N., et al. J. Am. Ceram. Soc., 1966, 49 (7): 369.

[5] Rahman I. A., Riley F. L. J. Europ. Ceram. Soc., 1989, 5 (1): 11.

[6] Zhang S., Cannon W. R. J. Am. Ceram. Soc., 1984, 67 (10): 691.

[7] Kubaschewski O., et al. Metallurgical Thermochemistry, 5th Edition, Pergamon Press, 1979: 378.

[8] Rochow E. G. Comprehensive Inorganic Chemistry [M]. Edited by J. C. Bailar, et al. Pergamon Press, 1973, 1: 1354.

(原载于 *Rare Metals*, 1991, 10 (2))

# O'-Sialon-ZrO$_2$-SiC 复合材料的摩擦磨损性能研究

张海军[1]  李文超[2]  姚熹[1]  张良莹[1]

(1. 同济大学功能材料研究所,上海 200092;
2. 北京科技大学理化系,北京 100083)

**摘 要**:研究了 O'-Sialon-ZrO$_2$-SiC 复合材料在室温和600℃的摩擦磨损行为,研究表明,在实验条件下,O'-Sialon-ZrO$_2$-SiC 复合材料的磨损率远低于不锈钢的磨损率,O'-Sialon-ZrO$_2$-SiC/不锈钢摩擦副间的磨损机制主要是粘着磨损。分形结果研究表明,分形维数与磨损存在对应关系,即分形维数越大,材料磨损越严重。复合材料磨损界面的分形维数在1.26~1.38之间。

**关键词**:O'-Sialon-ZrO$_2$-SiC;粘着磨损;分形维数;复合材料;摩擦;陶瓷

## Friction and Wear Properties of O'-Sialon-ZrO$_2$-SiC Composite

Zhang Haijun[1]  Li Wenchao[2]  Yao Xi[1]  Zhang Liangying[1]

(1. Functional Material Research Laboratory, Tongji University, Shanghai 200092;
2. Department of Physics Chemistry, University of Science and Technology Beijing, Beijing 100083)

**Abstract**: The friction and wear properties of O'-Sialon-ZrO$_2$-SiC composite were studied at room temperature and 600℃. Results show that the wear rate of O'-Sialon-ZrO$_2$-SiC is much less than that of stainless steel at the testing condition and main wear mechanism of the unlubricated friction is adhesive wear. It also indicates that the fractals dimension is related to the abrasion, the larger the fractals dimension, the greater the abrasion, the fractals dimension of wear ranges from 1.26 to 1.38.

**Keywords**: O'-Sialon-ZrO$_2$-SiC; adhesive wear; fractals dimension; composite material; friction; ceramic

## 1 引言

目前,化工、冶金等行业要求机械设备具有高度的可靠性、长寿命和低消耗。据统计

---

基金项目:国家自然科学基金资助项目 (59634130)。
张海军,1995~1999 年在北京科技大学师从李文超教授攻读博士学位。目前在武汉科技大学工作,湖北省"楚天学者"特聘教授。获得武汉市黄鹤英才、河南省杰出青年科学基金及河南省级学术与技术带头人等荣誉称号,发表学术论文 200 余篇,获省部级科技奖 6 项。

研究，摩擦磨损缩短机械寿命，降低机械性能，它不仅是机械零件的一种失效形式，也是引起后来失效的最初原因。陶瓷作为耐磨材料是近二十年来的事情，尤其最近一段时间发展更为迅速。这主要是由于其具有特殊优良的性能，包括高的耐磨及耐腐蚀能力、好的高温稳定性及高温抗氧化性、低的比重、低的摩擦系数、低的传热系数，而且在相当大的温度范围内具有较高的硬度。对 $Al_2O_3$、$Si_3N_4$ 和 SiC 陶瓷磨损性能的研究[1]发现，由于晶粒间各向异性的弹性和热扩散特性，在摩擦过程中晶间处受到微应力作用，当晶间强度较弱时，容易发生晶间断裂而使晶粒被拔出或挤出，从而导致磨粒磨损；相反，当晶间强度较强时，陶瓷材料发生的是轻微磨损。陈南平等认为，金属—陶瓷摩擦副的磨损机理主要有五种机制：粘着磨损机制、微切削机制、疲劳磨损机制、一次加载下的断裂磨损机制和磨蚀磨损机制。在各类磨损情况下，上述五类机制往往是几种同时作用，只是随着工况和材质的变化，对磨损起主要作用的机制可发生变化[2]。

Sialon、$ZrO_2$、SiC 等结构陶瓷具有高强度、高硬度、断裂韧性好、耐化学腐蚀等一系列优良的物理和化学性能，在工程应用中已表现为理想的耐磨损材料。这些材料在高温下的使用常常无法使用传统的润滑剂，故而研究其在未润滑情况下的摩擦磨损性能是十分必要的。本文重点研究了 O'-Sialon-$ZrO_2$-SiC 复合材料在室温及 600℃ 时随载荷变化的干摩擦磨损行为，并加以比较。

## 2 实验方法

研究用的 O'-Sialon-$ZrO_2$-SiC 复合材料的具体组成如表 1 所示。试样于 1700℃、20MPa 下热压成型，保温时间 1h。

表 1 制备 O'-Sialon-$ZrO_2$-SiC 材料的原始配方表
Table 1 Starting mixes composition of O'-Sialon-$ZrO_2$-SiC (wt%)

| $Y_2O_3$ | $SiO_2$ | $Si_3N_4$ | $Al_2O_3$ | $ZrO_2$ | SiC |
|---|---|---|---|---|---|
| 6 | 23.07 | 44.52 | 11.41 | 10 | 5 |

将合成的 O'-Sialon-$ZrO_2$-SiC 复合材料（硬度 HV3200；体密度 3.30g/cm³；吸水率 0.21%）切成直径为 24mm、高为 7.88mm 的圆柱体，经细磨、抛光后，光洁度达 ▽₉。金属试样由 GCr15 不锈钢加工而成，试样高为 16.5mm，直径为 20mm，经淬火及回火后，硬度 HRC57-61，表面光洁度为 ▽₉。试验在 SRV 试验机上进行，采用 O'-Sialon-$ZrO_2$-SiC 复合材料和不锈钢对磨，试验温度为室温和 600℃，钢块的滑动频率为 40Hz，行程为 2.0mm，磨损时间 30min。试验结束后，试样在超声波室中净化 30min，再用纯净的丙酮及乙醇清洗，然后在红外线灯下干燥，干燥器中冷却，最后用精确至 0.1mg 的电子分析天平称量。摩擦系数由试验机中的自动绘图仪给出，磨损量为试样试验前后的质量损失，通过扫描电镜观察摩擦副磨损后的表面形貌。

## 3 实验结果与分析

### 3.1 摩擦系数

图 1 为干摩擦条件下，O'-Sialon-$ZrO_2$-SiC 复合材料/不锈钢摩擦副的典型摩擦曲线。

可以看出：干摩擦下摩擦副的摩擦系数很不稳定。表 2 为实验条件下摩擦副的摩擦系数，可以看出，干摩擦磨损时，随负载的增加，摩擦系数呈下降趋势，这可能和负载增加引起表面温升有关，表面温度增加，降低了摩擦副表面的强度和硬度；另一方面负载面粗糙度下降，故而负载增加，摩擦系数降低。

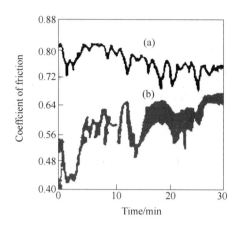

图 1　O'-Sialon-$ZrO_2$-SiC 复合材料/不锈钢摩擦副干摩擦下的摩擦曲线

Fig. 1　Friction curves of O'-Sialon-$ZrO_2$-SiC/stainless steel sliding couple under dry friction

（a）Room temperature，150N；（b）600℃，150N

表 2　O'-Sialon-$ZrO_2$-SiC 复合材料/不锈钢摩擦副的平均摩擦系数

Table 2　Mean friction coefficient of O'-Sialon-$ZrO_2$-SiC composite/stainless steel couples

| Temperature/℃ | Load/N | | | |
| --- | --- | --- | --- | --- |
| | 50 | 100 | 125 | 150 |
| Room temperature | 0.94 | 0.80 | 0.76 | 0.74 |
| 600 | 0.61 | 0.56 | 0.56 | 0.54 |

600℃时，经过磨合期后，摩擦曲线变得平稳，且其摩擦系数比室温时相应载荷下的要低。造成这种现象的原因是 600℃摩擦时，不锈钢表面形成一层铁和铬氧化膜，起到润滑的作用，从而使摩擦系数降低。这和大量的陶瓷—金属摩擦副摩擦磨损实验研究结果相一致[3,4]。文献［5］得出了陶瓷材料在滑动摩擦磨损时其摩擦系数，可用下式表示：

$$\mu = A \frac{\tau}{P^{1/3}} \left(\frac{3}{4E'}\right)^{2/3}$$

式中，$A$ 为与表面几何形状有关的常数；$P$ 为法向载荷；$E'$ 为有效弹性模量；$\tau$ 为界面极限剪切强度。

当陶瓷试样表面覆盖有氧化膜时，基体承受载荷，而摩擦则在氧化膜上发生[6]。由于 Fe 氧化产物的弹性模量和硬度比基体材料低得多，使得剪切强度 $\tau$ 减小。由式（1）可见，$\tau$ 减小，摩擦系数也相应降低。因此试样表面氧化膜的存在能降低摩擦系数，具有减磨作用。

## 3.2 磨损率

图2为不锈钢及O'-Sialon-ZrO$_2$-SiC复合材料试样在室温和600℃、干摩擦条件下磨损量随载荷的变化情况。可以看出，两种条件下，不锈钢试样的磨损量随载荷的增加都呈上升趋势，但变化过程稍有差别。室温时，不锈钢的磨损量随载荷增加而增大的幅度较大；600℃时，不锈钢的磨损量增加相对比较缓慢。这是由于不锈钢表面氧化膜的形成起到润滑作用，降低了金属表面的机械磨损。O'-Sialon-ZrO$_2$-SiC的磨损量变化规律与不锈钢试样相类似，但其磨损量比不锈钢试样要低一个数量级，其主要原因是：（1）相对O'-Sialon-ZrO$_2$-SiC材料而言，不锈钢的硬度低，容易剪切，粘着点断裂多发生在不锈钢表面；（2）不锈钢的粘着倾向较大，摩擦过程中不锈钢在O'-Sialon-ZrO$_2$-SiC材料表面形成大片的转移层，使不锈钢与O'-Sialon-ZrO$_2$-SiC材料之间的摩擦在一定程度上变成了不锈钢与不锈钢之间的摩擦；因此O'-Sialon-ZrO$_2$-SiC材料的磨损量要比不锈钢低得多。

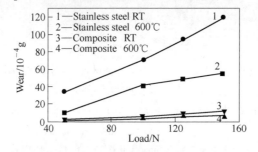

图2　不锈钢及O'-Sialon-ZrO$_2$-SiC在室温和600℃干摩擦条件下的磨损量对比

Fig. 2　Comparison of the wear rates of stainless steel and O'-Sialon-ZrO$_2$-SiC under dry friction condition at room temperature and 600℃ for 30min

## 4　磨损表面分析及显微结构

图3为不同载荷、不同温度下不锈钢试样表面磨损后的扫描电镜照片，从中可以看出

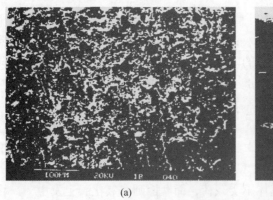
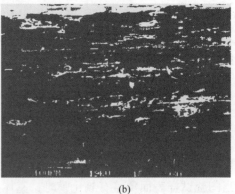

(a) 　　　　　　　　　　　　　　(b)

图3　干摩擦时金属表面扫描电镜照片

Fig. 3　SEM morphology of stainless steel under dry friction

(a) RT, 50N；(b) 600℃, 50N

粘着磨损的迹象。同时从图 4（a）中可以看到不锈钢表面存在大量的微裂纹，图 4（b）为微裂纹局部放大扫描电镜照片。可以看出裂纹已经在平行于和垂直于滑动方向上扩展，裂纹在这两个方向上的扩展必然会使不锈钢产生大片状的剥落，从而导致磨粒的形成，并在进一步的运动过程中导致磨粒磨损。

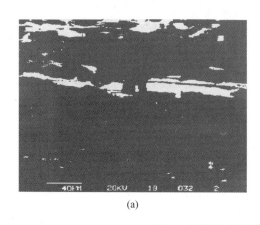
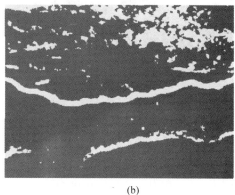

(a)　　　　　　　　　　　　　　　(b)

图 4　不锈钢表面裂纹的显微结构照片

Fig. 4　SEM pictures of microcracks on the surface of stainless steel

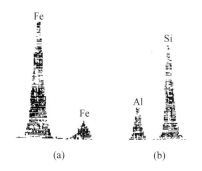

(a)　　　　　　(b)

图 5　图 4（a）不锈钢试样的 EDAX 能谱图

Fig. 5　EDAX patterns of Fig.4(a)

(a) The white adbesive layer;

(b) The black grain

图 5 为图 4（a）中不规则黑色颗粒与白色粘着物的能谱图，从能谱图推测白色粘着物应为 Fe，黑色颗粒应为 O'-Sialon，这表明在摩擦过程中，摩擦副表面既有铁发生转移，又有 O'-Sialon-$ZrO_2$-SiC 颗粒发生转移。日本 T. Sasada 认为，摩擦副接触面摩擦过程中，粘着磨损产生的磨屑不是由母体材料表面直接分离出来的磨粒，其磨屑的形成实际上经过了一个"元素传递"过程，从摩擦表面上粘着而新撕脱的磨粒，随摩擦发生运动，它或者会遇到其他磨粒，由于较高表面能及摩擦作用，分散的磨粒间很容易发生结合、积聚、长大，最后形成一个较大的游离磨屑离开摩擦面；或者同新撕脱的磨粒在摩擦过程中又黏结到母体金属上去，或者黏结到对偶摩擦面上去，这就是粘着磨损"元素传递"的过程。

在实验中，金属在 O'-Sialon-$ZrO_2$-SiC 表面上的转移是显而易见的（见图 6）。金属的转移主要是由其硬度和氧化活性决定的。Bowden[7] 很早就研究了在大气条件下各种金属的粘着系数与其硬度之间的关系，认为金属的粘着系数随其硬度的增加而降低，即硬度越低的金属越易发生粘着，不锈钢的硬度远低于 O'-Sialon-$ZrO_2$-SiC 的硬度，因此摩擦副间粘着点的断裂多发生在不锈钢表面，所以其在 O'-Sialon-$ZrO_2$-SiC 材料上的转移量大。此外 Hiratsuka[8] 等认为，金属在陶瓷（如 $Al_2O_3$、$ZrO_2$ 等）上的粘着倾向与该金属的氧化活性直接相关，越易氧化的金属在陶瓷上形成的粘着越强，粘着力越大。铁和铬都非常容易氧化，因此其粘着倾向也较大。

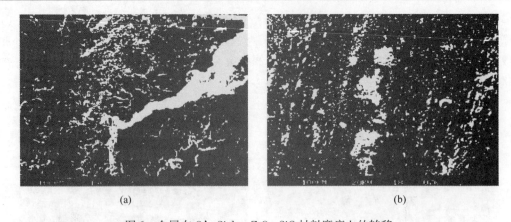

图 6 金属在 O'-Sialon-$ZrO_2$-SiC 材料磨痕上的转移

Fig. 6 SEM morphology of metallic transfer films on O'-Sialon-$ZrO_2$-SiC composite

(a) 125N, 600℃; (b) 150N, 600℃

图 7 (a) 表明，O'-Sialon-$ZrO_2$-SiC 试样表面黏附着不少磨粒，其对应能谱见图 7 (b)，通过能谱的分析推测可知，该颗粒应是从 O'-Sialon-$ZrO_2$-SiC 材料中剥落下来的。造成这种现象的原因可能是高温摩擦下，O'-Sialon 棒状颗粒的脆性断裂。

图 7 O'-Sialon-$ZrO_2$-SiC 试样表面的显微照片

Fig. 7 Morphology of O'-Sialon-$ZrO_2$-SiC surface

(a) 600℃; (b) The EDAX pattern of the white grain

综上所述，O'-Sialon-$ZrO_2$-SiC 材料/不锈钢摩擦副的磨损是以粘着磨损和磨粒磨损两种类型为主。

## 5 材料磨损的分形特征

借助于扫描电子显微镜（SEM）对磨损表面的观察，发现材料的磨损界面具有分形的特征，即具有统计意义上的自相似性。因此本文对不同条件下磨损后 O'-Sialon-$ZrO_2$-SiC 材料的界面进行分形维数计算，以确定材料的磨损机理与分形维数之间的关系。利用扫描仪将电镜照片扫描到计算机中，得到相应灰度图，再提取出材料磨损界面形貌的边界，然后将边界图二值化，利用 FDCP（Fractal Dimension Calculation Program）程序的盒计数法计算模块计算出不同载荷时材料分形维数,如图 8 所示。将不同载荷下材料的分形维数对载荷作图,得到分形维数随载荷变化曲线,如图 9 所示。

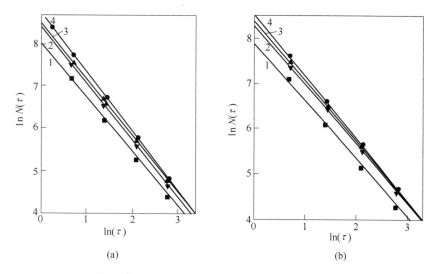

图 8  常温及 600℃ 时陶瓷材料磨损界面分形维数

Fig. 8  Box-dimension of wear interface of O'-Sialon-$ZrO_2$-SiC at room temperature and 600℃

(a) Room temperature; (b) 600℃

(a) 1—50N, $D=1.28$, $R=0.9913$; 2—100N, $D=1.32$, $R=0.9988$; 3—125N, $D=1.34$, $R=0.9980$; 4—150N, $D=1.38$, $R=0.9997$;

(b) 1—50N, $D=1.26$, $R=0.9990$; 2—100N, $D=1.30$, $R=0.9990$; 3—125N, $D=1.31$, $R=0.9978$; 4—150N, $D=1.37$, $R=0.9994$

从材料在不同载荷下分形维数的测定结果看,随磨损载荷的延长,O'-Sialon-$ZrO_2$-SiC 复合材料磨损界面的分形维数逐渐增大,表明材料磨损加剧,这和实验结果相一致。同时从图 9 中可以看出,相同载荷下,600℃ 时复合材料的磨损界面维数比室温时的要低,从另一方面证实了实验结果。

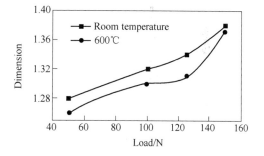

图 9  O'-Sialon-$ZrO_2$-SiC 材料分形维数随载荷的变化

Fig. 9  Box-dimension vs time curve of O'-Sialon-$ZrO_2$-SiC composite

## 6  结论

(1) 干摩擦条件下,O'-Sialon-$ZrO_2$-SiC 复合材料/不锈钢摩擦副的摩擦系数很不稳定,随负载的增加,摩擦系数呈下降趋势。600℃ 时 O'-Sialon-$ZrO_2$-SiC 材料/不锈钢摩擦副的摩擦系数比室温时相应载荷下的要低。

(2) 在干摩擦滑动条件下,O'-Sialon-$ZrO_2$-SiC 材料/不锈钢摩擦副间的磨损主要是粘着磨损,随载荷的增加,磨损量增大,且在相同条件下,O'-Sialon-$ZrO_2$-SiC 的磨损率要比不锈钢的低一个数量级。摩擦过程中,不仅金属向 O'-Sialon-$ZrO_2$-SiC 表面发生了转移,

O'-Sialon-$ZrO_2$-SiC 材料也向金属表面上发生了转移。

（3）分形研究表明，随实验载荷的增加，磨损的加剧，材料对应磨损界面的分形维数也变大，表明分形维数与磨损存在对应关系。在 O'-Sialon-$ZrO_2$-SiC 材料/不锈钢摩擦副的磨损实验过程中，磨损界面的分形维数在 1.26～1.38 之间。

## 参 考 文 献

[1] Zutshi A, Haber R A, Niesz D E, et al. J. Am. Ceram. Soc., 1994, 77: 883-890.
[2] 陈南平，刘家浚. 材料科学进展，1987, 1 (2): 3-9.
[3] 程荫芊，狄向东，汪复兴，等. 无机材料学报（Journal of Inorganic Materials），1990, 5 (4): 360-367.
[4] 赵兴中，刘家浚，朱宝亮，等. 硅酸盐学报，1996, 24 (5): 515-522.
[5] Liu H Y. J. Am. Ceram. Soc., 1993, 76 (1): 105-112.
[6] Page R A. Development of self-lubricating ceramics using surface a bulk oxidizing species [J]. Advances in Engineering Tribology. Illinois: STLE Publications, 1991, 145.
[7] 鲍登 F P，泰伯 D，著. 陈绍澧，袁汉昌，张绪寿，译. 固体的摩擦与润滑 [M]. 北京：机械工业出版社，1982.
[8] Hiratsuka K, Enomoto A, Sasada T. Wear, 1992, 153 (2): 361-373.

（原载于《无机材料学报》，2000, 15 (3): 480-486）

# 合成 β-SiAlON-AlON 复相材料的热力学分析的研究

黄向东　李文超　王福明　邵昱

(北京科技大学理化系，北京　100083)

**摘　要**：首先通过热力学分析了合成 β-SiAlON-AlON 复相材料的可行性，由热力学计算确定其合成的工艺条件，通过热压制备了 β-SiAlON-AlON 复相材料，XRD 分析表明所合成的材料为 β-SiAlON-AlON 复相材料，实验结果与热力学分析相吻合。

## 1 引言

随着冶金新技术的发展，如反向凝聚技术，尤其是已列入跨世纪的攀登计划的超纯净钢生产技术对我国的耐火材料提出了新的要求，超纯净钢的杂质含量要求在 ppm 数量级的范围内，现有的含碳耐火材料则不能完全胜任这样的要求，而只能通过研制出新的不含碳的耐火材料来满足冶金新技术的不断发展。因此，近几十年来国内外耐火材料界密切注意高技术陶瓷材料的发展，并正广泛开展高技术陶瓷材料在耐火材料中的应用与研究工作。赛隆 (Sialon) 是近二十年发展起来的一类高性能陶瓷材料。

SiAlON 陶瓷在工业上早就发挥着巨大的作用，如日本用 SiAlON 作汽车发动机的推拉杆，德国使用 SiAlON 陶瓷制作发动机的垫圈，而在耐火材料中应用则较晚，一般作耐火材料的结合相。欧洲一些国家[1]已在高炉上使用 SiAlON 结合 SiC 砖，与 $Si_3N_4$ 结合 SiC 相比，SiAlON 结合 SiC 材料具有优良的抗碱侵蚀性，因而更适合于推广应用。法国耐火材料首先推出 SiAlON 结合刚玉砖，自 1991 年到 1993 年在德国、法国、比利时和中国台湾等地已在六座高炉（直径为 9.3～14.9 米）上应用这种耐火材料。

AlON 由于其优良的抗氧化性、抗侵蚀性近年来受到耐火材料界的重视。T. Hosaka，M Kato[2] 报道了出铁沟材料中引入 AlON 材料的研究情况，在铁沟料中引入 AlON（5AlN·$9Al_2O_3$）材料后，发现铁沟材质的耐侵蚀性有大幅度的提高，抗热震性能也随着 AlON 含量的增加而提高，同时也发现 AlON 的引入并没有影响施工性能。K Murakami[3] 等为提高 $Al_2O_3$-C 材质滑动水口的耐用性，他们在 $Al_2O_3$-C 材质的滑动水口中引入部分 AlON 后，发现引入 AlON 的水口，其孔径扩大进度比未引入 AlON 材料的水口低 20%，而且使用次数也增加了，这一试验充分表明了引入 AlON 材料后耐用性的提高，K Takeda[4] 等将 AlON 材料引入 $Al_2O_3$-C 质浸入式水口中，同样提高了抗侵蚀性和抗剥落性。对于 β-SiAlON-AlON 是否能与 AlON 复合，以及如何获取 β-SiAlON-AlON 复合材料，至今尚未见报道，本文从热

---

本项目由国家自然科学基金资助。
黄向东，1997-2001 年于北京科技大学师从李文超教授攻读博士学位。目前英国定居。

力学角度出发，分析了合成的条件及可能性，并通过实验验证了分析结果。

## 2　β-SiAlON-AlON 的稳定性的热力学分析

### 2.1　Al-O-N 和 Si-O-N 两个体系的热力学参数状态图

β-SiAlON-AlON 的合成涉及 Al-O-N 和 Si-O-N 两个体系，从这两个体系的参数状态图入手，对合成的中间过程进行讨论。由于 β-SiAlON-AlON 及其他一些到目前为止，没有可用的热力学数据，本文采用文洪杰等人运用逆抛物面规则评估的热力学数据，进行计算。

表1和表2列出了 Al-O-N 和 Si-O-N 体系凝聚相及其平衡分压的关系（1800K）。

表1　Si-O-N 体系凝聚相及其平衡分压的关系（1800K）

| | |
|---|---|
| (1) $SiO_{2(s)} = Si_{(s)} + O_{2(g)}$ | $\lg(P_{O_2}/P^\ominus) = -16.81$ |
| (2) $\beta\text{-}Si_3N_{4(s)} = 3Si_{(s)} + 2N_{2(g)}$ | $\lg(P_{N_2}/P^\ominus) = -1.68$ |
| (3) $2SiO_{2(s)} + N_{2(g)} = Si_2N_2O_{(s)} + 3/2O_{2(g)}$ | $\lg(P_{N_2}/P^\ominus) = 3/2\lg(P_{O_2}/P^\ominus) + 21.18$ |
| (4) $2\beta\text{-}Si_3N_{4(s)} + 3/2O_{2(g)} = 3Si_2N_2O_{(s)} + N_{2(g)}$ | $\lg(P_{N_2}/P^\ominus) = 3/2\lg(P_{O_2}/P^\ominus) + 30.61$ |
| (5) $Si_2N_2O_{(s)} = 2Si_{(s)} + N_{2(g)} + 1/2O_{2(g)}$ | $\lg(P_{N_2}/P^\ominus) = 1/2\lg(P_{O_2}/P^\ominus) - 12.44$ |

表2　Al-O-N 体系凝聚相及其平衡分压的关系（1800K）

| | |
|---|---|
| (1) $2Al_{(s)} + 3/2O_{2(g)} = Al_2O_{3(s)}$ | $\lg(P_{O_2}/P^\ominus) = -21.19$ |
| (2) $Al_{(s)} + 1/2N_{2(g)} = AlN_{(s)}$ | $\lg(P_{N_2}/P^\ominus) = -6.94$ |
| (3) $2Al_7O_9N_{(s)} + 3/2O_{2(g)} = 7Al_2O_{3(s)} + N_{2(g)}$ | $\lg(P_{N_2}/P^\ominus) = 3/2\lg(P_{O_2}/P^\ominus) + 23.72$ |
| (4) $2Al_7O_9N_{(s)} = 14Al_{(s)} + 9O_{2(g)} + N_{2(g)}$ | $\lg(P_{N_2}/P^\ominus) = -9\lg(P_{O_2}/P^\ominus) - 199.3$ |

将表1和表2中的各式的关系作图，可得到 Si-O-N 系和 Al-O-N 系的叠加参数状态图（1800K），如图1所示。

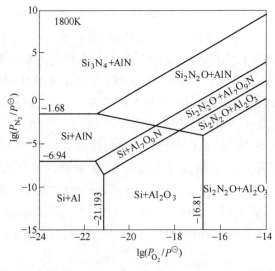

图1　Si-O-N 系和 Al-O-N 系的叠加参数状态图（1800K）

## 2.2 合成 β-SiAlON-AlON 的热力学分析

在合成 β-SiAlON-AlON 复相材料的过程中涉及以下可能的反应:

$$14Si_2N_2O_{(s)} + 2Al_7O_9N_{(s)} + 6N_2 = 7Si_4Al_2O_2N_{6(s)} + 9O_2 \quad (1)$$

$$\Delta_f G^\ominus_{Si_2N_2O(s)} = -951.68 + 0.2906T \ ^{[6]}$$

$$\Delta_f G^\ominus_{Al_7O_9N(s)} = -5364.12 + 1.0724T \ ^{[5]}$$

$$\Delta_f G^\ominus_{Si_4Al_2O_2N_6(s)} = -2598.08 + 0.8680T \ ^{[6]}$$

$$\Delta_r G_1^\ominus = 5865.2 - 0.1372T$$

$$42Si_4Al_2O_2N_6 + 24Al_7O_9N_{(s)} + 123O_{2(g)} = 14Si_{12}Al_{18}O_{39}N_{8(s)} + 82N_{2(g)} \quad (2)$$

$$\Delta_f G^\ominus_{Si_{12}Al_{18}O_{39}N_8(s)} = -22438.30 + 4.6325T \ ^{[7]}$$

$$\Delta_r G_2^\ominus = -76277.97 + 2.6614T$$

$$Si_4Al_2O_2N_6 + 2Al_7O_9N_{(s)} + 4N_2 = 4SiAl_4O_2N_{4(s)} + 6O_2 \quad (3)$$

$$\Delta_f G^\ominus_{SiAl_4O_2N_4(s)} = -1683.4 + 0.5236T \ ^{[7]}$$

$$\Delta_r G_3^\ominus = 6592.72 - 0.9184T$$

$$7Si_4Al_2O_2N_6 + 6Al_7O_9N_{(s)} + 4N_2 = 14Si_2Al_4O_4N_4 + 6O_2 \quad (4)$$

$$\Delta_f G^\ominus_{Si_2Al_4O_4N_4} = -3275.57 + 0.8487T \ ^{[7]}$$

$$\Delta_r G_4^\ominus = 4513.3 - 0.6286T$$

$$10N_2 + 2Si_{12}Al_{18}O_{39}N_{8(s)} + 6SiAl_4O_2N_{4(s)} = 15Si_2Al_4O_4N_{4(s)} + 15O_2 \quad (5)$$

$$\Delta_r G_5^\ominus = 5843.451 + 0.3239T$$

$$9Si_5AlON_{7(s)} + 12SiAl_4O_2N_{4(s)} + 12O_2 = 19Si_3Al_3O_3N_{5(s)} + 8N_2 \quad (6)$$

$$\Delta_r G_6^\ominus = -16153.59 + 2.2305T$$

(1) 在 $N_2$ 气氛下,取 $P_{O_2} = 10^{-18}$ atm、$P_{N_2} = 1$ atm。

$N_2$ 气氛下,1600℃:

$$\Delta_r G_1 = -201.4474 kJ;\ \Delta_r G_2 = 8106.682 kJ;$$
$$\Delta_r G_3 = 999.3803 kJ;\ \Delta_r G_4 = -537.2439 kJ;$$
$$\Delta_r G_5 = -3232.733 kJ;\ \Delta_r G_6 = -4229.509 kJ$$

$N_2$ 气氛下,1720℃:

$$\Delta_r G_1 = -590.1274 kJ;\ \Delta_r G_2 = 13513.07 kJ;$$
$$\Delta_r G_3 = 641.0244 kJ;\ \Delta_r G_4 = -860.8239 kJ;$$
$$\Delta_r G_5 = -3814.229 kJ;\ \Delta_r G_6 = -3465.553 kJ$$

说明在 $N_2$ 气氛中,无论在 1600℃ 或 1720℃ 下,$Si_2N_2O$ 可以和 AlON 反应生成 β-SiAlON,同时 β-SiAlON 与 AlON 反应使 β-SiAlON 的 Z 值升高。

(2) Ar 气氛下,取 $P_{O_2} = 10^{-18}$ atm、$P_{N_2} = 10^{-5}$ atm。

Ar 气氛下,1600℃:

$$\Delta_r G_1 = 874.4037 kJ;\ \Delta_r G_2 = -6597.116 kJ;$$
$$\Delta_r G_3 = 1716.739 kJ;\ \Delta_r G_4 = -180.1149 kJ;$$
$$\Delta_r G_5 = -1439.711 kJ;\ \Delta_r G_6 = -5664.04 kJ$$

Ar 气氛下,1720℃:

$\Delta_r G_1 = 554.6516 \text{kJ}$；$\Delta_r G_2 = -2132.779 \text{kJ}$；
$\Delta_r G_3 = 1404.343 \text{kJ}$；$\Delta_r G_4 = -97.50507 \text{kJ}$；
$\Delta_r G_5 = -1906.331 \text{kJ}$；$\Delta_r G_6 = -4991.992 \text{kJ}$

计算结果表明，Ar 气氛，无论在 1600℃ 或 1720℃ 下，β-SiAlON 与 AlON 反应生成 χ 相，并进而与 15R 反应生成 β-SiAlON；同时，也存在着 β-SiAlON 与 AlON 反应使 β-SiAlON 的 Z 值变大的过程。$N_2$ 气氛下，有 β-SiAlON 与 AlON 反应使 β-SiAlON 的 Z 值变大的反应。同时都存在低 Z 值的 β-SiAlON 与 15R 反应产生高 Z 值的 β-SiAlON 的过程。

基于以上分析，我们将 β-SiAlON 粉与预合成的 AlON 粉分别在 $N_2$ 或 Ar 中，于 1720℃ 热压烧结。所得产物及原料的 XRD 图谱如图 2～图 4 所示。XRD 分析表明，β-SiAlON 原料粉其主晶相为 β-SiAlON，也有一定量的衍射峰与 15R 很接近的相，可认为是制备过程中晶格发生畸变的 15R 相，根据 β-SiAlON 相的 Z 值与其晶格常数的关系，可以确定其 Z 值为 0.96，$N_2$ 气氛下热压产物中的 β-SiAlON 相的 Z 值为 3.5；Ar 气氛下，热压产物中的 β-SiAlON 相的 Z 值为 3.0；无论 $N_2$ 气氛下还是 Ar 气氛下，只有少量的 15R 存在，其相对含量较原料粉明显要少，都没有 χ 相产生。而且 $N_2$ 气氛下热压产物中的 AlON 的量较 Ar 气

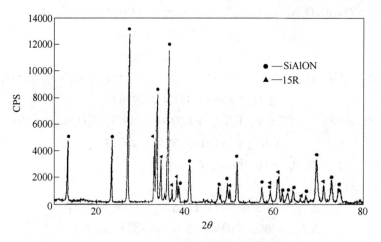

图 2　β-SiAlON 原料粉的 XRD 图谱

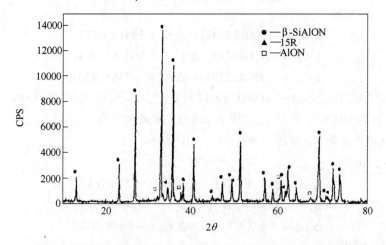

图 3　Ar 气氛下制备的 β-SiAlON-AlON 复相材料的 XRD 图谱

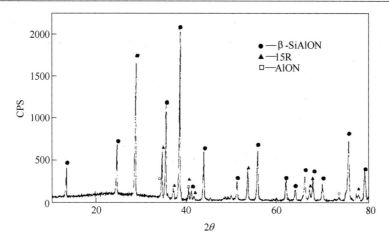

图 4  $N_2$ 气氛下制备的 β-SiAlON-AlON 复相材料的 XRD 图谱

氮下的要少。

从 β-SiAlON 与 AlON 作用使 β-SiAlON 的 $Z$ 值变大的反应来看,很明显由于 $N_2$ 气氛下,有利于使 β-SiAlON 的 $Z$ 值变大的反应,使得 $N_2$ 气氛下,所得产物中的 β-SiAlON 相的 $Z$ 值要比 Ar 气氛下的要大,并使 AlON 的量减少,同时,由于低 $Z$ 值的 β-SiAlON 与 15R 反应产生高 $Z$ 值的 β-SiAlON,因此,无论是在 Ar 气氛下还是 $N_2$ 气氛下,烧结产物中,15R 的含量较原料粉要少得多。尽管 Ar 气氛下,在反应过程中,β-SiAlON 与 AlON 作用形成 χ 相,由于形成的 χ 相进一步与 15R 反应最终生成 β-SiAlON,因而没有 χ 相的残留。

## 3  结论

对 β-SiAlON 与 AlON 复相材料合成过程的热力学分析表明:

(1) $N_2$ 气氛下,β-SiAlON 与 AlON 作用使 β-SiAlON 的 $Z$ 值变大,$N_2$ 有利于该过程的进行。Ar 气氛下,$N_2$ 分压降低该过程减缓。

(2) Ar 气氛下,存在着 β-SiAlON 与 AlON 作用形成 χ 相,χ 相进一步与 15R 反应最终生成 β-SiAlON 的过程。

(3) 无论是 Ar 气氛下还是 $N_2$ 气氛都存在低 $Z$ 值的 β-SiAlON 与 15R 反应产生高 $Z$ 值的 β-SiAlON 的过程。

(4) 控制一定的工艺条件,可合成 β-SiAlON-AlON 复相材料。

热力学分析与实验结果相吻合。

### 参 考 文 献

[1] J M Bauer. 欧洲高炉炉衬耐火材料的发展趋势 [J]. 国外钢铁, 1992, 9: 6-15.
[2] T Hosaka, M Kato. A Study of Compositional Modification of Trough using Aluminum oxynitride [J]. 耐火物, 1985, 37 (10): 22-26.
[3] K Murakami, A Iwasakil, Y akatsuka, et al. One result of sliding nozzle Refractories using Aluminum oxynitride [J]. 耐火物, 1986, 38 (1): 18-20.
[4] K Takeda, T Hosaka. Characteristics of New Raw Materials "AlON" for Refractory [J]. Interceram, 1989,

38（1）：18-20.

[5] H. X. Willems, et al. Thermodynimies of Alon I：Stability at Lower Temperatures ［J］. J, Eur, Ceram, Soc. ,1992,10：327-337.

[6] D. A. Gunn. A Theoretical Evaluation of Sialon-bonded Silicon Carbide in the Blast Furnace Environment ［J］. J, Eur, Ceram, Soc. , 1993, 11：35-41.

[7] 文洪杰. 红柱石合成 β-SiAlON 的基础研究 ［D］. 1998.

（原载于《第七届全国耐火材料青年学术报告会论文集》，1999）

# 热压烧结 Ta/β'-Sialon 系梯度功能材料的残余热应力分析

丁保华　李文超

（北京科技大学理化系，北京　100083）

**摘　要**：探讨了热压烧结制备 Ta/β'-Sialon 系梯度功能材料（FGM）的最佳参数，利用显微力学探针测量了 Ta/β'-Sialon 系 FGM 各层的杨氏模量。提出了用于评价复合材料物性的抛物线模型，并对其物性进行了估算。在此基础上，对 Ta/β'-Sialon 系 FGM 的残余热应力进行了分析，得出最佳成分分布指数 $P$ 为 0.89。

**关键词**：热压烧结；Ta/β'-Sialon FGM；残余热应力；抛物线模型；梯度功能材料

## Analysis of Residual Thermal Stress in Ta/β'-Sialon FGM Prepared by Hot-pressing Sintering

Ding Baohua　Li Wenchao

(Department of Physical Chemistry, University of Science and Technology Beijing, Beijing 100083)

**Abstract**: Preparation of Ta/β'-Sialon FGM by using hot-pressing sintering is introduced in this paper. The Young's modules of composites made up of different components are measured. A new model called parabolic model is put forward to estimate various physical properties. Based on these work, the residual thermal stress in Ta/β'-Sialon FGM has also been analyzed in this paper.

**Keywords**: hot-pressing sintering; Ta/β'-Sialon FGM; residual thermal stress; parabolic model; functionally gradient material

## 1　引言

目前，用于梯度功能材料（FGM）制备的方法主要有以下 5 大类[1~5]：物理气相沉积法（PVD）、化学气相沉积法（CVD）、等离子喷镀法（PS）、自蔓延高温合成法（SHS）与粉末冶金法（PM）。粉末冶金法能够制备出较大尺寸的 FGM，而且设备简单，工艺流程短，便于进行工业生产，因此粉末冶金法是应用于制备 FGM 最为常用的方法之一。金属/陶瓷系 FGM 的一侧是具有高强度、良好导热导电性能、易加工的金属材料，另一侧则是

---

本文得到国家自然科学基金资助。

丁保华，于北京科技大学理化系取得博士学位。主要从事梯度功能材料、高技术陶瓷材料及计算机辅助设计与模拟的研究工作。

具有良好高温性能的陶瓷材料。金属/陶瓷系 FGM 具有许多优异的性能，因此，成为梯度功能材料研究的热点之一。由于金属钽可以耐各种液态金属腐蚀，在核辐照下由于退火效应可以去除辐照破坏，因此 Ta/β'-Sialon 系 FGM 可望用于核工业的防护材料和热障结构材料。材料从烧结温度冷却至室温时产生的残余热应力应该尽可能小，否则可能会破坏该材料。因此，正确评估材料的残余热应力是梯度功能材料设计的关键。本文将着重研究 Ta/β'-Sialon 系 FGM 的残余热应力问题。

## 2 实验方法

热应力缓和型梯度功能材料设计是在设定构件的形状和使用条件的基础上，基于设计知识库，选择合适的材料组合；然后进行复合材料的性能推断，模拟温度分布与热应力分析，得出最佳成分分布指数 $P$。为了求解 FGM 内部的温度分布与热应力分布，必须获得多层混合系的任意混合比中的物性值。然而一般来说，材料物性不仅与组成有关，而且还随着制造过程的不同而有所改变。因此，利用实验的方法获得不同组成比的复合材料的物性值，是非常困难的。目前一般根据已掌握的构成 FGM 的基本材料物性值，利用一定的数学物理模型来推断其中间组成比的各种物性值。金属/陶瓷系圆片 FGM 的组成由 $C=(Z/H)^P$ 决定[6~8]。其中，$C$ 是梯度层中金属相的体积百分含量；$Z$ 是以纯陶瓷相一端为原点坐标的纵坐标；$H$ 是样品的厚度；$D$ 是试样的直径；$P$ 是组成分布指数，它决定了组成分布曲线的形状。

圆片 FGM 的成分设计模型如图 1 所示。材料的组成结构设计就是对不同 $P$ 值下的热应力进行计算，据此得出最佳的 $P$ 值。原料准确称量后，经过球磨 24h、干燥，采用 11 层等厚进行布料，试样总厚度为 22mm；预压成型然后在 1750℃、25MPa 压力、高纯 Ar 气氛保护下进行热压烧结。烧结后的试样切成 22mm×5mm×5mm 大小的样块，经粗磨、细磨、抛光、清洁与干燥处理，注意保持两面平行；然后利用显微力学探针（NANO Indenter Ⅱ）采用压痕法测得各层的弹性模量。

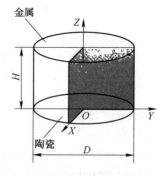

图 1　圆片状 FGM 成分设计模型
Fig. 1　Design model of disc FGM

## 3 实验结果与讨论

### 3.1 抛物线模型的提出

有关多相材料的各种物性值的复合准则，Kemer 等人较早开始研究。其后，不少学者也对复合材料的物性参数估算进行了研究。依据微观结构决定的"混合律"可半定量得到不同混合配比材料的物性参数。假定 FGM 由陶瓷 A 与金属 B 组成，$P_A$、$P_B$ 分别是纯陶瓷 A 与纯金属 B 的物性参数，$C_A$、$C_B$ 分别是陶瓷 A 与金属 B 在复合材料中的体积百分数（$C_A+C_B=1$）。复合材料物性参数模型大致有以下几种：

算术平均模型

$$P=C_AP_A+C_BP_B \tag{1}$$

调和平均模型

$$\frac{1}{P} = \frac{C_A}{P_B} + \frac{C_B}{P_A} \tag{2}$$

可调参数模型

$$P = k(C_A P_A + C_B P_B) + (1-k)\frac{P_A P_B}{C_A P_A + C_B P_B} \tag{3}$$

算术平均模型是最简单的模型，仅仅是使用线性插值的办法来估算材料的物性参数，因此计算误差较大；调和平均模型与算术平均模型都是已知两点物性参数来估算复合材料物性参数，称为两点模型。可调参数模型是在算术平均模型与调和平均模型基础之上提出的一种模型，它的实质是将算术平均模型与调和平均模型赋予不同的权值，然后进行加权平均以得到物性参数。式（3）中的 $k$ 是可调参数，满足 $0 \leq k \leq 1$。当 $k$ 取 1 时，可调参数模型就蜕变为算术平均模型；当 $k$ 取 0 时，可调参数模型就蜕变为调和平均模型。以上三种模型在一定程度上基于线性插值，而材料之间的相互作用是非线性关系。

考虑到上述三种模型的局限性，本文提出一种基于非线性插值的用于计算复合材料物性参数的模型。假定复合材料的物性符合抛物线规律，由于确定抛物线方程需要三个已知点的坐标，因此抛物线模型与可调参数模型都是三点模型。与两点模型相比，它需要三个已知条件。除纯物质 A 与纯物质 B 的物性必须已知以外，还需要已知某一个组成的物性。假设 $P_{0.5}$（$C_A = C_B = 0.5$ 的物性）已知，可以推导出抛物线模型的计算公式。

抛物线模型

$$P = aC_A^2 + bC_A + c \tag{4}$$

分别将三个已知点 $(0, P_B)$、$(0.5, P_{0.5})$ 与 $(1, P_A)$ 代入式（4），求得待定参数 $a$、$b$、$c$ 的解，如式（5）所示：

$$\begin{cases} a = 2P_A + 2P_B - 4P_{0.5} \\ b = 4P_{0.5} - P_A - 3P_B \\ c = P_B \end{cases} \tag{5}$$

因此，抛物线模型表达式如式（6）所示：

$$P = (2P_A + 2P_B - 4P_{0.5})C_A^2 + (4P_{0.5} - P_A - 3P_B)C_A + P_B \tag{6}$$

将式（6）推广，假设三个已知点为 $(0, P_B)$、$(C_M, P_M)$ 与 $(1, P_A)$，其中 $0 < C_M < 1$。同理求解得待定参数 $a$、$b$ 与 $c$，代入式（4），得到抛物线模型一般表达式：

$$P = \frac{C_M P_A - P_M + (1-C_M)P_B}{C_M - C_M^2}C_A^2 + \frac{P_M - C_M^2 P_A + (1-C_M^2)P_B}{C_M - C_M^2}C_A + P_B \tag{7}$$

为了对以上四种模型估算物性的准确性作一个比较，分别用这四个模型计算了金属 Ta 与 β'-Sialon 陶瓷不同组成比的热膨胀系数，如图 2 所示。曲线 1 是实验数据；曲线 2 为算术平均模型估算值；曲线 3 为调和平均估算值；曲线 4 为可调参数模型估算值；曲线 5

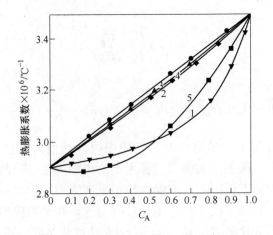

图 2 Ta 与 β'-Sialon 不同混合比热膨胀系数估算

Fig. 2 Estimative coeffecient of thermal expansion for various ratio of Ta to β'-Sialon

为抛物线模型估算值。由图可以看出：抛物线模型在已知三点的情况下，明显优于前三种模型。其中，抛物线模型选择 $C_M = 0.5$。这说明不同组成比的复合材料，其物性在一定程度上存在抛物线规律。用抛物线模型估算的各层物性参数如表 1 所示。

表 1 Ta 与 β'-Sialon 复合材料的物性值

Table 1 Physical properties of Ta/β'-Sialon composites

| Ta/% | 热膨胀系数 $\times 10^6/℃^{-1}$ | 杨氏模量/GPa | 泊松比 |
|---|---|---|---|
| 0 | 2.50 | 66.377 | 0.290 |
| 10 | 2.90 | 76.274 | 0.291 |
| 20 | 3.30 | 87.564 | 0.293 |
| 30 | 3.70 | 100.508 | 0.294 |
| 40 | 4.10 | 115.390 | 0.297 |
| 50 | 4.50 | 132.455 | 0.299 |
| 60 | 4.90 | 151.693 | 0.303 |
| 70 | 5.30 | 172.136 | 0.308 |
| 80 | 5.70 | 189.393 | 0.316 |
| 90 | 6.10 | 185.417 | 0.328 |
| 100 | 6.51 | 190.000 | 0.350 |

## 3.2 热应力分析

假定试样符合平板模型，即在不产生翘曲应力的前提下，可以用径向应力来反映材料的残余热应力。计算模型如下：设两平板试样在 $T_0$（高温）时，长度均为 $L_0$；当冷却至 $T_1$ 时，两试样自由膨胀（实际为收缩）后，长度分别为 $L_1$ 与 $L_2$，烧结体长度为 $L_S$。如图 3 所示。

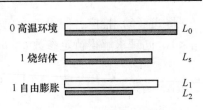

图 3 热应力计算平板模型

Fig. 3 Plate model of thermal stress calculation

并假定冷却过程只产生弹性形变。两材料的线性热膨胀系数、泊松比与杨氏弹性模量分别为：$\alpha_1$、$\alpha_2$、$\nu_1$、$\nu_2$、$E_1$ 与 $E_2$。

$$L_1 = L_0(1 + \alpha_1 \Delta T) \tag{8}$$

$$L_2 = L_0(1 + \alpha_2 \Delta T) \tag{9}$$

$$\sigma_1 = \frac{E_1(L_S - L_1)}{(1 - \nu_1)L_1} \tag{10}$$

$$\sigma_2 = \frac{E_2(L_S - L_2)}{(1 - \nu_2)L_2} \tag{11}$$

$$\Delta T = T_2 - T_1 \tag{12}$$

$\sigma_1$ 与 $\sigma_2$ 分别是上下两侧所受的应力，由于 $\sigma_1$ 与 $\sigma_2$ 大小相等、方向相反，联立式（8）至式（12）得

$$\sigma_1 = \frac{\Delta\alpha \Delta T G_1 G_2}{(1 + \alpha_1 \Delta T)G_2 + (1 + \alpha_2 \Delta T)G_1} \tag{13}$$

其中

$$\Delta\alpha = \alpha_2 - \alpha_1 \tag{14}$$

$$G_1 = \frac{E_1}{1 - \nu_1} \tag{15}$$

$$G_2 = \frac{E_2}{1 - \nu_2} \tag{16}$$

因此，两材料直接热应力可由式（13）计算出来。定义热应力缓和率

$$\eta = \left(1 - \frac{\sigma_i}{\sigma}\right) \times 100\% \tag{17}$$

在式（17）中，$\sigma_i$ 为第 $i$ 个界面层的热应力，$\sigma$ 为纯金属与纯陶瓷直接结合界面的热应力。

根据估算的物性值，对不同 $P$ 值的 Ta/$\beta'$-Sialon 系 FGM 同样进行了热应力分析，计算温度场从烧结温度冷却至室温 25℃，假设不发生塑性形变，结果如图 4 所示。

由图 4 可知，当取 $P = 0.5$、0.6 时，Ta/$\beta'$-Sialon 系 FGM 的层间热应力在纯 Ta 一侧缓和效果最好，即热应力在纯 Ta 一侧最小；随着 $P$ 值不断增大，纯 $\beta'$-Sialon 侧热应力缓和率逐渐提高；当 $P = 0.85$ 时，热应力最佳缓和位置过渡到材料中部；当 $P = 0.89$ 时，热应力在第三个层间界面缓和最好，表明热应力在第三层与第四层的层间界面达到最小值；随着 $P$ 值的进一步增大，热应力最佳缓和位置继续向纯 $\beta'$-Sialon 一侧转移。当 $P = 1.0$ 时，热应力缓和最佳位置为纯 $\beta'$-Sialon 一侧；当 $P$ 继续增大时，纯 Ta 侧的热应力缓和效果继续下降；达到最大热应力缓和时的 $P$ 值为 0.89，即 $P_0 = 0.89$。因此，随着 $P$ 值增大，纯 Ta 侧热应力缓和率不断降低，纯 $\beta'$-Sialon 侧热应力缓和率不断升高。$P = 0.89$ 时，整个 FGM 热应力缓和最

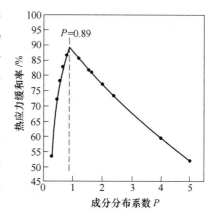

图 4　不同 $P$ 值的 Ta/$\beta'$-Sialon 系 FGM 热应力分析

Fig. 4　Analysis of thermal stress in Ta/$\beta'$-Sialon system

佳。由图 4 可知，$P=1.0$ 的最低热应力缓和率也高达 87.6%。因此可以取 $P=1.0$ 进行 Ta/β'-Sialon 系 FGM 的配制。

## 4 结论

（1）提出了用于估算复合材料物性参数的抛物线模型，并以 Ta/β'-Sialon 系 FGM 为例，验证了抛物线模型的可靠性。

（2）对 Ta/β'-Sialon 系 FGM 的热应力进行了估算，结果表明：热应力缓和效果与 $P$ 值有关，随着 $P$ 值的增大，热应力缓和最佳的界面从纯 Ta 一侧过渡到纯 β'-Sialon 一侧。最佳成分分布指数为 0.89，与 1.0 非常接近，因此可以选择 1.0 作为实验配方的成分分布指数。

（3）采用 $P=1.0$，利用热压烧结制备了 Ta/β'-Sialon 系 FGM；中间层热应力缓和率接近 90%，与本文提出的模型吻合较好。

## 参 考 文 献

[1] Seüchi Uemura, Yoshio Sohda, Yuk inori Kude, et al. Preparation and Evalication of SiC/C functionally gradient materials by chemical vapor deposition [J]. 粉体およひ粉末冶金，1990，37（2）：275-282.

[2] Makoto Sasaki, Toshiyuki Hashida, Toshio Hirai, et al. Thermal shock resistance of SiC/C functionally gradient material prepared by chemical vapor deposition [J]. 粉体およひ粉末冶金，1990，37（7）：966-967.

[3] Akira Kawasaki, Ryuzo Watanabe. Microstructural designing and fabrication of disk shaped functionally gradient material by powder metallurgy [J]. 粉体およひ粉末冶金，1990，37（2）：253-258.

[4] Toshikazu Takakura, Isso Tanaka, Yoshinan Miyamto, et al. Fabrication of $TiB_2$-Ni functionally gradient materials by a gas-pressure combustion simering [J]. 粉体およひ粉末冶金，1990，37（7）：933-936.

[5] 丁保华，李文超，王福明，等. 粉末冶金法制备 Ta/β'-Sialon 梯度功能材料的研究 [J]. 中国稀土学报，1998，16（8）：637-641.

[6] 丁保华，李文超. 梯度功能材料的研究现状与展望 [J]. 耐火材料，1998，32（5）：295-298.

[7] 丁保华，李文超，王福明，等. 分形在梯度功能材料中的应用 [J]. 材料导报，1998，12（5）：6-8，15.

[8] 沈强，张联盟，袁润章. Ni/Ni3Al-TiC 系梯度功能材料的组成结构设计与制备 [J]. Journal of the Chinese Ceramic Society，1997，25（4）：406-412.

（原载于《耐火材料》，2000，34（1）：27-30）

## 石英向α-方石英转化率研究

王金淑[1] 王俭[1] 李文超[1] 王彦平[2] 吕虎太[2]

(1. 北京科技大学,北京 100083;2. 邯郸陶瓷二厂,邯郸 056000)

## Study of Transformation from α-Quartze to α-Cristobalite

Wang Jinshu[1] Wang Jian[1] Li Wenchao[1] Wang Yanping[2] Lv Hutai[2]

(1. University of Science and Technology Beijing, Beijing 100083;
(2. The Second Factory of Procetain, Handan 056000)

**摘 要**:本文研究了1320℃下石英向方石英的转化率,讨论了金属氧化物添加剂对石英转化的影响,证实:氧化物中金属离子半径越小,价态越低,石英向方石英的转化率越高。

## 1 引言

众所周知,石英有三种形态,即石英、鳞石英和方石英;它们由α向β晶型转变的温度分别为573℃、117℃和200~275℃[1],1913年Ferner的研究表明:在870℃以下为石英的稳定区;870~1470℃时鳞石英是一种特殊的结构,在完全没有外加化学成分时便不可能形成[6,8]。因此,在标准大气压下,纯氧化硅系晶相转化仅限于石英和方石英[2]。

文献[3]中报道,在1470~1650℃石英转化为方石英。低于1470℃,此转化过程的实现,必须外加添加剂。Takao Suzuki等人的研究表明,$Al_2O_3$,可以促进石英的转化[4]。但转化时间长,且转化率低,不足40%。

生产α-方石英强化瓷,使石英在成瓷温度下转化为α-方石英是关键的工艺之一。因此,寻找高效添加剂,提高石英的转化效率,具有重要的实用价值。

## 2 实验方法

### 2.1 原料

其化学成分示于表1。

表1 石英的化学成分

| 组成 | $SiO_2$ | $Al_2O_3$ | $Fe_2O_3$ | CaO | MgO | $K_2O$ | $Na_2O$ | TiO | MnO | $P_2O_5$ |
|---|---|---|---|---|---|---|---|---|---|---|
| wt% | 99.71 | 0.54 | 0.03 | 0.03 | 0.04 | 0.96 | 0.03 | 0.008 | 0.007 | 0.006 |

注:ICP分析,微量元素28种(略)。

将原料球墨,使其粒度达到 2.95~10nm;然后分为若干批次,分别加入不同的金属氧化物添加剂,混合均匀,备用。

## 2.2 转化率试验

将试样分别装入刚玉坩埚,置于钼丝炉恒温区内,用 Pt-PtRh 热电偶测温,DWK-702 控温。当温度开至 1320±2℃时,保温 2 小时,而后降温;当温度降至 300℃时,为防止 α-方石英转变为 β-方石英,将坩埚取出急冷。整个过程模拟隧道窑的烧成工艺。

## 2.3 转化率分析

将转化后的试样制成 X 射线衍射试样,进行 X 射线衍射分析[6],计算其转化率。计算公式为:

$$\frac{C_Q}{C_{cr}} = \frac{I_Q/L_Q}{I_{cr}/L_{cr}}$$

式中,$C_Q$、$C_{cr}$ 分别为石英、α-方石英的含量,wt%;$I_Q$、$I_{cr}$ 分别为石英、α-方石英在(101)面的衍射强度;$L_{cr}/L_Q$ 为常数,可由物质的吸收系数、结构常数等计算而得,在上式中 $L_{cr}/L_Q = 0.61$。于是 $C_{cr} = I_{cr}/(I_{cr} + 0.61 I_Q)$。

## 2.4 形貌观察

将转化前后的粉末放入酒精中,经超声波处理,打碎团聚体,制成扫描电镜或透射电镜试样;经喷碳处理后,进行相应的观察。

# 3 试验结果分析

## 3.1 相变催化剂的选择

从相变热力学得知,在 1470℃石英开始向方石英转化。因此,纯石英在 1320℃是不可能发生转化的。然而,由于杂质的存在,约有 6.17% 的石英发生了转化(见图 1(a))。如何选择相变催化剂呢?文献中报道 CaO、MgO、CuO、PbO、$Al_2O_3$、$Na_2O$ 等均可采用,且认为 $Al_2O_3$ 较为理想[4]。为此,研究了不同相变催化剂含量,其离子半径对比石英转化率的影响见表 2 和图 2、图 3。

从实验结果可以看出,相变催化剂的离子半径和价态对石英转化率有明显的影响。在相同的加入量下,对于同一价态的相变催化剂,其离子半径越小,催化效果越明显,转化率也就越高,在相同加入量下,催化剂的价态越低,催化作用越强,转化率也越高(见图 2、图 3)。一般来讲,阳离子半径电荷越小,离子半径就越小,其金属性就越强;因此破坏 Si—O 键的能力就越强,使晶体内缺陷增多,有利于方石英成核。所以,相变催化剂的催化效果依次为:一价离子 > 二价离子 > 三价离子。

石英转化为方石英,发生硅氧键断裂、改键。由扎哈里亚生的网格学说[7],金属氧化物可使 Si—O 键断裂数目增多,并形成较多的晶格缺陷,有利于 α-方石英成核,从而降低了转化温度。

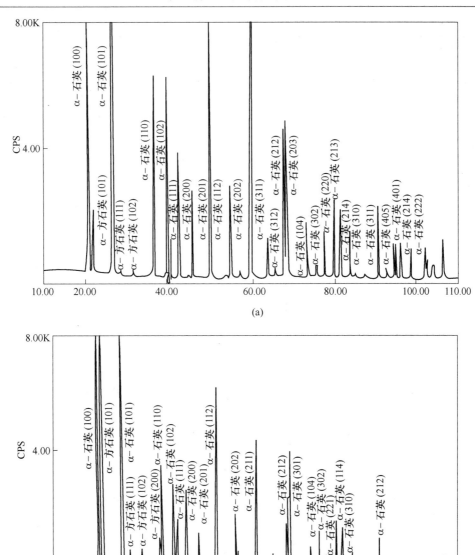

**图 1　1320℃下处理后石英的 X 射线衍射谱**

(a) 无添加剂；(b) 添加 1% $La_2O_3$

**表 2　相变催化剂的作用**

| 相变催化剂 | 转化率/%    含量/wt%    离子半径 | 0.5 | 1.0 | 1.5 | 2.0 |
|---|---|---|---|---|---|
| $La_2O_3$ | $La^{3+}_{0.61}$ | 29.89 | 28.24 | 17.51 | 43.52 |
| $Al_2O_3$ | $Al^{3+}_{0.05}$ | 28.20 | 29.00 | 26.50 | 27.30 |
| $MgO$ | $Mg^{2+}_{0.65}$ | 33.11 | 29.76 | 30.68 | 29.89 |
| $CaO$ | $Ca^{2+}_{0.80}$ | 33.70 | 39.60 | 36.46 | 35.79 |

续表2

| 转化率/% 含量/wt% 相变催化剂 离子半径 | | 0.5 | 1.0 | 1.5 | 2.0 |
|---|---|---|---|---|---|
| $Na_2O$ | $Na_{0.97}^{1+}$ | 95.35 | 98.70 | 99.39 | >99.39 |
| $CuO$ | $Cu_{0.69}^{2+}$ | 22.78 | — | — | — |
| $PbO$ | $Pb_{0.86}^{2+}$ | — | 35.00 | — | — |

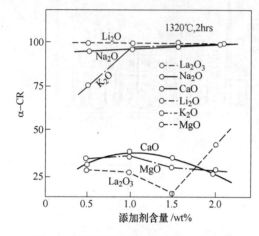

图2 不同添加剂及其含量对 α-方石英转化率的影响

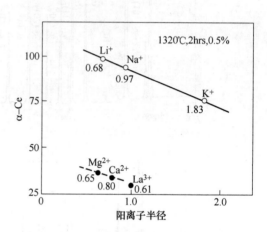

图3 添加剂的阳离子半径对 α-方石英转化率的影响

### 3.2 石英粒度对转化率的影响

实验证明,石英粒度越细,在相同条件下转化率越高。以添加 4% $Al_2O_3$ 为例,在 1320℃下,当石英粒度为 2.95nm,转化率为 42.2%;而当石英粒度为 9.17nm 时,其转化率仅为 29.1%。催化作用从表面或界面开始,石英颗粒越细,意味着比表面积大,催化效果好。

### 3.3 形貌观察和转化机理分析

透射电镜观察结果,示于图4。

实验观察到,石英转化为方石英颗粒长大,密度减小。如平均粒度为 2.95nm 的石英,转化为 α-方石英后平均粒度为 8.71nm;又如平均粒度为 9.17nm 的石英颗粒,转化为 α-方石英后平均粒度为 13.56nm。测得的密度分别为 $d_{石英} = 2.635 \times 10^{-3} kg/m^3$;$d_{方石英} = 2.32 \times 10^{-3} kg/m^3$。可以推断,α-方石英中的缺陷大于石英中的缺陷。

## 4 结论

(1) 在 1320℃下,通过加入金属氧化物相变催化剂可使石英提前转化为 α-方石英;催化剂的离子半径越小,越有利于转化率提高。

(2) 原料颗粒变细,会增大比表面能,改善界面催化的动力学条件,有利于转化率的

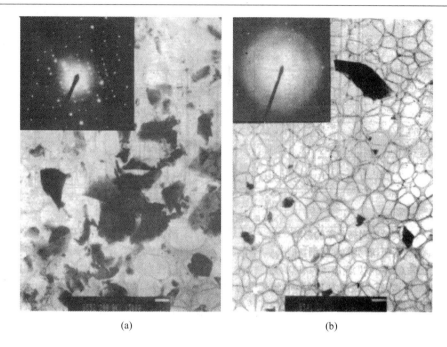

(a)　(b)

图 4　石英与方石英投射电镜形貌

(a) α-石英；(b) α-方石英

提高。

(3) 石英转化为 α-方石英，颗粒长大，密度下降，缺陷增多。

## 参 考 文 献

[1] E. M. Levin, H. F. Mcmurdie, F. P. Hall. Amer. Ceram. Soc., 1956：35.
[2] 莎尔满，舒尔茨，著. 黄熙柏，译. 陶瓷学 [M]. 北京：中国轻工业出版社，1989.
[3] A. O. Chakader, A. Roberts. J. Amer. Sci., 1961, 44.
[4] TadahisaArahori, Takao Suzuki. J. of Material Science, 1987, 22 (6)：2248.
[5] 李树堂. 金属 X 射线衍射与电子显微分析技术 [M]. 北京：冶金工业出版社，1980.
[6] D. M. Roy, R. Roy. Amer. Minerl., 1964, 49：952.
[7] 桥本谦一，滨野健也. 陶瓷基础 [M]. 北京：中国轻工业出版社，1986.
[8] M. Sanmoy. Trans Brit Ceram. Soc., 1977, 76：71.

（原载于《中国陶瓷》，1992，127 (6)：1-4）

# 刚玉强化日用瓷的理论分析

李文超[1]　王俭[1]　刘建华[1]　李国桢[1]　刘自强[2]　范社岭[2]　吴文亮[2]

（1. 北京科技大学，北京　100083；2. 河北邯郸第四陶瓷厂，邯郸　056200）

## Theoretical Analysis of Durable Porcelain with α-Al$_2$O$_3$ Addition

Li Wenchao[1]　Wang Jian[1]　Liu Jianhua[1]　Li Guozhen[1]
Liu Ziqiang[2]　Fan Sheling[2]　Wu Wenliang[2]

（1. University of Science and Technology Beijing，Beijing 100083；
2. No. 4 Porcelain Factory of Handan，Handan 056200）

**摘　要**：本文分析了刚玉强化日用瓷的物理化学基础，用相图、液固相反应动力学讨论了莫来石生成机理，推导了速度公式；理论推导与实验得到的经验公式一致。

$\ln(1.38 - m) = -0.408t - 0.336$，证实莫来石生成为一级反应，受扩散控制。

本文在单因素分析的基础上用模式识别处理了八个因素的影响；对53组实验按主成分分析法进行计算机信息处理，最后在二维空间投影，从而得到强化瓷的优化区为：

$$Y > 1.23X + 178 \quad Y < 110 \text{ 或 } Y > 110 + \sqrt{138^2 - (X-120)^2} \quad (X < 258)$$

在该区域内选取了三个点，在工厂进行了扩大实验，均得到了令人满意的结果。

本文还考察了结构与力性的关系，证实了均匀的刚玉相分布，莫来石交织成网状结构强化了玻璃相，较少的残余石英（<3%），以及均匀分布的气孔等是获得良好力性的基础。

# 1　引言

众所周知，随着科学技术的发展，机械化供餐、蒸洗消毒设备、家用电器等得到广泛的应用，因而对强化日用瓷的需求量与日俱增。近二十年来，强化瓷在国外已成为重点研究的课题。1975年苏联出现了高强度低温瓷[1]，1983年西德在陶瓷三组分的基础上，通过物质的替代，把日用瓷的强度提高了一倍[2]。日本于1972年开始研制，1988年有六家陶瓷公司生产强化瓷中间性商品，1989年近五十家陶瓷公司投入强化日用瓷的开发生产，其强度比传统的日用瓷提高了2~3倍[3,4]。我国台湾地区生产的大同强化瓷，通过添加Al$_2$O$_3$使日用瓷的强度提高了1~2倍。大陆先后有四个科研与生产厂家，研究并试生产了强化日用瓷，取得了满意的结果。本文拟对强化瓷的理论基础进行分析与探讨。

---

刘建华，1988~1991年于北京科技大学师从李文超教授攻读硕士学位。目前在北京科技大学冶金工程研究院工作，教授。发表论文150余篇，获省部级奖2项。

## 2 日用强化瓷的物理化学基础

物理化学是众多工程技术学科的理论基础,诸如:冶金、陶瓷、化工、石油等都建立在物理化学的基础上,物化热力学是不可违背的原理性理论;物化动力学是解释并指导工艺生产的现象定律,称之为结构性理论,研究试制刚玉强化瓷也自觉不自觉地要遵守这些原理。

### 2.1 刚玉强化瓷的相围分析

我们把历代青花和国内各单位研制的刚玉强化日用瓷瓷胎成分,以及我们实验室研究的若干配方都绘制在 $\Sigma SiO_2 + P_2O_5 + (\Sigma RO_2) - 2K_2O + Na_2O + CaO + MgO + (\Sigma R_2O + RO) - Al_2O_3$ 三元相图的等温截面图上(见图1)。

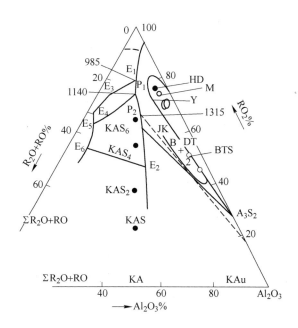

图1 $RO_2 - (R_2O + RO) - Al_2O_3$ 陶瓷三元系

由图可以看出:(1)传统青花与强化日用瓷以低共熔点 $E_1$ 与莫来石($A_3S_2$)连线为界分为两个区域;(2)传统与强化日用瓷各式分代表点均落在 $SiO_2$-$KAS_6$-$A_3S_2$ 基础三角形内。由杠杆原理和重心规则可以证明,在不改变传统工艺的条件下,强化瓷胎中莫来石量增加,而玻璃相与石英相相对减少;在准平衡的条件下,强化瓷相组成为莫来石、石英和白榴石;在快速冷却的条件下,相组成为莫来石、α-方石英和玻璃相。

综上所述,由相图分析得出,从强化瓷的化学成分在三元系的代表点看,在传统的工艺条件下不可能产生刚玉相。因此,外加并稳定刚玉相是研制刚玉强化日用瓷的关键所在。

### 2.2 强化瓷工艺过程中的动力学分析

强化瓷之所以强化,一是出现均匀分布的强化相刚玉;二是在玻璃相和长石残骸中生长出交织成网状的一次和二次莫来石,强化了坯体;三是石英相明显下降,甚至少于2%,因而石英与石英周围裂纹的断裂源相应消除,达到了强化的目的。下面我们对强化瓷中莫来石生成的动力学规律进行理论分析。

### 2.2.1 反应机理

反应机理可分为如下几步:

(1) 液相中的铝离子 $Al^{3+}$ 向固相表面(长石残骸等)迁移;

(2) 硅氧四面体向反应界面迁移;

(3) 在反应界面上发生界面化学反应生成莫来石。

$$6Al^{3+}_{(1)+2+5} = 3Al_2O_3 \cdot 2SiO_2(s) \tag{1}$$

由此可见,生成莫来石的反应为连串反应,整个过程受最慢一步控制。在高温下(1300℃)化学反应速度较快,可认为达到局部平衡;硅氧四面体浓度较高,且在生成莫来石的过程中消耗 $Al^{3+}$ 是 $SiO_4^{4-}$ 的三倍,由此得出整个过程应受铝离子的扩散所控制。

### 2.2.2 反应速度公式推导

由上述分析得知,铝离子的扩散速度即为莫来石的生成速度,即

$$V_{A_3S_2} = -V_{Al^{3+}}; \quad \frac{dm}{dt} = -\frac{d[\%Al^{3+}]}{dt}$$

$$\frac{dh_{Al^{3+}}}{dt} = A\frac{D_{Al^{3+}}}{\delta}(C_{Al^{3+}} - C^0_{Al^{3+}})$$

换算为重量百分浓度

$$\frac{d[\%Al^{3+}]}{dt} = \frac{AD_{Al^{3+}}}{V\delta}([\%Al^{3+}] - [\%Al^{3+}]^0) \tag{2}$$

式中,$[\%Al^{3+}]^0$ 为反应界面上铝离子浓度,可通过体相浓度来表示。而莫来石的生成浓度与铝离子浓度成正比,即 $dm = k_0 d[\%Al^{3+}]$ 分离变量,定积分得到:

$$-\int_{[\%Al^{3+}]^0}^{[\%Al^{3+}]} \frac{d[\%Al^{3+}]}{[\%Al^{3+}] - \psi} = \int_0^t \frac{AD_{Al^{3+}}}{V\delta} K_0 dt \tag{3}$$

或写为

$$-\int_{m^0}^{m} \frac{dm}{m - \psi} = \int_0^t \frac{AD_{Al^{3+}}}{V\delta} K_0'' dt$$

所以

$$\ln\frac{m^0 - \psi}{m - \psi} = kt \tag{4}$$

这与文献中不定积分得到的公式

$$\ln(m^0/m) = kt + C \tag{5}$$

是一致的。

经过处理本实验数据得到莫来石生长的动力学表达式为:

$$\ln(1.38 - m) = -0.408t - 0.336 \tag{6}$$

此为一级反应,由扩散控制(见图2)。

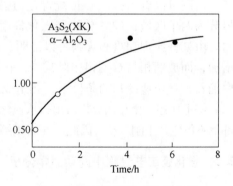

图 2 莫来石生长的动力学曲线

## 3 强化瓷研制过程中的模式识别应用

试验研究证明,影响强化瓷性能的因素很多,诸如:在最佳烧成条件下,随强化瓷中 $Al_2O_3$ 含量的增加,抗折强度增加;在相同配方条件下,抗折强度与烧成温度、原料的粒度、保温时间、相应含量的矿化剂等因素又都存在一个最佳值。在众多因素中,如何考虑各影响因素的交互作用呢?这用普通的正交设计是很难完成的。于是我们利用了计算机信息处理,进行了模式识别[5]。用主成分分析法编辑了计算机程序,其粗框图示于图3。其原理是对多维空间作 Karhuhen-Loeve 变换,把特征量重新组合为相互正交的特征量组 ($X_i$, $Y_i$),使高维空间点列投影式映照到,对应于最大变化幅度的平面上 ($X_i$, $Y_i$),以期得到最佳的分离效果。

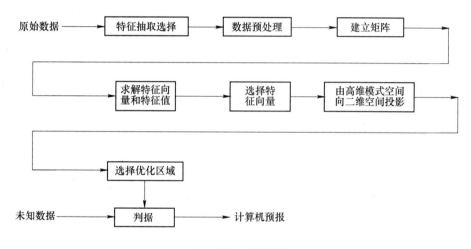

图 3 模式识别框图

我们根据单一因素实验结果,选取了矿化剂、保温时间、烧成温度等七个因素,共53个实验点作为训练点,进行了模式识别。经过反复训练,得到一个分辨率较高的平面,示于表1、表2和图4。

**表 1 模式空间向投影平面的投影方式**

| | | | |
|---|---|---|---|
| 5.73 | *$Al_2O_3$ | 7.05 | *$Al_2O_3$ |
| 73.55 | *CaO | 45.4 | *CaO |
| 88.66 | *MgO | 91.5 | *MgO |
| −41.48 | *$KNa_2O(K_2O + Na_2O)$ | 76.92 | *$KNa_2O(K_2O + Na_2O)$ |
| .255 | *T | −8.7E−04 | *T |
| 8.3 | *t | −2.12 | *t |
| −1.38 | *1d | −3.5E−02 | *1d |
| −244.21 | | −583.31 | |

表2 模式空间在投影平面上的投影

| 投影坐标 | $X_i$ | $Y_i$ | 投影坐标 | $X_i$ | $Y_i$ |
| --- | --- | --- | --- | --- | --- |
| $I=3$ | 235.63 | 49.67 | $I=16$ | 206.10 | 112.01 |
| $I=6$ | 136.01 | 128.08 | $I=17$ | 218.86 | 111.96 |
| $I=8$ | 169.39 | 80.90 | $I=20$ | 211.82 | 174.82 |
| $I=9$ | 156.63 | 80.94 | $I=21$ | 186.54 | 178.45 |
| $I=15$ | 193.34 | 112.05 | $I=22$ | 224.58 | 174.77 |
| $I=18$ | 186.31 | 174.90 | $I=23$ | 216.02 | 136.30 |
| $I=19$ | 199.06 | 174.86 | $I=24$ | 228.78 | 136.26 |
| $I=27$ | 254.30 | 136.17 | $I=25$ | 241.54 | 136.22 |
| $I=28$ | 336.82 | 131.30 | $I=26$ | 216.26 | 139.85 |
| $I=32$ | 258.86 | 112.92 | $I=29$ | 344.48 | 131.27 |
| $I=33$ | 263.96 | 112.90 | $I=30$ | 349.58 | 131.25 |
| $I=36$ | 292.40 | 104.49 | $I=31$ | 251.20 | 112.95 |
| $I=37$ | 275.63 | 108.68 | $I=34$ | 50.32 | 111.92 |
| $I=50$ | 258.96 | 112.92 | $I=35$ | 250.47 | 115.04 |
| $I=51$ | 259.00 | 112.93 | $I=38$ | 424.54 | 91.22 |
| $I=52$ | 301.69 | 136.14 | $I=39$ | 375.66 | 103.59 |
| $I=53$ | 300.75 | 149.64 | $I=40$ | 457.73 | 164.12 |
| $I=1$ | 210.62 | 17.82 | $I=41$ | 506.61 | 151.75 |
| $I=2$ | 248.39 | 49.63 | $I=42$ | 350.94 | 63.27 |
| $I=4$ | 315.29 | 36.63 | $I=43$ | 394.47 | 87.57 |
| $I=5$ | 302.53 | 36.72 | $I=44$ | 349.43 | 78.14 |
| $I=7$ | 123.25 | 128.12 | $I=45$ | 390.60 | 116.19 |
| $I=10$ | 152.39 | 149.09 | $I=46$ | 477.10 | 114.99 |
| $I=11$ | 165.15 | 149.04 | $I=47$ | 474.76 | 128.93 |
| $I=12$ | 177.91 | 149.00 | $I=48$ | 244.85 | 112.57 |
| $I=13$ | 190.67 | 148.95 | $I=49$ | 254.79 | 112.82 |
| $I=14$ | 180.58 | 112.10 | | | |

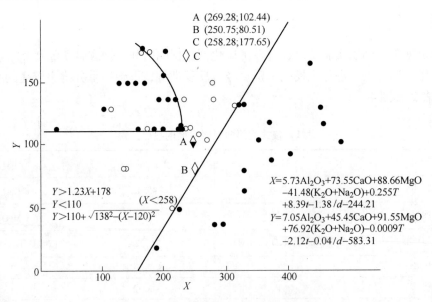

图4 模式空间向投影平面的投影方式

从表1、表2和图4可以看出,抗折强度大于980.665MPa的点分布在如下区域:

$$Y > 1.23X + 178$$
$$Y < 110 \quad (X < 258)$$
$$Y > 110 + \sqrt{138^2 - (X-120)^2}$$

为了验证本模式识别的可靠性，我们在该区域内选了三个代表点：A（269.28；102.44）、B（250.75；80.51）、C（258.28；177.65），以其代表的配方和工艺参数在现场进行了扩大试验，得到产品的抗折强度分别为：1321.94MPa、1119.92MPa 和 1423.93MPa，证实了在大量实验的基础上，模式识别处理预报未知是相当成功的。

$$X = 5.73Al_2O_3 + 73.55CaO + 88.66MgO - 41.48(K_2O + Na_2O) + 0.255T + 8.39t - 1.381d - 244.21$$

$$Y = 7.054Al_2O_3 + 45.46CaO + 91.55MgO + 76.92(K_2O + Na_2O) - 0.00087T - 2.12t - 0.0351d - 583.31$$

## 4 强化瓷的性能与结构的关系

强化瓷的力学性能主要取决于其显微结构（包括相分布、相间结合以及玻璃基质的数量等）。

### 4.1 相组成与力性的关系

我们对国内五种强化瓷 B、D、H、J、Q 进行了 X 射线衍射结构分析，并用相对强度式（7）计算了相组成比，结果见表3。

$$\frac{W_\alpha}{W_\beta} = K \frac{I_\alpha}{I_\beta} \tag{7}$$

相应的显微结构示于图5；相应的力性测试结果示于表4。

表3 国内各强化瓷相组成比

| 牌 号 | B | D | H | J | Q |
|---|---|---|---|---|---|
| $A_3S_2$:α-$Al_2O_3$:α-$Si_2O_3$ | 1:0.65:0.30 | 1:1.46:0 | 1:1.64:0 | 1:0.89:0.50 | 1:0.37:0 |

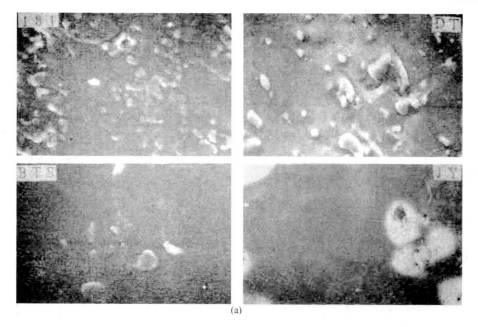

(a)

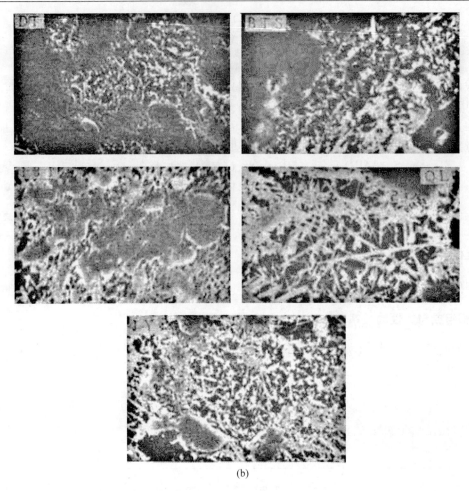

图 5　强化瓷腐蚀前后形貌

（a）刚玉相分布 735×，腐蚀前；（b）莫来石与刚玉分布 4290×，腐蚀后

表 4　国内各强化瓷力性比较

| 代 号 | 抗折强度 /MPa | 抗冲击 /J·m$^{-2}$ | 釉面硬度 HV/MPa |
|---|---|---|---|
| B | 86.89 | | |
| D | 110.82 | | 6574 |
| H | 132.19 | 3619 | 7090 |
| J | 76.88 | | 6580 |
| Q | 70.51 | 3628 | 6390 |

注：取各家成品样；分别按 GB 4741—84 和 GB 3297—82 标准测定。

由上述结果可以看出，刚玉相越多，力性越高；莫来石分布均匀，尺寸在 5～15 μm 范围内对提高力性有利。

### 4.2　保温过程显微结构变化与力性关系

我们用 X 射线衍射方法研究了不同保温条件下，α-$Al_2O_3$ 相的变化，同时用扫描电镜观察了显微结构变化，分别示于图 6、表 5。

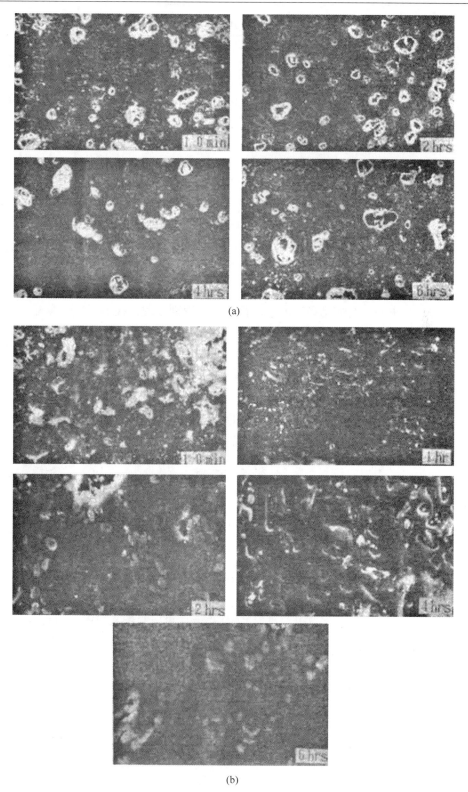

图 6 保温时间与显微结构
(a) 缩孔分布 (680×); (b) 刚玉相分布 (680×)

表 5  H 强化瓷不同保温时间下相组成比

| 保温时间/h | 0.17 | 1 | 2 | 4 | 6 |
|---|---|---|---|---|---|
| $A_3S_2:\alpha\text{-}Al_2O_3$ | 1:0.61 | 1:0.91 | 1:1.04 | 1:1.37 | 1:1.30 |

由上述结果可以看出，随着保温时间延长，莫来石逐渐长大，3 小时以后莫来石增长缓慢，满足抛物线规律，为一级反应，受扩散控制。从显微结构来看，保温 4 小时左右，相间结合好，缩孔最小，力性应为最佳。

## 5 结论

（1）利用物化相平衡分析表明：强化瓷现有的配方中，在一般条件下不易生成刚玉相；外加刚玉相并稳定刚玉相的存在是获得强化瓷的关键条件之一。

（2）物化动力学分析与实验研究表明：强化瓷中莫来石的生成受扩散控制，为一级反应；本实验条件下得到的经验表达式为：

$$\ln(1.38 - m) = -0.408t - 0.336$$

（3）用七个主要影响强化瓷的因素进行了模式识别处理，找出了强化瓷配方和工艺参数的区域为（抗折强度大于 980.67MPa 的点分布在如下区内）：

$$Y > 1.23X + 178 \quad Y < 110$$

或

$$Y > 110 + \sqrt{138^2 - (X - 120)^2} \quad (X < 258)$$

在此区域内，识别率为 88%。

（4）对显微结构进行了分析并与力性测试结果结合起来进行分析，证实：力性主要取决于显微结构；良好的显微结构对应于优异的力性。

### 参 考 文 献

[1] Порявкова С Москва，1975（3）．
[2] Dip-Ling，(FH) R. Rubin，Dr. ch. Hahu. Sprechseal，1983，116（10）：175.
[3] News，Industrial Ceramic，1987，7（2）：93.
[4] 王可鸣. 中国陶瓷，1989，(6)：45.
[5] 陈季镐. 统计模式识别 [M]. 北京：北京邮电科学院出版社，1989.

（原载于《中国陶瓷》，1991，117（2）：7-13）

# 用穆斯堡尔谱和吸收光谱研究汝瓷天青釉呈色机理

秦建武　李国桢　李文超　王　俭

（北京科技大学，北京　100083）

## Investigation of Colouration Mechanism of Ru Ware's Sky Blue Glaze by Mossbauer and Optical Spectrum

Qin Jianwu　Li Guozhen　Li Wenchao　Wang Jian

(University of Science and Technology Beijing, Beijing 100083)

**摘　要**：测定汝瓷天青釉中 $Fe^{57}$ 的 Mossbauer 效应，发现：天青釉中的 $Fe^{2+}$ 同时以四面体和八面体配位存在；少量的 $Fe^{3+}$ 以四面体配位存在。比较计算机拟谱后的各吸收峰面积，得出：$Fe^{2+}/\sum Fe$ 为 87%；$Fe^{3+}/\sum Fe$ 为 13%。再结合热力学参变数状态图和天青釉的吸收光谱，根据配位场理论讨论了铁离子对天青釉呈色的影响，为仿制宋代汝瓷提供了理论依据，并为解释青瓷呈色打下了基础。

## 1　前言

釉是施在坯体表面上，经烧成的多种氧化物组成的玻璃体，由于釉在熔融过程中受到坯体的限制，所以其微观结构与化学成分的均匀性均较玻璃差。

宋代五大名瓷之一——汝瓷，其釉色美似天青，属乳浊釉[1]。天青釉中存在大量钙长石微晶，使得釉色乳浊[2]。釉呈天青色的主要原因，是由于釉中存在不同价态的铁离子；然而，至今仍缺乏充足的实验依据和理论分析。本工作利用穆斯堡尔谱和吸收光谱，确定了铁离子在天青釉中的价态和其在玻璃熔体中的配位位置。在实验的基础上，结合物化热力学分析和配位场理论，进一步探讨了汝瓷天青釉的呈色机理。

## 2　实验方法

### 2.1　试样制备

将河南宝丰清凉寺出土的经考古鉴定的汝瓷残片，用切片机剥离釉层。在操作过程中，用流动水冷却，以防摩擦生热引起釉层 $Fe^{2+}$ 在空气中再氧化。取部分釉片作为吸收光谱试样；其余釉片经研磨，过 200 目筛。取一部分釉粉作为等离子光谱试样，对釉的化学成分进行全分析（除常规元素外，还分析了 28 个微量元素）。另一部分釉粉均匀洒在有机

---

秦建武，1987～1990 年于北京科技大学师从李文超教授攻读硕士学位。目前在南京亿达高科环保技术有限公司工作，任执行董事。

胶膜上，制成 Mossbauer 谱试样，供分析铁的价态与配位状态。

**2.2 穆斯堡尔谱和吸收光谱的测定**

（1）穆斯堡尔谱测定。实验设备采用国产 MS-1 型等速度穆斯堡尔谱仪，γ 射线由 NaI（Ⅱ）闪烁计数器检测，放射源为 57Co（铜基）。以 γ-Fe 为标准试样，在室温下测谱，定标常数为 0.8585（mm/s/chanel），中心道数为 128.675（chanel）。实验数据按 Lorentz 函数，用最小二乘法高斯—牛顿法与不用矩阵法交替处理；在 IBM-AT 计算机上拟合，最后得到有关的 Mossbauer 参数。

（2）吸收光谱测定。用日本理光分光光度计测定。

## 3 实验结果与讨论

### 3.1 化学分析

化学分析表明，汝瓷天青釉中铁是主要呈色元素，等离子光谱对汝瓷天青釉分析的部分结果示于表1。

表1 宋代汝瓷残片天青釉的化学成分

| 化学成分 | SiO$_2$ | Al$_2$O$_3$ | Fe$_2$O$_3$ | CaO | MgO | K$_2$O | Na$_2$O | TiO$_2$ | MnO | P$_2$O$_5$ | BaO | SrO | ZrO$_2$ | CuO | Li$_2$O | ZnO |
|---|---|---|---|---|---|---|---|---|---|---|---|---|---|---|---|---|
| wt% | 66.94 | 14.26 | 1.51 | 9.38 | 1.20 | 3.71 | 2.00 | 0.20 | 0.11 | 0.31 | 0.073 | 0.034 | 0.016 | 0.089 | 0.023 | 0.035 |
| mole% | 72.29 | 9.08 | 0.62 | 10.84 | 1.95 | 2.53 | 2.08 | 0.16 | 0.10 | 0.14 | 0.03 | 0.02 | 0.006 | 0.07 | 0.05 | 0.03 |
| 酸度系数：1.5577 | | | | 氧比：3.1 | | | | RO$_2$/R$_2$O$_3$：7.5 | | | | RO·R$_2$O:R$_2$O$_3$:RO$_2$ = 1:0.5523:4.1387 | | | | |

由表1可以看出，天青釉可形成硅铝酸盐质的玻璃熔体，其中 SiO$_2$、P$_2$O$_5$ 等为玻璃网络形成子，Al$_2$O$_3$、Fe$_2$O$_3$ 等为网络中间子，而 CaO、K$_2$O、Na$_2$O 等为网络修饰子。变价元素有 Fe、Ti、Mn、Cu 等，而以 Fe 的含量最高。可以推断铁为主要呈色元素，其他变价元素为辅助呈色。因此，研究铁离子在玻璃网络结构中的作用是从本质上解释天青釉呈色的关键。

### 3.2 天青釉中铁离子的价态和配位数

由拟谱所得到的 Mossbauer 参数，确定天青釉中铁离子的价态和配位数。

室温下测得的 Mossbauer 谱吸收曲线示于图1。

图1中两个吸收峰强度不相等，可解释为 Fe$^{3+}$ 的存在所引起的。

对实测谱进行计算机拟谱，结果示于图2。由图2可以看出，拟合曲线由两套双峰和一个单峰组成，从而得到表2中的 Mossbauer 参数。

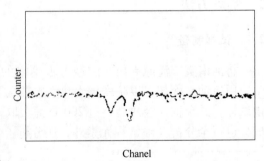

图1 室温下天青釉的穆斯堡尔谱

表2 计算机拟谱后得到的 Mossbauer 参数

| Subspec. No | I.S/mm·s$^{-1}$ | Q.S/mm·s$^{-1}$ | Area/% | Γa/mm·s$^{-1}$ |
|---|---|---|---|---|
| 1 | 1.3041 | 2.0063 | 67.7 | 0.258 |
| 2 | 0.9670 | 1.8786 | 19.3 | 0.258 |
| 3 | 0.2812 | — | 13.0 | 0.129 |

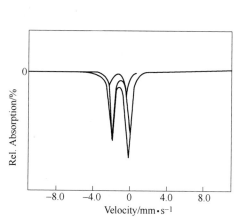

图2 穆斯堡尔谱的计算机拟合曲线

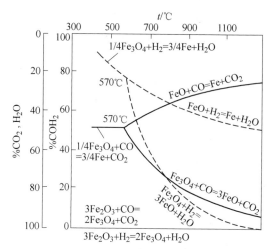

图3 天青釉中铁氧化物还原的热力学状态图

根据文献[3~9]研究结果，可以确定I.S值为1.304mm/s的双峰是由于$Fe^{2+}$八面体配位的吸收形成的；I.S为0.967mm/s的双峰应为$Fe^{2+}$的四面体配位，而I.S为0.2812mm/s的单峰则由$Fe^{3+}$的四面体配位引起的吸收峰。根据吸收峰的面积得到各相对含量为：$Fe_{tet}^{2+}$占19.3%；$Fe_{oct}^{2+}$占67.7%；$Fe_{tet}^{3+}$占13.0%。在还原气氛下存在三价铁离子，从热力学上分析是否合理呢？

## 3.3 汝瓷天青釉的热力学分析

天青釉生料中的$Fe_2O_3$在1240℃，4%~7% CO的还原气氛下，其热力学的稳定相为$Fe_3O_4$，参见图3。

由图3可以看出，在釉未熔时（1100℃），$Fe_2O_3$已全部转化为$Fe_3O_4$，随着CO、$H_2$浓度的增加$Fe_3O_4$→FeO的趋势增大，由此可见，从烧成到冷却过程中的再氧化，都证实在天青釉中应存在少量的三价铁离子。这与穆斯堡尔谱的分析是一致的。

## 3.4 用配位场理论分析天青釉呈色机理

配位场理论成功地解释了许多过渡金属离子在玻璃中的呈色；呈现的颜色除和元素的电子排列有关外，配位环境也有很大的影响[10]。天青釉中处于四面体和八面体配位的铁离子，其外层电子d轨道发生了不同方式的分裂。处于八面体配位的$d^6 Fe^{2+}$离子的电子成对能$P=19150cm^{-1}$，大于$Fe^{2+}$在八面体配位中d轨道的分裂参数$\Delta_0=9520cm^{-1}$。因此，$Fe^{2+}$处于弱八面体场，电子构型为高自旋状态$4T_{2g}$和$2E_{g0}$，当低能级轨道$4T_{2g}$的电子向高能级轨道$2E_g$跃迁时，要伴随有光的吸收；吸收光的能量$\Delta E$应等于轨道能级差。即配

场强度 10Dg（或 $\Delta_0$）。因此，吸收光的波长可根据式

$$\lambda = C/V = hc/\Delta E$$

来计算；计算结果为 $\lambda = 1051nm$，即 $Fe_{oct}^{2+}$ 应在近红外区 1051nm 处产生吸收。这一结论是否正确呢？我们用天青釉的吸收光谱进行了检验，参见图 4。实验证实，1051nm 处是 $Fe^{2+}$ 八面体配位引起的。吸收峰之所以延伸到可见光区，这主要是由于天青釉中含有相当量的 $CaO$、$K_2O$ 和 $Na_2O$ 等，碱度较大，使 $O^{2-}$ 离子的极化率增加，配位体畸变成低对称的多面体，从而导致 d 轨道的进一步分裂，即发生了 John-Teller 效应；Eg, Tg 分别分裂为两个能级，使 d 轨道能级分裂参数增大；由此引起了 1051nm 处的强吸收峰向短波方向移动，直延伸到可见光区，从而使釉呈现短波长光色——蓝色。

而处于四面体配位的 $Fe^{2+}$，引起的吸收峰位置在远红外区 4000nm 处，对可见光区的吸收没有影响，因此它对釉的呈色也影响不大。

同理可以计算出 $Fe^{3+}$ 在四面体配位时，产生的吸收光谱峰的位置在 280nm 处。由图 4 可以看出，此吸收峰较弱，这进一步说明了天青釉中 $Fe^{3+}$ 含量很少。总之，$Fe^{3+}$ 对光的吸收远小于 $Fe^{2+}$；由此可见，$Fe^{3+}$ 的着色强度也远小于 $Fe^{2+}$。

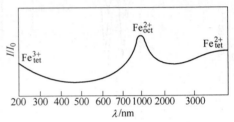

图 4 汝瓷天青釉的吸收光谱曲线

当天青釉料在氧化气氛下烧成时，由于反应 $4Fe^{2+} + O_2 + 14O^{2-} = 4[FeO_4]^{5-[11]}$ 的发生，使 $Fe^{2+}$ 变成四面体配位的 $Fe^{3+}$，则必将使 200nm 附近的吸收峰加强，直延伸到可见光区，此时釉泛黄色；这已被实验室模拟实验所证实。模拟实验还证实，较弱的还原气氛也将会使 $Fe^{3+}$ 相对增加，同样也加深了釉的黄色基调。

综上所述，合理选择温度和气氛是确保汝瓷天青釉色调再现的工艺关键所在。

## 4 结论

（1）用穆斯堡尔谱测定了汝瓷天青釉中 $Fe^{2+}$ 和 $Fe^{3+}$ 的相对含量分别为 87% 和 13%，这与热力学计算的结果基本一致。同时，还确定了在天青釉中 $Fe^{2+}$ 以四面体和八面体配位存在，而 $Fe^{3+}$ 仅以四面体配位存在。

（2）根据配位场理论和汝瓷天青釉的吸收光谱实验讨论了呈色机理；认为占据八面体配位的 $Fe^{2+}$ 在近红外区的强吸收，并延伸到可见光区是天青釉呈色的关键离子；而处于四面体配位的少量 $Fe^{3+}$ 对呈色作用不大，起辅助呈色的作用。因此，要获得天青釉除恰当地选择配方外，还必须正确地控制温度和气氛两个重要的工艺参数。以上结论均被实验室模拟实验所证实。

### 参 考 文 献

[1] 叶喆民. 中国古陶瓷科学浅说 [M]. 北京：中国轻工业出版社，1980.
[2] 郭演仪，等. 硅酸盐学报，1980，8（3）：232.
[3] C. R. Kurkjan. Phys. Chem. Glasses. , 1964, (5): 63.
[4] R. Gosselin. Phys. Chem, Glasses, 1967, (8): 56.

[5] K. Hirao, N. Soga. J. Non-Cryst. Solids, 1980, 40: 315.
[6] N. Iwanto. J. Non-cryst. Solids, 1978, 29: 347.
[7] A. Nolet. J. Non-cryst. Solids, 1980, 37: 99.
[8] K. Hirao, N. Soga. J. Amer. Ceramic Society, 1979, 62: 109.
[9] R. E. M. Hedge. Nature, 1975, 254: 501.
[10] C. R. Bamford. Phys. Chem. Glasses, 1962, (3): 189.
[11] H·舒尔兹, 著. 玻璃的本质结构和性质 [M]. 北京: 中国建筑工业出版社, 1984.

(原载于《中国陶瓷》, 1991, 118 (3): 59-62)

# $MgO\text{-}SiO_2\text{-}Al_2O_3$ 体系用后耐火材料合成新材料的研究

赛音巴特尔[1]　廖洪强[1]　岳昌盛[2]　张梅[2]

(1. 首钢环保产业事业部，北京　100041；
2. 北京科技大学冶金与生态工程学院，北京　100083)

**摘　要**：由于高温工业所产生的用后耐火材料总量逐年增加，基于资源充分利用及存量最少化的前提下，对用后耐火材料进行循环再利用将是合适的选择。$MgO\text{-}SiO_2\text{-}Al_2O_3$体系内材料在耐火材料领域中占有极其重要的地位，本文以热力学和化学匹配为基础，采用体系内的用后耐火材料，如黏土砖、硅砖、滑板砖和镁碳砖作为主要原料合成了纯度较高的莫来石材料和包括 SiC、β-SiAlON 和 MgAlON 在内的非氧化物材料。

**关键词**：$MgO\text{-}SiO_2\text{-}Al_2O_3$体系；用后耐火材料；合成；非氧化物；莫来石

# Synthesis of New Materials From Wasted Refractories of $MgO\text{-}SiO_2\text{-}Al_2O_3$ System

Sainbaatar[1]　Liao Hongqiang[1]　Yue Changsheng[2]　Zhang Mei[2]

(1. Shougang Environmental Protection Department, Beijing 100041;
2. School of Metallurgical and Ecological Engineering, University of Science and Technology Beijing, Beijing 100083)

**Abstract**: With an ever-increasing amount of wasted refractories in high-temperature industry, refractories recycling is one kind of necessary processing mode from the standpoint of comprehensive utilization of resources and waste minimization. Since the materials made from $MgO\text{-}Al_2O_3\text{-}SiO_2$ system have been widely used, based on thermodynamic analysis and chemical matching, high-purity mullite, non-oxide include SiC, β-SiAlON and MgAlON, have been synthesized from $MgO\text{-}Al_2O_3\text{-}SiO_2$ system wasted refractory such as wasted clay brick, silica brick, slide gate and MgO-C brick.

**Keywords**: $MgO\text{-}SiO_2\text{-}Al_2O_3$ system; wasted refractories; synthetic; non-oxide; mullite

# 1　引言

耐火材料是钢铁冶金行业的重要辅助材料，在钢铁企业有着广泛的应用。国外许多国

---

赛音巴特尔，2001~2005 年于北京科技大学师从李文超教授攻读博士学位。目前在首钢技术研究院工作，教授级高工，首钢技术专家。申请和获得国家专利 20 余项，发表论文 70 余篇，主编、参编专著 4 部，获省部级、首钢级等奖 6 项。

家,尤其是发达国家,对用后耐火材料的再利用非常重视,用后耐火材料的再利用率在60%以上[1-3]。我国用后耐火材料再利用率不足20%,近几年随着国家环保政策的推行力度加大,以及耐火材料原材料市场价格的提升,用后耐火材料的再利用逐步受到重视。

首钢总公司现在每年产生约5万吨的用后耐火材料,其中有许多具有再利用价值的用后耐火材料多被外单位收购,这些加工单位多以破碎成颗粒后外卖方式处理这些物料。但这些用后耐火材料若经严格的拣选、分类和特殊的工艺处理后,可以制备出利用价值更高的再生定型和不定型耐火材料产品。

$MgO-SiO_2-Al_2O_3$体系物相作为冶金耐火材料的一个极其重要的组成部分,体系中的$MgO$、$Al_2O_3$、$SiO_2$、莫来石、镁铝尖晶石等在高温工业中得到大规模广泛应用,或直接使用如硅砖、高铝砖、莫来石砖等,或结合第二相使用如镁碳砖、滑板砖、铁钩料等,其占用后耐火材料总量的70%以上,因此对体系内用后耐火材料进行再利用研究具有重大意义。

用后耐火材料可以通过加入适量有益氧化物以改善、优化高温性能制备优质改性料,如莫来石;或可通过高温还原反应制备转型料,如$SiC$、$\beta$-$SiAlON$、$MgAlON$等。氧化物与非氧化物复合材料由于具有强度、抗热冲击性能和抗侵蚀性能等优异的性能,将会成为新型高效耐火材料品而用于高温关键部位[4]。因此利用用后耐火材料合成莫来石、$SiC$、$\beta$-$SiAlON$、$MgAlON$等高附加值材料将可以实现资源和能源的有效利用,符合环保与可持续发展战略目标。

## 2 试验原料

试验原料采用首钢用后黏土砖、硅砖、滑板砖、镁碳砖,其原料组成成分如表1所示。合成采用用后耐火材料的粒径为0.044mm。

表1 用后耐火材料的化学组成
Table 1 Composition of wasted refractories (%)

| Code | $Al_2O_3$ | $SiO_2$ | MgO | $Fe_2O_3$ | $TiO_2$ | CaO | $R_2O$ | C |
|---|---|---|---|---|---|---|---|---|
| 黏土砖 | 33.36 | 59.66 | 0.45 | 2.5 | 1.42 | 0.53 | 0.82 | — |
| 硅砖 | 0.85 | 94.54 | 0.55 | 1.51 | 0.03 | 1.78 | 0.20 | — |
| 滑板砖 | 89.9 | 4.38 | 0.04 | 0.14 | 0.01 | 0.04 | 0.18 | 5.35 |
| 镁碳砖 | 5.25 | 2.66 | 75.55 | 1.27 | 0.04 | 1.55 | 0.07 | 13.42 |

实验在热力学分析基础上确定不同原料合成控制因素,采用首钢用后黏土砖、用后滑板砖合成莫来石,采用用后硅砖以C为还原剂合成SiC,采用用后滑板砖合成$\beta$-SiAlON,采用用后滑板砖、用后镁碳砖合成MgAlON,在XRD分析基础上研究合成复相材料的相组成,并对其理化性能进行分析。

## 3 结果与讨论

### 3.1 合成莫来石

莫来石为铝硅酸盐矿物,组成处在$2Al_2O_3 \cdot SiO_2$至$3Al_2O_3 \cdot 2SiO_2$之间,具有耐火度

高，抗热震性、抗化学侵蚀、抗蠕变性能好，荷重软化温度高，体积稳定性好，电绝缘性强等性能，是理想的高级耐火材料，在耐火材料行业中应用广泛，可以用作热风炉砖和窑具砖。大规模合成莫来石一般采用烧结法和电熔法，其中烧结法较为经济。

用后耐火材料中由于成分组成相异，故可对不同原料进行混合达到莫来石理论组成。用后滑板砖中含有残碳，高温下将不利于莫来石的合成，因此在1000℃下保温2h预先除碳。实验分别采用首钢用后黏土砖与用后滑板砖（编号为NHM）、用后硅砖与用后滑板砖（编号为GHM）作为主要原料，控制合成原料中氧化物 $Al_2O_3$ 与 $SiO_2$ 的组成接近于 $3Al_2O_3 \cdot 2SiO_2$，试样在空气中升温至1873K保温4h。图1示出了NHM与GHM的XRD图谱。

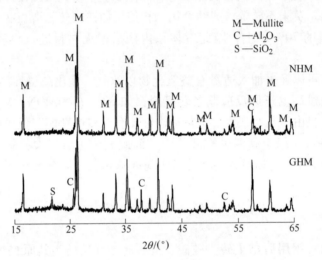

图1 用后耐火材料合成莫来石的XRD图
Fig. 1 XRD pattern of mullite from wasted refractory

由图1可以看出分别采用用后黏土砖、用后硅砖与用后滑板砖均可合成纯度较高的莫来石。其中采用黏土砖和滑板砖合成的试样NHM结果纯度较高（相对含量可达98%以上），而采用硅砖和滑板砖合成的试样GHM中含有部分 $SiO_2$ 与 $Al_2O_3$，分析其原因，用后硅砖中含有一定的玻璃相，同黏土砖中的 $SiO_2$ 相比玻璃相反应活性较低，因此反应不易完全进行，故有 $SiO_2$ 剩余，同时 $Al_2O_3$ 由于不能完全与 $SiO_2$ 反应，故出现 $Al_2O_3$ 剩余。NHM的常温抗折强度为52MPa，GHM为40MPa，可见当莫来石纯度较高时材料具有较高的抗折强度。

## 3.2 合成 SiC

SiC具有机械强度高、热导率好、耐磨性和抗侵蚀性能优异等特性，在耐火材料、高温冶金等工业领域应用广泛，例如高炉用 $Si_3N_4$ 结合SiC砖、铁钩料等。

实验采用用后硅砖与活性炭反应合成SiC。温度1723~1823K，加碳量为实际反应计算量的100%~150%，保温时间均设定4h，合成采用空气中埋碳控制气氛。表2示出了实验合成参数与试样的XRD结果。图2为SC2试样1823K、SC3试样1723~1823K时的XRD叠加图谱。

表2 试样的合成参数与相组成

Table 2 Synthetical parameters and phase composition of specimens

| Code | C 加入量 | 合成温度 | 合成 XRD 结果 |
| --- | --- | --- | --- |
| SC1 | 过量 0% | 1723K | $SiC > SiO_2$ |
| SC2 | 过量 25% | 1773K | $SiC > SiO_2 > Si_2N_2O$ |
| | | 1823K | $SiC > Si_2N_2O > SiO_2$ |
| SC3 | 过量 50% | 1723K | $SiC > SiO_2 > Si_2N_2O$ |
| | | 1820K | $SiC > Si_2N_2O > SiO_2$ |

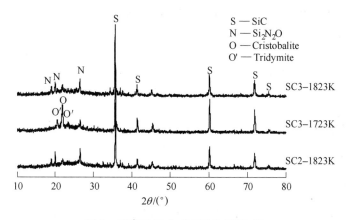

图2 用后硅砖合成 SiC 的 XRD 图

Fig. 2 XRD pattern of SiC from wasted silica brick

由表2与图2可以看出,高温反应下合成了以 β-SiC 为主晶相、$Si_2N_2O$ 为副晶相的材料,β-SiC 相对含量可达 90% 以上。对于同一配方 SC3 而言,当温度由 1723K 升至 1823K 时,方石英相与鳞石英相含量降低,而 $Si_2N_2O$ 相含量明显增加,但 SiC 的含量却有所减少,这表明温度的提高有利于降低合成相中 $SiO_2$ 的含量,但会转化为 $Si_2N_2O$ 相,不利于合成 SiC 相纯度的提高。对于同一温度而言,当碳加入量增加时,合成试样中 SiC 的含量有所增加,但变化较小,这表明碳加入量的增加并不能明显提高 β-SiC 的含量。

### 3.3 合成 β-SiAlON

β-SiAlON 是 SiAlON 材料中最稳定的晶相,具有热膨胀系数低、抗熔融金属及渣侵蚀能力好的优异性质,SiAlON/SiC 材料成为替代 $Si_3N_4$/SiC 耐火材料的第二代高炉关键部位的内衬材料,可有效地提高高炉的使用寿命[5,6]。SiAlON 陶瓷应用在很大程度上仍受到其高成本的限制,降低成本并保持其优异性能,成为 SiAlON 陶瓷发展的重要方向。

实验以用后滑板砖为主要原料,加入 Si 粉、Al 粉与残碳复合还原氮化合成 β-SiAlON。在热力学分析基础上[7],拟合成 $z=3$ 的 β-SiAlON($Si_3Al_3O_3N_5$),合成温度为 1800K,保温时间 4h,试验采用埋 $Si_3N_4$ 粉通氮控制合成气氛。图3示出了合成 β-SiAlON 的 XRD 图谱。

XRD 表明合成了以 β-SiAlON 为主相的复相材料,试样的抗折强度为 67MPa,具有较好的力学性能。反应相中存在部分 $Al_2O_3$ 和少量 SiC,分析其原因一方面用后耐火材料中的

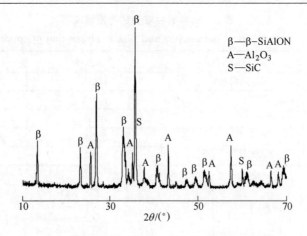

图3 用后耐火材料合成 β-SiAlON 的 XRD 图

Fig. 3 XRD pattern of β-SiAlON from wasted refractory

$Al_2O_3$ 活性较天然原料低,反应不易完全进行,加之合成 $z$ 值较高,故合成以 β-SiAlON 与 $Al_2O_3$ 的介稳形式共存。

## 3.4 合成 MgAlON

MgAlON 材料是一种具有尖晶石结构的新型陶瓷材料,具有优异的力学性能,耐化学侵蚀性好,对玻璃熔液和铁水的润湿性小,抗热冲击性能好,在耐火材料领域中具有广阔的应用前景,但合成成本较高制约了其大规模合成,如何降低成本成为工业化应用的关键[8-11]。

实验采用用后滑板砖和用后镁碳砖为主要原料,加入少量活性炭与残碳复合碳热还原氮化合成 MgAlON 材料。实验拟合成 MgAlON 的相组成为 $Mg_{0.3}Al_{1.3}O_{2.1}N_{0.1}$;合成温度为 1873K,保温时间为 6h,合成采用埋 $Si_3N_4$ 通氮气控制合成气氛。图4 示出了合成 MgAlON 的 XRD 图谱。

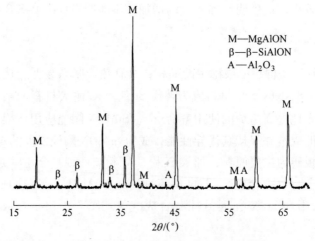

图4 用后耐火材料合成 MgAlON 的 XRD 图

Fig. 4 XRD pattern of MgAlON from wasted refractory

由图 4 可以看出，实验合成了以 MgAlON 为主晶相（含量较高）的材料，合成试样中含有少量 β-SiAlON 相，分析其原因，用后耐火材料原料中含有部分的含硅化合物如 $SiO_2$、SiC 等，高温下其与原料中 $Al_2O_3$ 发生还原氮化反应生成 β-SiAlON。

## 4 结论

（1）$MgO$-$SiO_2$-$Al_2O_3$ 体系物相在冶金耐火材料工业中占有极其重要的地位，对体系内用后耐火材料进行合成高附加值材料再利用的研究将有助于实现资源和能源的有效利用。

（2）分别以用后黏土砖与用后滑板砖、用后硅砖与用后滑板砖为主要原料，高温下合成了纯度较高的莫来石，具有较好的力学性能，其常温抗折强度分别可达到 40MPa 和 52MPa。其中采用黏土砖合成的试样纯度较高，莫来石含量可达 98% 以上。

（3）以用后硅砖与活性炭反应合成 SiC，XRD 表明高温反应下合成了以 β-SiC 为主晶相、$Si_2N_2O$ 为副晶相的材料，β-SiC 含量可达 90% 以上。

（4）以用后滑板砖为主要原料，利用金属/残碳复合还原氮化合成了以 β-SiAlON 为主晶相的复相材料，具有较好的力学性能。

（5）以用后滑板砖和用后镁碳砖为主要原料，加入少量活性炭与残碳复合碳热还原氮化合成了纯度较高的 MgAlON 材料，其抗折强度可达 32MPa，具有较好的力学性能。

感谢国家科技支撑计划资助项目（2006BAC21B02-1、2007BAB24B03），国家自然科学基金重点基金资助项目（50332010）对本课题的支持。

## 参 考 文 献

[1] 田守信. 用后耐火材料的再生利用 [J]. 耐火材料，2002，36（6）：339-341.
[2] 冯慧俊，田守信. 宝钢用后废弃 MgO-C 砖的再生利用 [J]. 宝钢技术，2006，1：17-19.
[3] 姜华，等. 宝钢用后耐火材料的技术研究与综合利用 [J]. 宝钢技术，2005，3：9-11.
[4] 钟香崇. 自主创新，发展新型优质耐火材料 [J]. 耐火材料，2005，39（1）：1-5.
[5] Jack K H. Review: Sialons and related nitrogen ceramics [J]. J Mater Sci，1976，11（16）：135-158.
[6] Gunn D A. Theoretical evaluation of the stability of sialon-bonded silicon carbide in the blast furnace environment [J]. J Eur Ceram Soc，1994，11（1）：35-41.
[7] 董鹏莉，王习东，张梅，等. β-SiAlON 及 β-SiAlON-SiC 复合材料合成的研究 [J]. 耐火材料，2006，40（2）：110-113.
[8] Weiss J，Greil P，Gauckler L J. The system Al-Mg-O-N [J]. J Am Ceram Soc，1982，65（5）：C68-C69.
[9] Granon A，Goeuriot F，Thevenot F，et al. Reactivity in the $Al_2O_3$-AlN-MgO system [J]. The MgAlON spinel phase. J Eur Ceram Soc，1994，13（4）：365-370.
[10] 王习东，王福明，李文超. MeAlON 陶瓷的合成热力学与相关性能 [J]. 无机材料学报，2003，18（1），83-90.
[11] 王习东，张作泰，张梅，等. 一种氮氧化铝镁/氮化硼复相耐火材料及其制备工艺 [P]. 中国，CN1603278A，2005-04-06.

（原载于《钢铁》，2008，43（增刊）：358-361）

# Recent Development of Andalusite Refractories in China

Wen Hongjie[1]    Li Wenchao[1]    Wang Jinxiang[2]    Miao Pu[2]

(1. University of Science and Technology Beijing, Beijing 100083;
2. Luoyang Institute of Refractories Research, Luoyang 471039)

**Abstract:** Exploitation and research of the chinese andalusite have made a great progress in 1990s. In this paper, some aspects in recent development of andalusite refractories in China are introduced. And the further development is also discussed.

## 1 Introduction

Andalusite is one of the aluminium silicate minerals and, with the ability of transforming into high performance mullite, it can be used for high grade refractories and for ceramic kiln furniture. As the refractories used in steelplant, its application abroad has spread almost to various fields, which include the blast furnaces and hot blast stoves, the iron or steel ledles, the roofs of electric arc furnaces, and so on[1].

Andalusite refractories were hardly ever developed in China until the year 1978, which seemed due to the abundant resource of high alumina bauxites. The general surey nationwide started in 1978 for the sillimanite minerals, which include the sillimanite, kyanite and andalusite. And the results indicated that there are more than 40 million tons of andalusite reserves which occur in about ten provinces of the country[2]. Since then, with primary seperating process, thousands of tons of andalusite concentrates have been produced a year. Meanwhile, the application as refractories started in home. For example, the thousand tons of low creep bricks had been produced and used for the blast furnace stoves in Angang, Baosteel and Wugang[3].

At the begining of 1990s, the National Key Project on the research of concentration technique of andalusite, properties of its concentrates and its applications with the chinese feature started, and by now, a great progress has been made in this field. This article will focus on it.

## 2 Development of concentrating processes

It is proved that China has rich resources of andalusite minerals which are of high quality[4]. For example, Xixia's andalusite deposit, which lies in Xixia county in Henan province, was considered

---

文洪杰，1992~1998 年于北京科技大学师从李文超教授攻读博士学位。目前在中国中钢集团公司科技管理部工作，任部门副总经理职务。发表论文30篇，获奖3项。

as the most valueable one because of the large size of its andalusite grains[5], and andalusite reverse is over 10 million tons in it. There are also other ore locations of andalusite minerals in China, such as those both in Wulian in Shandong province and in Fengcheng in Liaoning province, etc[2].

For Xixia's andalusite, because of weathering the adjoining rocks of some andalusite ores which occur on the surface of the deposit are usually soft and friable and easy to separate andalusite crystals from them. Therefore, a few of concentration plants were set up in 1980s to produce andalusite concentrates from the weathering ores by primary dressing processes, i.e., only by desliming process. Later, gravitational (table)-magnetic or electrostatic seperation had been developed[6]. However, the production has been limited obviously since the limited quantities of the weathering ore. To improve the recovery of andalusite and qualities of the concentrates, some more efficient processes, including flotation, have been tested in this period. Table 1 shows the typical results of them[7].

Table 1  Andalusite seperating processes and qualities of concentrites in 1980s

| Mining areas | Dressing processes① | Andalusite /% | $Al_2O_3$ /% | Particle size /mm | Recovery /% |
|---|---|---|---|---|---|
| Fengcheng | G-M-F | — | >57 | 0.2-0.4 | 84.3 |
| Wulian | D-G-M | — | 55.80 | — | — |
| Xixia | G-M-G | 89.56 | 55.71 | 5-0 | 84.54 |
|  | G-M-G | 93.28 | 58.72 | 2-0 | 79.17 |
| Xixia | D-G-M | 92.06 | 59.5 | 3.2-0.5 | 23.04 |
|  | D-G-M-F | 88.20 | 55.17 | <0.5 | 37.90 |
| A area | M-D-F | 99.32 | 56.88 | <0.1 | 76.07 |
| B area | M-F | — | 55.96 | <0.1 | 64.41 |

①D, F, G and M denote desliming, flotation, gravitational and manetic seperation, respectively.

Considering available conditions of the mineral resources and needs of refractories, the National Key Project on research of seperating processes of Xixia's andalusite has been carried since 1991. The new gravitational-magnetic-flotation process, as shown in Fig. 1[8], has come to success both in small-scale test and in enlarge-scale one. It is the first time in China that the heavy media seperation with hydrocyclones has been applied to andalusite seperating process, which makes it more efficiently to separate large grains of andalusite, especially from the primary ore.

And the results show that by this process the products of large grains can be obtained as well as those of fines from the primary ore in Xixia, and that the qualities of them, as shown in Table 2, are close to those of K-57 grade andalusite produced in South Africa[9].

A trial plant equiped with this new technique is to be set up in Xixia before long, the annual production of which is around three thousand tons of andalusite concentrates. Moreover, a new concentration plant with the same production capability of andalusite has already been operated at Fengcheng in Liaoning province recently.

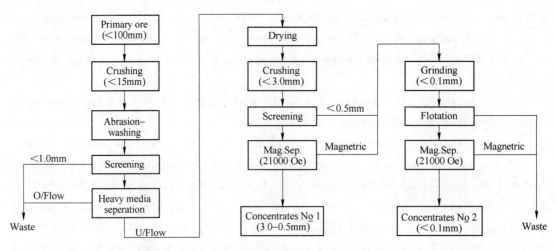

Fig. 1  Trial flow diagram of andalusite seperation

Table 2  Properties of andalusite concentrates by new seperating process

| Particle size/mm | Andalusite/% | $Al_2O_3$/% | $Fe_2O_3$/% | Recovery/% |
|---|---|---|---|---|
| 3.0–2.0 | 86.89 | 56.85 | 1.07 | |
| 2.0–1.0 | 90.11 | 56.6 | 0.88 | |
| 1.0–0.5 | 92.85 | 58.07 | 0.66 | |
| 3.0–0.5 | | | | 26.42 |
| <0.1 | 85.88 | 55.29 | 1.44 | 36.85 |
| Σ | | | | 63.27 |
| Primary ore | 9.42 | 18.34 | 4.13 | 100.00 |

## 3  Researches on andalusite characteristics

### 3.1  Thermal expansion characteristics

Andalusite is the anhydrous aluminium silicate, which makes it possible directly to use without precalcined as being done to bauxites. Moreover, small everlasting expansion which is raised by mullitization during the manufacture or the use at high temperature leads it to be of high volume stability. Therefore, the thermal expansion of andalusite is one of the important property and has been investigated comprehensively. Fig. 2 shows the typical thermal expansion curves of andalusite with various chemical composition and particle size[10].

There are two peaks which appear in each curve as the temperature increases, and as the

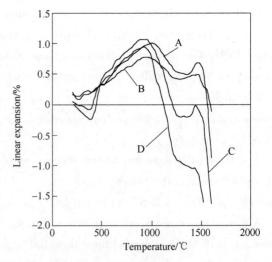

Fig. 2  Typical thermal expansion curves of andalusite[8]

chemical composition and particle size changes, the differences of the second peaks which occur at higher temperature become more obvious.

According to the linear dependances of andalusite's lattice constants ($a$, $b$ and $c$) on temperature which changes from 0℃ to 1000℃[11], only has the physical expansion taken place up to 1000℃.

Otherwise, for the reaction of andalusite's mullitization, illustrated as following[12]:

$$3(Al_2O_3 \cdot SiO_2) \longrightarrow 3Al_2O_3 \cdot 2SiO_2 + SiO_2$$
$$\text{(andanusite)} \qquad \text{(mullite)} \qquad \text{(quartz)}$$
$$\Delta_r G^0 = 48.25 - 3.78 \times 10^{-2} T \quad kJ(298-3000K) \tag{1}$$

when $\Delta_r G^0 = 0$, $T = 1275K$ or 1002℃, that is, the starting temperature of this reaction is 1002℃ in theory. In fact, this temperature is up to 1200℃[13] or 1240℃[10], which measured by DAT and by XRD respectively. All these state that the reaction can not takes place until the temperature rises to over 1002℃ theoretically or over 1200℃ practically. On the other hand, the expansion peak which occurs above 1200℃ corresponds to the mullitization of andanusite.

The effects of the chemical composition and particle size of andalusite on its thermal expansion are very considerable, especially at high temperature. From Fig. 2 the peaks led by mullitization become higher both with the content of $Al_2O_3$ increases (A and C, with 60% and 56% $Al_2O_3$ respectively) and with the particle size increases (B, C and D, with 0.56mm, 0.10mm and 0.071mm respectively). It is concluded from these experiments that the positive expansion of andalusite at high temperature is not always abtained, and that by changing reasonablly the chemical composition or the particle size distribution the thermal expansion property of andalusite refractories can be adjusted.

## 3.2 Kinetic characteristics of mullitization

Research on isothermal conversion of andalusite at 1300℃, 1350℃ and 1400℃ has been done. The dependences of conversion ratio ($\alpha$) on time ($t$), as shown in Fig. 3[14], basically correspond with the Avrami's kinetic mechanism, that is

$$\alpha = 1 - \exp(-kt^n) \tag{2}$$

where the reaction rate constant and the reaction order, denoted $k$ and $n$ espectively, shown in Table 3.

The activation energy is about 467.6kJ/mol. Therefore, the dependence of the rate constant ($k$) of the reaction on the reaction temperature ($T$) is shown as following:

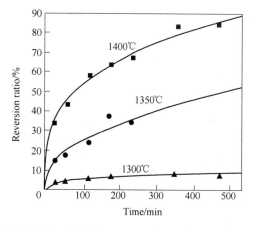

Fig. 3 The dependence of conversion ratio on time

$$k = 2.03 \times 10^{-3} \exp(-5.56 \times 10^4/T) \tag{3}$$

Table 3  Kinetic data

| $T/°C$ | $k/\min^{-n}$ | $n$ |
| --- | --- | --- |
| 1300 | $0.919 \times 10^{-3}$ | 0.4 |
| 1350 | $2.74 \times 10^{-2}$ | 0.5 |
| 1400 | $7.60 \times 10^{-2}$ | 0.5 |

According to this mechanism, the course of mullitization of andalusite relies greatly on the formation of crystal nucleus of mullite conversed from andalusite, and on the amount of activated points which caused especially by surface defects and impurities in the reactants effects considerably on the reaction rate. This result has also corresponded well with that of the thermal expansion experiment mentioned above.

The effect of some oxides, such as $TiO_2$, $Fe_2O_3$, CaO and $Na_2O$, etc., on the decomposition of andalusite at 1400℃ has been investigated with cross experiment. By the variance analysis on the test results, the effect has followed:

$$CaO > Na_2O \geqslant Fe_2O_3 > TiO_2$$

The phase composition in the clinkers measured by XRD shows that the content of mullite transformed from andalusite in CaO-contained andalusite clinker is much less than that in others. The result indicates that andalusite materials have the lower resistance to CaO-bearing impurities or slag.

## 3.3 Sintering characteristics

The sinterability of andalusites, alone or combined with other rich alumina bearing materials, has been investigated largely. The results can be generalized as following.

(1) The sintering temperatures of the pure andalusit may be within the extent of 1500℃ to 1600℃; when combined with industrial alumina or bauxite the temperatures will increase by about 100℃ to 200℃, which seems due to the secondary mullitization of the mixture materials during sintering[10,15-17].

(2) Some oxides which occur originally or may be added as sintering agents in andalusite have great effects on their sintering. The effects of $TiO_2$, $La_2O_3$ and CaO show as following:

$$TiO_2 > CaO > La_2O_3$$

Considering the microstructure and the thermomechanical properties of the sintered materials, both $TiO_2$ and $La_2O_3$ are the more efficient agents for sintering andalusite or its mixtures combined with other high alumina materials[17].

(3) The microstructures of sintered pure andalusites and its mixtures which combined with other rich alumina materials are different obviously. In the sintered pure andalusites, there are large amounts of glass phases with which the mullites are surrounded. While the mixture samples have been improved noticeably. With the amounts of glass phases decrease the microstructure has been characterized by the higher degree of crystal-to-crystal bonding. Thus, it can be predicted that by combined with other rich alumina bearing materials the thermomechanical properties of andalusite

materials could be improved.

# 4 Development of andalusite refractories

## 4.1 Andalusite-bearing refractories

With adding 10% to 25% andalusite grains to conventional clay bricks or bauxite-based high alumina bricks, the properties of them, such as thermal shock resistance, slag resistance and creep resistance, etc., are improved considerably, as shown in Table 4[18-22]. These improved refractory bricks have already been produced in industrial-scale and used successfully for hot blast stoves, torpedo ladles, or arc furnace roofs.

Table 4  Properties of andalusite-bearing refractory bricks

| Properties | Bricks | | | |
|---|---|---|---|---|
| | A | B | C | D |
| $Al_2O_3$/% | 55.36 | 58.30 | 54.28 | — |
| $Fe_2O_3$/% | 1.91 | 1.33 | 1.74 | — |
| Bulk density/g·cm$^{-3}$ | 2.48 | 2.58 | 2.48 | 2.62 |
| Porosity/% | 13.7 | 10.34 | 14.6 | 16.9 |
| Crushing strength/MPa | 110.8 | 85.7 | 108.5 | 38.0 |
| Refractrieness/℃ (under 0.2MPa) | (0.6%) 1560 | (0.6%) 1600 | 1620 | 1520 |
| Reheat liner change /% | (1450℃ ×2h) +0.07 | — | (1550℃ ×2h) +0.08 | (1500℃ ×3h) +0.3 |
| Creep/% (under 0.2 MPa) | (at1300℃ ×50h) 0.7 | — | — | — |
| Number of cycles/times (25℃　1100℃) | >30 | 30 | — | — |
| Main starting raw materials | andalusite + sintered clay | andalusite + sintered bauxite | andalusite + sintered clay | andalusite + sintered bauxite |
| Preparing processing | burned | burned | burned | heated at 400℃ |
| Application | blast stove | torpedo ladle | arc furnace roof | arc furnace roof |

Andalusite-bearing monolithic refractories have been produced in the past years. The low-cement andalusite castable which is characterized by high thermal shock resistance has been used successfully in the converter in nickle making in 1992[22], and the dry ramming monolithics of andalusite used in the induction furnace in copper making[23].

The practice shows that andalusite has never been looked down for improving the properties of the high alumina refractories though it seems to be taken as the second role because of the less amount of andalusite than that of fire clay or high alumina bauxite used in the production as mentioned above. This is due to the more abundant resources of fire clay and bauxite in China.

## 4.2 Fabrication of synthetic materials with andalusite

### 4.2.1 Synthetic mullite and its use for kiln furniture

Combined with other rich alumina materials, andalusite can be applied to prepare synthetic mullite. And the experiments on the synthetizing process have been done. The sarting materials in the tests involved industrial alumina and raw or sintered bauxites besides fine andalusite concentrates[17]. The content of mullite could reach up to 90 percent in these synthetic mateials, and the relective density will be over 90 percent when sintered at above 1700℃. Some oxide additions can improve the sinterability of synthetic mullite.

With the mullite matrix caused by adding fine andalusite and sillimanite concentrates, the SiC-mullite kiln furniture, a kind of plate with 470mm×430mm×32mm in size, has high thermal shock resistence and long operating life, which has been used for more than 260 times in a tunnel kiln operated at 1280 – 1300℃[24]. It is considered reasonablly that such high performance is attributed to the form of continuous interlocking network made of mullite crystals observed in the microstrurcture.

### 4.2.2 Fabrication of β-Sialon

Since decomposed in the condition of reduction, andalusites as high alumina refractories are usually used at air or less oxidition atmosphere. By the carbon-thermic reduction, however, andalusite could be used for fabricting β-Sialon material. Recently, the tests concerned with has been done[25].

By mixing fine andalusite with graphite powders, shapping into pellets and then, heating at 1400-1700℃ with a certain amount of nitrogen, the synthetic β-Sialon has been prepared. The crystal phases that occur in the samples have been investigated by XRD, as shown in Fig. 4.

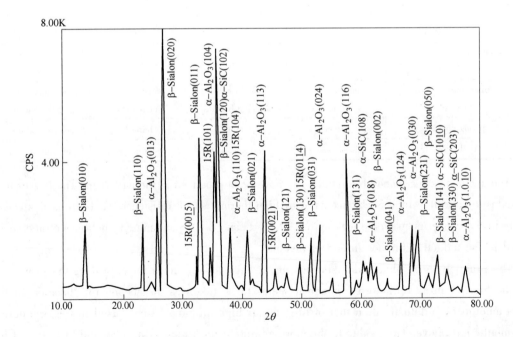

Fig. 4 Typical XRD pattern of synthetic β-Sialon

From Fig. 4, it is obvious that there are also $\alpha\text{-}Al_2O_3$ and $\alpha\text{-}SiC$ in the synthetic $\beta$-Sialon, which implies that by changing the ratio of $\beta$-Sialon to $\alpha\text{-}Al_2O_3$ or $\alpha\text{-}SiC$ the $\alpha\text{-}Al_2O_3/\beta$-Sialon or $\alpha$-SiC/$\beta$-Sialon composite refractories or ceramics could be prepared.

## 5 Prospects in development of andalusite

Although a great progress has been made in last five years, as mentioned above, we'll have a lot to do. In general, some aspects of further developments of andalusite in China could be generized as follows.

(1) For some andalusite ore, such as Wulian's andalusite ore in Shandong province, the seperable tests done previously show that it is very difficult to seperate single andalusite crystals from rich-iron-bearing impurity minerals by the available methods, even though all the primary ore grinded into the powder with the particle size of less 0.07mm. Therefore, the more efficient dressing process is expected to be developed.

(2) In the past years large grains of andalusite (over 0.5mm in size) have been used widely because of their perfect crystal structure and high density. And the fine andalusite concentrites with the size of less 0.1mm, such as the products by floration, seemed to be ignored since they could be replaced of sintered bauxites that have the higher content of $Al_2O_3$ and may be used at the less cost in China, and of the silliminites that have the higher conversion temperature and can maintain the performance of themselves at higher temperature. However, the amounts of the fine andalusite could be more than 60 percent of all the andalusite concentrates even in Xixia (see Table 2) which is well known for large grain concentrates. Therefore, the exploitation of the fine andalusite may be carried out so as to utilize the resources reasonablly.

(3) Some new fields of the utilization of andalusite may be found. If the synthetized $\beta$-Sialon materials, as illustrated previously in this paper, come to practice successfully, the fields will be extended from that of high alumina refractories alone to that of Si-Al-O-N materials.

## 6 Summary

Since 1991 the exploitation and research of andalusite have developed considerably in China. With the new dressing processes come to use, both the quality and the quantity of refined andalusite will be raised greatly. The results of the studies on various properties of andalusite can make it more efficient to use andalusite. Although the utilization in metellurgic industries increased obviously, it is still not suitable to the relatively abundant resources of andalusite minerals. Some researches may enlarge the extents of using andalusite and make the fine andalusite concentrates used fully.

**References**

[1] H. J. S. Kriek. Andalusite as a Refractory Material in Steel Making [C]. Proceedings of the First International Conference on Refractories. Tokyo, Japan, 1983: 576-587.

[2] Liu Hongquan. Exploitage of Kyanite, Sillimanite and Andalusite in China [C]. Proceedings of Chinese-Korea Symposium on Refractories. Zibo, Shandong, China, 1995: 74-77.

[3] LIRR. Exploitation and Application of Kyanit, Sillimanite and Andalusite Refractories (in Chinese). 1989: 67-77.

[4] Huang Wenjing, et al. Properties and Applications of China's Sillimanite Group Minerals [C]. Proceedings of International Symposium on Refractories- Refractory Raw Materials and High Performance Refractory Products. Hangzhou, China, 1988: 156-166.

[5] Wong Da, et al. Exploitage of Xixia andalusite [C]. Proceedings of Chinese- Korea Symposium on Refractories. Zibo, Shandong, China, 1995: 83-89.

[6] Miao Feng. Study on Seperation Technique of Andalusite Shale [J]. Nonmetallic Minerals (in Chinese). 1987, (1): 26.

[7] Miao Feng. Situation of Test on Andalusite Seperation [J]. ibid, 1988, 6 (5): 15-18.

[8] Liu Shiqian, et al. Report of enlarged Test on Seperation of Xixia's Andalusite (in Chinese). Central Laboratory Henan. MGM, 1994.

[9] 富原忠臣. Cullinan Andalusite K-57 (in Japanese) [J]. 耐火物, 1987, 39 (1): 34-36.

[10] Wen Hongjie, Miao Pu. Thermal Expansion Characteristics of Andalusite (in Chinese) [C]. Proceedings of 30th LIRR's Birthday Celebration symposium. Luoyang, China, 1993: 13-22.

[11] M. K. Derrrill. The $Al_2SiO_5$ Polymorphs [J]. Reviews in Mineralogy. Ed. by H. R. Paul. 1990, 22: 21-25.

[12] 山口明良, 著. 张文杰, 译. 实用热力学及其在高温陶瓷中的应用 [M], 武汉: 武汉工业大学出版社, 1993: 127-185.

[13] Yu Qing. LIRR's Master Thesis, 1990.

[14] Wen Hongjie, Miao Pu, Li Wenchao, Mullitization Dynamics of Andalusite [J]. Refractories (in Chinese), 1995, 29 (3): 140-141, 148.

[15] Shi Gan, Sun Gengchen. The Mullitization Behaviour of Three Sillimanite in Materials [J]. ibid, 1992, 26 (6): 311-314.

[16] Lin Binyin, et al. The Study for the Sintering Property of Andalusite [J]. ibid, 1992, 26 (1): 28-32.

[17] Wen Hongjie, Miao Pu, Li Wenchao. Sinterability of concentrated Andalusite and Synthetic Mullite [J]. ibid, 1995, 29 (5): 247-250.

[18] Zhu Liming, Meng Fanjun, Zhang Dajie. Effect of Sillimanite or Andalusite on Properties of $Al_2O_3$-$SiO_2$ System Refractory Products [J]. ibid, 1990, 24 (4), 5-7, 16.

[19] Zhu Liming, Chen Xiuping, et al. Development and Production of Andalusite Brick for Hot Blast Stove [J]. ibid, 1991, 25 (3): 166-167.

[20] Li Shucai, Xie Zhupei. Preparation and Application of Andalusite Brick for Torpedo Ladles [J], Refractories (in Chinese), 1992, 26 (5): 302-303.

[21] Tian Xiuzhou, Test and Production of Andalusite Brick for Roof of Arc Furnace [J]. ibid, 1995, 29 (4): 244.

[22] Wu Qinshun, Li Xiaoming, Wu Yunyun. Chinese Sillimanite Group and its Application in Refractory [C]. Proceedings of Chinese-Korea Symposium on Refractories, 1995, Zibo, Shandong, China, 1995: 78-82.

[23] Hao Yunyu, Preparation and Application of Dry Ramming Refractories [J]. Refractories (in Chinese), 1992, 26 (5): 305.

[24] Wen Hongjie, Miao Pu. Preparation of Low-cement Andalusite Castable for the Converter in Nickle Making (in Chinese) [R]. LIRR's Research Report, 1992.

[25] Wen Hongjie, Li Wenchao, Wang JinXiang. Fabrication of β-Sialon with Andalusite. (to be published).

(原载于'96 *International Symposiumon Refractories*, 1996: 12-15)

# 红柱石分解过程的分形研究

文洪杰[1]　李文超[1]　王金相[2]　苗圃[2]

（1. 北京科技大学理化系，北京　100083；2. 洛阳耐火材料研究院，洛阳　471003）

**摘　要**：本文在对红柱石的热分解进行实验研究的基础上，运用周长—面积法测定了分解颗粒的分形维数，并利用分形维数与反应时间的关系探讨了红柱石分解反应的历程。

**关键词**：红柱石；反应动力学；分形

## Study on the Fractal of Andalusite Grains Decomposed

Wen Hongjie[1]　Li Wenchao[1]　Wang jinxiang[2]　Miao Pu[2]

(1. University of Science and Technology Beijing, Beijing 100083;
2. Luoyang Institute of Refractory Research, Luoyang 471003)

**Abstract:** Based on the experiment of decompostion dynamics of andalusite, the fractal dimensions of the decomposed grains have been measured by means of the perimeter-area mathod. The reaction mechanisium has also been discussed with the relation of the fractal dimension and the reaction time.

**Keywords:** andalusite; reaction dynamics; fractal

## 1　引言

分形理论诞生于 20 世纪 70 年代，它是以自然界中的非线性过程为研究对象，以新的观念、新的手段，透过过程中无序的混乱现象和不规则的形态，揭示隐藏在复杂现象背后的规律的科学[1]。分形理论的应用非常广泛，几乎包括了自然科学和社会科学的各个领域。其在材料科学中的应用也较为突出[2,3]。

红柱石是一种硅酸盐矿物，可用于生产高性能耐火材料。有关红柱石的一些基本性能，如分解温度、热膨胀性能以及烧结性能等，已有过一些报道[4,5]。本文作者也曾对红柱石分解过程动力学进行了实验研究。本文将在此基础上，运用分形理论，通过测定红柱石颗粒分解后的分形维数及其与反应时间的变化关系对红柱石分解过程的分形特征进行研究，并对该分解反应的历程做进一步探讨。

---

本工作由国家自然科学基金项目（594734020）资助。

文洪杰，1992~1998 年于北京科技大学师从李文超教授攻读博士学位。目前在中国中钢集团公司科技管理部工作，任部门副总经理职务。发表论文 30 篇，获奖 3 项。

## 2 周长—面积法测定颗粒的分形维数

分形维数是对分形图形的表征。有关分形维数的测定方法很多[6,7],周长—面积法是根据测度关系求分形维数的一种方法。

各种测度与其相应维数之间有如下关系式:

$$L \sim S^{1/2} \sim V^{1/3} \sim X^{1/D} \tag{1}$$

式中的 $L$、$S$、$V$ 和 $X$ 分别代表一维测度、二维测度、三维测度和 $D$ 维测度的量。用上式可以测定类似于岛屿海岸线的分形的维数。如图1所示,设小岛的面积为 $S$,海岸线的长度为 $X$。因岛的面积明显是具有二维测度的量,所以,根据 $S^{1/2} \sim X^{1/D}$ 即可求得海岸线的分形维数 $D^{[6]}$。

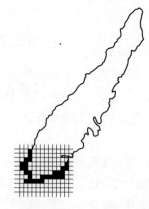

图 1 周长—面积方法测定分形维数

Fig. 1 The perimeter-area method for measuring fractal dimension

## 3 红柱石分解动力学实验结果

图2为红柱石分解转化率 $\alpha$ 与分解反应时间 $t$ 的关系。对反应机理的分析认为,莫来石的晶核化是该分解反应的控速环节。

用耶费洛米(Avrami)的晶核化模型对实验数据进行处理[8],如图3所示。结果表明,理论分析与实验模型处理吻合得很好。由上图可求得不同温度下反应的动力学参数,并得到反应动力学的半经验方程如下:

$$1300℃: \alpha = 1 - \exp(-9.19 \times 10^{-4} t^{0.4})$$
$$1350℃: \alpha = 1 - \exp(-2.74 \times 10^{-2} t^{0.5})$$
$$1400℃: \alpha = 1 - \exp(-7.6 \times 10^{-2} t^{0.5})$$

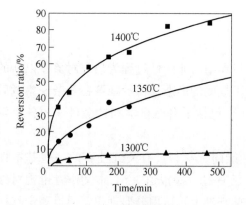

图 2 红柱石的转化率与时间的关系

Fig. 2 The dependence of conversion ratio on time

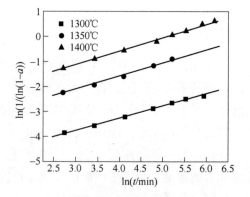

图 3 用耶费洛米动力学模型处理得到的动力学曲线

Fig. 3 Reaction dynemic curve by Avrami model

## 4 分解颗粒分形维数的测定

如前所述,采用周长—面积法测定红柱石分解颗粒边界的分形维数。测定过程如图4

所示。

图 5 是对在同一个 SEM 视域中的 6 个红柱石颗粒提取的分形图形。该试样经过了 1400℃ 保温 360 分钟的分解处理。

利用图像分析仪分别测定各个颗粒的面积 $S$ 和周长 $X$，并作 $\ln(S/S^0) \sim \ln(X/X^0)$ 关系曲线，详见图 6。由图中的线性关系可知，红柱石的分解颗粒具有一定的分形特征。从该直线的斜率可求出该条件下红柱石颗粒的分形维数是 1.18。

用同样的方法可以得到经不同时间分解的红柱石颗粒的分形维数。图 7 显示了红柱石颗粒的分形维数与其在 1400℃ 时的分解时间的关系。

将图 7 与图 2 比较，二者具有相似的变化趋势。在反应速率较大的初期，分形维数的变化也较大；而在中后期，反应速率明显降低，此时颗粒的分形维数的变化也趋于平缓。由此可知，用分形维数与反应时间的变化关系同样可以描述红柱石的分解过程。

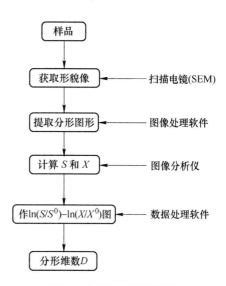

图 4　分形维数测定过程

Fig. 4　The flow of mesurement of fractal dimension

图 5　红柱石颗粒的分形图形（在 1400℃ ×360min 分解）

Fig. 5　The fractal graph of andalusite grains (decomposed at 1400℃ ×360min)

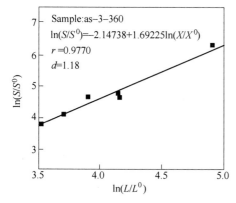

图 6　$\ln(S/S^0) \sim \ln(X/X^0)$ 关系曲线

Fig. 6　$\ln(S/S^0) \sim \ln(X/X^0)$ curve

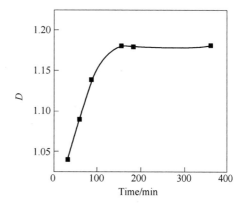

图 7　红柱石颗粒分形维数与反应时间的关系（1400℃）

Fig. 7　The relation of fractal dimension and reaction time（1400℃）

红柱石分解颗粒分形维数的变化反映了其表面形貌规整性的变化。在未发生反应时，颗粒表面主要是一些规整的解理面，此时的分维近似为1。根据对反应机理的研究表明，红柱石的分解反应开始于颗粒表面，并受分解产物莫来石的晶核化的控制。莫来石晶核在颗粒表面的大量生成，原来颗粒表面的规整性被破坏，表面变得粗糙不平，这使得颗粒的分形维数增大。当颗粒表面完全被产物所覆盖，反应的界面达到最大。这时颗粒的分形维数也应达到最大。从图7可知，此时的分形维数为1.18，反应已进行了约150分钟。此后，反应的继续进行，对颗粒分形维数的影响已不大。

## 5 结论

对红柱石分解颗粒分形特征的研究表明：红柱石的分解颗粒具有一定的分形特征。其分形维数与反应时间的变化关系可以描述红柱石的分解过程。当颗粒的分形维数达到最大时，对应于反应界面也最大，并由此可以推测反应达到这一阶段所经历的时间。对反应过程的这一精确描述是传统的动力学研究方法难以做到的。

### 参 考 文 献

[1] B. B. Mandelbrot. The Fractal Geometry of Nature [M]. New York: W. H. Freeman and Company, 1982.
[2] 丁保华, 等. Mo/β'-Sialon 梯度功能材料的显微结构及分形计算 [C]. 现代技术陶瓷——第九届全国高技术陶瓷学术年会论文专辑, 1996, (增刊): 4-065, 344-347.
[3] 李文超, 等. 分形及其在耐火材料中的应用 [J]. 耐火材料, 1997, 31 (2): 113.
[4] 林彬荫, 等. 红柱石的烧结性能研究 [C]. 全国优质硅酸铝制品学术会议论文, 1988, 7: 9-12.
[5] 文洪杰, 苗圃. 红柱石的热膨胀特性 [C]. 三十周年院庆学术报告会论文集, 冶金工业部洛阳耐火材料研究院, 1993: 13-22.
[6] 文洪杰, 苗圃, 李文超. 红柱石的莫来石化动力学 [J]. 耐火材料, 1995, 29 (3): 140-141, 148.
[7] 张济忠. 分形 [M]. 北京: 清华大学出版社, 1995.
[8] 谢和平, 等. 分形应用中的数学基础与方法 [M]. 北京: 科学出版社, 1997.

(原载于《北京科技大学学报》, 1998, 20 (1): 85-87)

# Fractal Calculation of Mo/β'-Sialon Functionally Gradient Materials by Powder Metallurgy

Ding Baohua  Li Wenchao  Wang Fuming  Zhong Weibin  Li Liuliu

(University of Science and Technology Beijing, Beijing 100083)

**Abstract:** The preparation of disk shaped Mo/β'-Sialon functionally gradient material (FGM) is studied. Random fractal of diffusion interface in Mo/β'-Sialon FGM is also studied by variation method.

## 1 Introduction

As the aerospace industry is developing at high speed, a new kind of material is needed to meet the requirements of high temperature and large difference of temperature. But neither current single metal material nor single ceramic material can work well in above-mentioned environment. Therefore, the concept of functionally gradient materials (FGM) was put forward by Masakuki Niino and Toshio Hirai in 1984[1]. Metal/Ceramics FGM is made of pure metal material on one end, and pure ceramic material on another end. The component of this material is in graded distribution from pure metal end to pure ceramics end. FGMs can be prepared by five methods as follows: Chemical Vapor Deposition (CVD), Phycical Vapor Deposition (PVD), Plasma Spraying (PS), Self-propagating High-temperature Synthesis (SHS) and Powder Metallurgy (PM). Powder metallurgy is now the most usual method in preparation of FGMs. There are some reports of FGMs prepared by powder metallurgy, such as $MgO/Ni$[2], $TiC/Ni_3Al$[3], $Ni/Ni_3Al$-$TiC$[4], PSZ/stainless steel[5] and $TiB_2/Ni$[6].

Sialon ceramics have high performance, e.g., O'-Sialon's good non-oxidizability, β'-Sialon's high strength at high temperature. Therefore, they will have good performance if metal/sialon FGMs are prepared successfully. There is few report about preparation of metal/sialon FGMs, especially in study of reaction mechanism and microstructure. Applying fractal theory in functionally gradient materials will form a new method for material research. The fractal's dimension can demonstrate the mechanism of sintering process from another point of view.

## 2 Experiments and methods

### 2.1 Calculation of fractal's dimension

Dimension is defined in Euclidean geometry by:

---

Sponsored by National Nature Science Foundation of China.

$$D = -\ln N/\ln r \tag{1}$$

$N$ is the multiple of length, area or cubic content of graph, $r$ is the multiple of measurement of the graph. For example, when square's $r$ is 1/2, its $N$ will be 4. So, the dimension of square is:

$$D_{\text{square}} = -\ln N/\ln r = -\ln 4/\ln 0.5 = 2 \tag{2}$$

Regular fractal's dimension can be calculated by a mathematical representation. Viscek graph[7] is a typical regular fractal, showing in Fig. 1.

Fig. 1  Viscek graph

The dimension of Viscek graph is:

$$D_{\text{Viscek}} = -\ln 5/\ln(1/3) = 1.4650 \tag{3}$$

Dimension of random fractal is calculated by variation method. Overlay the graph with varied height but the same width rectangles, moving it all over the graph. $N(r)$ is the sum of all the rectangles' area. Changing the width of rectangles to repeat the process, will get a series of $\ln N$ and $\ln r$. Plotting $\ln N$ with $\ln r$, Will get a line whose slope coefficient is negative. Otherwise, this graph is not a fractal graph. Therefore, fractal's dimension is expressed by:

$$N \propto r^{-D} \tag{4}$$

It accords with formula (1).

## 2.2  Preparation of Mo/β'-Sialon FGM

Starting materials: Mo powder (Purity: 99%), β'-Sialon powder (sintered by SHS, 10μm, 99.5%).

The volume percentage of substance A in A/B disk shaped FGM is decribed by: $C = (x/d)^{P}$ [3-5]. It is shown in Fig. 2. The task of FGMs' design is to seek for the best $P$ value. Set pure substance B side as original point, $x$ as vertical coordinate, as $x$ increases from 0 to $d$, the volume percentage of substance A increases from 0 to 100%. Mo/β'-Sialon FGM is prepared by powder metallurgy in 11 equal thickness layers. It was sintered in 1750℃, 17.3MPa pressure, high purity of $N_2$ atmosphere, 1h holding time by hot pressed sintering.

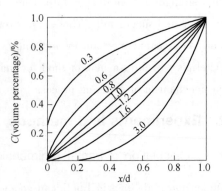

Fig. 2  Thethotical component distribution curve

## 2.3 Samples for SEM test

The sintered samples were cut to the size of 20mm × 5mm × 5mm. After coarse grinding, fine grinding, politure, cleaning and weathering, we observe the pattern's photo and analyse the component by SEM.

## 3 Results and discussion

### 3.1 Component distribution

We analysed the component of Mo/β'-Sialon FGM with EDAX. From pure Mo side to pure β'-Sialon side, the component distribution is described by Fig. 3.

### 3.2 Application of fractal theory in Mo/β'-Sialon FGM

The SEM photo can be input into computer by scanner. Put Mo/β'-Sialon FGM's SEM interface photo into computer by Microtek ScanMaker II scanner. The photos are saved as photo files with a certain format. And then transform this grayscale photo to binary image. The border-

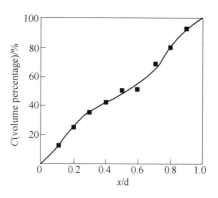

Fig. 3 Component distribution of Mo in Mo/β'-Sialon FGM

line of binary image can be found and can be written to another file, which is the random fractal graph. The flow of fractal graphs' extraction is listed in Fig. 4.

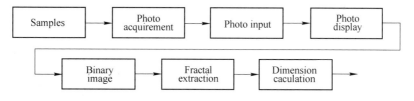

Fig. 4 Flow of random fractal graph's extraction

After extracting the fractal graph of Mo/β'-Sialon FGM's interface, the dimension can be calculated by variation method. Fig. 5 (a) is the fractal graph of Mo/β'-Sialon FGM's interface, and Fig. 5 (b) is fractal's dimension calculation by variation method.

Results of experiments show that the sintering process is controlled by diffusion. The dimension of Mo/β'-Sialon FGM's fractal graph is 1.518, so the sintering process is charged by diffusion[8]. The holding temperature and holding time should be controlled strictly in sintering process of FGMs.

## 4 Conclusions

Mo/β'-Sialon FGM can be prepared by powder metallurgy. Its sintering process is controlled by diffusion. The dimension of Mo/β'-Sialon FGM's interface is 1.518, it also reflects that the sintering process is controlled by diffusion.

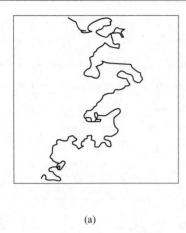
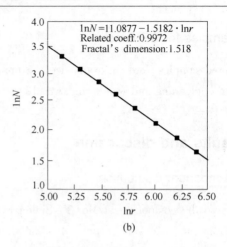

(a)　　　　　　　　　　　　　　(b)

Fig. 5　Fractal graph and dimension calculation

(a) Fractal graph；(b) Fractal's dimension calculation

## References

[1] Yu Maoli, Wei Mingkun. Research development of functionally gradient materials [J]. Functional Materials, 1992, 23 (3): 184-189.

[2] Zhang Lianmeng, Tang Xinfeng, Chen Fuyi, et al. Design and preparation of MgO/Ni functionally gradient materials [J]. Chin. Ceram. Soc., 1993, 21 (5): 406-412.

[3] Zhang Lianmeng, Tu Rong, Zhang Qingjie. Thermal insulation's stress and component distribution's optimization of TiC/Ni$_3$Al functionally gradient materials in stable state [J]. Chin. Ceram. Soc., 1996, 24 (4): 418-422.

[4] Shen Qiang, Zhang Lianmeng, Yuan Runzhang. Design and fabrication of Ni/NiAl-TiC functionally graded materials [J]. Chin. Ceram. Soc., 1997, 25 (4): 406-412.

[5] Akira Kawasaki, Ryuzo Watanabe. Microstructural designing and fabrication of disk shaped functionallty gradient material by powder metallurgy [J]. Micromeritics and Powder Metallurgy, 1990, 37 (2): 253-258.

[6] Toshikazu Takakura, Isao Tanaka, Yoshinari Miyamoto, et al. Fabrication of TiB$_2$-Ni Functionally Gradient Materials by a Gas-Pressure Combustion Sintering [J]. Micromeritics and Powder Metallurgy, 1990, 37 (7): 933-936.

[7] Zhang Jizhong. Fractal [M]. Beijing: Tsinghua University Press, 1995.

[8] Ding Baohua, Wen Hongjie, Zhong Weibin, et al. Microstructure and fractal calculation of Mo/β'-sialon functionally gradient materials [J]. Advanced Ceramics (Supplement), 1996, 17 (3): 344-347.

(原载于 *China's Refractories*, 1998, 7 (2): 27-29)

# 第三部分

## 功能陶瓷材料
### Functional Ceramics Materials

随着国民经济的高速发展，在生命、能源、环境，以及军工国防和高科技领域需要研发一些具有一定性能的功能陶瓷材料，诸如各种陶瓷传感材料、生物材料、绿色电源材料、催化与净化材料等属极迫切的材料研究领域。

本文集利用物理化学原理和计算机辅助研究的手段，选择了具有代表性的若干功能材料进行研发，可供相关领域研究者参考。

# Effects of Preparing Conditions on Controllable One-step Electrodeposition of ZnO Nanotube Arrays

## Lu Hui  Zheng Feng  Zhang Mei  Guo Min

(State Key Laboratory of Advanced Metallurgy, School of Metallurgical and Ecological Engineering, University of Science and Technology Beijing, Beijing 100083)

**Abstract:** By using a one-step electrodeposition method, well oriented single crystalline ZnO nanotube arrays (ZNTAs) with average diameter less than 200 nm were synthesized on FTO substrate under different conditions. The effects of preparing conditions such as pretreatment of substrate, pH value of electrolyte, electrodeposition potential, concentration of electrolyte, reaction temperature and growth time on the electrodeposition of ZNTAs were systematically investigated by scanning electron microscopy (SEM), and ultraviolet visible absorbance spectroscopy to make the formation mechanism clear. It is indicated that the electrodeposition parameters have significant influence on the morphology, average diameter and growth density of as-prepared ZNTAs. The aligned ZNTAs can be only obtained on the FTO substrate pre-treated with acid solution when the applied potential is controlled from -0.8V to -1.0V and the pH value of electrolyte is from 3.10 to 3.90, simultaneously. The growth temperature and the electrolyte pH value have great impact on the morphology of as-prepared ZnO nanostructure. The tube diameters can be monitored to some extent only by changing the concentration of the zinc precursors or growth time, in addition, the growth time is closely related to the band gap energy ($E_g$) of as-prepared ZNTAs.

**Keywords:** ZnO nanotube arrays; one-step electrodeposition; controlled growth; growth mechanism

## 1 Introduction

As a direct wide band gap (3.37eV) semiconductor with a large excitation binding energy (60 meV), the wurtzite ZnO has attracted more attention for its fundamental research and potential applications such as solar cells[1], optoelectronic[2], catalysis[3], and sensors[4], etc. Although many efforts still focus on synthesizing randomly oriented 1D ZnO nanostructure including nanorods[5], nanowires[6], nanobelts[7], nanosheets[8] and nanotubes[9], it has been realized that the construction of well-aligned ZnO nanotube arrays (ZNTAs) on substrates is expected to represent important building blocks for nanodevices due to their larger surface area and unique optical and electrical properties[10]. To date, various kinds of techniques have been proposed to synthesize ZNTAs including the sol-gel method[11], hydrothermal method[12], template method[13], chemical vapor deposition (CVD)[14], vapor-solid mechanisms[15] and electrodeposition

---

张梅,1996~2000年于北京科技大学师从李文超教授攻读博士学位。目前在北京科技大学工作,教授,博士生导师。获得中国金属学会冶金青年科技奖、新世纪优秀人才荣誉称号,发表论文140余篇,获省部级二等奖5项。

method[16-20]. Among these approaches, synthesizing ZNTAs by electrodeposition method is considered as one of the promising routes to meet the demand of low-cost large-scale production for industrial needs. Till now, two-step method [21-23], namely, electrodeposition of ZnO nanorods on conducting substrates firstly and then etching the top of nanorods by chemical or electrochemical approach, were usually used to prepare ZNTAs. Zhang et al. [21] deposited ZnO nanorods on FTO substrate firstly and then reversed the polarity of the working electrode changing the nanorods into nanotubes. The obtained ZnO nanotubes were mainly 300-500nm in diameter. Tena-Zaera et al. [22] put the ZnO nanowire arrays prepared by electrodeposition method in a KCl solution, resulting in the formation of the tubular structures. The mean value of the diameter was about 200-500nm and the influences of KCl concentration, temperature and immersion time were systematically investigated. In 2012, Lin et al. [23] had reported that a large scale and highly oriented ZNTAs with an average diameter of 400 nm were synthesized on ITO substrate by a two-step route in alkaline solution. From above we know that the diameter, density and orientation consistence of the as-prepared ZNTAs mainly depended on that of the nanorod arrays formed at the first step, and generally the erosion degree of the ZnO nanorods can't be controlled, leading to semirod/semitube structures appeared. Therefore, how to fabricate the single-crystalline ZNTAs by one-step approach attracted more attention[24,25]. Tang et al. [24] firstly synthesized ZNTAs with larger diameters directly on the FTO substrate by one-step method, and they proposed that the surface condition of the substrate played a key role in the nanotube formation. Hwang et al. pointed out that the formation of the ZnO nanotubes was governed by the self-etching process with $H^+$ generated during the growth process by direct electrodeposition. However, how to control the morphology, the size of diameter (especially smaller than 200nm), growth density of well-oriented ZNTAs are still the most challenging issues by using one-step electrodeposition approach. But detailed works on these issues are inadequate up to now. Therefore, it is necessary to explore systematically the effect of preparing conditions on the direct electrodeposition of well-oriented ZNTAs.

In this paper, ZNTAs with average diameter less than 200nm were synthesized on FTO substrate under different electrodeposition parameters. The effects of preparing conditions such as pretreatment of substrate, pH value of electrolyte, electrodeposition potential, reaction temperature, growth time and concentration of electrolyte on the electrodeposition of ZNTAs were systematically investigated. It is shown that the morphology features, diameter and growth density of ZNTAs can be effectively controlled by using suitable growth conditions. Moreover, the band gap energy of ZNTAs can be controlled by adjusting electrodeposition time.

## 2 Experimental

All chemicals were of analytical reagent grade and used without further purification, and all aqueous solutions were prepared using de-ionized water. The FTO glass ($10-15\Omega \cdot cm^{-2}$) was used as substrate and tailored into rectangular shape with dimensions of $1 \times 3 cm^2$. Before electrodeposition, the FTO glass was cleaned with ultrasonic in acetone and absolute ethyl alcohol for 10min respectively, after that, the substrate was immersed in diluted hydrochloric acid (pH = 4.00) for

2min, then rinsed with de-ionized water. The electrolyte was $0.005\ mol \cdot dm^{-3}$ $ZnCl_2$ aqueous solution bubbling with saturated oxygen ($20mL \cdot min^{-1}$), and $0.1\ mol \cdot dm^{-3}$ KCl aqueous solution was introduced into the electrolyte as the supporting solution insuring the good conductivity of the electrolyte.

Electrodeposition of ZNTAs was performed in a standard three electrodes system with the FTO substrate as working electrode, a Pt wire as the counter electrode, and a saturated calomel electrode (SCE) as the reference electrode, respectively. The electrolyte pH value was adjusted by $0.01\ mol \cdot dm^{-3}$ hydrochloride acid and measured by using PHS-25C of Shanghai Kangyi in order to control at 3.70. The deposition temperature was fixed at 80℃ by a water bath and the electrodeposition potential was carried out at $-1.3V$ (vs. SCE, named as nucleation potential) for 10s and $-1.0V$ (vs. SCE, named as growth potential) for 1800s using CHI660C potentiostat.

The as-prepared ZNTAs were characterized as follows: The surface morphology and size distribution of the nanotubes was investigated by scanning electron microscope (SEM) (Zeiss supra 55) operated at 10kV. The Ultraviolet and Visible (UV-Vis) absorption spectrum was measured to obtain the band gap energy of ZNTAs by UV-VIS-NIR recording spectrophotometer (Shimadzu, UV-3100) using $BaSO_4$ as background and creating the baseline from 240 to 800nm.

## 3 Results and discussion

Generally, well aligned ZNTAs on conducting substrates were prepared by a two-step electrodeposition/chemical reaction approach, which means the oriented ZnO nanorod arrays (ZNRAs) were firstly electrodeposited on conducting substrates, then the nanorods whose top surface ((001) crystal face) was more unstable than other surfaces, could be converted to ZNTAs by selectively being eroded in acid or alkali solution due to its amphoteric property. The relevant equations can be written as follows:

$$O_2 + 2H_2O + 4e^- \longrightarrow 4OH^- \qquad (1)$$

$$Zn^{2+} + xOH^- \rightleftharpoons Zn(OH)_x^{2-x} \qquad (2)$$

$$Zn(OH)_x^{2-x} \rightleftharpoons ZnO + H_2O + (x-2)OH^- \qquad (3)$$

$$ZnO + 2H^+ \longrightarrow Zn^{2+} + H_2O\ (\text{in acidic solution}) \qquad (4)$$

$$ZnO + 2OH^- \longrightarrow ZnO_2^{2-} + H_2O\ (\text{in alkaline solution}) \qquad (5)$$

In our previous work[26], it is indicated that we could directly prepare ZNTAs on conducting substrates by one-step electrodeposition method only adjusting the electrolyte pH value lower than 7, which means keeping the two competitive reactions, namely, ZnO crystal growth (reaction (3)) and selective self-etching by $H^+$ (reaction (4)) in balance during the electrochemical deposition process.

### 3.1 The influence of substrate pretreatment

It is known that substrate surface characteristics is an important factor for formation of ZNTAs by using different preparing methods including hydrothermal and Pulse Laser Deposition approaches[12,27]. In our previous works [28], we investigated the relationship between the substrate surface characteristics and the growth of ZNTAs by a hydrothermal approach. It is shown that the ZNRAs with different size, morphology and distribution could be obtained on ITO substrates pretreated by highly pure nitrogen and hydrochloric acid, respectively. Fig. 1 showed the typical SEM images of ZnO nanostructures on FTO substrates pretreated by different ways including immersing the substrates into deionized water, $1 \times 10^{-4}$ mol·dm$^{-3}$ potassium hydroxide solution and $1 \times 10^{-4}$ mol·dm$^{-3}$ hydrochloric acid, respectively. It can be clearly seen that ZNTAs with uniform size and large scale were successfully prepared only on the substrate pretreated with hydrochloric acid, while just ZNRAs were obtained on the substrate pretreated with alkali solution, suggesting that substrates pretreated with acid are more effective for electrodeposition of ZNTAs. It is worth noting that ZnO nanorods with partially etched top surfaces appeared solely on the substrate pretreated by immersing in de-ionized water. The foregoing results indicated that ZnO with different nanostructure (nanorods/nanotubes) could be selectively electrodeposited on the FTO substrates pretreated specifically with different ways.

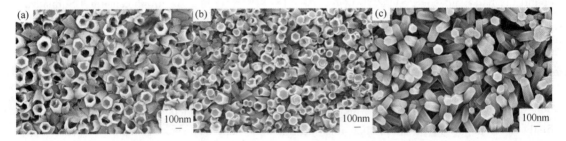

Fig. 1 SEM images of ZnO nanorod/nanotube arrays electrodeposited under different substrate pretreatments:
(a) immersing into hydrochloric acid solution; (b) immersing into deionized water;
(c) immersing into potassium hydroxide solution

Preparing conditions: $E = -1.3$ V 10s, ($-1.0$ V 1800s, $T = 80$ ℃, pH = 3.70, $C_{Zn^{2+}} = 0.005$ mol·dm$^{-3}$)

Such differences might be attributable to the different surface structure of the respective substrates pretreated in different ways. When the substrate was immersed in $1 \times 10^{-4}$ mol·dm$^{-3}$ hydrochloric acid for 2 min, a lot of hydrogen ions would be adsorbed uniformly onto the substrate surface. The positive surface charge would result in a double layer in which the $Zn^{2+}/OH^-$ ratio was less than in the bulk solution. Reduced $Zn^{2+}/OH^-$ concentration ratio may favor the formation of tapered ZnO nanorods with relatively more high-index polar crystal faces [29]. These unstable polar faces would be apt to be selectively etched and dissolved in the electrolyte solutions, leading to the formation of ZnO nanotubes, as observed in Fig. 1 (a). Alternatively, when the substrate was immersed in $1 \times 10^{-4}$ mol·dm$^{-3}$ potassium hydroxide solution for 2 min, a lot of hydroxyl ions would be adsorbed uniformly on the substrate surface. The electrostatic effects would work in the opposite sense, increasing the $Zn^{2+}/OH^-$ concentration ratio in the vicinity of the polar surface, thereby favoring growth of ZnO nanorods with flat and uniform stable polar surfaces, and sustaining ZnO nanorods growth, as shown in Fig. 1 (c).

## 3.2 The influence of electrolyte pH value

On one hand, based on the growth habits of ZnO crystal in aqueous solution, we know that the ZnO nanorods can be obtained only when the pH values of bulk solution is controlled nearly neutral (reaction 3)[30]. On the other hand, as an amphoteric oxide, the obtained ZnO nanorods whose top surface ((001) crystal face) was more unstable than other surfaces, can be converted to ZNTAs by selectively being eroded in acid solution (reaction 4). Therefore, how to control the electrolyte pH value in order to keep the two competitive reactions (ZnO crystal growth and selective self-etching by $H^+$) in balance is the key role in formation of ZNTAs during the electrodeposition process.

Fig. 2 gave the SEM images of the ZnO nanostructures electrodeposited on pre-treated FTO substrates when the electrolyte pH values were controlled from 3.00 to 6.90. It can be seen that well aligned ZNRAs with uniform size and flat tip were obtained on FTO substrates when the pH value was under 6.90 as shown in Fig. 2 (a). With the pH value reducing to 4.00, the formed ZNRAs became uneven and the tops of nanorods were selectively etched. Further decreasing the electrolyte pH values from 3.90 to 3.10, oriented ZNTAs could be synthesized on FTO substrates as illustrated in Fig. 2 (c) – (g), confirming that the top surfaces of ZnO nanorods were more unstable than other lateral surfaces, and especially the two competitive reactions including ZnO crystal growth and selective self-etching by $H^+$ were indeed kept in balance during this period of the electrolyte pH values, which would lead to the formation of ZNTAs. When the pH value was lower than 3.00, the synthesized ZnO was dissolved completely and disappeared, even the surface of substrate would be etched to some extent (Fig. 2 (h)).

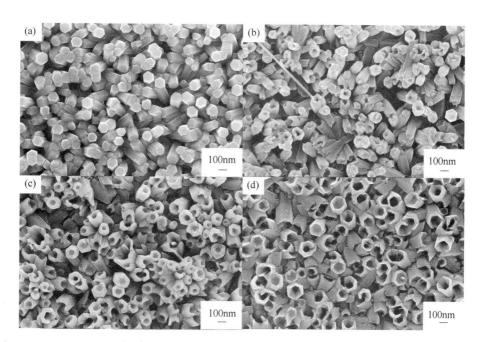

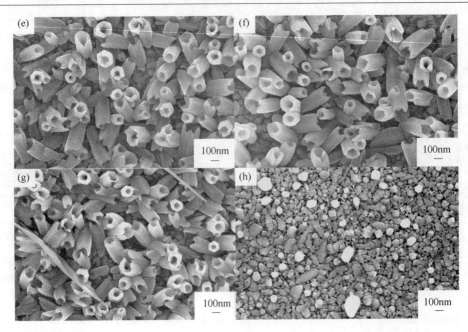

Fig. 2 SEM images of ZnO nanorod/nanotube arrays electrodeposited under different electrolyte pH values:
(a) 6.90; (b) 4.00; (c) 3.90; (d) 3.70; (e) 3.50; (f) 3.30; (g) 3.10; (h) 3.00
(Preparing conditions: $E = -1.3V$ 10s, $-1.0V$ 1800s, $T = 80℃$, $C_{Zn}^{2+} = 0.005 mol \cdot dm^{-3}$, FTO substrate immersing in diluted hydrochloric acid (pH = 4.00) for 2min)

From the chemical reaction point of view, it is suggested that two reactions (reactions (3) and (4)) were most important for the selective etching to form the ZNTAs. In fact, reaction (3) and reaction (4) conducted at the same time. It is believed that the faster rate of reaction (3) could lead to form the ZNRAs, otherwise, morphology of ZNTAs would be favored. If the pH value approached to neutrality (4.00-6.90), the concentration of $H^+$ ions in the solution was too low, leading to the reaction (4) conducted slowly. On the other hand, reaction (3) occurred rapidly and the seed of ZnO was easier to get. In this case, the growth of ZnO nanorods were predominant due to the speed of the growth was far greater than the rate of erosion. Finally the ZnO nanorods were observed on the substrate. However, when the pH value approached to 3.00, $H^+$ ions would be overmuch and ZnO nanostructures were dissolved. The effect of pH value on the growth of ZNTAs showed that the tubular structure could be obtained especially when the pH values of electrolyte solution were controlled from 3.10 to 3.90, and it had a complete tubular structure at 3.70.

## 3.3 The influence of electrodeposition potential

The electrodeposition of ZnO is based on the electrochemical reaction involving oxygen (reaction (1)) which generate electrochemically hydroxide ions leading to the formation of ZnO at the electrode surface, and the electrochemical reaction (reaction (1)) is mainly related to the applied reduction potential of oxygen (more negative than the theoretical value 0.16V vs. SCE due to the high overpotential). Fig. 3 showed the SEM images of as-prepared ZnO nanostructures under the same nucleation potential (-1.3V) and different growth potential (-1.0, -0.8 and

−0.6V). From Fig. 3 it can be seen that ZNTAs were prepared on FTO substrate only when the applied potential was controlled at −1.0V, while the ZNRAs with etched tops were obtained when the applied potential were at −0.8 and −0.6V, indicating the electrodeposition potential played an important role in formation of ZNTAs. This phenomenon may be explained as follows: the relatively more negative applied potential would lead to generate more hydroxide ions according to reaction (1), and meanwhile, zinc ions in the electrolyte could easily diffuse to or adsorb on the cathode surfaces due to the stronger electric field intensity. This would promote the electrodeposition proceeding, and the growth rate of the electrodeposition of ZNRAs would increase. Considering ZnO growth (reaction (3)) and selective self-etching (reaction (4)) occurred simultaneously during the deposition process, therefore, the electrodeposition of ZNRAs at −1.0V would firstly reach equilibrium, and the rate of dissolution was faster than the rate of formation. Under this circumstance, the polar surfaces having the higher energy would be dissolved in priority to decrease the system energy and the tubular structures were formed. These results indicated that the polarity of ZnO and the applied electrodeposition potential were necessary factors for the ZNTAs during the whole growth process.

Fig. 3　SEM images of ZnO nanorod/nanotube arrays electrodeposited at −1.3V 10s then under different applied potentials for 1800s: (a) −1.0V; (b) −0.8V; (c) −0.6V

(Preparing conditions: $T=80℃$, pH = 3.70, $C_{Zn^{2+}} = 0.005 mol \cdot dm^{-3}$,

FTO substrate immersing in diluted hydrochloric acid (pH = 4.00) for 2min)

### 3.4　The influence of growth time

As for the electrodeposition of ZNTAs, multi-potential steps technique was used: firstly, the applied deposition potential was fixed at −1.3V for 10s, then the potential was controlled at −1.0V for different time. Fig. 4 gave the variation of the current as the function of the deposition time for 1800s at 80℃, and Fig. S1 (ESI) also illustrated the variation of the current as the function of other deposition time (300s, 900s and 1500s) at 80℃. It can be clearly seen that the process of electrodeposition of ZNTAs for different time included two stages: nucleation process at higher applied potential (−1.3V), and growth process at relatively lower applied potential (−1.0V). We noted that the average current decreased abruptly from about 14 to 7mA, then became stable at 7mA during the first 10s shown in the inset parts of Fig. 4 and Fig. S1 (ESI), indicating reaction (1) occurred rapidly and the overmuch of $OH^-$ were instantly formed on/near the surface of cathode (FTO substrates) at −1.3V. Because of large amount of zinc ions in the electrolyte in the initial stage, they could easily diffuse to or adsorb on the cathode surfaces leading to generate Zn

$(OH)_x^{2-x}$, and the heterogeneous nucleation occurred lead to the formation of ZnO nano-grains on the surface of the substrate. Till now, the nucleation process for electrodeposition of ZNTAs at higher applied potential completed. When the growth process started at $-1.0V$, the current increased sharply from about $-4mA$ to $2mA$ for about 25s, then the current remained relatively constant at $2mA$ when the growth time increased from 25s to 300, 900, 1500 and 1800s as illustrated in Fig. S1 (ESI) and Fig. 4, suggesting that the growth of ZnO nanorods from the ZnO seeds formed during the nucleation process went into a fast step. During the ZnO nanorods growth stage, the precursor could continuously provide liberally reactants for the electrodeposition by diffusion with no limitation, and the concentration variation of hydroxides and zinc ions could be negligible in a certain period from 25s to 300s at the substrate surface. However, with the deposition proceeding, the concentration of $Zn^{2+}$ decreased more seriously than that of $H^+$ in the electrolyte solution, the reaction (2) would be limited and growth of the ZnO nanorods occurred slower. In this case, the selective self-etching of the formed ZnO nanorods became the main process. Therefore, the well aligned tubular structures were finally formed as shown in Fig. S2 (b-d) (ESI). Further increasing electrodeposition time, the previously formed ZnO nanotubes with smaller size would be easier to dissolve owing to its high specific surface energy, leading to relatively larger average diameters, lower density and even poor orientation of as-prepared ZNTAs.

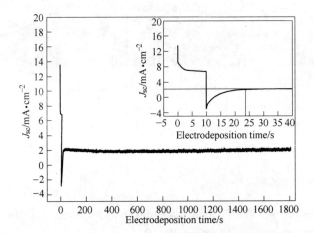

Fig. 4  The variation of current as the function of time: $-1.3V$ 10s, $-1.0V$ 1800s

(Preparing conditions: pH = 3.70, $T = 80°C$, $C_{Zn^{2+}} = 0.005 mol \cdot dm^{-3}$,

FTO substrate immersing in diluted hydrochloric acid (pH = 4.00) for 2min.

The insets correspond to their magnified images from 0 to 40s.)

The corresponding diameter distributions of as-prepared ZnO nanostructures were illustrated in Fig. S3 (ESI), and the average diameter and growth density were summed up in Table 1. It is indicated that with increasing the electrodeposition time from 300s to 1800s, the average diameters (outer diameter for ZnO nanotube) of ZnO nanostructures increased from $133 \pm 10nm$ to $204 \pm 10nm$, whereas the growth density decreased from $3.48 \times 10^9$ to $1.76 \times 10^9$ rods or tubes $cm^{-2}$ correspondingly. In addition, it is worth noting that the average inner diameters of synthesized ZNTAs also increased from 112 ± 10nm to 178 ± 10nm, while the wall thickness of the ZnO nanotubes decreased from about 30nm to 15nm with the time prolonging from 900s to 1800s.

Table 1  The influence of the electrodeposition time on growth density and average diameters of ZnO nanorod/nanotube arrays

| Group | Deposition time/s | Density /rods or tubes cm$^{-2}$ | Average diameter/nm | Wall thickness/nm |
| --- | --- | --- | --- | --- |
| a | 300 | $3.48 \times 10^9$ | 133 ± 10 | — |
| b | 900 | $3.04 \times 10^9$ | 167 ± 10 | 30 |
| c | 1500 | $2.25 \times 10^9$ | 186 ± 10 | 22 |
| d | 1800 | $1.76 \times 10^9$ | 204 ± 10 | 15 |

Moreover, the UV-Vis absorption spectra of ZnO nanostructures electrodeposited for different time were also measured and the result agreed well with that literature [26] reported. The absorbance variation trend of ZnO nanostructures was similar, the absorption wavelength threshold of samples red shifted, and the relevant band gap energies decreased from 3.59eV to 3.48, 3.37 and 3.31eV, respectively by increasing the time from 300s to 1800s, implying that the electrodeposition time has a little impact on the band gap energy of ZNTAs. This case may be mainly due to the different crystalline degree of the formed ZNTAs electrodeposited for different time.

### 3.5 The influence of growth temperature

As for electrodeposition of ZnO nanostructures, it is known that the zinc hydroxide may be transited into zinc oxide by dehydration reaction when the temperature was controlled at 34℃ and above. However, the formation of zinc oxide nucleus and crystal growth rate by dehydration reaction was relatively low when the growth temperature was controlled between 34℃ and 50℃ [31], which may result in the delay time of transition between zinc hydroxide and zinc oxide, and consequently, the as-prepared ZnO were not well-defined crystallized and dispersed randomly on the FTO substrates. Fig. 5 (a) gave the SEM image of ZnO structure electrodeposited at 50 ℃. It can be seen that none of the ZnO nanotubes appeared on the substrate while the agglomerated ZnO particles with poor crystallization, and a few larger ZnO plates were observed. However, with the electrodeposition temperature increasing to 70℃, the ZNTAs with larger size distribution were formed on the substrates, and in the meantime, nearly all the ZnO large plates were dissolved and disappeared as shown in Fig. 5 (b). Further increasing the temperature to 80℃, the uniform ZNTAs with narrower size distribution and well crystallization were gained (Fig. 5 (c)). The average diameter of as-deposited ZNTAs was summed up in Table 2. It can be seen that the average diameters of the ZNTAs decreased from 194 ± 10nm to 189 ± 10nm with the electrodeposition temperature increasing from 70℃ to 80℃. It is indicated that the growth temperature has a little effect on diameter distributions of as-deposited ZNTAs. This phenomenon may be explained as follows: firstly, with the deposition temperatures increasing, not only the diffusion rate of zinc ions in the electrolyte improved, but also the dehydration reaction rate of $Zn(OH)_2$ to ZnO accelerated, which may result in higher growth rate of ZNRAs. Therefore, the growth of ZnO nanorods could quickly reach equilibrium. Secondly, the diffusion rate of $H^+$ in the electrolyte improved, so, the rate of dissolution (reaction (4)) was faster than the rate of formation, and the selective self-etching of the formed nanorods became the main process. The previously formed ZnO plates and smaller ZnO nanotubes would be easier to dissolve in the acid electrolyte solution due to its high specific surface energy. Therefore, the ZNTAs with uniform size were finally obtained.

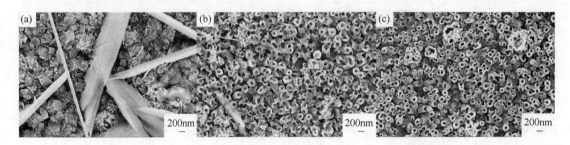

Fig. 5 SEM images of ZnO nanostructures electrodeposited under different growth temperatures:
(a) $T=50℃$; (b) $T=70℃$; (c) $T=80℃$
(Preparing conditions: $E=-1.3V$ 10s, $-1.0V$ 1800s, pH = 3.70,
$C_{Zn}^{2+} = 0.005$ mol · dm$^{-3}$, FTO substrate immersing in diluted hydrochloric acid (pH = 4.00) for 2min)

Table 2 The influence of growth temperature on growth density and average diameters of ZnO nanorod/nanotube arrays

| Group | Growth temperature/℃ | Density/rods or tubes cm$^{-2}$ | Average diameter/nm |
| --- | --- | --- | --- |
| a | 50 | — | — |
| b | 70 | $2.98 \times 10^9$ | $194 \pm 10$ |
| c | 80 | $3.24 \times 10^9$ | $189 \pm 10$ |

### 3.6 The influence of zinc concentration

The SEM image of the ZnO nanorod/nanotube arrays electrodeposited on FTO substrates in different zinc concentration solutions was shown in Fig. 6. It is indicated that well oriented ZNRAs with flat tops were obtained by using 0.010mol · dm$^{-3}$ ZnCl$_2$ solution (Fig. 6 (a)), while ZNTAs appeared on substrates when the zinc concentrations decreased to 0.005 and 0.001 mol · dm$^{-3}$ as illustrated in Fig. 6 (b) and (c), suggesting that the lower precursor concentration was needed for preparation of ZNTAs by electrodeposition. The detailed statistical results of the influence of zinc concentration on average diameter and growth density of as-deposited ZnO nanorods/nanotubes were shown in Table 3. It can be seen that with the zinc concentrations decreasing from 0.010 to 0.001mol · dm$^{-3}$, the average diameters of the nanorods/nanotubes decreased obviously from $196 \pm 10$nm to $144 \pm 10$nm, whereas the growth density increased from $3.53 \times 10^9$ to $6.38 \times 10^9$cm$^{-2}$.

Fig. 6 SEM images of ZnO nanorod/nanotube arrays electrodeposited under different concentrations:
(a) $C_{Zn}^{2+} = 0.01\text{mol} \cdot \text{dm}^{-3}$; (b) $C_{Zn}^{2+} = 0.005\text{mol} \cdot \text{dm}^{-3}$, (c) $C_{Zn}^{2+} = 0.001\text{mol} \cdot \text{dm}^{-3}$
( Preparing conditions: $E = -1.3\text{V}$ 10s, $-1.0\text{V}$ 1800s, $T = 80$ ℃, pH = 3.70, FTO substrate immersing in diluted hydrochloric acid (pH = 4.00) for 2min)

Table 3 The influence of the zinc concentration on growth density and average diameters of ZnO nanorod/nanotube arrays

| Group | Zinc concentration/mol·dm$^{-3}$ | Density/rods or tubes cm$^{-2}$ | Average diameter/nm |
|---|---|---|---|
| a | 0.010 | $3.53 \times 10^9$ | 196 ± 10 |
| b | 0.005 | $5.32 \times 10^9$ | 178 ± 10 |
| c | 0.001 | $6.38 \times 10^9$ | 144 ± 10 |

Based on the growth mechanism of ZNTAs, it is known that the average diameter and density of the ZNTAs were mainly determined by the ZNRAs firstly electrodeposited on the FTO substrates in the early stage. And the obtained ZnO nanorods seemed to have a specific average diameter and a specific density when the growth of ZnO nanorods reached equilibrium under the preparing conditions. When the concentration of $Zn^{2+}$ was in a higher level, e.g., $0.010\text{mol} \cdot \text{dm}^{-3}$, it would lead to generate more hydroxide ions, and in the meantime, zinc ions in the electrolyte easily diffused to or absorbed on the cathode surfaces due to the stronger electric field intensity. All this phenomena would promote the electrodeposition proceeding, and the growth rate of the electrodeposition of ZnO would increase and the rods with the larger diameter and lower density could be formed. With the deposition proceeding, the concentration of $Zn^{2+}$ decreased according to reaction (2), but its critical concentration was not lower than that of $H^+$ in the electrolyte solution. So the growth of ZnO nanorods had been the master step during all the deposition time, leading to the formation of ZNRAs on the FTO substrate with larger diameter and lower growth density (e.g, Table 3).

Contrarily, when the concentration of $Zn^{2+}$ was in a lower level (0.005 or $0.001\text{mol} \cdot \text{dm}^{-3}$), the growth of ZnO nanorods with smaller diameter and larger density could be formed. With the deposition proceeding, the concentration of $Zn^{2+}$ decreased more seriously than that of $H^+$ in the electrolyte solution. The growth of ZnO nanorods reached equilibrium. So, the rate of dissolution was faster than the rate of formation, and the selective self-etching of the formed nanorods became the main process. Therefore, the ZNTAs with smaller diameter and larger density were finally obtained (Fig. 6 (b) – (c)). The results are in accordance with our previous works[30].

## 4 Conclusion

Well aligned single crystalline ZNTAs were prepared on FTO substrate under different conditions by using a one-step electrodeposition method. The effects of electrodeposition parameters played important roles in controlling the morphology, the average diameter and density of as-prepared ZnO nanostructures. The oriented ZNTAs had been prepared when the applied potential was controlled from of -0.8V to -1.0V and the pH value of electrolyte from 3.10 to 3.90 at the same time on the acid pretreated FTO substrates. The growth temperature and the electrolyte pH value had great impact on the morphology of as-prepared ZnO nanorod/nanotube arrays. When the growth temperature was 70℃ to 80℃, the ZNTAs with well crystallization and small diameter (<200nm) could be formed. In addition, the tube average diameters can be monitored to some extent by mainly changing the concentration of the zinc precursors or growth time. What's more, the band gap energy (Eg) of ZNTAs could be controlled by changing the electrodeposition time.

**Acknowledgements:** The work is supported by the National Nature Science Foundation of China (No. 51272025 and 50872011), and 973 Program of China (No. 2014CB643401).

## References

[1] F. Dehghan Nayeri, E. Asl Soleimani, F. Salehi. Synthesis and characterization of ZnO nanowires grown on different seed layers, the application for dye-sensitized solar cells [J]. Renewable Energy, 2013, 60: 246.

[2] S. S. Shinde, A. P. Korade, C. H. Bhosale, et al. Influence of tin doping onto structural, morphological, optoelectronic and impedance properties of sprayed ZnO thin films [J]. Journal of Alloys and Compounds, 2013, 551: 688.

[3] M. Ahmad, E. Ahmed, Z. L. Hong, et al. A facile one-step approach to synthesizing ZnO/graphene composites for enhanced degradation of methylene blue under visible light [J]. Applied Surface Science, 2013, 274: 273.

[4] Y. H. Shi, M. Q. Wang, C. Hong, et al. Multi-junction joints network self-assembled with converging ZnO nanowires as multi-barrier gas sensor [J]. Sensors and Actuators B, 2013, 177: 1027.

[5] J. Y. Zhu, J. X. Zhang, H. F. Zhou, et al. Microwave-assisted synthesis and characterization of ZnO-nanorod arrays [J]. Trans. Nonferrous Met. Soc. China, 2009, 19: 1578.

[6] O. Lupan, V. M. Guérin, I. M. Tiginyanu, et al. Well-aligned arrays of vertically oriented ZnO nanowires electrodeposited on ITO-coated glass and their integration in dye sensitized solar cells [J]. Journal of Photochemistry and Photobiology A: Chemistry, 2010, 211: 65.

[7] X. R. Qu, S. C. Lv, J. J. Wang, et al. Preparation and optical property of porous ZnO nanobelts [J]. Materials Science in Semiconductor Processing, 2012, 15: 244.

[8] S. J. Chen, Y. C. Liu, C. L. Shao, et al. Structural and optical properties of uniform ZnO nanosheets [J]. Adv. Mater, 2005, 17: 586.

[9] K. Yang, G. W. She, H. Wang, et al. ZnO nanotube arrays as biosensors for glucose [J]. J. Phys. Chem. C, 2009, 113: 20169.

[10] X. F. Ma, M. J. Gao, J. B. Zheng, et al. Conversion of large-scale oriented ZnO rod array into nanotube array under hydrothermal etching condition via one-step synthesis approach [J]. Physica E, 2010, 42: 2237.

[11] X. H. Liu, J. Q. Wang, J. Y. Zhang, et al. Sol-gel-template synthesis of ZnO nanotubes and its coaxial nano-

composites of LiMn$_2$O$_4$/ZnO [J]. Materials Science and Engineering A, 2006, 430: 248.

[12] Y. Sun, N. G. Ndifor-Angwafor, D. J. Riley, et al. Synthesis and photoluminescence of ultra-thin ZnO nanowire/nanotube arrays formed by hydrothermal growth [J]. Chemical Physics Letters, 2006, 431: 352.

[13] F. Ochanda, K. Cho, D. Andala, et al. Synthesis and optical properties of Co-doped ZnO submicrometer tubes from electrospun fiber templates [J]. Langmuir, 2009, 25: 7547.

[14] C. C. Wu, D. S. Wuu, P. R. Lin, et al. Three-step growth of well-aligned ZnO nanotube arrays by self-catalyzed metal organic chemical vapor deposition method [J]. Crystal Growth & Design, 2009, (9): 4555.

[15] Y. J. Xing, Z. H. Xi, Z. Q. Xue, et al. Optical properties of the ZnO nanotubes synthesized via vapor phase growth [J]. Appl. Phys. Lett., 2003, 83: 1689.

[16] L. F. Xu, Q. Liao, J. P. Zhang, et al. Single-crystalline ZnO nanotube arrays on conductive glass substrates by selective dissolution of electrodeposited ZnO nanorods [J]. J. Phys. Chem. C, 2007, 111: 4549.

[17] F. Xu, J. Chen, L. Y. Guo, et al. In situ electrochemically etching-derived ZnO nanotube arrays for highly efficient and facilely recyclable photo catalyst [J]. Applied Surface Science, 2012, 258: 8160.

[18] C. L. Cheng, J. S. Lin, Y. F. Chen. A simple approach for the growth of highly ordered ZnO nanotube arrays [J]. J. Alloys Compd., 2009, 476: 903.

[19] G. W. She, X. H. Zhang, W. S. Shi, et al. Controlled synthesis of oriented single-crystal ZnO nanotube arrays on transparent conductive substrates [J]. Appl. Phys. Lett., 2008, 92: 0531111.

[20] X. Ren, C. H. Jiang, D. D. Li, et al. Fabrication of ZnO nanotubes with ultrathin wall by electrodeposition method [J]. Materials Letters, 2008, 62: 3114.

[21] G. W. She, X. H. Zhang, W. S. Shi, et al. Electrochemical/chemical synthesis of highly-oriented single-crystal ZnO nanotube arrays on transparent conductive substrates [J]. Electrochemistry Communications, 2007, (9): 2784.

[22] J. Elias, R. Tena-Zaera, G. Y. Wang, et al. Conversion of ZnO nanowires into nanotubes with tailored dimensions [J]. Chem. Mater., 2008, 20: 6633.

[23] J. Y. Yang, Y. Lin, Y. G. Meng, et al. A two-step route to synthesize highly oriented ZnO nanotube arrays [J]. Ceramics International, 2012, 38: 4555.

[24] Y. W. Tang, L. J. Luo, Z. G. Chen, et al. Electrodeposition of ZnO nanotube arrays on TCO glass substrates [J]. Electrochemistry Communications, 2007, (9): 289.

[25] C. J. Xu, B. S. Kim, J. H. Lee, et al. Seed-free electrochemical growth of ZnO nanotube arrays on single-layer grapheme [J]. Mater. Lett., 2012, 72: 25.

[26] H. Lu, F. Zheng, M. Guo, et al. One-step electrodeposition of single-crystal ZnO nanotube arrays and their optical properties [J]. J. Alloys Compd., 2014, 588: 217-221.

[27] T. Ohshima, R. K. Thareja, T. Ikegami, et al. Preparation of ZnO thin films on various substrates by pulsed laser deposition [J]. Surface and Coatings Technology, 2003, 517: 169-170.

[28] Z. N. Liu, X. Guo, M. Zhang, et al. Effect of substrate pretreatments on the growth of ZnO micro/nanotube arrays by a hydrothermal method [J]. Journal of University of Science and Technology Beijing, 2011, (6): 756.

[29] P. Li, H. Liu, Y. F. Zhang, et al. Synthesis of flower-like ZnO microstructures via a simple solution route [J]. Materials Chemistry and Physics, 2007, 106: 63.

[30] M. Guo, C. Y. Yang, M. Zhang, et al. Effects of preparing conditions on the electrodeposition of well-aligned ZnO nanorod arrays [J]. Electrochimica Acta, 2008, 53: 4633.

[31] A. Goux, T. Pauporte, J. Chivot, et al. Temperature effects on ZnO electrodeposition [J]. Electrochim. Acta, 2005, 50: 2239.

（原载于 *Electrochim. Acta*, 2014, 132: 370-376）

# Oxygen Sensitivity of Nano-$CeO_2$ Coating $TiO_2$ Materials

Zhang Mei  Wang Xidong  Wang Fuming  Li Wenchao

(Department of Physical Chemistry, University of Science and Technology Beijing, Beijing 100083)

**Abstract:** The oxygen sensitivity of pure $TiO_2$, nano-$CeO_2$ coating micro-$TiO_2$ and nano-$CeO_2$ coating nano-$TiO_2$ materials was investigated in the present work. The experiments showed that the oxygen sensitivity of the nano-$CeO_2$ coating nano-$TiO_2$ materials was much better than that of the pure $TiO_2$ and $CeO_2$ coating micro-$TiO_2$ materials. Thermodynamic analyses and XPS experimental results indicated that $CeO_2$ was easier to be reduced into nonstoichiometric compound than $TiO_2$ under the same change of partial oxygen pressure. With large special surface area and capacity of storing and releasing oxygen while the atmosphere changes, nano-$CeO_2$ is a very good catalyst for the oxygen sensitivity of $TiO_2$.

## 1 Introduction

Titanium dioxide is receiving increased attention in view of its advantageous properties [1-9] such as high dielectric constant, excellent optical transmittance, high refractive index, high chemical stability and suitable band gap. It can be processed into dielectric capacitor [1], optical coating layer [2], catalyst supports, photo-catalysts [3-5] and semi-conductor sensors.

$TiO_2$ (rutile in most cases) can not only be used as humidity-sensitive and pressure-sensitive materials but also used as the gas sensors such hydrogen, oxygen, carbon monoxide sensors [6-8]. In order to improve its sensitivity, a proper catalyst, such as Pt, is often required.

$CeO_2$ has been widely used as a catalyst and it has the electrochemical properties of storing or releasing oxygen while the atmosphere changes [9-11]. Thus it is reasonable to assume that $CeO_2$ will be a suitable catalyst for the oxygen sensitivity of $TiO_2$.

It is known that while a sensor is made of nano-materials, its responding rate and sensitivity will be improved and its working temperature will be reduced due to its large specific surface area [10,11]. In addition, nano-catalyst can also improve its catalysis. Therefore, nano-$CeO_2$ and nano-$TiO_2$ composite sensitive materials have been investigated in the present work. Nano-$CeO_2$ coating nano-$TiO_2$ materials have been experimentally prepared (in order to improve the interaction between $TiO_2$ and $CeO_2$, each nano-$TiO_2$ particle was synthesized surrounded by several nano-$CeO_2$ particles). Oxygen

---

张梅，1996～2000年于北京科技大学师从李文超教授攻读博士学位。目前在北京科技大学工作，教授，博士生导师。获得中国金属学会冶金青年科技奖、新世纪优秀人才荣誉称号，发表论文140余篇，获省部级二等奖5项。

sensitivities of nano-$CeO_2$ coating nano-$TiO_2$ synthesized at varied conditions have been measured. A comparison of the oxygen sensitivity of nano-$CeO_2$ coating nano-$TiO_2$ with that of pure $TiO_2$ and nano-$CeO_2$ coating micro-$TiO_2$ has been done and discussed.

## 2 Experimental procedures

The preparation of the nano-$CeO_2$ coating nano-$TiO_2$ materials was shown in Fig. 1. To prepare nano-$CeO_2$ coating nano-$TiO_2$, the key point is to put nano-$TiO_2$ into the solution of $Ce(NO_3)_3 \cdot 6H_2O$ and urea. Thus, nano-$TiO_2$ provides a seed crystal or a site for $CeO_2$ to crystal. Therefore, a $TiO_2$ particle is covered by several nano-$CeO_2$ particles, A TEM analysis photo of the synthetic material is shown in Fig. 2.

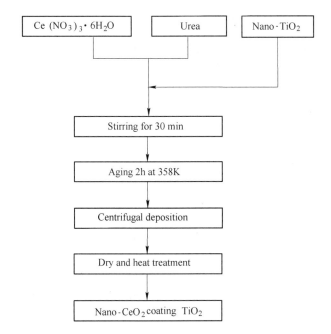

Fig. 1　Preparation procedures of nano-$CeO_2$ coating nano-$TiO_2$

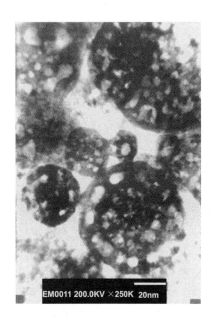

Fig. 2　Morphology of nano-$CeO_2$ coating $TiO_2$

After nano-$CeO_2$ coating nano-$TiO_2$ was prepared, it was processed into $\phi 15 \times 2mm^3$ thin cylinders and each end was coated Pt electrodes connected to Pt wires. At the same time, pure $TiO_2$ and nano-$CeO_2$ coating micro-$TiO_2$ had been processed in a similar way (the procedures to prepare $CeO_2$ coating micro-$TiO_2$ is almost the same as Fig. 1, the only different step is that micro-$TiO_2$ instead of nano-$TiO_2$ was put into the solution of $Ce(NO_3)_3 \cdot 6H_2O$ and urea). The electric resistance of these materials was measured under different oxygen pressure (In order to get different oxygen pressure, air was mixed with pure $N_2$ gases).

In order to understand the mechanism of conductivity of nano-$CeO_2$ coating nano-$TiO_2$ materials, the XPS spectra of different elements of the nano-$CeO_2$ coating nano-$TiO_2$ materials under the same treating conditions has been analyzed.

## 3  Experimental results and discusses

### 3.1  Comparison of oxygen sensitivity of nano-$CeO_2$ coating $TiO_2$ and that of pure $TiO_2$

The electric resistance of nano-$CeO_2$ coating nano-$TiO_2$ (5% $CeO_2$) and pure nano-$TiO_2$ was measured under different oxygen pressure. The average size of nano-$CeO_2$ coating nano-$TiO_2$ is about 70nm and the average size of nano-$TiO_2$ is about 13.6nm what was observed from TEM photo.

The oxygen sensitivity of nano-$CeO_2$ coating nano-$TiO_2$ (5% $CeO_2$) was compared with that of pure nano-$TiO_2$, the results is shown in Fig. 3. As can be seen, the slop of nano-$CeO_2$ coating $TiO_2$ is greater than that of pure $TiO_2$, which indicates that oxygen sensitivity of nano-$CeO_2$ coating $TiO_2$ is better than that of pure $TiO_2$.

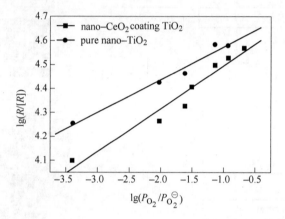

Fig. 3  Oxygen sensitivity compare of nano-$CeO_2$ coating nano-$TiO_2$ with pure nano-$TiO_2$

### 3.2  Comparison of oxygen sensitivity of the nano-$CeO_2$ coating nano-$TiO_2$ with that of nano-$CeO_2$ coating micro-$TiO_2$

After the $CeO_2$ coating micro-$TiO_2$ with different $CeO_2$ content was prepared, the oxygen sensitivities have been measured at different temperatures. The average size of micro-$TiO_2$ is 2.0$\mu$m. Oxygen sensitivity of nano-$CeO_2$ coating nano-$TiO_2$ and nano-$CeO_2$ coating micro-$TiO_2$ is shown in Fig. 4 and Fig. 5.

As can be seen in Fig. 4 and Fig. 5, the slope of the resistance curve of nano-$CeO_2$ coating nano-$TiO_2$ was obviously larger than that of nano-$CeO_2$ coating micro-$TiO_2$ materials. Therefore, nano-$CeO_2$ coating nano-$TiO_2$ responded greater than that of $CeO_2$ coating micro-$TiO_2$ materials under the same change of oxygen pressure.

### 3.3  XPS analysis of nano-$CeO_2$ coating $TiO_2$ material

Fig. 3 showed that the oxygen sensitivity was obviously improved after $TiO_2$ coated by $CeO_2$. Fig. 4 and Fig. 5 showed that the oxygen sensitivity of the nano-$CeO_2$ coating nano-$TiO_2$ was better than

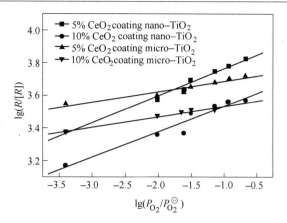

Fig. 4  Comparison of resistance with different content and grain size of CeO$_2$ coating TiO$_2$ at 1073K

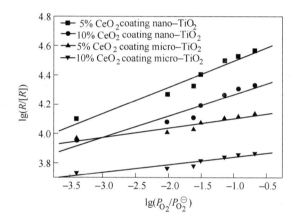

Fig. 5  Comparison of resistance with different content and grain size of CeO$_2$ coating TiO$_2$ at 973K

that of the nano-CeO$_2$ coating micro-TiO$_2$ materials. In order to understand the conductive mechanism, XPS analyses of nano-CeO$_2$ coating nano-TiO$_2$ have been carried out.

Under the same treating conditions, the XPS spectra of different elements of nano-CeO$_2$ coating nano-TiO$_2$ materials are shown in Fig. 6 (a) – (c). Spectra of Ti$_{2p}$ (Fig. 6 (b)) and O$_{1s}$ (Fig. 6 (c)) showed that their chemical valence didn't change on the whole treating procedure. This indicated that ionic Ti$^{4+}$ was not reduced into Ti$^{X+}$ ($X = 3.96, 3.9, \cdots$). On the contrary, the content of Ce$^{3+}$ increased obviously after the materials had been bombarded (Fig. 6 (a) -2), and the ionic Ce$^{3+}$ has been re-oxidized into Ce$^{4+}$ while the partial oxygen pressure increased (Fig. 6 (a) -3). These confirmed that CeO$_2$ was easily reduced into lower chemical valence than that of the Ti$^{4+}$. These results were in consistent with the following thermodynamic analyses.

Some standard molar Gibbs energies of Ce- and Ti- oxides are listed in Table 1. On the basis of Table 1, standard molar Gibbs energies of the relative reactions can be calculated and the results are listed in Fig. 7.

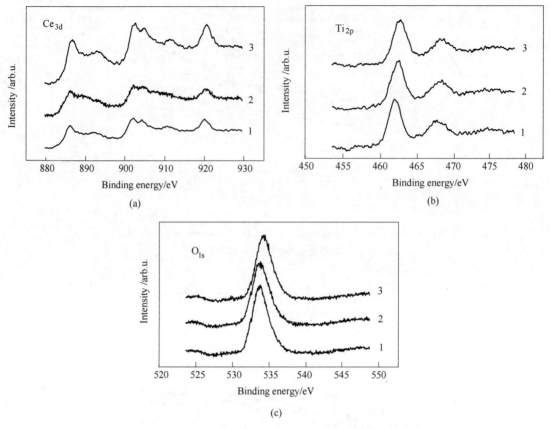

Fig. 6  XPS analysis of nano-$CeO_2$ coating $TiO_2$ material (823K)

(a) Spectra of $Ce_{3d}$;  (b) Spectra of $Ti_{2p}$;  (c) Spectra of $O_{1s}$

1—$4 \times 10^{-4}$Pa, 30min; 2—$Ar^+$ bombarding, 20sec; 3—$4 \times 10^{-4}$Pa, 20min

**Table 1  Standard molar Gibbs free energies of different reaction**

| compound | $\Delta_f G^\ominus - T/J \cdot mol^{-1}$ | Temperature/K |
|---|---|---|
| $CeO_2$ | $-1085700 + 211.3T$ | 298 – 1077 |
| $Ce_6 O_{11}$ | $-5843920 - 1001.0T$ | 298 – 2000 |
| $Ce_{18} O_{31}$ | $-1688260 - 2971.6T$ | 298 – 2000 |
| $TiO_2$ | $-943500 + 179.1T$ | 298 – 1940 |
| $Ti_{50} O_{99}$ | $-46781500 + 8880T$ | 298 – 2000 |
| $Ti_{20} O_{39}$ | $-1739980 - 2914.1T$ | 298 – 2000 |

From Fig. 7, it can be seen that $TiO_2$ is more stable than $CeO_2$ and reaction (1) in Fig. 7 is the easiest to react under the same partial oxygen pressure. Hence, $CeO_2$ can be reduced into non-stoichiometric compound.

When the partial oxygen pressure decreased, $CeO_2$ was firstly reduced into non-stoichiometric compound, it can be written as follow:

$$4Ce_{Ce} + 2O_O \rightleftharpoons O_2 + 4Ce'_{Ce} + 2V_O^{\cdot\cdot} \tag{1}$$

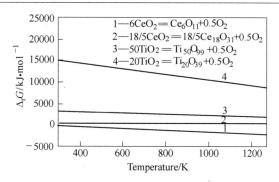

Fig. 7 Standard Gibbs energies of some reactions

The oxygen produced in equation (1) escaped into the outer environment, thus a great deal of oxygen ion vacancy ($V_O^{\cdot\cdot}$) existed in $TiO_2$ ($CeO_2$) lattice. Those charged oxygen ion vacancy can conduct electricity [6,7], hence, the resistance of the $CeO_2$ coating $TiO_2$ materials was decreased more than that of the pure $TiO_2$ as oxygen pressure decreased. On the other hand, if the partial oxygen pressure increased, equation (1) would react backward. Thus, $Ce^{3+}$ combined the oxygen and was oxidized into $Ce^{4+}$ and the amount of the $V_O^{\cdot\cdot}$ decreased obviously, so the resistance of the material increased greatly. Therefore, the oxygen sensitivity of the $CeO_2$ coating $TiO_2$ materials was better than that of the pure $TiO_2$.

It is known that special surface area of nano-materials is much bigger than that of micro-material. Hence, The oxygen sensitivity of nano-$CeO_2$ coating nano-$TiO_2$ materials was better than that of nano-$CeO_2$ coating micro-$TiO_2$ materials.

## 4 Conclusions

(1) Thermodynamic analyses and XPS experimental found that $CeO_2$ was easier to be reduced into nonstoichiometric compound than $TiO_2$ under the same partial oxygen pressure.

(2) The oxygen sensitivity of nano-$CeO_2$ coating nano-$TiO_2$ materials was much better than that of the pure $TiO_2$ and $CeO_2$ coating micro-$TiO_2$ materials.

(3) Nano-$CeO_2$ can improve the oxygen sensitivity of $TiO_2$ for its large special surface area and capacity of the storing and releasing oxygen while the atmosphere changes. Nano-$CeO_2$ shows good catalyst for the oxygen sensitivity of $TiO_2$.

## References

[1] Shu-Chuan Huang, Tong-Fong Lin, Shi-Yuan Lu, et al. J. of Materials Science, 1999, 34: 4293-4304.
[2] Jun-Hua Huang, Lian Gao. J. Inorganic Materials, 1996, 11 (1): 51-57.
[3] M. Graetzel. Comments Inorg. Chem., 1991, 12: 93-111.
[4] H. Tang, H. Berger, P. E. Schmid, et al. Solid State Commun., 1994, 92: 267-271.
[5] H. Tang, K. Prasad, R. Sanjines, et al. J. Appl. Phys., 1994, 75: 2042-2047.
[6] K. D. Schreibaum, U. K. Kirner, J. F. Geigeret al. Sensors and Actuators B, 1991, 4: 87-94.
[7] Yulong Xu, Kui Yao, Xiaohua Zhou, et al. Sensors and Actuators B, 1993, 14: 492-494.
[8] T. Y. Tien, H. L. Stagler, E. F. Gibbons et al. Am. Ceram. Soc. Bull., 1975, 54: 280-282.

[9] M. T. Wu, X. Yao, Z. H. Yuan, et al. Sensors and Actuators B, 1993, 491: 13-14.
[10] A. Tschoepe, J. Y. Ying. Nonaphase Materials: Properies Applications. ed. G. C. Hadjipanayis et al. Netherlands, Kluwer Academic, 1994: 781-785.
[11] K. B. Sundaram, P. F. Wahid. J. Vac. Sci. Technol. A, 1997, 15 (1): 52-56.

(原载于 *Sensors and Actuators B*, 2003, 92 (1-2): 167-170)

# Thermodynamic Analysis of Combustion Synthesis of $Al_2O_3$-TiC-$ZrO_2$ Nanoceramics

Dong Qian[1,2]　Tang Qing[1]　Li Wenchao[2]　Wu Dongya[2]

(1. Institute of Chemical Metallurgy, Chinese Academy Sciences, Beijing 100080;
2. University of Science and Technology Beijing, Department of Physical-Chemistry, Beijing 100083)

**Abstract:** Combustion synthesis of $Al_2O_3$-TiC-$ZrO_2$ nanoceramics by reactions in $TiO_2$-Al-C-$ZrO_2$ system is a new method with advantages of simplicity and efficiency. In the present work, the effect of $ZrO_2$ nanoparticles on the thermodynamics of combustion synthesis of $TiO_2$-Al-C system is studied in order to obtain desired phases. The result of thermodynamic analysis shows that the adiabatic temperature $T_{ad}$ of the system keeps at about 2327K (the melting point of $Al_2O_3$) with the addition of $ZrO_2$ in the range of 0–15wt%, and the fraction of molten $Al_2O_3$ varied from 100% to 78%. The possible combustion products are discussed with an approach of overlapped Phase Stability Diagram (PSD) of Al-O-N, Ti-O-N, Zr-O-N and C-O-N systems at 2300K. It has been shown that the combustion product is a mixture of $Al_2O_3$-TiC-$ZrO_2$, which coincides with the results obtained by XRD and TEM.

## 1 Introduction

Combustion synthesis or self-propagating high-temperature synthesis (SHS) is a new process for producing high-performance ceramics[1-3]. $Al_2O_3$-TiC multiphase composites can be used to produce ceramic cutting-tool, but its strength and toughness still need to be improved. The combustion reaction under study is an aluminothermic reaction:

$$3TiO_2 + 4Al + 3C = 3TiC + 2Al_2O_3 \quad (1)$$

Which has previously been used to produce $Al_2O_3$-TiC ceramic[4-6]. Additives, especially nanometric particles in combustion system are one kind of important method to control combustion process and even the properties of materials[7].

On the other hand, the search for the optimum conditions for the synthesis of products of desired chemical and phase composition is necessary. Along with the methods of combustion theory and structural macrokinetics, those of combustion chemistry, particularly thermodynamic analysis of the combustion synthesis plays an important role in the solution of this search. The development of the thermodynamic branch is based on the estimation of both temperature and equilibrium compositions of the combustion product. Thermodynamic analysis is especially important for the multicomponent

---

董倩，1998~2001年于北京科技大学师从李文超教授攻读博士学位。目前在挪威DNV GL公司工作，任高级工程师。

systems, when the composition of combustion product is not apparent because of competing side reactions and transformations.

In the current study, the effect of $ZrO_2$ nanometric additive on the thermodynamics of the combustion synthesis system and phase composites of the products will be discussed. In addition, the purposes of present study are to describe the microstructure of an $Al_2O_3$-TiC-$ZrO_2$ sintered body and to provide new information that will contribute to researching the mechanism of combustion synthesis.

## 2  Experimental procedures

The starting materials utilized in this study are as follows: commercial anatase (average particle size is $2\mu m$, 99% purity), aluminum (average particle size is $100\mu m$, 99% purity), graphite (average particle size is $10\mu m$, >98% purity) and synthesized tetragonal $Y_2O_3$ stabilized $ZrO_2$ (average particle size is 10nm, >99%) powders. The reactant powders with the content of $ZrO_2$ ranging from 0 to 15wt% were mixed with a C/Ti atomic ratio of 0.9 based on the following reaction:

$$4Al + 3TiO_2 + 2.7C + wZrO_2 \longrightarrow 3TiC + 2Al_2O_3 + wZrO_2 \quad (2)$$

where TiC is off-stoichiometric composites.

The mixture of reactant powders was alcoholic wet-milled in a nylon jar for 24 hours. The well-blended powders were then filled in a paper column to form a green compact with the density of 0.75 to 1.1 $g/cm^3$. The combustion synthesis processes were carried out in a reaction chamber at room temperature in argon atmosphere. The combustion reactions were initiated with an electrified tungsten coil on the upper surface of the sample, i.e., the "wave propagation" combustion mode. Once ignited, combustion wave was established and it self-propagated across the entire sample to complete the reaction. Combustion temperature was measured by a fine tungsten-rhenium (W-3% Re and W-25% Re) thermocouple with the diameter of 0.25 millimeter, which was embedded into the center of each sample through a small hole in the samples. An IBM compatible computer with data acquisition software recorded the temperature data at every 2.23-millisecond interval. The as-combusted samples were ground into powders, which were used for the phase analysis. Phases in the as-combusted products were analyzed with a D/MAX-RB X-ray diffractometer (XRD). The morphology of the combustion-synthesized products was observed with Cambridge S-250MK3 scanning electron microscope (SEM) and JEM-2010 analytical transmission electron microscope (TEM) with an X-ray energy-dispersive spectrometer (EDS, Model ISIS).

## 3  Results and discussion

### 3.1  Calculation of the adiabatic temperature

The adiabatic temperature ($T_{ad}$) is an important parameter in the thermodynamics of combustion synthesis. $T_{ad}$ can be used to determine simply whether the combustion synthesis can be self-sustaining or not and the state of various phases. According to criteria provided by Reference[8] and[9], assuming that no heat loses to surrounding, i. e., adiabatic conditions, the adiabatic temperature and the fraction of molten $Al_2O_3$ can be calculated from the following equation based on:

$$\Delta_r H_{298}^0 = 2\left(\int_{298}^{800} C'_{p_{Al_2O_3}} dT + \Delta_{tr} H_{Al_2O_3} + \int_{800}^{T_{ad}} C''_{p_{Al_2O_3}} dT + v\Delta_m H_{Al_2O_3}\right) + 3\int_{298}^{T_{ad}} C_{p_{TiC}} dT \\ + w\left(\int_{298}^{1478} C'_{p_{ZrO_2}} dT + \Delta_{tr} H_{ZrO_2} + \int_{1478}^{T_{ad}} C''_{p_{ZrO_2}} dT\right) \quad (3)$$

where $\Delta_r H_{298}^0$ is the enthalpy of the reaction (2) at 298K, $C_{p_i}$ is heat capacity of product $i$, $\Delta_{tr} H$ is the enthalpy of phase transitions of product $i$, $\Delta_m H_i$ is the enthalpy of fusion of the product $i$ at its melting point, $v$ is the fraction of the product $i$ which has melted, for this case, $0 \leq v \leq 1.0$.

According to thermodynamic data from Reference[10], the result of thermodynamic calculation shows that the adiabatic temperature ($T_{ad}$) keeps at about 2327K (the melting point of $Al_2O_3$) with the addition of $ZrO_2$ in the range of 0-15wt%. On the other hand, the experimental combustion temperature ($T_c$) approximately keeps constant with the increasing content of $ZrO_2$ from 0 to 15wt%, and the fraction of molten $Al_2O_3$ varied from 100% to 78.28%. The comparison between the calculated adiabatic temperature and the experimental combustion temperature of $TiO_2$-Al-C-$ZrO_2$ system is gives in Fig. 1. It can be found that the adiabatic temperature is higher than the experimental one. The deviation between theoretical calculation and experiments is probably attributed to some side-reactions related in the system and thermal loss. As shown in Fig. 2 shows the combustion temperature profiles of a $TiO_2$-Al-C sample with 5wt% $ZrO_2$ additive collected by a computer data acquisition system at different positions of the combustion compact.

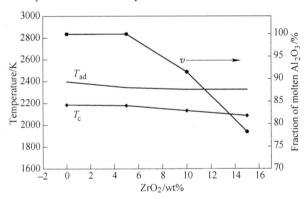

Fig. 1 Influence of addition of $ZrO_2$ on the adiabatic temperature ($T_{ad}$), the combustion temperature ($T_c$) and fraction of molten $Al_2O_3$ ($v$) in $TiO_2$-Al-C-$ZrO_2$ system

## 3.2 Phase stability diagrams

The system of Al-Ti-Zr-C-O-N under consideration is complex. In order to obtain the desired phases and suitable processing conditions, the isothermal phase stability diagrams (PSD) or so-called predominance area diagrams are used. The result from thermodynamic calculation has shown that the adiabatic temperature $T_{ad}$ keeps at about 2327K with the addition of $ZrO_2$ in the range of 0-15wt%. Therefore the phase stability diagrams of Al-O-N, Ti-O-N, Zr-O-N and C-O-N at 2300K are calculated and simply overlapped together as shown in Fig. 3. The phases that can exist

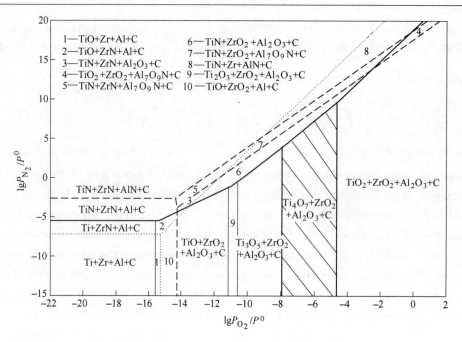

Fig. 2 Combustion temperature profiles of a $TiO_2$-Al-C sample with 5wt% $ZrO_2$ additive in each phase stability area are indicated, and the partial pressure relations of various equilibrium systems between $O_2$ and $N_2$ at 2300K are shown in Table 1-3.

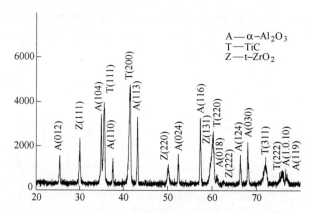

Fig. 3 Overlapped phase stability diagram for Al-O-N, Ti-O-N, Zr-O-N and C-O-N systems at 2300K

Table 1  The partial pressure relations of Al-O-N equilibrium systems between $O_2$ and $N_2$ at 2300K

| Reactions | The partial pressure relations |
|---|---|
| $Al_{(l)} + 1/2N_2 = AlN_{(s)}$ | $\lg P_{N_2}/P^0 = -2.70$ |
| $2Al_{(l)} + 2/3O_2 = Al_2O_{3(s)}$ | $\lg P_{O_2}/P^0 = -14.21$ |
| $2Al_7O_9N_{(s)} + 3/2O_2 = 7Al_2O_{3(s)} + N_2$ | $\lg P_{N_2}/P^0 = 3/2 \lg P_{O_2}/P^0 + 17.58$ |
| $7/3AlN_{(s)} + 3/2O_2 = 1/3Al_7O_9N_{(s)} + N_2$ | $\lg P_{N_2}/P^0 = 3/2 \lg P_{O_2}/P^0 + 18.79$ |
| $14Al_{(l)} + 9O_2 + N_2 = 2Al_7O_9N_{(s)}$ | $\lg P_{N_2}/P^0 = -9 \lg P_{O_2}/P^0 - 131.63$ |

Table 2  The partial pressure relations of Ti-O-N equilibrium systems between $O_2$ and $N_2$ at 2300K

| Reactions | The partial pressure relations |
|---|---|
| $Ti_{(l)} + 1/2N_2 = TiN_{(s)}$ | $\lg P_{N_2}/P^0 = -5.46$ |
| $Ti_{(l)} + 1/2O_{2(g)} = TiO_{(s)}$ | $\lg P_{O_2}/P^0 = -15.37$ |
| $2TiO_{(s)} + 1/2O_2 = Ti_2O_{3(s)}$ | $\lg P_{O_2}/P^0 = -10.97$ |
| $3/2Ti_2O_{3(s)} + 1/4O_2 = Ti_3O_{5(s)}$ | $\lg P_{O_2}/P^0 = -10.40$ |
| $4/3Ti_3O_{5(s)} + 1/6O_2 = Ti_4O_{7(s)}$ | $\lg P_{O_2}/P^0 = -7.73$ |
| $Ti_4O_{7(s)} + 1/2O_2 = 4TiO_{2(s)}$ | $\lg P_{O_2}/P^0 = -4.55$ |
| $TiN_{(s)} + 1/2O_2 = 1/2N_2 + TiO_{(s)}$ | $\lg P_{N_2}/P^0 = \lg P_{O_2}/P^0 + 9.91$ |
| $2TiN_{(s)} + 3/2O_2 = N_2 + Ti_2O_{3(s)}$ | $\lg P_{N_2}/P^0 = 3/2 \lg P_{O_2}/P^0 + 15.40$ |
| $3TiN_{(s)} + 5/2O_2 = 3/2N_2 + Ti_3O_{5(s)}$ | $\lg P_{N_2}/P^0 = 5/3 \lg P_{O_2}/P^0 + 17.13$ |
| $4TiN_{(s)} + 7/2O_2 = 2N_2 + Ti_4O_{7(s)}$ | $\lg P_{N_2}/P^0 = 7/4 \lg P_{O_2}/P^0 + 17.78$ |
| $TiN_{(s)} + O_2 = 1/2N_2 + TiO_{2(s)}$ | $\lg P_{N_2}/P^0 = 2 \lg P_{O_2}/P^0 + 18.91$ |

Table 3  The partial pressure relations of Zr-O-N equilibrium systems between $O_2$ and $N_2$ at 2300K

| Reactions | The partial pressure relations |
|---|---|
| $Zr_{(l)} + 1/2N_2 = ZrN_{(s)}$ | $\lg P_{N_2}/P^0 = -7.00$ |
| $Zr_{(l)} + O_2 = ZrO_{2(s)}$ | $\lg P_{O_2}/P^0 = -15.29$ |
| $ZrN_{(s)} + O_2 = 1/2N_2 + ZrO_{2(s)}$ | $\lg P_{N_2}/P^0 = 2 \lg P_{O_2}/P^0 + 23.57$ |

The primary partial pressures of nitrogen and oxygen in reaction chamber are estimated as 0.5Pa and 1Pa, respectively. So the equilibrium point of the combustion system is placed in the shadow area under the experimental condition. It can be seen from the overlapped PSD of Al-$TiO_2$-C-$ZrO_2$ system that the phases in the shadow area include $Al_2O_3$, $Ti_4O_7$, $ZrO_2$ and C. In order to estimate that which phase can react preferentially with carbon under the given conditions, Gibbs free energies of the following reactions that probably occurred have been calculated (as shown in Table 4).

Table 4  Results of thermodynamic calculation ($T = 2300K$)

| No. | Reaction | $\Delta G^0$/kJ |
|---|---|---|
| 1 | $C + 2/9 Al_2O_3 = 1/9 Al_4C_3 + 2/3CO$ | -3.43 |
| 2 | $C + 1/11Ti_4O_7 = 4/11TiC + 7/11CO$ | -74.93 |
| 3 | $C + 1/3 ZrO_2 = 1/3 ZrC + 2/3CO$ | -41.05 |

As can be seen from Table 4, reaction (1) - (3) can occur at 2300K because of $\Delta G^0 < 0$, and the trend of reaction (2) is the most obvious, which means that TiC is more stable than ZrC and $Al_4C_3$. Based on the calculations above, if the dynamic effects are ignored, it can be considered that the equilibrium compositions of the as-combustion products are $Al_2O_3$, TiC and $ZrO_2$. And trace of ZrC and $Al_4C_3$ probably can be found.

### 3.3 XRD analysis results

In order to confirm the reliability of thermodynamic analysis, XRD result of the as-combusted products is shown in Fig. 4, which indicates that the powder is composed of $Al_2O_3$, TiC and tetragonal $ZrO_2$, and no residual $TiO_2$, aluminum, carbon or other phases were detected. It can be considered that the combustion reaction processed at a relatively complete degree. This result is accordance with the theoretical predictions.

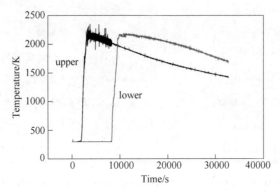

Fig. 4 X-ray diffraction patters of the combustion-synthesized products for $TiO_2$-Al-C-$ZrO_2$ system

It is noteworthy that the right shift in the peak position and the obvious increase in the d-value of TiC can be found in the XRD pattern of $Al_2O_3$-TiC-10wt%$ZrO_2$, which is expected some other phases dissolved in the crystal lattice of TiC. Calculation indicates the experimental crystal-lattice parameter of TiC ($a = 4.3409$nm) is bigger than the value ($a = 4.3274$nm) given by JPDSC card[11]. The above thermodynamic analysis shows that it is possible to form ZrC phase. In addition, the lattice structure of TiC and ZrC is centred cubic (NaCl type) and the radius difference between $Ti^{4+}$ and $Zr^{4+}$ is less than 30%. So it is not difficult to form a solid solution between TiC and ZrC at high temperature. The existence of ZrC could not be confirmed by XRD analysis, maybe the amount of ZrC was too little to be detected by XRD analysis. But the EDS results indicate that titanium dissolved a small amount of zirconium.

### 3.4 Morphology of the powder

Fig. 5 (a) shows a typical morphology of combustion-synthesized $Al_2O_3$-TiC-$ZrO_2$ ceramic products observed with SEM. It can be seen that the combustion-synthesized products are loosening and there are numerous pores.

Fig. 5 (b) shows the TEM micrograph of combustion-synthesized $Al_2O_3$-TiC-$ZrO_2$ specimen. It can be seen that most of large grains are spherical and the size are in the range of 0.3-0.6μm. Electron-diffraction patterns from one of the grains are shown in Fig. 6 (a) and (b), which was identified as $Al_2O_3$ phase. Since the $T_{ad}$ of the combustion synthesis system keeps at about the melting point of $Al_2O_3$, it could be considered that $Al_2O_3$ grains crystallize from molten phases. The molten regions around $Al_2O_3$ particles are indicated as amorphous region by the halos of the selected area diffraction pattern in TEM (as shown in Fig. 6 (c)). EDS results of these areas shows the presence of aluminum, titanium and a small amount of zirconium. It is well known that combustion synthesis is a typical non-equilibrium process due to its rapid velocity and high temperature gradient. It is rather difficult for every phase to reach equilibrium completely, which lead to form amorphous or metastable phases in the final combustion-synthesized product.

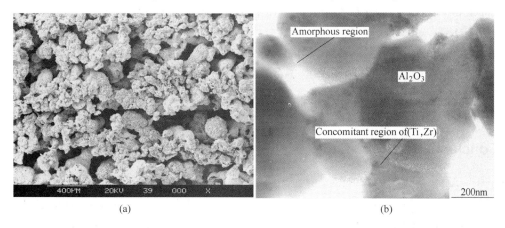

Fig. 5  SEM and TEM micrograph of as-combusted $Al_2O_3$-TiC-$ZrO_2$ composites

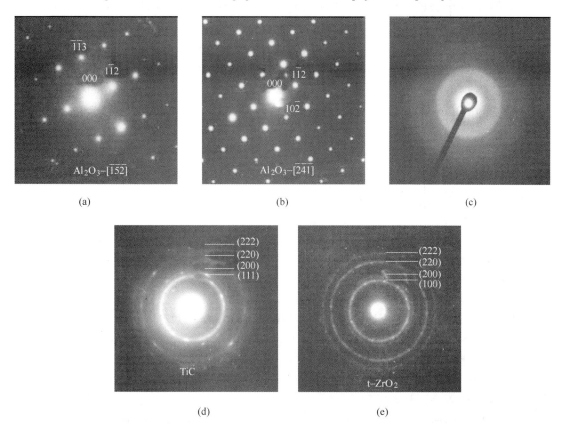

Fig. 6  The selected area diffraction patterns of $Al_2O_3$-TiC-$ZrO_2$ composites

On the other hand, some dark gray particles (marked by an arrow in Fig. 5 (b)) are randomly dispersed in the $Al_2O_3$ particles and amorphous regions. These are likely to be the original agglomerates of fine crystallites with the diameter of less than 20nm. Analysis results obtained by EDS in the TEM showed that these small darker regions were Ti-rich and Zr-rich. Considering the XRD and EDS result, the gray particles may consist of TiC and $ZrO_2$. The results of selected area electron diffraction patterns of such particles confirmed the presence of fine crystals of $ZrO_2$ and TiC (as shown in Fig. 6 (d) and Fig. 6 (e)). There is no obvious morphological difference between TiC and $ZrO_2$.

Based on the above analysis, it shows that combustion synthesis is feasible to produce the nanoceramics. Therefore, it is expected that the bulk specimen with in-situ homogeneous distribution of fine particles can be easily obtained by hot pressing using the combustion-synthesized powder as starting materials. Investigations of mechanical properties of the composite are in next paper.

## 4 Conclusions

$Al_2O_3$-TiC-$ZrO_2$ nanoceramics was synthesized by combustion-synthesis process. The adiabatic temperature ($T_{ad}$) of this system keeps constant at about 2327K (the melting point of $Al_2O_3$) with the addition of $ZrO_2$ in the range of 0 to 15wt%, and the fraction of molten $Al_2O_3$ varied from 100% to 78%. The combustion temperature ($T_c$), which is lower than $T_{ad}$, approximately keeps constant with the increasing content of $ZrO_2$. Thermodynamics analysis of combustion synthesis of $TiO_2$-Al-C-$ZrO_2$ system by phase stability diagram (PSD) indicates that the combustion products are mainly $Al_2O_3$, TiC and tetragonal $ZrO_2$. It coincides with the results obtained by XRD and TEM. $Al_2O_3$ grains crystallized from the amorphous phases. And TiC and $ZrO_2$ particles with a diameter of approximately 20nm are distributed randomly in the $Al_2O_3$ particles and some amorphous regions.

**Acknowledgements:** This work was financially supported by the National Natural Science Foundation of China (NSFC) No. 59774019.

## References

[1] A. G. Merzhanov. Combustion and plasma synthesis of high-temperature materials [M]. Edited by Z. A. Munir, J. B. Holt. New York: VCH Publishers Inc., 1990.
[2] Z. A. Munir. Ceram. Bull., 1988, 67: 342.
[3] S. Yin. Combustion synthesis [M]. Beijing: Metallurgical Industry Press, 1999.
[4] R. A. Cutler, K. M. Rigtrup. J. Am Ceram. Soc., 1992, 75: 36.
[5] C. R. Bowen, B. Derby. J. Mater. Sci., 1996, 31: 3791.
[6] Q. Tang. The combustion synthesis and the SHS-densification of TiC-Ni-Mo hard alloy and TiC-$Al_2O_3$ ceramics [D]. Univ. Sci. & tech. Beijing, 1993.
[7] Q. Tang, D. J. Zhang, F. Li, et al. The relationship on properties, processing and misconstrue of advanced ceramics [M]. Edited by J. K. Guo, X. Yao. Shanghai: Shanghai scientific and technology literature publishing

house, 2000.
[8] J. B. Holt, Z. A. Munir. J. Mater. Sci., 1986, 21: 251.
[9] J. Subrahamanyam, M. Vijayakumar. J. Mater. Sci., 1992, 27: 6249.
[10] Y. J. Liang, Y. C. Che. Handbook of thermodynamic data of inorganic materials [M]. Shenyang: Northeast University Press, 1993.
[11] JCPDS card: 32-1383.

(原载于 *Journal of Materials Research*, 2001, 16 (9): 2494-2498)

# Microwave Plasma Sintered Nanocrystalline $Bi_2O_3$ - $HfO_2$ - $Y_2O_3$ Composite Solid Electrolyte

Zhen Qiang[1,3]   Girish M. Kale[2]   He Weiming[3]   Liu Jianqiang[3]

(1. Nano-science and Nano-technology Research Center, School of Materials Science and Engineering, Shanghai University, Shanghai 200444, China; 2. Institute for Materials Research, University of Leeds, Leeds LS2 9JT, United Kingdom; 3. School of Materials Science and Engineering, Shanghai University, Shanghai 200072, China)

Processing and characterization of nanocrystalline $Bi_2O_3$-$HfO_2$-$Y_2O_3$ composite solid electrolyte having high density and conductivity has been investigated and reported in this paper. Nanopowders of mixed bismuth oxide, hafnia and yttrium oxide in $Bi_2O_3$:$HfO_2$:$Y_2O_3$ = 3:1:1 molar ratio have been prepared by a reverse titration chemical coprecipitation from $Bi^{3+}$, $Hf^{4+}$, and $Y^{3+}$ containing aqueous solution. The high density, nanocrystalline $Bi_2O_3$-$HfO_2$-$Y_2O_3$ solid electrolyte has been synthesized by microwave plasma and pressureless sintering. From the XRD results, the growth behavior indicates that growth of both $\delta$-$Bi_2O_3$ and c-$HfO_2$ crystallites obeys the parabolic rate law, expressed as $(D - D_0)^2 = kt$, during the sintering process. After the samples were sintered in microwave plasma at 700℃ for 30min, the relative density was found to be greater than 96%. Moreover, the sintered specimens exhibit considerably finer microstructure and greater densification compared to that of samples sintered by conventional pressureless sintering. Results of AC impedance spectroscopy of the solid electrolyte indicates that the conductivity of nanocrystalline $Bi_2O_3$-$HfO_2$-$Y_2O_3$ electrolyte was more than $10^{-6}$ S·cm$^{-1}$ at 350℃ and $10^{-2}$ S·cm$^{-1}$ at 550℃, which is significantly higher than that of microcrystalline $HfO_2$-based solid electrolyte.

## 1 Introduction

Functional ceramics based on stabilized $\delta$-$Bi_2O_3$ phase with high oxide ($O^{2-}$) ion conductivity are promising solid electrolyte materials for solid oxide fuel cells (SOFC), high purity oxygen generators, and electrochemical sensors[1-6]. The oxide ion conductivity of microcrystalline $Bi_2O_3$-based solid electrolyte, e.g. $Y_2O_3$-doped $Bi_2O_3$, is several orders of magnitude higher than that of the corresponding $ZrO_2$-based material at temperatures below 800℃[7]. $HfO_2$ has similar crystal structure and physical and chemical properties akin to $ZrO_2$. Compared to $ZrO_2$-based solid electrolyte, $HfO_2$-based solid electrolyte has superior mechanical properties and much lower

---

甄强，1996～2000年于北京科技大学师从李文超教授攻读博士学位。目前在上海大学工作，现任科技发展研究院书记兼主持工作院长职务。获得"赣鄱英才"省级特聘教授荣誉称号，发表论文60余篇，荣获省部级科技进步奖4项。

limiting oxygen partial pressure for the onset of electronic conduction ($P_{e'} < 10^{-12}$ Pa) which implies that $HfO_2$-based solid electrolyte can exhibit high $O^{2-}$ ion transference number at much lower oxygen partial pressure[8]. Therefore, a new dense nanocrystalline composite solid electrolyte in the $Bi_2O_3$-$HfO_2$-$Y_2O_3$ system is expected to have ionic conductivity superior to $Y_2O_3$-stabilized $HfO_2$ and mechanical properties superior $Y_2O_3$-stabilized $Bi_2O_3$.

Systematic investigations for the production and properties of nanocrystalline $Bi_2O_3$-$HfO_2$-$Y_2O_3$ solid electrolyte have been scanty although there is enough literature available on synthesis, densification, and measurement of physical as well as electrical properties of microcrystalline $Bi_2O_3$-$Y_2O_3$ ceramics and $HfO_2$-$Y_2O_3$ related ceramics[1-10]. Recently, nanopowders of materials containing $Bi_2O_3$ or $HfO_2$ have been prepared by using various methods, such as reverse chemical titration[1], chemical precipitation[11-13], sol-gel[14], spontaneous combustion[15], and solid-state reaction method[16]. The bulk density of sintered ceramics approaching close to the theoretical density could be easily achieved at fairly low temperatures when the particle size of the powders is in the nanometer range compared to the conventional ceramic powders in submicron range[13]. Besides this, the nanostructured high-density compact also exhibits significantly improved mechanical properties[14]. Application of nanocrystalline ceramics can lead to the development of electrochemical devices that have considerably low operating temperatures and can exhibit significantly improved ionic conductivity[10]. Therefore, nanocrystalline $Bi_2O_3$-$HfO_2$-$Y_2O_3$ could be a promising solid electrolyte material for the fabrication of gas sensors for in-line monitoring of automotive emission, air-to-fuel ratio in automotive combustion process, designing planar fuel cells, and monitoring the atmospheric pollutants.

The fabrication of nanocrystalline $Bi_2O_3$-$HfO_2$-$Y_2O_3$ solid electrolyte involves two key steps: first the preparation of mixed nanopowders and second their transformation into nanocrystalline $Bi_2O_3$-$HfO_2$-$Y_2O_3$ solid electrolyte powder. Among the methods mentioned earlier for the synthesis of nanopowders of materials containing $Bi_2O_3$ or $HfO_2$, the chemical precipitation process incorporating surface modifiers has been one of the most effective methods for the preparation of nanopowders of functional ceramics. The nanopowders of functional ceramic materials can prove to be technologically useful if they are transformed into dense structures by sintering at elevated temperatures without allowing excessive grain growth. The main advantage of microwave plasma sintering process is that the process is rapid and can prevent excessive grain growth during sintering process in comparison to the other conventional sintering processes. Further, the coupling of microwave with the cations and anions in the ceramic materials during the heating process can appreciably enhance the mass transfer within the ceramic compact which can in turn influence the sintering and grain growth kinetics of the material compared to that observed in the conventional sintering processes of ceramic materials. There is a very limited understanding of the influence of microwave sintering process on the grain growth and kinetics of sintering process in ceramic materials due to the limited number of investigations in this area of research.

Hence, the present work has been focused on preparing mixed nanopowders of $Bi_2O_3$ + $HfO_2$ + $Y_2O_3$ by reverse chemical titration coprecipitation[1] and fabrication of dense nanocrystalline $Bi_2O_3$-$HfO_2$-$Y_2O_3$ solid electrolyte by a novel microwave plasma sintering process. The molar ratio of $Bi_2O_3$ : $HfO_2$ : $Y_2O_3$ = 3:2:1 is maintained for two main reasons, first to produce $\delta$-$Bi_2O_3$ and c-$HfO_2$ single phases successfully in a composite and also to enable sintering of the composite solid electrolyte at lower temperatures with the assistance of $\delta$-$Bi_2O_3$ acting as a sintering aid while c-$HfO_2$ acting as a toughening agent. In addition to this, the grain growth kinetics of $\delta$-$Bi_2O_3$ and c-$HfO_2$, as well as the electrical properties of the dense nanocrystalline $Bi_2O_3$-$HfO_2$-$Y_2O_3$ solid electrolyte, have been measured in this study.

## 2 Experimental methods

### 2.1 Sample preparation

Mixed nanopowders of $Bi_2O_3$-$HfO_2$-$Y_2O_3$ in $Bi_2O_3$ : $HfO_2$ : $Y_2O_3$ = 3:2:1 molar ratio has been prepared by reverse chemical titration coprecipitation[1]. First, 550mL of aqueous ammonia solution having a concentration of 1.26mol/L and a pH $\approx$ 12 was prepared by mixing analytically pure aqueous ammonia and distilled water. The analytically pure Bi($NO_3$)$_3$ · $5H_2O$, $HfOCl_2$ and Y($NO_3$)$_3$ · $6H_2O$, in a molar ratio of 3:1:1, were dissolved in dilute nitric acid to prepare a nitrate solution containing a total metal ion ($Bi^{3+}$, $Hf^{4+}$ and $Y^{3+}$) concentration of 0.125mol/L. The 50mL prepared nitrate solution was added dropwise into the aqueous ammonia solution in a reaction vessel at room temperature. The solution was continuously stirred using a magnetic needle and was maintained at pH $\approx$ 12. During the chemical titration, 1wt.% PEG6000 (poly (ethylene glycol) of average molecular weight equal to 6000) dissolved in aqueous ammonia solution was added as a dispersing agent. The obtained suspension was further agitated in an ultrasonic bath to prevent any formation of agglomerates and was washed repeatedly with distilled water to washout all the chloride anions. The precipitate containing mixed hydroxide of $Bi^{3+}$, $Hf^{4+}$, and $Y^{3+}$ was oven dried at 80℃ for 12h and calcined at 700℃ for 1h to yield mixed nanopowders of $Bi_2O_3$-$HfO_2$-$Y_2O_3$.

The mixed oxide nanopowders synthesized as above were pressed uniaxially into disk-shaped pellets having an outer diameter of 10mm and a thickness of 2mm at a relatively low pressure of 15MPa. Approximately 3 wt.% polyvinyl alcohol (PVA) was used as a binder and a 20 wt.% stearic acid mixed with ethyl alcohol as a lubricant. The samples were sintered at temperature of 600, 700, and 800℃ in microwave plasma under oxygen atmosphere of 40mmHg for different lengths of sintering time in order to study the microstructure development and grain growth kinetics. Fig. 1 shows the schematic diagram of microwave plasma equipment used in this study. This equipment consists of microwave generator, microwave transmission and conversion system, gas control system, reaction chamber, vacuum system, and pressure monitor. In this work, the microwave power was maintained in the range of 0-1000 W at 2.45GHz frequency. The samples' temperature was measured with an infrared optical temperature monitor. The reaction chamber consists of a quartz bell-jar ($\phi$120mm), quartz current equalizer, quartz platform ($\phi$76mm), and quartz sample holder. The

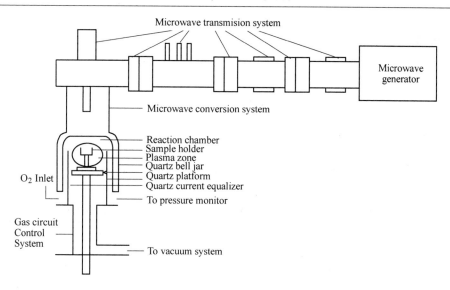

Fig. 1 A schematic diagram of microwave plasma equipment

samples were placed in the center of the quartz holder in order to keep them in the center of plasma during the sintering period so that the samples could be heated uniformly and rapidly by plasma.

Duplicate specimens were sintered by a conventional pressureless sintering process at the same temperature for comparison with the microwave plasma sintering process. The heating rate of all the samples during the sintering stage was maintained at 10℃/min. After the selected time interval, the samples were withdrawn from the furnace and were allowed to cool naturally in ambient atmosphere.

## 2.2 Materials characterization

The phases present in the calcined nanopowder and sintered samples were analyzed by X-ray diffraction (Rigaku D/max2550V) using CuKα radiation with a nickel filter at room temperature. The mean diameter of nanocrystalline grains was determined according to Scherrer formula, $t = 0.9\lambda/\beta\cos(\theta)$, from the X-ray diffraction peak width at half-maxima. The morphology of nanopowder and sintered samples were also determined using transmission electron microscopy (Hitachi H-800) and high-resolution scanning electron microscopy (JSM-6700F). The bulk density of the sintered specimens was measured at ambient temperature with an accuracy of $100\mu g \cdot cm^{-3}$ using an electronic densimeter (SD-120L, Mirage JICC Co) that employed the Archimedes principle. In all cases, distilled water was used as an immersion medium.

AC impedance spectroscopy using a two-probe method with Ag electrodes was carried out using a PAR M273A frequency response analyzer over a frequency range of 0.01Hz to 99kHz. The frequency response analyzer was interfaced to a computer using ZPlot software. Due to the small sample impedances at high temperature, a nulling technique was necessary to remove any artifacts caused by inductive responses of the test leads and the equipment. The impedance of the leads without a sample was obtained and subtracted from the measurements involving the sample and the electrical leads. The measurements were conducted between 200℃ and 600℃ in ambient atmosphere.

## 3 Results and discussion

### 3.1 Powder preparation

The uniformity of the powders is significantly influenced by the preparation process and conditions. Since the solubility product of $Bi(OH)_3$, $Hf(OH)_4$, and $Y(OH)_3$ are $4.0 \times 10^{-31}$, $4.0 \times 10^{-26}$ and $3.2 \times 10^{-26}$, respectively[17], the pH range for the commencement and completion of the precipitation of different metal ions ($Bi^{3+}$, $Hf^{4+}$ and $Y^{3+}$) has been determined separately and is listed in Table 1.

Table 1   pH range of $Bi^{3+}$, $Hf^{4+}$, and $Y^{3+}$ metal ion precipitation

| Initial concentration | pH value when precipitation commences | pH value when precipitation ends /residual ion concn < $10^{-14}$ mol/L |
|---|---|---|
| $Bi^{3+}$ initial concentration /0.075 mol·L$^{-1}$ | 4.24 | 8.53 |
| $Hf^{4+}$ initial concentration /0.025 mol·L$^{-1}$ | 8.4 | 11.2 |
| $Y^{3+}$ initial concentration /0.025 mol·L$^{-1}$ | 6.37 | 10.5 |

It is well-known that a series of precipitates, such as $BiONO_3$ and $Bi(OH)_3$, can be present simultaneously during the hydrolysis process of $Bi(NO_3)_3$[11,18,19]. It was noticed that during the forward titration process of $Bi(NO_3)_3$ with aqueous ammonia, $BiONO_3$, which could interfere in the successive reactions, was first precipitated and as a result the final product of the precipitation process was a multiphase mixture of precipitates of $Bi^{3+}$ salts mentioned earlier. Analogously, $HfO(OH)_2$ precipitates can also be present simultaneously during the hydrolysis process of $HfOCl_2$[20], Therefore, it was difficult to obtain the pure $Bi(OH)_3$, $Hf(OH)_4$ and $Y(OH)_3$ mixed powders by the forward titration process. In order to overcome this difficulty and obtain pure $Bi(OH)_3$, $Hf(OH)_4$ and $Y(OH)_3$ mixed powders, the reverse chemical titration coprecipitation process was adopted in this research work[1]. In this process, nitrate solution with three metal ions $Bi^{3+}$, $Hf^{4+}$ and $Y^{3+}$ was simultaneously titrated with the aqueous ammonia solution having pH ≈ 12 to obtain uniform $Bi(OH)_3$, $Hf(OH)_4$, and $Y(OH)_3$ mixed powders without the formation of $BiONO_3$ and $HfO(OH)_2$. The suitable pH value of aqueous ammonia solution in the reverse chemical titration coprecipitation process was greater than 11.2 from the data presented in Table 1.

X-ray diffraction pattern of mixed hydroxide powders calcined at 700℃, 1h is shown in Fig. 2. The result indicates that the calcined powder mainly consists of δ-$Bi_2O_3$ and c-$HfO_2$ with only traces of β-$Bi_2O_3$ (tetragonal) because it is believed that during the calcination process at 700℃ the following solid solution reactions occur between the component oxides[1]:

$$Bi_2O_3 + xY_2O_3 = Bi_2Y_{2x}O_{3+3x} \qquad (1)$$

$$HfO_2 + xY_2O_3 \Longrightarrow HfY_{2x}O_{2+3x} \tag{2}$$

As a result of the above reactions between $Bi_2O_3$ and $Y_2O_3$ as well as $HfO_2$ and $Y_2O_3$ the β-$Bi_2O_3$ (tetragonal) phase transforms into δ-$Bi_2O_3$ (cubic) and the m-$HfO_2$ (monocline) phase transforms into c-$HfO_2$ (cubic) with $Y_2O_3$ acting as a stabilizer. No noticeable residual peaks due to unreacted component oxides could be observed in the XRD trace.

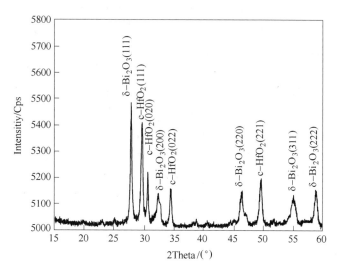

Fig. 2 XRD pattern of mixed powder of δ-$Bi_2O_3$ and c-$HfO_2$ calcined at 700℃ for 1h

Fig. 3 shows the TEM micrograph of the powder after calcination. The particle size of powder was about 40nm without significant agglomeration as seen in Fig. 3. According to the Scherrer formula and from the X-ray diffraction data, the mean diameter of δ-$Bi_2O_3$ and c-$HfO_2$ has been found to be approximately 40 and 35nm, respectively. This is in qualititative agreement with the results shown in the TEM image in Fig. 3, suggesting that the particles do not exhibit the tendency of forming agglomerates. Further, the particles of δ-$Bi_2O_3$ and c-$HfO_2$ are equiaxed, suggesting that the structure of both the solid solution composition is cubic. It was difficult to determine the exact composition of δ-$Bi_2O_3$ (cubic) and c-$HfO_2$ (cubic) with $Y_2O_3$ acting as a stabilizer due to less than 40nm grains by EDAX analysis of individual grains as the spot size of the electron beam in SEM is comparable with the grain size. However, on the basis of the binary T-X phase diagram of $Bi_2O_3$-$Y_2O_3$ and $HfO_2$-$Y_2O_3$ system reported in the literature and the initial cation concentration reported in Table 1, the concentration of $Y_2O_3$ in δ-$Bi_2O_3$ (cubic) and c-$HfO_2$ (cubic) is estimated as 25 (±2) mol% and 12.5 (±2) mol%, respectively.

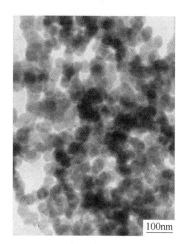

Fig. 3 TEM image of nanoparticles of δ-$Bi_2O_3$ and c-$HfO_2$ mixed powder after calcining at 700℃ for 1h

## 3.2 XRD analysis

The XRD patterns of nanocrystalline solid electrolyte samples synthesised from mixed nanopowders of $\delta$-$Bi_2O_3$ and c-$HfO_2$ by the microwave plasma sintering process at 600℃, 700℃, and 800℃ for different lengths of sintering time are shown in Fig. 4. It is clear from Fig. 4 that the major phase constitution of the sintered composite material is $\delta$-$Bi_2O_3$ and c-$HfO_2$ phase with only traces of $\beta$-$Bi_2O_3$ coexisting during the sintering process. The presence of traces of $\beta$-$Bi_2O_3$ is manifested by the presence of a broad peak around 33° of 2-theta in Fig. 2 and in Fig. 4. As the temperature and time of sintering is increased, the peak doublet at 33° of 2-theta in Fig. 4 also increased in intensity and became a singlet. This in our opinion is due to the transformation of traces of impurity phase consisting of $\beta$-$Bi_2O_3$ to $\delta$-$Bi_2O_3$. On sintering for 30min and 60min at 800℃, the intensity of the $\delta$-$Bi_2O_3$ peak at 27.5° of 2-theta increased whereas that at 33° decreased and the peak intensity of c-$HfO_2$ at 31° of 2-theta decreased too. This is possible due to the partial formation of ternary solid solution between $Bi_2O_3$-$HfO_2$-$Y_2O_3$. Further, the X-ray diffraction peak width at half-maxima diminished with increase in temperature and sintering time because of the slight increase in the grain diameter of $\delta$-$Bi_2O_3$ and c-$HfO_2$ grains during the microwave plasma sintering process.

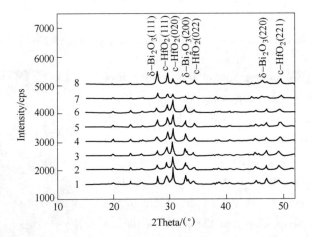

Fig. 4  XRD patterns of the samples sintered by using microwave plasma
1—600℃, 30min; 2—600℃, 60min; 3—700℃, 5min; 4—700℃, 30min;
5—700℃, 60min; 6—800℃, 5min; 7—800℃, 30min; 8—800℃, 60min

## 3.3 Grain growth

The variation of crystallite size of $\delta$-$Bi_2O_3$ and c-$HfO_2$ of the nanocrystalline $Bi_2O_3$-$HfO_2$-$Y_2O_3$ solid electrolyte samples sintered by the microwave plasma process as a function of time at 600℃, 700℃, and 800℃ is presented in Fig. 5 and Fig. 6, respectively. It can be seen that the variation of the grain size of both $\delta$-$Bi_2O_3$ and c-$HfO_2$ with sintering time follows a parabolic rate law at different sintering temperatures for the samples sintered by the microwave plasma process. It is clear from Fig. 5 and Fig. 6 that the crystallites grew rapidly in the early sintering stage at 600℃ because both $\delta$-$Bi_2O_3$

and c-$HfO_2$ have a high specific surface area in the initial stages of sintering as a result of smaller crystallite size. However, the growth rate reduced as the grain size increased with increase in sintering time. A similar grain growth behavior has been observed at 700℃ for δ-$Bi_2O_3$ and c-$HfO_2$, but as expected the rates of grain growth were higher at 700℃ than that at 600℃ for both the solid solution composition. At 800℃ the rate of grain growth of δ-$Bi_2O_3$ and c-$HfO_2$ increased dramatically as seen in Fig. 5 and Fig. 6, respectively. We believe that this phenomenon is expected as the rate of diffusion of ionic species and as a result the mass transfer across the grain boundaries in the sintered specimens is likely to increase within the solid solution phase as the temperature increases. The interesting point is that there is a window of temperature and time, typically between 600 and 700℃ for 30 to 60min, within which the grain growth is limited to below 100nm even though the densification of the sample is greater than 95% for microwave-sintered specimen and greater than 90% for pressureless sintering process as will be seen later.

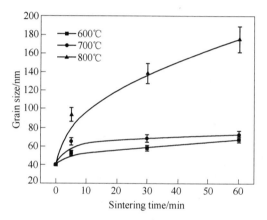

Fig. 5  Rate of change of δ-$Bi_2O_3$ grain size of nanocrystalline $Bi_2O_3$-$HfO_2$-$Y_2O_3$ samples at different temperature during microwave plasma sintering process

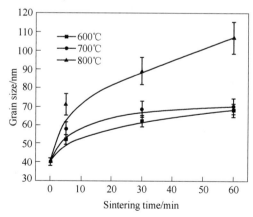

Fig. 6  Rate of change of c-$HfO_2$ grain size of nanocrystalline $Bi_2O_3$-$HfO_2$-$Y_2O_3$ samples at different temperature during microwave plasma sintering process

As a first approximation, the oxide nanoparticles can be considered close to spheres for a cubic solid solution of $Bi_2O_3$-$Y_2O_3$ and $HfO_2$-$Y_2O_3$. According to the spherical model[21], the relationship between the variation of grain size ($D - D_0$) in a sintering process, and sintering time ($t$) at a fixed temperature is given by the expression:

$$(D - D_0)^2 = kt \tag{3}$$

where $k$ is a constant; $D$ is average grain size after sintering, and $D_0$ is the average initial grain size. Therefore, using the data in Fig. 5 and Fig. 6, the grain growth in the $Bi_2O_3$-$HfO_2$-$Y_2O_3$ solid solution system at the three different sintering temperatures can be represented by the following equations. The value of $D_0$ for δ-$Bi_2O_3$ and c-$HfO_2$ used in the calculation is 40 and 35 nm, respectively, obtained from the average size of δ-$Bi_2O_3$ and c-$HfO_2$ mixed nanopowders calcined at 700 ℃ for 1h. Therefore, from the results obtained in this study, the following relationships have been derived at 600, 700 and 800 ℃, respectively.

The following relationships have been obtained at different sintering temperatures for the δ-$Bi_2O_3$ grains,

$$600\ ℃: (D - D_0) = 3.36 \times t^{\frac{1}{2}} \tag{4}$$

$$700\ ℃: (D - D_0) = 4.66 \times t^{\frac{1}{2}} \tag{5}$$

$$800\ ℃: (D - D_0) = 17.66 \times t^{\frac{1}{2}} \tag{6}$$

and for the c-$HfO_2$ grains,

$$600\ ℃: (D - D_0) = 3.83 \times t^{\frac{1}{2}} \tag{7}$$

$$700\ ℃: (D - D_0) = 4.54 \times t^{\frac{1}{2}} \tag{8}$$

$$800\ ℃: (D - D_0) = 8.81 \times t^{\frac{1}{2}} \tag{9}$$

Based on the above dynamic equations, the calculated rate constant for the growth of δ-$Bi_2O_3$ and c-$HfO_2$ grains as a function of time at three different sintering temperatures is given in Table 2.

Table 2  Grain growth rate constants ($k$) of δ-$Bi_2O_3$ and c-$HfO_2$

| Material | 600 ℃ | 700 ℃ | 800 ℃ |
| --- | --- | --- | --- |
| δ-$Bi_2O_3$ | 11.31 | 55.99 | 311.94 |
| c-$HfO_2$ | 14.75 | 30.41 | 77.43 |

According to the Arrhenius equation, the above rate constant ($k$) in eq. (3) can be expressed as a function of the reciprocal of absolute temperature by the expression:

$$k = k_0 \exp\left(\frac{-\Delta E}{R \times T}\right) \tag{10}$$

where $k$ is a rate constant at a fixed temperature of grain growth, $k_0$ is a preexponential constant at the same temperature, $\Delta E$ is a apparent activation energy, $R$ is the universal gas constant, and $T$ is the absolute temperature. Using the data from Table 2, eq. (10) yields the apparent activation energy ($\Delta E$) for the growth of nanocrystalline δ-$Bi_2O_3$ and c-$HfO_2$ grains to be equal to 125.87 kJ·mol$^{-1}$ and 62.77 kJ·mol$^{-1}$, respectively within 1% of the uncertainty limits.

The variation of grain size of δ-$Bi_2O_3$ and c-$HfO_2$ of the composite nanocrystalline $Bi_2O_3$-$HfO_2$-

$Y_2O_3$ solid electrolyte samples sintered by microwave plasma and by pressureless sintering as a function of time at 700 ℃ is compared in Fig. 7 and Fig. 8, respectively. It can be seen from Fig. 7 and Fig. 8 that the rate of grain growth of both $\delta\text{-}Bi_2O_3$ and $c\text{-}HfO_2$ using the microwave plasma sintering process is significantly lower compared to that of pressureless sintering.

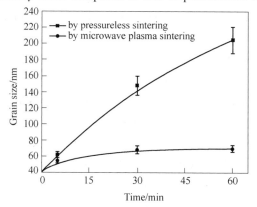

Fig. 7  Rate of change of $\delta\text{-}Bi_2O_3$ grain size of nanocrystalline $Bi_2O_3\text{-}HfO_2\text{-}Y_2O_3$ samples at 700 ℃ by different sintering process

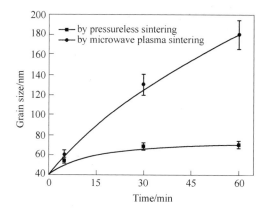

Fig. 8  Rate of change of $c\text{-}HfO_2$ grain size of nanocrystalline $Bi_2O_3\text{-}HfO_2\text{-}Y_2O_3$ samples at 700 ℃ by different sintering process

From the results shown in Figs. 5 – 8 it is apparent that the appropriate temperature and time for microwave plasma sintering of nanocrystalline composite solid electrolyte materials made from mixed nanopowders of $\delta\text{-}Bi_2O_3$ and $c\text{-}HfO_2$ is probably between 600 and 700 ℃ for 60 min or more. Under this condition, the grain growth in nanocrystalline $Bi_2O_3\text{-}HfO_2\text{-}Y_2O_3$ solid electrolyte is at a minimum during the densification process. It is well-known that prevention of excessive grain growth is essential to reap the benefits of nanostructured materials on the physical, electrical, and mechanical properties of the sintered high density solid electrolyte.

## 3.4  Density variation

The variation of the relative density of $Bi_2O_3\text{-}HfO_2\text{-}Y_2O_3$ nanocrystalline samples sintered by microwave plasma and pressureless sintering process as a function of sintering temperature between 600℃ and 800℃ for 30min is shown in Fig. 9. It can be seen from Fig. 9 that both processes exhibit a high densification rate probably due to the smaller grain size, high surface area, and smaller diffusion distances. However, at all the three temperatures selected, the relative density of samples sintered by microwave plasma is much higher than that by the pressureless sintering process. This is probably due to the effect of microwave plasma sintering on the rate of diffusion of ions within the ceramic material, enhanced mass transfer at lower temperatures and improved sintering and grain growth kinetics[22,23]. The nanocrystalline particles have a large surface area, grain boundaries, and a concentration of defects, which when coupled with microwave, significantly enhances the diffusion of charged particles ($Bi^{3+}$, $Hf^{4+}$, $Y^{3+}$ and $O^{2-}$) present in the composite solid electrolyte leading to the enhancement of the rate of sintering processes in the field of microwave plasma. Consequently, the samples having greater than 96% of the relative density could be easily obtained at 700℃ within 30min of sintering time with negligible grain growth during microwave plasma sintering as shown in Fig. 5 and Fig. 6 compared to the conventional pressureless sintering process.

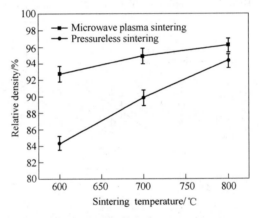

Fig. 9 Relative density of $HfO_2$-$Bi_2O_3$-$Y_2O_3$ materials by different sintering process in same sintering time of 30min

## 3.5 Microstructure

The microstructure at 40,000 magnification of the nanocrystalline $Bi_2O_3$-$HfO_2$-$Y_2O_3$ composite solid electrolyte sintered by microwave plasma and by pressureless sintering is shown in Fig. 10 (a) and (b), respectively. The sample sintered by microwave plasma at 700℃ for 60min has dense microstructure with a small amount of residual porosity and is virtually crack free as seen in Fig. 10 (a).

However, the sample sintered by the pressureless sintering process at an identical temperature and for the same length of time (700℃, 60min) shows that there is significant porosity and has relatively larger grain size, greater than 100nm, as seen in Fig. 10 (b). From Fig. 10 (a), it can be also seen that both δ-$Bi_2O_3$ and c-$HfO_2$ grains have an average size between 60nm and

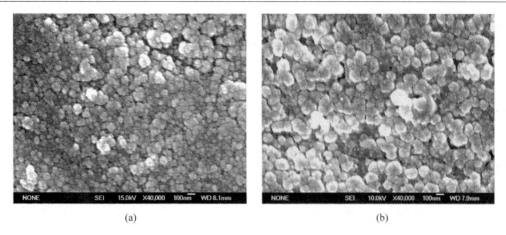

Fig. 10 Morphology of $Bi_2O_3$-$HfO_2$-$Y_2O_3$ materials by different sintering process.
(a) by microwave plasma at 700℃ 60min; (b) by pressureless sintering at 700℃ 60min

70nm and have equiaxed morphology, which is excellent for superior mechanical properties such as high fracture toughness and electrical properties such as high ionic conductivity.

## 3.6 Conductivity

Conductivity of the $Bi_2O_3$-$HfO_2$-$Y_2O_3$ composite solid electrolyte sintered by the microwave plasma process at 700℃ over a period of 1h has been measured using a two-probe AC impedance spectroscopy method. A typical complex impedance diagram at 400℃ is shown in Fig. 11. The measured impedance as a function of signal frequency was modeled using commercial software, and the intercept of the arc at lower frequency was treated as a measure of impedance of grain and grain boundary whereas the intercept at high frequency was treated as the impedance of grain.

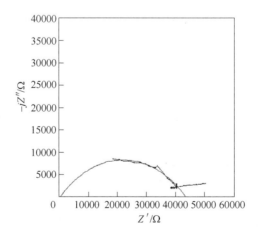

Fig. 11 Typical complex impedance diagram of $Bi_2O_3$-$HfO_2$-$Y_2O_3$
composite solid electrolyte at 400℃. The frequency of ac
signal increases from right to left in the diagram

The grain and grain boundary conductivity of the composite nanostructured solid electrolyte at different temperature computed from the measured impedance data is given in Table 3. The conductivity of nanocrystalline $Bi_2O_3$-$HfO_2$-$Y_2O_3$ is compared with the cubic form of microcrystalline $HfO_2$-$Y_2O_3$ solid electrolyte[24] and the delta form of $(Bi_2O_3)_{0.75}(Y_2O_3)_{0.25}$ solid solution[25] in Fig. 12. It can be seen from Fig. 12 that the conductivity of nanocrystalline $Bi_2O_3$-$HfO_2$-$Y_2O_3$ solid electrolyte obeys an Arrhenius type relationship and is 2 to 4 orders of magnitude higher than that of the microcrystalline $HfO_2$-based solid electrolyte reported in literature[24].

Table 3  Calculated grain, grain boundary and total conductivity of $Bi_2O_3$-$HfO_2$-$Y_2O_3$ composite solid electrolyte from the complex impedance diagrams at different temperature of measurement

| Temp/℃ | Conductivity of grains ($\sigma_b$) /$\Omega^{-1} \cdot cm^{-1}$ | Conductivity of grain boundary ($\sigma_{gb}$) /$\Omega^{-1} \cdot cm^{-1}$ | Conductivity of total ($\sigma$) /$\Omega^{-1} \cdot cm^{-1}$ |
|---|---|---|---|
| 200 | $5.10 \times 10^{-8}$ | $1.22 \times 10^{-8}$ | $9.86 \times 10^{-9}$ |
| 250 | $2.59 \times 10^{-7}$ | $1.10 \times 10^{-7}$ | $7.74 \times 10^{-8}$ |
| 300 | $1.65 \times 10^{-6}$ | $5.90 \times 10^{-7}$ | $4.35 \times 10^{-7}$ |
| 350 | $6.84 \times 10^{-6}$ | $6.16 \times 10^{-6}$ | $3.24 \times 10^{-6}$ |
| 400 | $1.11 \times 10^{-4}$ | $3.02 \times 10^{-5}$ | $2.38 \times 10^{-5}$ |
| 450 | $2.32 \times 10^{-4}$ | $1.92 \times 10^{-4}$ | $1.05 \times 10^{-4}$ |
| 500 | $2.83 \times 10^{-3}$ | $4.85 \times 10^{-4}$ | $4.14 \times 10^{-4}$ |
| 550 | — | $1.10 \times 10^{-2}$ | $1.10 \times 10^{-2}$ |
| 600 | — | $6.36 \times 10^{-2}$ | $6.36 \times 10^{-2}$ |

Further, from Fig. 12 it can be seen that the conductivity of the composite solid electrolyte is approximately 2 orders of magnitude lower than that of δ-$Bi_2O_3$ consisting of 25 mol% $Y_2O_3$[25]. This clearly suggests that the nanocrystalline $Bi_2O_3$-$HfO_2$-$Y_2O_3$ composite solid electrolyte exhibits average conducting properties compared with the pure $HfO_2$-$Y_2O_3$ solid electrolyte[24] and $(Bi_2O_3)_{0.75}(Y_2O_3)_{0.25}$ solid solution[25].

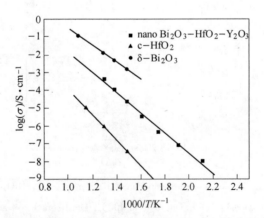

Fig. 12  Conductivity of the $Bi_2O_3$-$HfO_2$-$Y_2O_3$ and $HfO_2$-based solid electrolyte materials. The error in the measured data is approximately ±2%

We assume that the composite solid electrolyte material investigated in this study is predominantly an oxygen ion conductor; therefore, the activation energy for oxygen ion ($O^{2-}$) conduction in $Bi_2O_3$-$HfO_2$-$Y_2O_3$ solid electrolyte obtained using the Arrhenius relationship is 1.134eV, which is less than that of the $HfO_2$-based microcrystalline solid electrolyte[24], 1.431eV, and very similar to that of $\delta$-$(Bi_2O_3)_{0.75}$-$(Y_2O_3)_{0.25}$ solid solution[25], 0.87eV. This suggests that the activation energy for the oxygen ion conduction in the composite solid electrolyte also tends to exhibit an average behaviour akin to the conductivity as shown in Fig. 12. These values compare reasonably well with the data reported in the literature by Kale et al[26-28], for ceramic oxide materials bearing similar crystal structure. As a result, the nanocrystalline $Bi_2O_3$-$HfO_2$-$Y_2O_3$ electrolyte can be successfully used as an oxide ion conduction membrane in devices such as solid-state gas sensors, low-temperature single chamber SOFC, and oxygen separation membranes. However, this remains to be proved, and needs further investigation into the chemical stability of the composite electrolyte in oxygen potential gradients and determination of the ionic conduction domain as a function of temperature and oxygen potentials.

## 4 Conclusion

In conclusion, mixed nanopowders of $\delta$-$Bi_2O_3$ and c-$HfO_2$ were successfully prepared by a reverse chemical titration coprecipitation process using $Bi(NO_3)_3 \cdot 5H_2O$, $HfOCl_2$, and $Y(NO_3)_3 \cdot 6H_2O$ as starting materials and PEG6000 as dispersing agent. After calcination at 700℃ for 60min, the mixed oxide powders were found to exhibit insignificant agglomeration and were composed of $\delta$-$Bi_2O_3$ and c-$HfO_2$ having an average particle size of approximately 40nm and 35nm, respectively. The average composition of $Y_2O_3$ in $\delta$-$Bi_2O_3$ is estimated at 25 mol% and that in c-$HfO_2$ is estimates at 12.5 mol%. The dynamic equations of grain growth obtained in this investigation at the three different temperatures follow the parabolic rate law. Moreover, under microwave plasma sintering condition, grain growth rate is much lower than that under conventional pressureless sintering. The apparent activation energy for the grain growth ($\Delta E$) of nanocrystalline $\delta$-$Bi_2O_3$ and c-$HfO_2$ grains has been found to be 125.87kJ·$mol^{-1}$ and 62.77kJ·$mol^{-1}$, respectively. There is a window of temperature and time, typically between 600℃ and 700℃ for 30min to 60min, within which the grain growth is limited to less than 100nm while the density of the sample is greater than 95% for microwave-sintered specimen relative to 90% for a pressureless sintering process probably because of the greater driving force for sintering of the nanopowders of $\delta$-$Bi_2O_3$ and c-$HfO_2$ admixture and possible enhancement of mass transfer within the ceramic compact at lower temperatures due to the coupling of microwave with the cations and anions in the $\delta$-$Bi_2O_3$ and c-$HfO_2$ ceramic composite. The conductivity of nanocrystalline $Bi_2O_3$-$HfO_2$-$Y_2O_3$ composite solid electrolyte has been found to be an average of the two end members. The activation energy for oxygen ion conduction of $Bi_2O_3$-$HfO_2$-$Y_2O_3$ composite solid electrolyte has been found to be 1.134eV, which is less than that of $HfO_2$-based solid electrolyte[24], 1.431eV and greater than $\delta$-$Bi_2O_3$, 0.87eV[25].

**Acknowledgements:** The authors are grateful to the National Natural Science Foundation of China

(Grant ref. no. 20101006), Nano Technology Special Foundation of Shanghai Science and Technology Committee (Grant ref no. 0452 nm073), and Institute for Materials Research at the University of Leeds for supporting this research work.

## References

[1] Zhen Q., Kale G. M., Li R., et al. Solid State Ionics, 2005, 176: 2727-2733.
[2] Doshi R., Richards Von L., Carter J. D., et al. J. Electrochem. Soc. 1999, 146, 1273-1278.
[3] Hirabayashi D., Hashimoto A., Hibino T., et al. Electrochem Solid State Letts., 2004, 7: A108-A110.
[4] Asahara, S., Michiba D., Hibino, M., et al. ElectrochemSolid State Letts., 2005, 8: A449-A451.
[5] Li X., Xiong, W., Kale G. M. Electrochem. Solid-State Letts., 2005, 8: H27-H30.
[6] Xiong W., Kale G. M. Sensors and Actuators B, 2006, 114, 101-108.
[7] Watanabe A., Kikuchi T. Solid State Ionics 1986, 21, 287-291.
[8] Wang, C. Z. Solid Electrolyte and Chemical Sensors [M], Beijing: Metallurgical Industry Press, 2000.
[9] Karch, J., Birringer, R., Gleiter H. Nature, 1987, 330: 556-558.
[10] Ferrari, A. Development, Industrialization of Nanocomposite Ceramic Material [R]. Proceedings of Nanostructure Materials and Coating' 95, Atlanta, Georgia, USA, 1995.
[11] Kruidhof, H., Seshan Jr K., Lippens B. C., et al. J. Materials Research Bulletin, 1987, 22, 1635-1643.
[12] Bhattacharya A. K., Mallick K. K. Solid State Communications, 1994, 91: 357-360.
[13] Hirano, T., Namikawa T. IEEE. Transcations on Magnetics 1999, 35: 3487-3489.
[14] Joshi P. C., Krupanidi, S. B. J. Appl. Phys, 1992, 72: 5827-5833.
[15] Zeng, Y., Lin, Y. S. J. Mater. Sci., 2001, 36: 1271-1276.
[16] Li W. Q., Li J., Xia X., et al. Acta Chimica Sinica, 1999, 57, 491-495.
[17] Stephen H., Stephen T. Solubilities of Inorganic Compounds [M]. Oxford: Pergamon Press, 1963.
[18] Pourbaix M. Atlas of Electrochemical Equilibria in Aqueous Solutions [M]. Oxford: Pergamon Press, 1966.
[19] He W. M., Zhen Q., Liu J. Q., et al. J. Funct. Mater. (Chinese), 2003, 34, 702-706.
[20] Ivanova E. A., Konakov V. G., Solovyeva E. N. Rev. Adv. Mater. Sci, 2003, 4, 41-47.
[21] Kingery W. D., Bowen H. K., Uhlman D. R. Introduction to Ceramics [M]. London: John Wiley and Sons Inc., 1967.
[22] Tiegs T. N., Kiggams J. O., Kimrey H. D. Mat. Res. Soc. Symp. Proc, 1990, 189, 243-245.
[23] Janney M. A., Kimrey H. D. Mat. Res. Soc. Symp. Proc. 1990, 189, 215-217.
[24] Zhuiykouv, S. J. Euro. Ceram. Soc., 2000, 20, 769-976.
[25] Takahashi, T., Iwahara H. Mater. Res. Bull., 1978, 13, 1447-1453.
[26] Kurchania, R., Kale, G. M. J. Mater. Res, 2000, 15, 1576-1582.
[27] Wang, L., Pan, L., Sun, J., et al. J. Mater. Sci., 2005, 40: 1717-1723.
[28] Xiong, W., Kale, G. M. Intl. J. Appl. Ceram. Technol., 2006, 3: 210-217.

(原载于 *Chemistry of Materials*, 2007, 19: 203-210)

# 一种新型湿化学方法合成 Ba(Mg$_{1/3}$Ta$_{2/3}$)O$_3$ 纳米粉末的研究

连 芳  徐利华  王福明  李文超

(北京科技大学材料科学与工程学院，北京 100083)

**摘 要**：介绍了一种工艺简便且成本较低的 Ba(Mg$_{1/3}$Ta$_{2/3}$)O$_3$ (BMT) 纳米粉末的湿化学制备方法。实验过程以 Ta$_2$O$_5$ 作为起始物，通过热碱反应与钽离子浓度控制法合成 Ta$_2$O$_5$·$n$H$_2$O 胶体，然后与 Ba、Mg 醋酸盐按化学计量比混合，经热处理制得 BMT 纳米粉末。实验结果表明该方法制备 BMT 粉体所需的合成温度仅为 800℃，比传统固相反应法合成温度降低 400℃ 左右，平均粒径仅为 70nm。此外，所制得的纳米陶瓷材料具有较佳的低温烧结性能和微波介电特性，比目前报道的醇盐系湿化学制备技术更具有实用性。

**关键词**：Ba(Mg$_{1/3}$Ta$_{2/3}$)O$_3$ 纳米粉；Ta$_2$O$_5$·$n$H$_2$O 胶体；湿化学法；微波介电特性

## A New Wet-Chemical Approach to Synthesis Ba(Mg$_{1/3}$Ta$_{2/3}$)O$_3$ Nanometric Powder

Lian Fang  Xu Lihua  Wang Fuming  Li Wenchao

(Department of Inorganic Material, University of Science and Technology Beijing, Beijing 100083)

**Abstract**: This paper introduced an easily-controlled and lower-cost wet-chemical process for the synthesis of nanometric Ba(Mg$_{1/3}$Ta$_{2/3}$)O$_3$ (BMT) powder. Tantalum pentaoxide was dissolved in molten sodium hydroxide, forming gelatinous Ta$_2$O$_5$·$n$H$_2$O at appropriate pH value and concentration. Then it was mixed with barium acetate and magnesium acetate as a certain proportion, nanometric BMT powders were finally prepared after heat treatment. Furthermore the experimental results have indicated that the heating temperature was only 800℃ for getting better powders with the average particle size of 70nm, and it has successfully reduced near 400℃ compared with traditional solid-reaction method. Also, the nano-ceramics prepared by BMT powder have low-temperature sinteribility and good microwave dielectric properties. So it is more practical than the present wet-chemical method in alcohol salt system reported in existing literature.

**Keywords**: Ba(Mg$_{1/3}$Ta$_{2/3}$)O$_3$ nanometric powder; Ta$_2$O$_5$·$n$H$_2$O colloid; wet-chemical process; microwave dielectric property

---

基金项目：清华大学新型陶瓷与精细工艺国家重点实验室开放课题基金（KF0002）。

连芳，1998~2001 年于北京科技大学师从李文超教授和徐利华教授攻读硕士学位，2003 年该篇论文收录于第七届北京青年科技论文评选获奖论文集。目前在北京科技大学材料科学与工程学院工作，任教授、博士生导师。获得 2008 北京市科技新星荣誉称号，发表论文 70 余篇，获奖 2 项。

## 1 引言

近年来，具有复合钙钛矿结构的 Ba($Mg_{1/3}Ta_{2/3}$)$O_3$陶瓷材料，简称BMT，以其优异的微波介电性能，成为毫米波通信，卫星直播电视以及国防军事工程上使用的微波谐振器的候选材料。就BMT粉末的合成技术而言，通常的制备工艺是固相反应法，即各类氧化物按比例混合进行高温固相反应。由于传统工艺存在合成温度高、保温时间长、合成的粉末颗粒较粗、粒径分布宽、且颗粒易团聚等缺点，目前已逐渐转向湿化学方法[1-2]。据文献报道，现有的湿化学合成技术，一般采用多金属醇盐为前驱体的溶胶凝胶方法。此醇盐系湿化学法的操作工艺复杂，生产周期长（要求在保护气氛下回流48h），附加成本高[3-5]。单醇盐法[6]与八羟基喹啉（Oxine）法[7,8]，在简化制备过程上有较大的进展，但仍涉及$TaCl_5$作为起始物制备前驱体（如Ta($OC_2H_5$)$_5$）的工艺步骤。$TaCl_5$的价格昂贵，并且化学性能不稳定，易水解，合成过程操作难度大。因此，这就制约了该制备技术的普及应用。为此针对上述问题，本文拟采用原料来源较广泛的$Ta_2O_5$为起始物，运用热碱反应合成Ta的胶状体（$Ta_2O_5 \cdot nH_2O$），然后与Mg、Ba醋酸盐混合反应合成BMT粉末。同时对BMT粉末的形态特征、形核机理以及烧结性能进行研究。

## 2 实验过程

### 2.1 设计思路与工艺过程

根据金属钽氧化物的化学属性可知，$Ta_2O_5$具有化学惰性，除HF外其他酸都不能侵蚀。但可以和熔融碱金属硫酸氢盐、碱金属碳酸盐或苛性碱作用[9]，得到多钽酸根离子，通过调节pH值及浓度，将得到可变水量的水合氧化物的胶状物。利用$Ta_2O_5$的特殊性质，以$Ta_2O_5$为原料，采用过量热碱熔融反应，用冰醋酸中和剩余碱液并调节pH值，制备氧化钽（$Ta_2O_5 \cdot nH_2O$）白色胶体。为了防止团聚，应调节$Ta^{5+}$至合适的浓度。$Ta_2O_5 \cdot nH_2O$经反复冲洗、过滤后，再与Ba($CH_3COO$)$_2$、Mg($CH_3COO$)$_2 \cdot 4H_2O$的醋酸混合溶液反应，其中$Ba^{2+}:Mg^{2+}:Ta^{5+}=3:1:2$，并调节pH值。具体工艺流程如图1所示。

### 2.2 测试方法

粉末中的合成相用X射线衍射（XRD）分析。为了研究粉末在煅烧过程中的物理化学变化，确定最佳煅烧温度，进行差热（DAT）及热重（TGA）分析，升温速度为5℃/min。粉末试样的颗粒度及形态特征采用透射电镜（TEM）观察。粉末试样经干压成型后置于硅钼棒电炉进行无压烧结，所制得陶瓷采用排水法测定体积密度，并通过HP4194和HP4291B型频谱仪测试相应的介电性能。

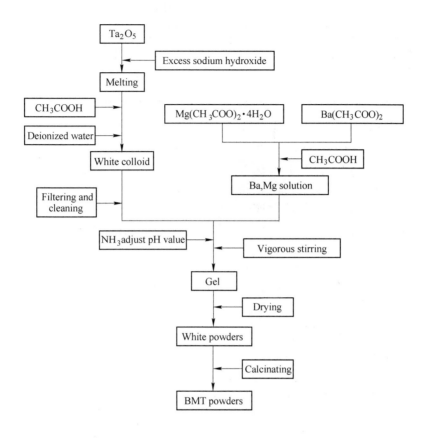

图1 制备 Ba(Mg$_{1/3}$Ta$_{2/3}$)O$_3$ 粉末的工艺流程图

Fig. 1 Scheme of the preparation of Ba(Mg$_{1/3}$Ta$_{2/3}$)O$_3$ powders

## 3 结果和讨论

### 3.1 BMT 纳米粉体合成过程与原理分析

Ta$_2$O$_5$·$n$H$_2$O 胶体的制备工艺是 BMT 纳米粉体合成的关键环节,本节着重讨论 Ta$^{5+}$浓度对胶状 Ta$_2$O$_5$·$n$H$_2$O 的胶体行为的影响及其成核生长机制。同时根据 BMT 干凝胶的综合热分析结果,以及煅烧粉末的相组成分析结果,对 BMT 粉体合成中的物理化学反应过程与 B 位有序化过程作了较详细的研究。

#### 3.1.1 Ta$_2$O$_5$ 优质胶体的制备工艺与形成机理

根据胶体化学中溶度积规则可知,热碱熔融反应物经去离子水稀释后,溶液中 Ta$^{5+}$浓度大小控制直接影响了 Ta$_2$O$_5$·$n$H$_2$O 胶体的质量优劣。表1为实验过程 Ta$^{5+}$浓度对胶体性质的影响。当 Ta$^{5+}$浓度较低时(≤0.1mol/L),胶体颗粒度还未达到胶粒尺寸,即 1nm~1μm,溶液中未出现白色胶状体;而当浓度过大时(≥0.4mol/L),Ta$_2$O$_5$·$n$H$_2$O 胶体颗粒的粒度过大将自发结成硬块,形成团聚和沉淀,这对后道工序——均质纳米颗粒合成极为不利。因此,控制适当的 Ta$^{5+}$浓度(如 0.25mol/L)是获得优质胶体的保障。

表1 $Ta^{5+}$ 浓度对反应的影响

Table 1 Influence of $Ta^{5+}$ concentration on reaction

| Concentration of $Ta^{5+}$ /mol·L$^{-1}$ | 0.1 | 0.25 | 0.4 |
|---|---|---|---|
| Phenomena | Xo gel | White gel | Precipitation |

上述实验现象可借助于溶液中晶体生长模型加以分析,如图2中的Lamer曲线所示[10]。在阶段Ⅰ,溶质的浓度$C$不断的积累,但此阶段并无晶核形成,相当于实验中$Ta^{5+}$浓度低而无胶体形成阶段;当浓度达到形核所需的最低过饱和浓度$C_{min}^*$时,进入阶段Ⅱ,即形核阶段。在这种状态下溶质的浓度$C$仍稍有增加,之后由于快速形核的大量消耗而使$C$急剧下降。当$C$降回至$C_{min}^*$,形核阶段结束,并进入阶段Ⅲ,即生长阶段。生长阶段一直延续到浓度接近溶解度$C_s$。该过程相当于实验中随着钽离子含量的变化,$Ta_2O_5·nH_2O$胶体颗粒的成核与生长阶段。

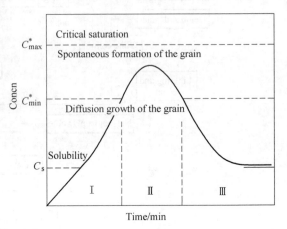

图2 Lamer 晶体生长模型

Fig. 2 Lamer model of crystal growth

当溶质的浓度过高时,如实验中$Ta^{5+}$浓度达到0.4mol/L,形核过程和晶核的长大过程同时发生,晶粒的大小难以控制,必然导致晶粒的团聚和沉淀的生成,此时氧化钽胶状体表现为自发结块团聚行为。但是从图2可知,当浓度处在稍高于$C_{min}^*$的范围内,经过短时间的形核,溶质浓度可以降至$C_{min}^*$以下,必将形核与核的长大过程分开,使已形成的晶核同步长大,且在长大过程中不再有新核生成。为了抑制团聚,并在尽量短的时间内形成凝胶,必须控制形核速度。因此在实验中,可通过同时控制钽离子浓度和稀释速率(成核时间)两个主要工艺参数,实现$Ta_2O_5$优质胶体的制备。

### 3.1.2 BMT有序化的形成过程及表征

为研究BMT相的合成过程,将实验中制备的粉末进行热分析实验。图3是差热分析(DTA)和热重分析(TGA)结果。

从图中可以判断出不同温度段物质的物理化学变化过程。在133℃附近有一个小的吸热峰,这是由于残余水分的蒸发。223℃左右的吸热峰是结晶水挥发造成的。在414℃处的放热峰尖锐,而且伴随很大的失重现象,失重率约达到20%,这是由于醋酸、醋酸盐的分解以及残余物的燃烧。562℃处的这个小的放热峰尤为重要,它显示出在此温度下立方钙钛矿结构晶体开始形成。接着DTA曲线上710～1000℃范围内出现一段较宽的放热峰,但TGA曲线变化很小,几乎没有失重现象,这表明晶体在进一步结晶、完善,B位离子排列有序化程度提高。

以上的推断可以在经过不同温度下煅烧粉末的X射线衍射图(XRD)(见图4)中得到证实。在600℃ BMT相已经成为主晶相,同时存在第二相$BaTa_2O_6$。经过800℃煅烧后

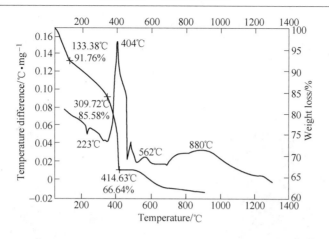

图 3 BMT 干凝胶的 DTA/TGA 分析图

Fig. 3 Results of DTA/TGA of the BMT aerogel

的试样中 $BaTa_2O_6$ 相已经不存在，说明此相随着温度的升高，转化为立方钙钛矿结构的 BMT 晶体。温度进一步升高，(100) 晶面出现。B 位原子的有序化程度往往用衍射峰强度之比来表示[4]，即：

$$S = [(I_{100}/I_{110,012})_{obs}]^{\frac{1}{2}} \times [(I_{100}/I_{110,012})_{ord}]^{-\frac{1}{2}}$$

其中 obs 表示试样实际的衍射峰强度，ord 表示 JCPDS 卡片上的标准衍射峰强度。因此，(100) 晶面衍射峰强度的增大表明钙钛矿结构中 B 位的 Mg 和 Ta 排列进一步有序化。

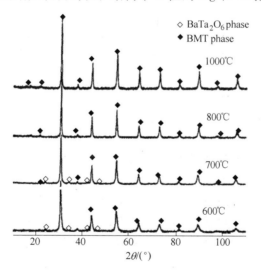

图 4 BMT 粉末不同温度下煅烧 1h 后的 XRD 分析图

Fig. 4 XRD patterns of BMT powders calcined at different temperatures for 1h

## 3.2 BMT 纳米粉体粒度表征与瓷体介电性能评估

为评价该制备工艺对合成材料性能的影响，分别将不同工艺条件下制备的陶瓷粉体进行显微结构分析。图 5 分别是 BMT 在 800℃、1000℃煅烧后的 TEM 图。从图中可以看出，800℃煅烧后得到颗粒平均直径为 70nm 的粉末，颗粒分布均匀。煅烧温度提高到1000℃时，颗粒直径达到100nm，并出现一定程度的团聚。因此由上可知，使用此种湿化学方法制备的粉体，经过800℃的预烧可以得到颗粒尺寸分布均匀的BMT纳米粉末。而传统固相反应法中单相BMT合成温度达到1200℃以上，得到的颗粒粒径为1.5um左右[11]。

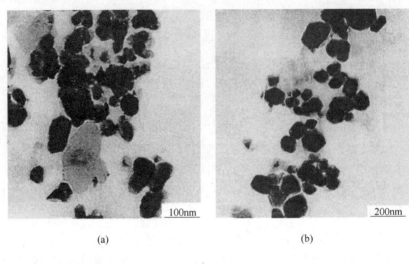

| hkl | (102, 110) | (202) | (212, 300 +) | (204, 220) | (410) |
|---|---|---|---|---|---|
| $dA_{JCPDS}$ | 2.887 | 2.045 | 1.669 | 1.444 | 1.094 |
| $dA_{Meas}$ | 2.884 | 2.022 | 1.662 | 1.421 | 1.066 |

图 5 经过 800℃ (a), 1000℃ (b) 煅烧 1h 的 BMT 粉末的 TEM 图
Fig. 5 TEM micrographs of BMT powder calcined at (a) 800℃, (b) 1000℃ for 1h

为进一步考察纳米粉末的烧结性能以及介电特性，实验中对BMT纳米粉体进行干压成型与烧结。结果表明经1400℃保温1h的无压烧结，即可得到较为致密（相对密度为93.4%）的BMT陶瓷。由此可见该工艺制备的纳米粉体具有低温烧结特性。

此外，作为介电陶瓷重要技术指标——品质因数（介电损耗倒数），其值的高低是判断该材料能否用作微波器件的重要标志。从表2的测试数据可以看出，在微波频段（1.52~1.56GHz），经1400℃烧结的致密陶瓷具有极高的品质因数$Q$，其数值达到20684，与其他湿化学方法制备的BMT陶瓷[4,8]的微波介电特性相接近。

图 6 经过 800℃ 煅烧 1h 的 BMT 粉末的选区衍射图及标定
Fig. 6 Selected area diffraction pattern and indexing of BMT powder calcined at 800℃ for 1h

**表 2  Ba(Mg$_{1/3}$Ta$_{2/3}$)O$_3$ 陶瓷的介电性质**

**Table 2  Dielectric properties of Ba(Mg$_{1/3}$Ta$_{2/3}$)O$_3$ ceramic sintered at different temperatures for 1h**

| Sintering temperature/℃ | Q | F/GHz |
| --- | --- | --- |
| 1000 | 19375 | 1.52 |
| 1200 | 10421 | 1.52 |
| 1400 | 20684 | 1.56 |

## 4 结论

以 Ta$_2$O$_5$ 替代价格昂贵的 TaCl$_5$ 作为起始物，利用湿化学法成功制备了 BMT 纳米粉。实验表明在 800℃，煅烧 1h 后，得到平均粒径为 70nm 的单相 BMT 粉末，且具有低温烧结性。随着粉末煅烧温度升高，晶粒长大，钙钛矿结构中 B 位上的 Mg 和 Ta 排列进一步有序化。通过热碱反应合成 Ta$_2$O$_5\cdot n$H$_2$O 胶体是获取高质量粉体的关键，从晶体生长模型可知必须同时控制 Ta$^{5+}$ 离子浓度与稀释速率。致密的 BMT 陶瓷材料在微波频段介电损耗值为 $10^{-5}$，与目前文献报道值相接近，具有一定的实用性。

### 参 考 文 献

[1] Wersing W. Electronic Ceramics [M]. London and New York: Electronic Material press, 1991.
[2] Trinogga L A, Kaizhou G, Hunter I C. Practical Microstrip Circuit Design [M]. New York: Ellis Horwood press, 1991.
[3] Matsumoto K, Hiuga T, Ichimura H. J. Am. Ceram. Soc., 1991, 12 (3): 98-101.
[4] Renoult O, Jean-Pierre Boilot, et al. J. Am. Ceram. Soc., 1992, 75 (12): 3337-3340.
[5] Bradley D C, Wardlaw W, Alice. Whitley. Jpn. J. Appl. Phys., 1994, 27 (5): 149-153.
[6] Mlatsumoto K, Hiuga T, Takadal K. Proc. Ferroelectr. Mater. Appl., 1986, 23: 118-121.
[7] Katayama S, Yoshinaga I, Yamada N. J. Am. Ceram. Soc., 1996, 79 (8): 2059-2064.
[8] 杨传仁, 秦广宇, 周大雨, 等. 硅酸盐学报, 1999, 27 (4): 437-444.
[9] 黄佩丽, 田荷珍. 基础化学元素 [M]. 北京: 北京师范大学出版社, 1992.
[10] Lamer V K. J. Am. Ceram. Soc., 1950, 72: 4847-4850.
[11] 连芳, 牛雅芳, 徐利华, 等. 北京科技大学学报, 2000, 22 (5): 460-463.

(原载于《无机材料学报》, 2002, 17 (2): 247-252)

# Effect of Ti Content on the Martensitic Transformation in Zirconia for Ti-$ZrO_2$ Composites

Teng Lidong  Li Wenchao  Wang Fuming

(Department of Physical Chemistry, University of Science and
Technology Beijing, Beijing 100083)

**Abstract:** Ti-$ZrO_2$ metal-ceramic composites with a series of graded compositions have been prepared by powder metallurgical process. The effect of Ti on the martensitic transformation in $ZrO_2$ from tetragonal to monoclinic structure has been studied by means of X-ray diffraction (XRD) and transmission electron microscope (TEM). The experimental results show that the phases of the sintered Ti-$ZrO_2$ system composites are α-Ti, tetragonal zirconia (t-$ZrO_2$) and monoclinic zirconia (m-$ZrO_2$). The volume fraction of m-$ZrO_2$ in $ZrO_2$ increases with the increase of Ti content when the volume fraction of Ti is more than 12.5 vol.% in Ti-$ZrO_2$ composites. The nucleation of m-$ZrO_2$ occurs preferentially at the Ti/$ZrO_2$ interface. The interfacial stress arising from the thermal expansion mismatch and the plastic deformation of Ti enhance the driving force for the t-$ZrO_2 \rightarrow$ m-$ZrO_2$ transformation. It is noted that the content of $Y_2O_3$ stabilizing agent in $ZrO_2$ should be increased as the Ti content increases in the preparation of Ti-$ZrO_2$ FGMs to appropriately control the volume fraction of m-$ZrO_2$.

**Keywords:** Ti-$ZrO_2$ system; martensitic transformation; microstructure; metal-ceramics; powder metallurgy

## 1 Introduction

In the past decade, $ZrO_2$-metal functionally graded materials (FGMs) have been studied intensively for their potential uses as thermal barrier materials[1]. Up to now, $ZrO_2$-stainless steel[2], $ZrO_2$-Ni[3,4], $ZrO_2$-Mo[5] and $ZrO_2$-Cu[6] FGMs have been investigated. In the studies mentioned above, 3mol% $Y_2O_3$-stabilized $ZrO_2$ (3Y-TZP) was usually chosen as the candidate composition of $ZrO_2$ ceramics, and the research interest is often focused on the structure, residual stress and thermal properties of the materials. In fact, the phase stability of t-$ZrO_2$ is very important for the preparation of $ZrO_2$-metal FGMs[7]. J. C. Zhu and his colleagues have ever mentioned that Ni might have influence on the transformation of t-$ZrO_2 \rightarrow$ m-$ZrO_2$ in $ZrO_2$-Ni FGMs[3], but further studies of the effect of metal content on the martensitic transformation of $ZrO_2$ in $ZrO_2$-metal FGMs have not been reported. The martensitic transformation of t-$ZrO_2 \rightarrow$ m-$ZrO_2$ in $ZrO_2$ has received considerable attention in recent years[8-12]. It is well-known that the kinetics of martensitic transformation in $ZrO_2$

---

腾立东，1998~2002 年于北京科技大学师从李文超教授攻读博士学位。目前在 ABB 工作，任首席高级工程师。发表论文 60 余篇，获奖 5 项。

are nucleation controlled, and the transformation is athermal, diffusionless, and involves both a shear strain and a volume change[10]. Classical and nonclassical nucleation models have been considered for martensitic transformations in $ZrO_2$. It is reasonable to assume that the relatively high stress near the $ZrO_2$/metal interface arising from thermal expansion mismatch might affect the nucleation of m-$ZrO_2$.

In this paper, a novel FGM based on Ti-$ZrO_2$ system has been studied. 3mol.% $Y_2O_3$-stabilized $ZrO_2$ (3Y-TZP) was also chosen as the candidate ceramic material. 3Y-TZP is one of the most important transformation-toughened engineering ceramic, which has excellent thermal barrier property, thermal shock resistance, anti-corrosion and wear resistance[13]. Pure titanium and its alloys are important aerospace materials that have high-temperature strength, thermal shock resistance, anti-corrosion and good heat-transfer properties[14]. This Ti-$ZrO_2$ thermal barrier graded material was chosen on the basis of the potential applications in aerospace industries. Based on authors' previous work[15], it is known that there is no reaction product in the Ti/$ZrO_2$ interfaces, and the bonding state between Ti and $ZrO_2$ is physical in the composite. A series of Ti-$ZrO_2$ composites with different titanium content have been prepared by powder metallurgical method to further study the effect of titanium content on the t-$ZrO_2$→m-$ZrO_2$ transformation in Ti-$ZrO_2$ FGMs.

## 2 Experimental procedures

The starting materials were 3mol.% $Y_2O_3$-stabilized tetragonal zirconia polycrystals (3Y-TZP) and pure titanium powders. The original particle size distributions of super-fine $ZrO_2$ and Ti powders are determined by particle sizer (England Malver 2200). The mean particle size of 3Y-TZP is 0.17μm, and that of Ti is 18.42μm. The X-ray diffraction pattern of 3Y-TZP powder is shown in Fig. 1. The results indicate that the phase composition of 3Y-TZP consists of tetragonal zirconia (t-$ZrO_2$) and monoclinic zirconia (m-$ZrO_2$), and that the volume fraction of m-$ZrO_2$ in 3Y-TZP is about 28.13%. The X-ray diffraction pattern of Ti powder indicates that the phase composition of Ti is hexagonal titanium (α-Ti).

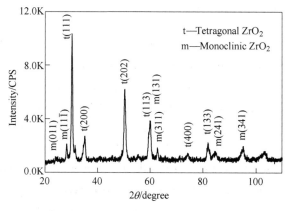

Fig. 1  X-ray diffraction spectroscopy of 3mol.% $Y_2O_3$-stabilized $ZrO_2$ (3Y-TZP) powders

The compositions of the designed Ti-ZrO$_2$ composites are listed in Table 1. The powder mixtures with different compositions were ball-milled for 5.0h with 3Y-TZP milling media, dried and granulated. The green bodies of each composition were fabricated at an optimized pressure (150.0MPa) using a strip-type steel die (47.5mm ×7.5mm ×6.5mm). When the content of the metal was above 75.0vol.%, 1.0mass.% polyvinyl alcohol (PVA) was added for forming green bodies. The compacted strip-type green bodies of each composition were embedded with special powders to control the atmosphere in a alumina crucible and pressureless sintered in the temperature range 1400 – 1650℃ for 2.0 – 3.0h in argon atmosphere and at 5.0℃ per minute on heating and cooling to relax the residual stress. According to the thermodynamic calculation, the sintering atmosphere for Ti-ZrO$_2$ composites must meet the following conditions to prevent oxidizing or nitriding of titanium:

$$7.26 \times 10^{-24} < p_{O_2} < 1.59 \times 10^{-23} \text{MPa}$$

$$\lg(p_{N_2}/p^0) < \lg(p_{O_2}/p^0) + 32.67$$

$$\lg(p_{H_2}/p^0) < 0.5\lg(p_{O_2}/p^0) + 13.96$$

Table 1  Sample number and composition of the designed Ti-ZrO$_2$ composites

| Sample No. | TZ-1 | TZ-2 | TZ-3 | TZ-4 | TZ-5 | TZ-6 | TZ-7 | TZ-8 | TZ-9 |
| --- | --- | --- | --- | --- | --- | --- | --- | --- | --- |
| Ti (vol.%) | 0.0 | 12.5 | 25.0 | 37.5 | 50.0 | 62.5 | 75.0 | 87.5 | 100.0 |
| ZrO$_2$ (vol.%) | 100.0 | 87.5 | 75.0 | 62.5 | 50.0 | 37.5 | 25.0 | 12.5 | 0.0 |

X-ray diffraction patterns were used to determine the phase composition by a X-ray diffractometer with Mo $K_\alpha$ radiation (Japan D/max-1200), and the working conditions were fixed to semi-quantitatively compare the diffractive peak intensities of various samples. The crystalline microstructures and the interface characteristics between Ti and ZrO$_2$ phases were studied by a transmission electron microscope (Japan H-800).

## 3  Results and discussion

### 3.1  Determination of phase compositions

The X-ray diffraction patterns of different compositions in Ti-ZrO$_2$ composites are shown in Fig.2. The phase compositions corresponding to various samples were α-Ti, t-ZrO$_2$ and m-ZrO$_2$, the relative content of each phase depends on the compositions of Ti-ZrO$_2$ composites. The volume fraction of m-ZrO$_2$ ($V_m$) in ZrO$_2$ was calculated by the following equation[10]:

$$V_m = \frac{I_{m(11\bar{1})} + I_{m(111)}}{I_{m(11\bar{1})} + I_{m(111)} + I_{t(111)}} \times 100\%$$

where $I$ is the relative intensity corresponding to the diffraction peaks of XRD. The subscript t and m represent the tetragonal and monoclinic zirconia, respectively. The calculated values of $V_m$ are shown in Fig.3. When the titanium content is less than 12.5vol.%, $V_m$ decreases with the increase in Ti content. When the titanium content is more than 12.5vol.%, $V_m$ increases with the increase of Ti content. The XRD results show that the volume fraction of Ti in the composites has considerable effect on the phase transformation from t-$ZrO_2$ to m-$ZrO_2$.

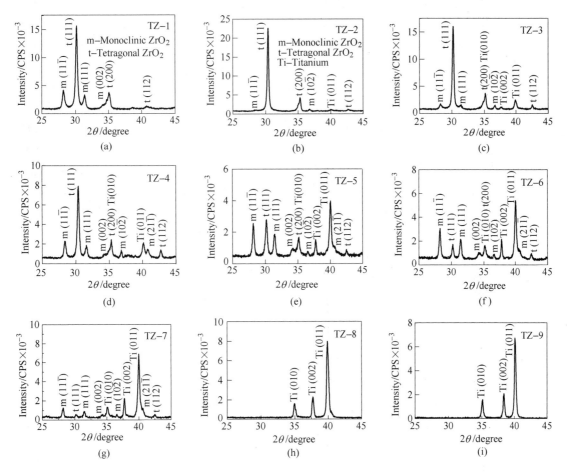

Fig. 2 (a) – (i) X-ray diffraction spectroscopies of Ti-$ZrO_2$ composites with different compositions
(a) 0.0 vol.%Ti; (b) 12.5 vol.%Ti; (c) 25.0 vol.%Ti; (d) 37.5 vol.%Ti; (e) 50.0 vol.%Ti;
(f) 62.5 vol.%Ti; (g) 75.0 vol.%Ti; (h) 87.5 vol.%Ti; (i) 100 vol.%Ti

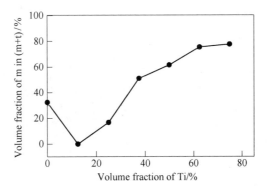

Fig. 3 Effect of Ti content on the volume fraction of m-$ZrO_2$ in Ti-$ZrO_2$ composites

In Ti-ZrO$_2$ metal-ceramic composites, the average thermal expansion coefficient of Ti (8.2 × 10$^{-6}$℃$^{-1}$) is less than that of t-ZrO$_2$ (9.6 × 10$^{-6}$℃$^{-1}$) and t-ZrO$_2$ is thus in a state of residual quasi-hydrostatic tension due to thermal expansion mismatch in the cooling process from the sintering temperature (1873K) to room temperature. The presence of the residual stress can decrease the activation energy for nucleation of m-ZrO$_2$ and then promote the transformation of t-ZrO$_2$→m-ZrO$_2$. The interfacial area of Ti/ZrO$_2$ and the residual stress will change with the increase of Ti content in Ti-ZrO$_2$ composites, and accordingly this change will affect the process of martensitic transformation in ZrO$_2$. When titanium content is less than 12.5vol.%, the addition of Ti in ZrO$_2$ can increase the density of the composites and prevent the growth of t-ZrO$_2$ grains[16], and the effect of residual stress on the martensitic transformation is relatively small. Therefore, the martensitic transformation is restrained and the volume fraction of m-ZrO$_2$ decreases with the increase of Ti content, as shown in Fig. 4. When titanium content is more than 12.5vol.%, the interfacial area of Ti/ZrO$_2$ increases as the increase of Ti content, and at this time the residual stress effect predominates the transformation process. Therefore, the martensitic transformation is accelerated and the volume fraction of m-ZrO$_2$ increases with the increase of Ti content. The effect of Ti on the martensitic transformation of t-ZrO$_2$→m-ZrO$_2$ has been demonstrated by the results of TEM analysis.

### 3.2 Characterization of Ti/ZrO$_2$ interface

TEM examination shows that the crystals of m-ZrO$_2$ are often observed in the Ti/ZrO$_2$ interface areas as shown in Fig. 4 (a). Fig. 4 (b) is the [1$\bar{3}$2] diffraction pattern from m-ZrO$_2$ area. Fig. 4 (c) is the [243] diffraction pattern from α-Ti area. The m-ZrO$_2$ has a twinned structure with lath crystals in ZT-5 composites sintered at 1650℃ for 2h. The formation of this martensitic twinned structure plays important role to reduce the strain energy accompanying the transformation of t-

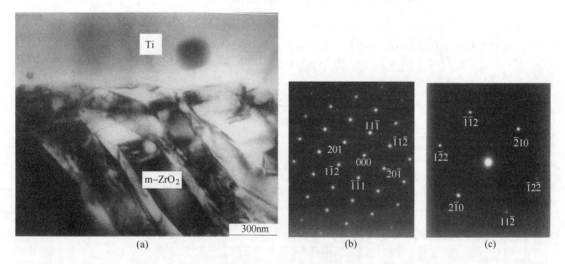

Fig. 4 (a) the morphology of the interfaces between m-ZrO$_2$ and α-Ti phases in ZT-5 sample sintered at 1650 °C for 2 h; (b) the [1$\bar{3}$2] diffraction pattern from m-ZrO$_2$ area;
(c) the [2 4 3] diffraction pattern from α-Ti

$ZrO_2 \rightarrow$ m-$ZrO_2$[17]. The transformations of t-$ZrO_2 \rightarrow$ m-$ZrO_2$ are nucleation controlled[18-20]. The free energy of transformation depends on the chemical free energy, strain free energy, and surface free energy. In the free energy function describing the formation of a martensite nucleus, the strain free energy is one of the main factors for the thermo-elastic martensitic transformation. Therefore, the transformation is controlled by the grain size, composition and the stress state of t-$ZrO_2$.

Fig. 5 is the morphology of m-$ZrO_2$ precipitated from the Ti/$ZrO_2$ interface in ZT-5 composites and the [101] diffraction pattern of twinned m-$ZrO_2$. TEM observation shows that the m-$ZrO_2$ near the Ti/$ZrO_2$ interface has grown more perfectly than the internal m-$ZrO_2$, and the m-$ZrO_2$ crystals grow perpendicularly to the Ti/$ZrO_2$ interface. It is reasonable to assume that the nucleation of m-$ZrO_2$ occurs preferentially at the Ti/$ZrO_2$ boundaries, due to the stress concentration existing at such regions. The plastic deformation of Ti near the Ti/$ZrO_2$ interface can also relax the strain arising from the volume change (+5%) accompanied by the t-$ZrO_2 \rightarrow$ m-$ZrO_2$ transformation. It was proposed that the presence of Ti/$ZrO_2$ interface lowers the activation energy for nucleation of m-$ZrO_2$ and enhances the t-$ZrO_2 \rightarrow$ m-$ZrO_2$ transformation. It was deduced that the increase of Ti content in Ti-$ZrO_2$ composites would increase the areas of t-$ZrO_2$/Ti interface, and then promote the transformation of t-$ZrO_2 \rightarrow$ m-$ZrO_2$. The analysis could readily explain the XRD results shown in Fig. 3.

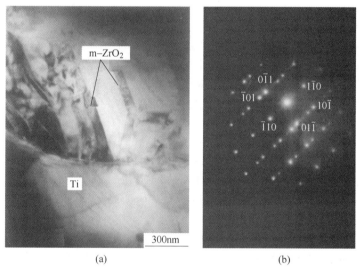

Fig. 5 (a) The morphology of m-$ZrO_2$ precipitated from the Ti / $ZrO_2$ interface in ZT-5 composites; (b) The diffraction pattern of twinned m-$ZrO_2$, B = [101]

The influence of Ti content on the volume fraction of m-$ZrO_2$ in Ti-$ZrO_2$ composites should be seriously considered in the preparation of Ti-$ZrO_2$ FGMs. Because of the graded change in the composition of Ti-$ZrO_2$ FGMs, the stabilizing agent of $Y_2O_3$ in $ZrO_2$ should increase with the increase of Ti content to control the volume fraction of m-$ZrO_2$. In this sense, the yttria content in Ti-$ZrO_2$ FGMs must play a crucial role in strengthening the mechanical properties of these materials.

## 4 Conclusions

Ti-$ZrO_2$ metal-ceramic composites with a series of graded compositions were fabricated by powder metallurgical process. The phases of the sintered Ti-$ZrO_2$ composites are α-Ti, tetragonal zirconia and monoclinic zirconia. The volume fraction of m-$ZrO_2$ in $ZrO_2$ increases with the increase of Ti content when the volume fraction of Ti is more than 12.5vol.% in Ti-$ZrO_2$ composites. The nucleation of m-$ZrO_2$ occurs preferentially at the Ti/$ZrO_2$ interface. The interfacial stress arising from the thermal expansion mismatch between Ti and $ZrO_2$ and the plastic deformation of Ti enhance the driving force for the t-$ZrO_2$→m-$ZrO_2$ transformation. It is noted that the stabilizing agent of $Y_2O_3$ in $ZrO_2$ should increase as the Ti content increases in the preparation of Ti-$ZrO_2$ FGMs to control the volume fraction of m-$ZrO_2$.

**Acknowledgements:** The authors wish to thank the National Nature Science Foundation of China for the financial support on No. 59872002 to this work.

## References

[1] A. Mortensen, S. Suresh. International Materials Reviews, 1995, 40 (6): 239-265.
[2] Y. G. Jung, S. C. Chol, U. G. Paik. J. Mater. Sci., 1997, 32: 3841-3850.
[3] J. C. Zhu, Z. D. Yin, Z. H. Lai. J. Mater. Sci., 1996, 31: 5829-5834.
[4] Q. L. Fan, X. F. Hu, J. K. Guo. Science in China (Series A), 1995, 25 (7): 777-784.
[5] X. F. Tang, L. M. Zhang, R. Z. Yuan. J. Chinese Ceram. Soc., 1994, 22 (1): 44-49.
[6] S. NaKashima, H. Arikawa, M. Chigasaki, et al. Surface and Coating Technology, 1994, 66: 330-333.
[7] Y. R. He, V. Subramanian, J. J. Lannutti. J. Mater. Res., 1997, 12 (10): 2589-2593.
[8] I. W. Chen, Y. H. Chiao. Acta Metall., 1983, 31 (10): 1627-1638.
[9] A. H. Heuer, M. Rühle. Acta Metall., 1985, 33 (12): 2101-2112.
[10] H. Toraya, M. Yashimura, S. Somiya. J. Amer. Ceram. Soc., 1984, 71: 198-202.
[11] M. Matsui, T. Soma, I. Oda. J. Amer. Ceram. Soc., 1986, 69 (3): 198-202.
[12] M. Yashima, M. Kakihana, M. Yoshimura. Solid State Ionics, 1996, 86-88: 1131-1149.
[13] R. A. Miller. Surface and Coatings Technology, 1987, 30: 1-11.
[14] R. Boyer, G. Welsch, E. W. Collings. Materials Properties Handbook: Titanium Alloys [M]. The Materials Information Society, OH44073-0002, 1994.
[15] Lidong Teng, Fuming Wang, Wenchao Li. Materials Science and Engineering A, 2000, 293 (1-2): 130-136.
[16] Lidong Teng, Fuming Wang, Wenchao Li. The Chinese Journal of Nonferrous Metals (English edition), 2000, 10 (4): 506~510.
[17] W. M. Kriven. Advances in Ceramics, Science and Technology of Zirconia. American Ceramics Society, Columbus, Ohio, 1981, 3: 168.
[18] A. H. Heuer. Advances in Ceramics. Edited by A. H. Heuer and L. W. Hobbs. Westerville: The American Ceramic Society, 1981, 3: 98.
[19] M. Ruhle, A. H. Heuer. Advances in Ceramics. Edited by N. Claussen, M. Ruhle, A. H. Heuer. Westerville: The American Ceramic Society, 1984, 12: 14.
[20] I. W. Chen, Y. H. Chiao. Advances in Ceramics. Edited by N. Claussen, M. Ruhle, A. H. Heuer. Westerville: The American Ceramic Society, 1984, 12: 33.

(原载于 *Journal of Alloys and Compounds*, 2001, 319: 228-232)

# Preparations and Characterizations of New Mesoporous ZrO₂ and Y₂O₃-stabilized ZrO₂ Spherical Powders

Zhang Hui[1]   Lu Hu[1]   Zhu Yawei[1]   Li Fan[2]
Duan Renguan[3]   Zhang Mei[4]   Wang Xidong[5]

(1. School of Science, Beijing Jiaotong University, Beijing 100044, China;
2. College of Environmental and Energy Engineering, Beijing University of Technology, Beijing 100124, China; 3. Core Engineering, Etch, Applied Materials, Inc., California 94085, USA;
4. Department of Physical Chemistry, School of Metallurgical and Ecological Engineering,
University of Science and Technology Beijing, Beijing 100083, China;
5. Department of Energy and Resources Engineering, College of Engineering,
Peking University, Beijing 100871, China)

**Abstract:** In the absence of template and thermal stabilizer, novel mesoporous $ZrO_2$ and $Y_2O_3$-stabilized $ZrO_2$ (YSZ) spherical powders were synthesized via the improved spray reaction (SR) technique. The synthesized materials were characterized by thermo-gravimetry/differential thermal analysis (TG/DTA), scanning electron microscopy (SEM), transmission electron microscopy (TEM), X-ray diffraction (XRD), and Nitrogen adsorption/desorption isotherm techniques. Results reveal that the microstructure of $ZrO_2$ precursor significantly depends on the aging process, and the particle size, surface area and phase composition of prepared $ZrO_2$-based materials significantly depend on the calcination temperature. For pure $ZrO_2$ spherical particles, the size varies from 1.63μm, 1.58μm to 1.60μm in correspondent to the calcined temperature 700℃, 800℃ and 1000℃, respectively. And also the major cubic $ZrO_2$ (c-$ZrO_2$) is formed when calcined at 700℃ whereas the predominant monoclinic $ZrO_2$ (m-$ZrO_2$) is synthesized when calcined at 800-1000℃. For YSZ spherical particles, the size changes from 1.65μm, 1.59μm, 1.62μm to 1.45μm in correspondent to the calcined temperature 500℃, 600℃, 700 ℃ and 1000℃, respectively. The synthesized YSZ spheres exhibit the similar phase constitution of predominant tetragonal $ZrO_2$ (t-$ZrO_2$) at all experimental temperatures but show the different surface area and pore diameter, ~14.72m²/g surface area and 10.7nm pore diameter at 500℃ as well as ~34.39m²/g surface area and 8.3nm pore diameter at 600℃.

**Keywords:** mesopores; $ZrO_2$-based spherical powders; aging process; pore structure; phase composition; particle size

# 1 Introduction

As early as 1992, mesoporous MCM-41 silica molecular sieve had been successfully synthesized

---

张辉,1999~2003 年于北京科技大学师从李文超教授攻读材料学学位,获得北京科技大学优秀博士学位论文荣誉称号。目前在北京交通大学工作,副教授。发表论文30余篇,获奖3项。

by scientists from Mobil Research and Development Corporation of USA[1]. This kind of mesoporous inorganic solid has found great utilities in catalysis, healthcare, electronics, energy and etc. because of their large internal surface area, perfect pore structure, excellent biocompatibility and other special features [2-4]. From that time on, the preparations and applications of the silica-based molecular sieves, transition-metal-doped silica-based mesoporous materials, transition metal oxide porous materials and so on, have been receiving increased attentions [5-9]. Of above mesoporous materials, zirconia has been applied extensively as catalysts or catalyst supports. Zirconia is the unique material that has both acidic and alkaline sites, coupled with its remarkable ion-exchange, charge-transfer, chemical stability and mechanical properties [10,11], which make it an ideal multi-functional catalytic materials. In recent years, a lot of research works have been published on preparation of mesoporous $ZrO_2$ by using self-assembled surfactant or structure-directing agent as template [8,12-15]. But $ZrO_2$ powders synthesized by template method aren't spherical. Moreover their pore structures will mostly collapse after surfactant is removed, hence the phosphate or sulfate as thermal stabilizer has to be introduced to improve thermal stability [16-25]. Also the removal of organic template will bring about environmental pollution.

In this paper, without using any template and thermal stabilizer, pure $ZrO_2$ and YSZ spherical powders with new mesoporous structure have been successfully synthesized by modified SR technique[9]. The synthesized materials properties have also been measured and discussed in detail. As we know, in the field of ceramics or catalysis, spherical $ZrO_2$-based powders with porous structure are important because they possess nice mobility, low thermal conductivity and high surface area, they are also facile to be densified and so on. Therefore these porous $ZrO_2$-based spherical particles prepared herein will find potential applications in oxygen sensor, solid oxide fuel cell as electrode materials, auto exhaust-conversion as an efficient catalyst support for the oxidation of CO, thermal barrier coating and etc..

## 2 Experimental

### 2.1 Preparations of mesoporous $ZrO_2$ and YSZ spherical powders

Two solutions were made by solving $ZrOCl_2 \cdot 8H_2O$ (analytical reagent purity) and $ZrOCl_2 \cdot 8H_2O/YCl_3 \cdot 6H_2O$ (analytical reagent purity) (molar ratio of Zr/Y: 97/6) in DI water, respectively. Ammonia gas (technical grade) stored in cylinder was used as-received.

SR technique proposed by us is, in fact, based on the reaction between droplets of zirconium or zirconium/yttrium salt solution, generated by atomizer, and $NH_3$ molecules. Because of the imperfections existed in original process, we made some improvements on it. Therefore a modified SR process was used to prepare $ZrO_2$ and YSZ spherical particles, and a schematic is shown in Fig. 1. In comparison with the original technique[9], the improved technique exhibits the following advantages: (1) new reactor with three $NH_3$ gas inlets in uniform alternating arrangement and droplets inlet is located near the bottom of the reactor; (2) installing a vacuum meter in the outlet of reactor and an air-flow meter prior to atomizer; (3) aging the precursor spheres; (4) minor adjustments to operating procedure. The adjusted operating procedure is that firstly, $NH_3$ gas from the cylinder with a pressure regulator flowed into the reactor. After the reactor was full of $NH_3$ gas, the valve was closed, and then the atomizer filled with aqueous solution of $ZrOCl_2 \cdot 8H_2O$ or $ZrOCl_2 \cdot 8H_2O/YCl_3 \cdot 6H_2O$ was connected to reactor. After that, the vacuum device, atomizer and ammonia cylinder were opened sequentially. Thus, the aerosols of metal salt solution were carried into the reactor where liquid-gas reaction took place at room temperature between droplets of zirconium or zirconium/yttrium salt solution and $NH_3$, forming the spherical $ZrO_2$ or YSZ precursor particles. Due to a reduced pressure in reactor, the synthesized precursor spheres were carried into particle collector, which was filled with distilled water. The specific process parameters for synthesizing $ZrO_2$ and YSZ precursors are shown in Table 1.

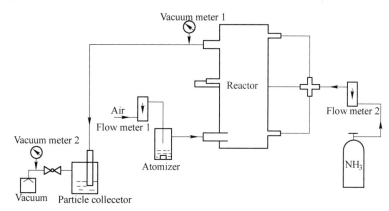

Fig. 1 A schematic diagram of the production system for $ZrO_2$ and YSZ precursor spheres

**Table 1 Synthesizing parameters for precursor spherical particles**

| Samples | $c_{ZrOCl_2 \cdot 8H_2O}$ /mol·L$^{-1}$ | $c_{YCl_3 \cdot 6H_2O}$ /mol·L$^{-1}$ | $V_{s(NH_3)}$ /m$^3$·s$^{-1}$ | $P_{vacuum\ 1}$ /kPa | $V_{s(air)}/V_{s(NH_3)}$ | Aging time /h |
|---|---|---|---|---|---|---|
| $ZrO_2$ precursor | 1.01 | 0.00 | $1.67 \times 10^{-5}$ | 81.3 | 1 | 72 |
| YSZ precursor | 1.01 | 0.06 | $1.67 \times 10^{-5}$ | 81.3 | 1 | 72 |

The collected precursor spheresas-synthesized were immediately aged for 72 hrs without other alkaline substance was added as $NH_3$ molecules were saturated in the aqueous solution in particle collector throughout collecting process. Then the aged precursor spheres were washed several times with distilled water until no chloride ions could be detected by dropping a small amount of silver nitrate solution into the separated-out supernatant solution and watching no more precipitation formed. Afterwards, the cleaned $ZrO_2$ or YSZ precursor spheres were washed twice with anhydrous ethanol to remove absorbed water molecules. Finally, the precursor was dried with microwave at "low fire" for ~2min, and calcined in air at 350℃ for 15min and at 500 – 1000℃ for 3hrs.

## 2.2 Characterizations

Particle size and morphology were evaluated usingHITACHI H-7650 TEM and HITACHI S-4800

SEM. Average particle sizes were estimated by counting several hundred spheres from representative SEM images of samples. The morphology and microstructure of spheres were observed on a JEOL JEM2010 TEM. Specimens for TEM were prepared by dropping 1-2 drops of suspension of spherical particles onto the grid for morphology observation or embedding spheres in copper film via electrical chemical reaction, fixing the copper film wrapping spheres on the grid and ion-milling for microstructure observation. Thermal analysis of precursor powder was conducted on HENVEN HCT-2 TG/DTA at a heating rate of 10.0℃/min. The phases in the powders were identified by Rigaku Dmax/2400 XRD using CuKα1 radiation. Pore structure in spherical particles was analyzed with Micrometritics ASAP 2010 $N_2$ adsorption apparatus.

## 3  Results and discussion

### 3.1  Effect of aging on the structure of precursor spheres

In our work, synthesized $ZrO_2$ precursor spheres and $ZrO_2$ spheres possess three kinds of interior structure: the homogenous structure, the structure with a loose core and dense shell, and the hollow structure. Reason for the formation of the above second and third structures has been discussed in detail in our previous publication[9]. Briefly, it is, really as liquid-gas reaction, because some of large droplets didn't react fully with $NH_3$ molecules in a shorter resident time (~0.365 sec) in reactor. It could be proposed that precursor sphere formation should contain one process and four reactions as listed below and formation speed should depend on the rates of both chemical reactions (2)-(5) and migration of reactants. At the beginning of precursor sphere formation, formation speed chiefly depended on that of process (1), probability of $NH_3$ molecules colliding successfully with droplets, which was smallest comparing with reactions (2)-(5). Once the shell of precursor spheres was formed, the rate of particle formation primarily depended on that of diffusion of the aqueous ammonia molecules/$OH^-$ ions into interior of spheres through dense shell, which was lowest among the processes of reactant migrations and chemical reactions. The diffusion rate of aqueous ammonia molecules/$OH^-$ ions was so slow after the shell of precursor spheres was formed that some of large droplets didn't build up dense structure in a shorter resident time (~0.365 sec) in reactor. Therefore an aging process for the collected precursor particle suspension was utilized to make those hollow and non-dense $ZrO_2$ precursor spheres completely precipitate by a slow migration of $OH^-$ ions in aqueous solution in particle collector into spheres through pores in the shell of spheres in a longer period of time.

$$NH_3 \xrightarrow{colliding} droplets \qquad (1)$$

$$NH_3 + H_2O \longrightarrow NH_3 \cdot H_2O \qquad (2)$$

$$NH_3 \cdot H_2O \longrightarrow NH_4^+ + OH^- \qquad (3)$$

$$Zr^{4+} + OH^- \longrightarrow Zr(OH)_4 \downarrow \qquad (4)$$

$$Y^{3+} + OH^- \longrightarrow Y(OH)_3 \downarrow \qquad (5)$$

TEM images (Fig. 2) of $ZrO_2$ precursor spheres before and after aged for 72 hrs clearly show that pre-aged non-dense precursor spheres mostly transform into post-aged dense ones. Therefore, this aging process was employed in synthesizing both pure $ZrO_2$ and YSZ precursor spheres in this work.

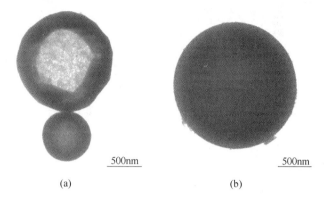

Fig. 2 TEM images of $ZrO_2$ precursor spheres (a) before and (b) after aged for 72 hrs

## 3.2 Properties of $ZrO_2$ spherical powder

On the basis of our previous work[9], it is known that the crystallization of $ZrO_2$ particles made by SR method takes place above 500℃. By this knowledge, precursor was calcined at 700℃, 800℃ and 1000℃ respectively to obtain crystallized $ZrO_2$ powder in this paper.

SEM images (Fig. 3 (a)-(c)) show that the $ZrO_2$ particles calcined at 700-1000℃ are quite spherical shape. The excellent spherical shape of $ZrO_2$ particles results from the great precursor spherical shape that is due to surface tension of droplets during reaction.

Average diameters of $ZrO_2$ microspheres calcined at different temperatures are 1.63μm (standard deviation: 0.55, 700℃), 1.59μm (standard deviation: 0.47, 800℃) and 1.60μm (standard deviation: 0.53, 1000℃), respectively, as shown in Fig. 3 (a')-(c'). It can be noticed that with calcination temperature rose, there are small changes in the sphere diameter but particle aggregation enhances. Small changes of $ZrO_2$ sphere diameters indicate that shrinkage of spheres has possibly been finished after 700℃. It can be noticed as well from Fig. 3(a) -(c) that the surface of some of spheres flakes and a few spheres crack due to existing shrinkage difference between surface and interior of spheres.

In order to study the interior microstructure of $ZrO_2$ spheres produced by using an aging process, we conducted TEM observation of cross-section of the $ZrO_2$ spheres calcined at 800-1000℃ as well. The representative TEM micrograph (Fig. 4) evidently shows that $ZrO_2$ microspheres attained using aging treatment on precursor are still composed of peanut-like nano-grains or nanocrystallites and irregular pores among grains, which is much similar to TEM characterization result of cross-section of the $ZrO_2$ spheres with homogenous structure produced without using aging treatment [9]. And the pore diameter < ~15nm in $ZrO_2$ spheres obtained can be estimated from Fig. 4. The formation of peanut-like nano-grains in $ZrO_2$ spheres originated from the peanut-shaped aggregates of precursor,

Zr(OH)$_4$ molecules [9], whose generation remains further research in our future work.

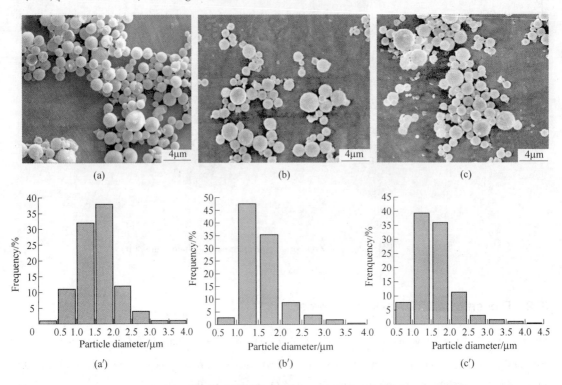

Fig. 3 (a) – (c) SEM images and (a′) – (c′) size distributions of ZrO$_2$ spherical particles attained at (a) 700℃, (b) 800℃ and (c) 1000℃

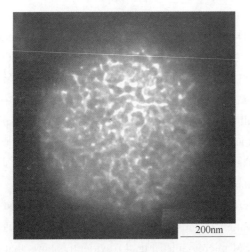

Fig. 4 Typical TEM micrograph of cross-sections of the ZrO$_2$ microspheres prepared using aging precursor process

XRD patterns (Fig. 5) of the spherical ZrO$_2$ powders reveal the phase composition after calcination at 700-1000℃ is major c-ZrO$_2$/minor t-ZrO$_2$ at 700℃ and major m-ZrO$_2$/minor t-ZrO$_2$ at 800-1000℃. For samples heated at 800-1000℃, the relative volume contents of m-ZrO$_2$ may be calculated using the most intensive diffraction peaks, ($\bar{1}11$) and (111) of monoclinic phase and (111) of tetragonal phase, by formula $V_m = \dfrac{I_m(\bar{1}11) + I_m(111)}{I_t(111) + I_m(\bar{1}11) + I_m(111)}$ [26]. Furthermore in 800-1000℃ calcination temperature range, m-$ZrO_2$ content increases with the increase of calcination temperature (Table 2). The grain size was calculated on the basis of the full width at half maximum of three most intensive diffraction peaks in X-ray diffraction patterns (Scherrer's formula) (Table 2). It is noted that with increasing the calcination temperature from 800 to 1000℃, grain size grows up. Grain growth of the same phase is driven by neighbouring grains that possess different energy levels due to the curvature of energetic grain boundaries and/or different amounts of accumulated strain energy. Surface tension across a curved boundary between two grains gives rise to a pressure difference, with the grain that possesses the concave boundary having a higher energy state. Atoms that receive sufficient kinetic energy at increased heating temperature tend to move from the high energy side of the boundary to the low energy side. This leads to grain boundary migration as the low energy crystal consumes the high energy crystal [27]. The crystallite sizes in both directions of m ($\bar{1}11$) and m (111), or c (111) and c (311) planes are larger than that calculated for the direction of m (020) or c (220) plane, which can lead to an aspect ratio (length to width), suggesting a strip-shaped $ZrO_2$ grain such as peanut-like shape as observed in Fig. 4.

**Table 2  Properties of $ZrO_2$ spherical particles**

| Calcination temperature /℃ | Particle size /μm | Phase composition /vol% | Crystallite size/nm | | | | BET area /$m^2 \cdot g^{-1}$ |
|---|---|---|---|---|---|---|---|
| | | | m ($\bar{1}11$) | m (111) | m (020) | Average | |
| Precursor | 4.43 | Amorphous | — | — | — | — | 233.66 |
| 700① | 1.63 | Cubic: major  Tetragonal: minor | 18.6② | 11.5③ | 18.6④ | 16.2 | 30.85 |
| 800 | 1.58 | Monoclinic: 87.38  Tetragonal: 12.62 | 22.9 | 23.1 | 17.4 | 21.1 | — |
| 1000 | 1.60 | Monoclinic: 97.65  Tetragonal: 2.35 | 29.5 | 26.0 | 17.4 | 24.3 | — |

①Sample prepared without aging precursor and calcined at 700℃ for 2.5hrs.
②Correspond to c (111).
③Correspond to c (220).
④Correspond to c (311).

BET analyses (Fig. 6) illustrate a high surface area for the mesoporous precursor spheres as-dried and a normal one for the mesoporous $ZrO_2$ spheres produced at 700℃ and without using aging process, which mainly results from that there was such a large shrinkage on the surface of $ZrO_2$ spheres after calcination that lots of pores on the surface closed.

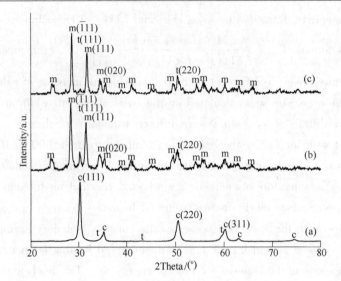

Fig. 5  XRD patterns of $ZrO_2$ spherical powders calcined at (a) 700℃, (b) 800℃ and (c) 1000℃

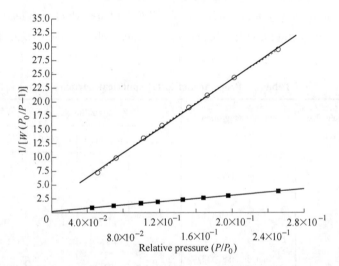

Fig. 6  BET surface area plots of spherical particles of $ZrO_2$ precursor (■) and $ZrO_2$ (○) calcined at 700℃

### 3.3  Properties of YSZ spherical powder

Thermal analysis (Fig. 7) of YSZ precursor illustrates its decomposition and phase transformation during heating. The DTA curve reveals an endothermic peak at ~86℃, which corresponds to the volatilization and vaporization of the absorbed ethanol and water in the powder. The TG curve shows a significant weight loss at around the corresponding peak temperature of ~86℃. Additionally, the DTA curve exhibits two exothermic peaks at ~296℃ and ~370℃, which correspond to the crystallizations of $ZrO_2$ and $Y_2O_3$, respectively. Meanwhile the TG curve shows a slow weight loss in the temperature range of ~250-400℃. In the DTA curve there is an exothermic peak at ~450℃, which is caused by complete crystallizations of both $ZrO_2$ and $Y_2O_3$, and another exothermic peak at ~660℃ results from that t-$ZrO_2$ partially transformed into m-$ZrO_2$. Correspondingly, the TG curve reveals constant weight after ~400℃, which suggests nothing was lost after ~400℃ but only phase transformations.

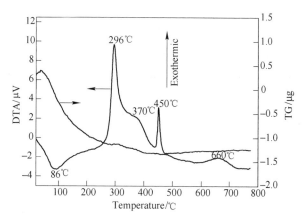

Fig. 7 TG/DTA curves of YSZ precursor powder

SEM micrographs (Fig. 8 (a)-(d)) of the YSZ particles calcined at 500-1000℃ show that the YSZ particles after calcinations still restain pretty spherical. Average diameters of the YSZ microspheres prepared at different temperatures are 1.65μm (standard deviation: 0.46, 500℃), 1.59μm (standard deviation: 0.41, 600℃), 1.62μm (standard deviation: 0.52, 700℃) and 1.45μm (standard deviation: 0.48, 1000℃), respectively, as shown in Fig. 8 (a)-(d).

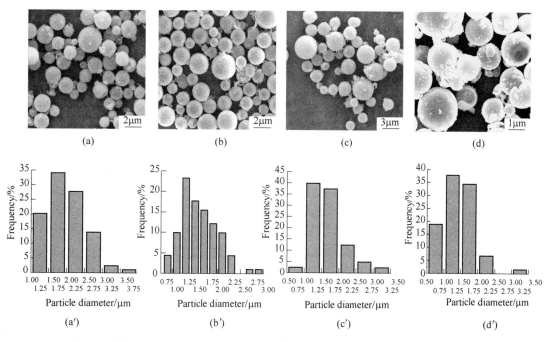

Fig. 8 (a)-(d) SEM images and (a')-(d') size distributions of YSZ spherical particles prepared at (a) 500℃, (b) 600℃, (c) 700℃ and (d) 1000℃

And with calcination temperature elevated, sphere diameter reduces, particle aggregation enhances, and sphere surface flakes increasingly because shrinkage difference increases between surface and interior of spheres.

In XRD patterns (Fig. 9) of the YSZ powders heated at 500-1000℃, almost all the main diffraction peaks can be indexed as t-$ZrO_2$ except for few small m-$ZrO_2$ peaks. The crystallite sizes were calculated on three diffraction maxima t (111), t (220) and t (131) from the half-width of diffraction peaks using Scherrer's formula, as presented in Table 3. It can also be seen that with increasing the calcination temperature the crystallite size increases. In addition, the crystallite sizes in the direction of t (111) (t (220) for the sample of 1000℃) plane are larger than those calculated for t (220) (t (111) for the sample of 1000℃) and t (131) planes, which will also result in an aspect ratio (length to width), suggesting a strip-shaped YSZ grain such as peanut-like shape as well.

**Table 3  Properties of YSZ spherical particles**

| Calcination temperature /℃ | Particle size /μm | Phase composition /vol% | Crystallite size/nm | | | | BET area /$m^2 \cdot g^{-1}$ | Pore volume /$cm^3 \cdot g^{-1}$ | Pore size /nm |
|---|---|---|---|---|---|---|---|---|---|
| | | | t (111) | t (220) | t (131) | Average | | | |
| 500 | 1.65 | Monoclinic: 7.50  Tetragonal: 92.5 | 25.9 | 13.8 | 12.8 | 17.5 | 14.72 | 0.050 | 10.7 |
| 600 | 1.59 | Monoclinic: 8.98  Tetragonal: 91.02 | 18.8 | 11.0 | 14.4 | 14.8 | 34.39 | 0.087 | 8.3 |
| 700 | 1.62 | Monoclinic: 8.00  Tetragonal: 92.00 | 37.6 | 20.1 | 12.8 | 23.5 | — | — | — |
| 1000 | 1.45 | Monoclinic: 2.66  Tetragonal: 97.34 | 31.8 | 49.0 | 28.9 | 36.6 | — | — | — |

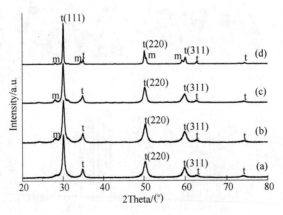

Fig. 9  XRD patterns of YSZ powders obtained at (a) 500℃, (b) 600℃, (c) 700℃ and (d) 1000℃

In order to describe the pore structure in YSZ particles prepared, for example, pore shape, size and its distribution, and volume, small angle XRD, $N_2$ adsorption/desorption isotherm, pore size distribution and BET surface area analyses for YSZ calcined at 500℃ and 600℃ were performed as shown in Fig. 10-Fig. 12.

Small angle XRD patterns (Fig. 10) show that the pore structure in samples calcined at both 500℃ and 600℃ is disordered because of no diffraction peaks appearing. The $N_2$ adsorption/desorption isotherm profiles for samples obtained at both 500℃ and 600℃ display a typical isotherm of mesoporous materials, a type Ⅲ isotherm with a large type H3 hysteresis loop according to the IUPAC classification.

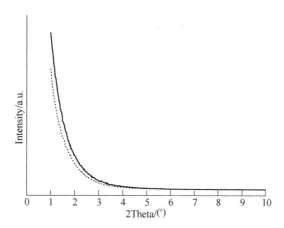

Fig. 10　Small angle XRD patterns of YSZ spherical particles calcined at 500℃ (solid line) and 600℃ (dash line)

The pore size distribution analyses (Fig. 11) reveal that YSZ particles heated at 500℃ and 600℃ possess a multimodal pore size distribution. For the samples calcined at 500℃ and 600℃ major and average pore diameters are ~3.8nm and ~10.7nm, ~3.3nm and ~8.3nm, respectively. And major pore sizes in samples calcined at 500℃ and 600℃ are close to those in samples prepared at same temperatures by template process. Increasing calcination temperature shifts the

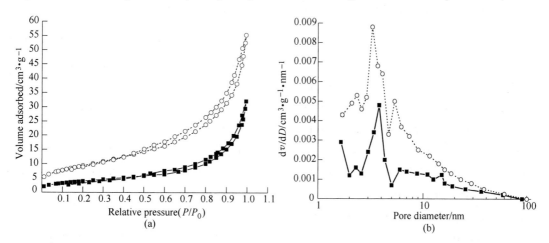

Fig. 11　(a) $N_2$ adsorption/desorption isotherms and (b) pore size distributions of YSZ spherical particles calcined at 500℃ (■) and 600℃ (○)

pore size distribution to smaller size, which results from enhancing of shrinkage of microspheres when calcination temperature increased ranging from 500 to 600℃.

The BET analyses (Fig. 12) exhibit a general specific surface area for the samples prepared at 500℃ and 600℃, which is not consistent with the internal microstructure of spheres observed in Fig. 4. And the specific surface area increases with calcination temperature increasing from 500 to 600℃. Reasons for above results are possibly that shrinkage of sphere shell was larger than that of sphere core during calcinations, which caused that pores in sphere shell closed partially then not so higher specific surface area was tested. Furthermore spheres heated at 600℃ cracked and peeled off more badly than those heated at 500℃, which resulted in more internal pores were exposed when sample heated at 600℃ than heated at 500℃, thus raising the specific surface area of sample obtained at 600℃.

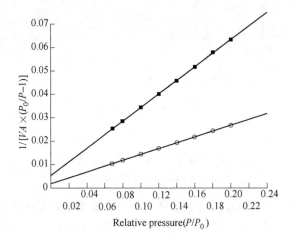

Fig. 12 BET surface area plots of YSZ spherical particles calcined at 500℃ (■) and 600℃ (○)

### 3.4 Formation and thermal stabilization of mesopores in $ZrO_2$-based spheres

For SR method for preparing mesoporous $ZrO_2$-based spheres, formation of mesopores is because that the droplets of zirconium or zirconium/yttrium salt solution could serve as many micro-reactors and the precipitation reaction took place inside these micro-reactors, forming a large number of nano-scaled precursor aggregates in droplets just as what happened in common "large" reactor, among which there were lots of micro- or meso- pores forming simultaneously [9]. After precursor nano-aggregates were heated, micro-or meso-pores became bigger and turned into great mesopores due to $H_2O$ molecules' loss while precursor nano-aggregates transformed into nano-crysllites interconnected. Moreover arrangement of pores formed is disordered as precursor aggregates are arranged in a three dimensional disorder way in precursor spheres.

Through above discuss it could beexplained that formation mechanism of mesopores in $ZrO_2$-based spheres by SR process is distinguished conspicuously from that by removal of self-assembled surfactant templates (liquid crystal template mechanism), where pore arrangement is crystallographically ordered while atomic arrangement in $ZrO_2$-based framework is disordered, bringing about that pore structures will be mostly destroyed after template is removed and framework is crystallized. Therefore for template process, the phosphate or sulfate as thermal stabilizer has to be introduced to improve thermal stability of pores as sulfate and phosphate species reinforce the bonding between zirconia fragments and delay the crystallization [16-25].

In our previous[9] and present works, thermal stabilization of pores in $ZrO_2$-based spheres produced by SR technique has been investigated partly. It is demonstrated that after heating at 500 – 1000℃, pores are preserved quite well, probably resulting from the three dimensional disordered porous structure and mechanically strong crystallized skeleton. It is generally concluded that the disordered pores possess better thermal stabilization than ordered ones [28]. However the specific surface area of samples obtained by SR method, due to fair uneven shrinkage between surface and interior of spheres during calcinations, is not so large ( ~ 30 – 35$m^2$/g) as that of samples prepared by template process (up to >500$m^2$/g ) at calcination temperatures below 700℃ [12].

## 4 Conclusion

Without using any template agent and thermal stabilizer, the novel mesoporous $ZrO_2$ and YSZ spherical powders were prepared by improved SR technique. Also it was the first time synthesizing YSZ spherical powder via SR route. The obtained $ZrO_2$-based spherical powders were characterized by TG/DTA, SEM, TEM, XRD and Nitrogen adsorption/desorption isotherm analyses. It is concluded that the amount of non-dense structure in $ZrO_2$ spheres is dramatically reduced by aging precursors for 72 hrs. Pure $ZrO_2$ spherical powders obtained in the temperature range of 700 – 1000℃, whose phase compositions are major c-$ZrO_2$ (700℃) and m-$ZrO_2$ (800 – 1000℃), and nanocrystallite size is less than or equal to 24.3nm, are less than or equal to 1.63μm in size, and ~30.85$m^2$/g (700℃) in BET surface area. YSZ spherical powders calcined in the temperature range of 500-1000℃, whose phase constitutions are mainly t-$ZrO_2$ and nanocrystallite size is less than or equal to 36.6nm, are less than or equal to 1.65μm in size, ~14.72$m^2$/g (500℃) and ~34.39$m^2$/g (600℃) in BET surface area, and have a narrow pore size distribution with the average pore diameters of ~10.7nm (500℃) and ~8.3nm (600℃).

**Acknowledgements:** This work has been financed by the Scientific Research Foundation for the Returned Overseas Chinese Scholars, State Education Ministry, Beijing Natural Science Foundation (No. 2073032), and National Science Foundation of China (No. 51072022).

## References

[1] C. T. Kresge, M. E. Leonowicz, W. J. Roth, et al. Ordered mesoporous molecular sieves synthesized by a liquid-crystal template mechanism [J]. Nature 1992, 359: 710-712.

[2] J. S. Beck, J. C. Vartuli, W. J. Roth, et al. A new family of mesoporous molecular sieves prepared with liquid crystal templates [J]. J. Am. Chem. Soc., 1992, 114: 10834-10843.

[3] Q. S. Huo, D. I. Margolese, U. Ciesla, et al. Generalized synthesis of periodic surfactant/inorganic composite

materials [J]. Nature, 1994, 368: 317-321.

[4] W. R. Zhao, J. L. Gu, L. X. Zhang, et al. Fabrication of uniform magnetic nanocomposite spheres with a magnetic core/mesoporous silica shell structure [J]. J. Am. Chem. Soc. , 2005, 127: 8916-8917.

[5] M. Guan, W. Liu, Y. L. Shao, et al. Preparation, characterization and adsorption properties studies of 3-(methacryloyloxy) propyltrimethoxysilane modified and polymerized sol-gel mesoporous SBA-15 silica molecular sieves [J]. Microporous Mesoporous Mater. , 2009, 123: 193-201.

[6] D. Y. Zhao, Y. Wan. Chapter 8 the synthesis of mesoporous molecular sieves [J]. Stud. Surf. Sci. Catal. , 2006, 168: 241-300, I - III.

[7] S. Li, J. T. Zheng, W. Y. Yang, et al. A new synthesis process and characterization of three-dimensionally ordered macroporous $ZrO_2$ [J]. Mater. Lett. , 2007, 61: 4784-4786.

[8] X. M. Liu, G. Q. Lu, Z. F. Yan. Synthesis and stabilization of nanocrystalline zirconia with MSU mesostructure [J]. J. Phys. Chem. B. , 2004, 108: 15523-15528.

[9] (a) H. Zhang, Z. T. An, F. Li, et al. Synthesis and characterization of mesoporous c-$ZrO_2$ microspheres consisting of peanut-like nano-grains [J]. J. Alloys Compd. , 2008, 464: 569-574.

(b) H. Zhang, Z. T. An, D. J. Zhang, et al. Effect of calcination temperature on the structure of the $ZrO_2$ microspheres synthesized by quasi-gaseous state reaction [J]. Acta Chim. Sinica, 2005, 63: 2098-2102.

[10] T. S. Wu, K. X. Wang, L. Y. Zou, et al. Effect of surface cations on photoelectric conversion property of nanosized zirconia [J]. J. Phys. Chem. C, 2009, 113: 9114-9120.

[11] A. Taguchi, F. Schuth. Ordered mesoporous materials in catalysis [J]. Microporous Mesoporous Mater. , 2005, 77: 1-45.

[12] H. R. Chen, J. L. Shi, L. Z. Hua, et al. Parameter control in the synthesis of ordered porous zirconium oxide [J]. Mater. Lett. , 2001, 51: 187-193.

[13] J. Zárate, H. Juárez, M. E. Contreras, et al. Experimental design and results from the preparation of precursory powders of $ZrO_2$ (3% $Y_2O_3$)/(10-95)% $Al_2O_3$ composite [J]. Powder Technol. , 2005, 159: 135-141.

[14] S. G. Liu, X. L. Zhang, J. P. Li, et al. Preparation and application of stabilized mesoporous MgO-$ZrO_2$ solid base [J]. Catal. Commun. , 2008, 9 : 1527-1532.

[15] X. L. Liu, D. L. Cui, Q. Wang, et al. Preparation of $ZrO_2$ porous nanosolid and its composite fluorescent materials [J]. Mater. Chem. Phys. , 2007, 105: 208-212.

[16] M. Mamak, N. Coombs, G. Ozin. Self-assembling solid oxide fuel cell materials: mesoporous yttria-zirconia and metal-yttria-zirconia solid solutions [J]. J. Am. Chem. Soc. , 2000, 122: 8932-8939.

[17] E. Zhao, S. E. Hardcastle, G. Pacheco, et al. Aluminum-doped mesoporous zirconia obtained from anionic surfactants [J]. Microporous Mesoporous Mater. , 1999, 31: 9-21.

[18] J. S. Reddy, A. Sayari. Nanoporous zirconium oxide prepared using the supramolecular templating approach [J]. Catal. Lett. , 1996, 38 : 219-223.

[19] (a) U. Ciesla, S. Schacht, G. D. Stucky, et al. Ein poröses zirconiumoxophosphat sehr hoher oberfläche durch eine tensidunterstützte synthese [J]. Angew. Chem. , 1996, 108: 597-600. (b) Formation of a porous zirconium oxo phosphate with a high surface area by a surfactant-assisted synthesis [J]. Angew. Chem. Int. Ed. Eng. , 1996, 35: 541-543.

[20] U. Ciesla, M. Fröa, G. Stucky, et al. Highly ordered porous zirconias from surfactant-controlled syntheses: zirconium oxide-sulfate and zirconium oxo phosphate [J]. Chem. Mater. , 1999, 11: 227-234.

[21] F. Schüth, U. Ciesla, S. Schacht, et al. Ordered mesoporous silicas and zirconias: control on length scales be-

tween nanometer and micrometer [J]. Mater. Res. Bull. , 1999, 34: 483-494.
[22] M. Lindén, J. Blanchard, S. Schacht, et al. Phase behavior and wall formation in Zr $(SO_4)_2$/CTABr and TiO-$SO_4$/CTABr mesophases [J]. Chem. Mater. , 1999, 11: 3002-3008.
[23] M. S. Wong, D. M. Antonelli, J. Y. Ying. Synthesis and characterization of phosphated mesoporous zirconium oxide [J]. Nanostruct. Mater. , 1997, 9 : 165-168.
[24] M. S. Wong, J. Y. Ying. Amphiphilic templating of mesostructured zirconium oxide [J]. Chem. Mater. , 1998, 10: 2067-2077.
[25] Y. Huang, W. M. H. Sachtler. Preparation of mesostructured lamellar zirconia [J]. Chem. Commun. , 1997: 1181-1182.
[26] R. C. Garvie, P. S. Nicholson. Phase analysis in zirconia system [J]. J. Am. Ceram. Soc. , 1972, 55: 303-305.
[27] C. V. Thompson. Grain growth in thin films [J]. Annu. Rev. Mater. Sci. , 1990, 20: 245-268.
[28] H. Chon, S. K. Ihm, Y. S. Uh (Editors). Synthesis and hydrothermal stability of a disordered mesoporous molecular sieve, Progress in zeolite and microporous materials [J]. Stud. Surf. Sci. Catal. , 1997, 105: 45-52.

(原载于 *Powder Technology*, 2012, 227: 9-16)

# Catalytically Highly Active Top Gold Atom on Palladium Nanocluster

Zhang Haijun[1,2]　　Tatshuya Watanabe[1]　　Mitsutaka Okumura[2,3]
Masatake Haruta[2,4]　　Naoki Toshima[1,2]

(1. Department of Applied Chemistry, Tokyo University of Science Yamaguchi,
SanyoOnoda, Yamaguchi 756-0884, Japan;
2. Core Research for Evolutional Science and Technology (CREST), Japan Science
and Technology Agency, Kawaguchi, Saitama 332-0012, Japan;
3. Graduate School of Science, Osaka University, Machikaneyama, Toyonaka, Osaka 560-0043, Japan;
4. Department of Applied Chemistry, Graduate School of Urban Environmental Sciences,
Tokyo Metropolitan University, Minami-osawa, Hachioji,
Tokyo 192-0397, Japan)

Catalysis using gold is emerging as an important field of research in connection with "green" chemistry[1-3]. Several hypotheses have been presented to explain the markedly activities of Au catalysts [4-10]. To date, the origin of the catalytic activities of supported Au catalysts can be assigned to the perimeter interfaces between Au nanoclusters (NCs) and the support [11]. However, the genesis of the catalytic activities of colloidal Au based bimetallic NCs is unclear. Moreover it is still a challenge to synthesize Au based colloidal catalysts with high activity. Here we now present a novel concept of "crown-jewel" (CJ, Fig. S1) for preparation of catalytically highly Au-based colloidal catalysts. Au/Pd colloidal catalysts containing an abundance of top (vertex or corner) Au atoms were synthesized according to the strategy on a large scale. Our results suggest that the genesis of the high activity of the catalysts could be ascribed to the presence of negative charged top Au atoms.

Pd/Au is one of the most attractive systems in catalysis research[1,12,13]. We prepared the crown-jewel structured Au/Pd NC (CJ-Au/Pd NC, Fig. S1) catalysts by a galvanic replacement reaction method [14]. It is reasonable to anticipate that this new catalyst would have excellent activities based on the followings reasons. First, Au atoms locating at the top position of a $Pd_{147}$ NC (cf. Fig. S1a, mother clusters) possess a high degree of coordinative unsaturation which leaves them more free than other atoms to engage in chemical reactions [15-17]. Second, the top Au atoms, which are surrounded by several Pd atoms, would be very interesting from the viewpoint of catalysis of metal NCs because the catalytic performance of a metal is often improved by the adjacent second

---

张海军，1995～1999年于北京科技大学师从李文超教授攻读博士学位。目前在武汉科技大学工作，湖北省"楚天学者"特聘教授。获得武汉市黄鹤英才、河南省杰出青年科学基金及河南省省级学术与技术带头人等荣誉称号。发表学术论文200余篇，获省部级科技奖6项。

and third elements [18,19]. In this catalyst, ideal $Pd_{147}$ NC serve as the crown, while the Au atoms formed by the spontaneous replacement reaction serve as jewels decorating the top position of the $Pd_{147}$ NCs (as shown in Fig. 1). Considering the surface free energies of the top, edge, and face atoms of the $Pd_{147}$ NCs, the replacement reaction on the surface of the $Pd_{147}$ NCs first occurs from the top Pd atoms, which guarantees that the top Pd atoms have a preferential probability to react with the $Au^{3+}$ ions to in-situ form the top Au atoms.

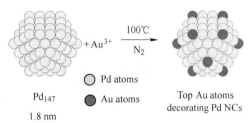

Fig. 1 Schematic illustration for the deposition of top Au atoms on Pd mother clusters by replacement reaction method

According to Mie theory, the position and intensity of the absorption bands of UV-Vis spectra are strongly influenced by the particle size, shape, concentration, composition, and dielectric properties of metallic NCs and their local environment. The UV-Vis absorption spectra of the aqueous dispersions of the $Pd_{147}$ (mother clusters, 1.8 ±0.6nm, prepared by alcohol reduction) and Au NCs (1.4 ±0.5nm, prepared by rapid injection of $NaBH_4$), and a series of CJ-Au/Pd NCs share a similar curve (Fig. S3a) and exhibit a featureless and monotonically increasing absorbance toward higher energies. The spectrum of the Au NCs dispersion exhibits a small peak at 520nm attributed to the surface plasmon resonance of the metallic Au. The prepared CJ-Au/Pd NCs show a similar size and geometrical shape with that of $Pd_{147}$ mother clusters and Au NCs (Fig. S3b). However, the UV-Vis spectra of CJ-Au/Pd NCs exhibit a higher intensity (Fig. S3a) than that of the $Pd_{147}$ NCs though the concentrations of CJ-Au/Pd NCs dispersions are lower than that of $Pd_{147}$ and Au NCs (Fig. S3b). These intensified absorptions suggest that the surface compositions of the CJ-Au/Pd NCs are different from that of the $Pd_{147}$ mother clusters. The absence of the surface plasmon peak of Au at 520nm indicates that Au atoms are deposited only in several areas across the surface of the $Pd_{147}$ NCs.

TEM images and size distribution histograms (Fig. S4) of the CJ-Au/Pd NCs, i.e., CJ-1, CJ-2, and CJ-3 with different Au contents (cf. Table S1 for the detailed preparation conditions) show that all the NCs are spherical and well-isolated, and mainly distributed within the range from about 0.75 to 4nm. The average sizes based on the TEM images are 1.8 ±0.5nm for CJ-1, 1.7 ±0.5nm for CJ-2, and 1.6 ±0.5nm for CJ-3.

Using electron tomography to obtain an unambiguous image of the three-dimensional structure of the ultra small NCs is challenging[20] because those ultra-small NCs are intrinsically unstable and may interact with the incident electron beam [21]. The high-angle annular dark-field scanning TEM (HAADF-STEM) was measured to further investigate the presence of the top Au atoms of the CJ-Au/Pd NCs. A representative HAADF-STEM image in Fig. 2 (a)-(1) clearly shows the existence of columnar atoms in a CJ-Au/Pd NCs of ~2nm diameter, the dotted yellow hexagon draws the whole shape of the cluster. The inset gives a corresponding fast Fourier-transform (FFT) pattern, which indicates that the surface of the NC is enclosed by both {111} (yellow color) and {100} (red color) facets (cf. Fig. 2 (a)-(2)). The atoms located at the center as well as corners

1 and 2 possess orderly alignments, suggesting that they are unreacted Pd atoms (green circle). However, the NC in Fig. 2 (a) does not form orderly atom alignments around corners 3 and 4, which can be ascribed to the size miss-match between the Au and Pd atoms since the interatomic distances of Au and Pd are 0.292nm and 0.276nm, respectively. Moreover, clear vacancies resulting from the replacement reaction are also observed in corners 3, 4, 5, and 6 (red circle), and similar phenomena are also observed in the other images of Figure S5. These vacancies as well as the disorder atom alignments can provide reasonable evidence to prove that the replacement reaction for the Au atoms formation does occur at the top sites of the $Pd_{147}$ NCs.

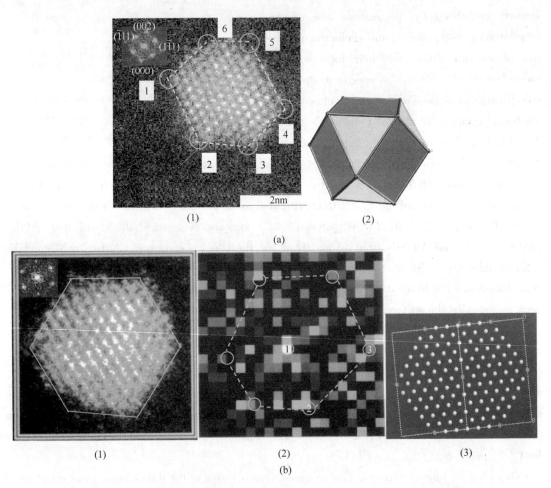

Fig. 2 Experimental characterization and theoretical modeling of CJ-Au/Pd NCs
(a) HAADF-STEM image of a single CJ-Au/Pd NCs; (b) HAADF-STEM image, EELS mapping, and the surface atomic configurations of a CJ-Au/Pd NCs recorded along the [110] zone axis
(The dotted blue hexagon in
Fig. 2(b)-(2) draws a shape of the cluster estimated based on EELS results. The red circles in Figure 2 (b)-(2) indicate the presence of the top Au atoms.)

EELS has been widely used to study the surface composition [22]. The elemental distribution of Pd and Au was obtained by sorting out a specific atom by means of EELS at various locations of a CJ-Au/Pd NC. Since the size of the electron beam in EELS is about 0.2nm ( < Au atom diameter,

0.292nm), the mapping of a NC by elements can be carried out in detail. Fig. 2 (b) show a representative HAADF-STEM image (Fig. 2 (b)-(1)), Au EELS map (Fig. 2 (b)-(2), 4.32 × 4.32nm, 20 × 20pixels, and 0.22nm/pixel) of a CJ-Au/Pd NC accompanied by its schematic atomic configurations (Fig. 2 (b)-(3)) in the {110} plane. In the Au EELS map (Fig. 2 (b)-(2)), the bright areas show the presence of Au atoms. The brighter the area, the more probable it corresponds to Au atom. It shows that the distribution of Au is not very orderly in the NC, and the Au atoms seem to move in a certain area. This phenomenon can be ascribed to the following two factors. First, the observed CJ-Au/Pd NC is not an ideal $Pd_N$ clusters, hence the replacement reaction also occurs at some defect sites. Second, the top Au atoms are unstable and easily move during the electron beam irradiation. We think the latter factor is more reasonable because the easy movement of top atoms were supported by other HAADF-STEM images of the CJ-Au/Pd NCs (Fig. S6). A set of images recording the moving process of a top atom is screened from several successive HAADF-STEM frames in Fig. S6, clearly indicating that the top atom (red arrow) finally disappears during the electron beam irradiation process. In fact, the electron beam intensity used and the scanning time required for an EELS mapping are usually much higher and longer than that for a single HAADF-STEM image, which could result in easier movement of the top Au atoms at certain areas in the EELS mapping as shown in Fig. 2 (b). However, the most striking observation is that some Au atoms do locate at the corner of the CJ-Au/Pd NC, such as atom 1, atom 2 and atom 3 shown in Fig. 2 (b)-(2). Based on the EELS results as well as the above-mentioned characteristic of the top atoms, we can approximately outline the distribution of the top Au atoms of the CJ-Au/Pd NC as shown in Fig. 2(b)-(2). Therefore, based on the results of the HAADF-STEM and EELS, we can conclude that the prepared Au atoms or, at least, parts of them locate at the top position of the Pd NC.

What is of interest in this novel CJ structured catalyst is its unprecedented high activity. Pure Pd NCs usually show a low activity towards glucose oxidation [23]. Thus, the catalytic activities of the Au atoms in the present CJ NCs could be calculated. (The activity of the Pd mother clusters was subtracted from that of the CJ NCs. The detail calculations are listed in Table S2.) The initial catalytic activity of the CJ-Au/Pd NCs for the glucose oxidation, normalized by Au mass, is much higher than that of Au, as shown in Figure 3. The maximum catalytic activity (TOF) of the top Au atoms is about 194,980 mol-glucose · $h^{-1}$ · mol-$Au^{-1}$. The specific activity is higher than that of the Au (1.4 ±0.5nm in diameter, even if the activity was normalized to the surface atoms) and Pd mother clusters (1.8 ±0.6nm) by a factor of 20-30, and that of the Pd/Au alloy NCs by a factor of 8-10, although all of those NCs possess almost the same particle size (2.4 ±0.8nm for Pd/Au (8/2) alloy NCs and 1.9 ±0.8nm for Pd/Au (3/7) alloy NCs, prepared by rapid injection of $NaBH_4$.). This result demonstrates that the CJ-Au/Pd NCs possess a special structure different from that of the Pd and Au, and Pd/Au alloy NCs, and provides an indirect proof for the presence of the CJ structure in the CJ-Au/Pd NCs. To the best of our knowledge, this activity is the highest reported to date for the glucose oxidation catalyzed by colloidal gold catalysts. Moreover, this strategy is also applicable to other systems. Our preliminary results have shown that the activity for glucose oxidation of the top Au atoms on the Ir/Pd NCs is as high as 463,900 mol-glucose · h$^{-1}$ · mol-Au$^{-1}$; this TOF value is the highest reported to date for the glucose oxidation catalyzed by all supported and unsupported catalysts.

It is interesting to note that CJ-3 NCs with increasing number of Au atoms show appreciable inferior catalytic performance to those of CJ-1 and CJ-2 NCs (Fig. 3 and Table S2) and catalyst activity for glucose oxidation significantly depends on the sites of Au atoms in the CJ NCs (cf. cluster model inserts of Fig. 3). Based on the results of HAADF-STEM observation, all the Au atoms formed by replacement reaction can be assumed to predominantly locate at the top position on the surface layer of CJ-1 and CJ-2 NCs. In contrast, in the case of CJ-3 NCs containing a lot of Au atoms, parts of the Au atoms formed by the replacement reaction could locate at the edge sites as well as the top sites. The low activity of CJ-3 NCs suggests that the presence of edge Au atoms will play a negative effect on the catalytic performance for glucose oxidation and that the top Au atoms are mainly responsible for the high activity, which is in agreement with the previous reports that the reactivity of a catalyst is dominated completely by a very small fraction of the surface active sites[24,25]. In other words, the activity of the Au atoms on the surface of Pd$_{147}$ mother clusters should decrease in the following order: top Au atom > edge Au atom > face Au atoms (Fig. S7). This correlation is an important guiding principle for designing novel metal cluster catalysts.

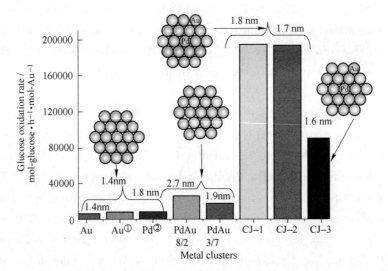

Fig. 3 Comparison of the catalytic activity of CJ-Au/Pd, Au, Pd, and Pd/Au alloy NCs for aerobic glucose oxidation

(Schematic inserts and numbers shown at the top of each bar indicate the structure models and the average particle sizes, respectively, of the NCs; Au①, the activity was normalized by the number of surface Au atoms in NCs; Pd②, the activity was normalized by the number of surface Pd atoms, activity of 8,290 mol-glucose · h$^{-1}$ · mol-surface Pd$^{-1}$.)

A question arises as to how the high activities of the CJ-1 and CJ-2 NCs can be explained. We think it should be relative to not only the unique geometrical structure but also the electronic properties of the top Au atoms. In order to examine the electronic properties of the catalytically active

CJ-Au/Pd NCs, another CJ-Au/Pd NCs (CJ-4, Table S3) protected by the low PVP content were prepared and investigated by XPS. As shown in Table S3, the electron apparent binding energy (BE) of Au $4f_{7/2}$ in the CJ-4 NCs was about 1.5 eV lower than that of the bulk Au, and about 0.25eV lower than that of the PVP-protected Au NCs with the mean diameter of 1.1 nm [26]. The negative shift of the Au 4f BE suggests that a negative charge is deposited on the Au atoms of the CJ-Au/Pd NCs, and provides evidence that the Pd atoms donate electrons to the Au atoms. In order to further confirm the electron donation of the Pd atoms to the Au atoms, DFT calculations were carried out to study the electronic states of the top Au atoms in the CJ-Au/Pd NCs. At the present stage, it is still impossible for us to make such a huge calculation using the $M_{147}$ ideal model, hence the DFT calculations for $Pd_{55}$ and the CJ structured $Pd_{43}Au_{12}$ model NCs were examined as a first step. The Mulliken charges of $Pd_{55}$ depicted in Fig. 4 and Table S4 showed that the top Pd atoms are weakly negatively charged (−0.0006). A more important result was that the negatively charge density of the top atoms (Fig. 4) significantly increased when all of the top atoms of $Pd_{55}$ clusters were replaced with Au atoms (−0.116). This result provides more evidence that the Pd atoms donate electrons to the Au atoms. On the basis of these above-mentioned facts, it can be concluded that the top Au atoms in the crown jewel-structured CJ-Au/Pd NCs are indeed negatively charged, and the negatively charged top Au atoms are formed due to the charge transfer from the palladium atom to the top gold atoms (Fig. S8).

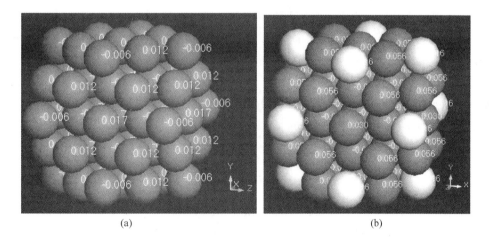

(a)　　　　　　　　　　　　　　　(b)

Fig. 4　DFT calculations of electronic structure of (a) $Pd_{55}$ and (b) crown jewel-structured $Pd_{43}Au_{12}$ NCs
(Yellow: Au atoms; Blue: Pd atoms)

These results of the catalytic activity, XPS, and DFT calculations led us to conclude that the anionic charge on the top Au atoms is the direct cause for the high reactivity. Tsukuda et al. reported similar results of the small PVP protected Au (<2nm)[26] and Ag/Au NCs[27] for the aerobic oxidation of p-hydroxybenzyl alcohol. Another similar correlation was observed in the gas-phase Au clusters, smaller cluster anions ($Au_n^-$) that exhibit a higher reactivity toward $O_2$ than the larger cationic clusters [28]. In all the cases, a hydroperoxo-like molecular oxygen species ($HO_2$) generated on the cluster surface under quasi-homogeneous conditions is believed to play a key role as an active species in these oxidation reactions. Based on the chemical nature of the present CJ-Au/Pd NCs, we propose that electron transfer from the anionic top Au atoms to $O_2$ also generates a hydroperoxo-like species that plays a crucial role in the oxidation of the glucose. In another way, it was accepted that during the process of aerobic glucose oxidation by colloidal gold nanoparticles, gluconic acid was produced together with hydrogen peroxide, which later quickly decomposes[29]. As shown in Fig. S9, the present CJ-Au/Pd NCs also show a much higher catalytic activity toward $H_2O_2$ decomposition than Pd, Au, and the Pd/Au alloy NCs. We believe the activity for $H_2O_2$ decomposition plays another positive role in the glucose oxidation by the CJ-Au/Pd NCs. Therefore, we can conclude that the highest catalytic activity of the top Au atoms could be associated with its high negatively charged density and its unique structure.

In summary, we have demonstrated for the first time a simple and versatile route for the large-scale assembly of Au atoms as an effective catalyst. The Au atoms can be controllably assembled at the top position on the surface of the mother clusters, and then exhibit a high catalytic activity. We anticipate that our finding will initiate attempts for understanding the structure-activity relations of a catalyst on an atomic level, which might lead to commercial exploitation of the high activity of the gold NCs. In addition, this concept potentially provides a general design with a reactive metal embedded in less reactive materials.

Supporting information available. HAADF-STEM images, TEM images, UV-Vis spectra, XPS results, preparation conditions and catalytic activities of the CJ-Au/Pd NCs catalysts. An example of the titration curve is also included.

**Acknowledgements:** This work was financially supported by Grants-in-Aid from the Core Research for Evolutional Science and Technology (CREST) program sponsored by the Japan Science and Technology Agency (JST).

## References

[1] Enache D. I., Edwards J. K., Landon P., et al. Solvent-free oxidation of primary alcohols to aldehydes using titania-supported gold-palladium catalysts [J]. Science, 2006, 311: 362-365.

[2] Comotti M., Della Pina C., Falletta E., et al. Is the biochemical route always advantageous? The case of glucose oxidation [J]. J. Catal., 2006, 244: 122-125.

[3] Hughes M. D., Xu Y. J., Jenkins P., et al. Tunable gold catalysts for selective hydrocarbon oxidation under mild conditions [J]. Nature, 2005, 437: 1132-1135.

[4] Herzing A. A., Kiely C. J., Carley A. F., et al. Identification of active gold nanoclusters on iron oxide supports for CO oxidation [J]. Science, 2008, 321: 1331-1335.

[5] Haruta M. Gold as a novel catalyst in the 21st century: Preparation, working mechanism and applications [J]. Gold Bull., 2004, 37: 27-36.

[6] Wang J. G., Hammer, B. Oxidation state of oxide supported nanometric gold [J]. Top. Catal., 2007, 44:

49-56.

[7] Lopez N., Janssens T. V. W., Clausen B. S., et al. On the origin of the catalytic activity of gold nanoparticles for low-temperature CO oxidation [J]. J. Catal., 2004, 223: 232-235.

[8] Sanchez A., Abbet S., Heiz U., et al. When gold is not noble: Nanoscale gold catalysts [J]. J. Phys. Chem. A, 1999, 103: 9573-9578.

[9] Xu Y., Mavrikakis M. Adsorption and dissociation of $O_2$ on gold surfaces: Effect of steps and strain [J]. J. Phys. Chem. B, 2003, 107: 9298-9307.

[10] Haruta M., Daté, M. Advances in the catalysis of Au nanoparticles [J]. Appl. Catal. A: Gen., 2001, 222: 427-437.

[11] Haruta M. When gold is not noble: Catalysis by nanoparticles [J]. Chem. Record, 2003 (3): 75-87.

[12] Chen M. S., Kumar D., Yi C. W., et al. The promotional effect of gold in catalysis by palladium-gold [J]. Science, 2005, 310: 291-293.

[13] Edwards J. K., Solsona B., Ntainjua N. E., et al. Switching off hydrogen peroxide hydrogenation in the direct synthesis process [J]. Science, 2009, 323: 1037-1041.

[14] Sun Y. G., Xia Y. N. Shape-controlled synthesis of gold and silver nanoparticles [J]. Science, 2002, 298: 2176-2179.

[15] Campbell C. T. The active site in nanopaticle gold catalysis [J]. Science, 2004, 306: 234-235.

[16] Lemire C., Meyer R., Shaikhutdinov S., et al. Do quantum size effects control CO adsorption on gold nanoparticles? [J] Angew. Chem. Int. Ed., 2004, 43: 118-121.

[17] Zanella R., Giorgio S., Shin, C. H., et al. Characterization and reactivity in CO oxidation of gold nanoparticles supported on $TiO_2$ prepared by deposition-precipitation with NaOH and urea [J]. J. Catal., 2004, 222: 357-367.

[18] Toshima N., Kanemaru M., Shiraishi Y., et al. Spontaneous formation of core/shell bimetallic nanoparticles: A calorimetric study [J]. J. Phys. Chem. B, 2005, 109: 16326-16331.

[19] Toshima N., Ito R., Matsushita T., et al. Trimetallic nanoparticles having a Au-core structure [J]. Catal. Today, 2007, 122: 239-244.

[20] Li Z. Y., Young N. P., DiVece M., et al. Three-dimensional atomic-scale structure of size-selected gold nanoclusters [J]. Nature, 2008, 451: 46-48.

[21] Ajayan P. M., Marks L. D. Experimental-evidence for quasimelting in small particles [J]. Phys. Rev. Lett., 1989, 63: 279-282.

[22] Egerton R. F. Electron energy loss spectroscopy in the electron microscope [M]. New York: Plenum Press, 1996.

[23] Comotti M., Della Pina C., Matarrese R., et al. The Catalytic Activity of "Naked" Gold Particles [J]. Angew. Chem. Int. Ed., 2004, 43: 5812-5815.

[24] Zambelli T., Wintterlin J., Trost, J., et al. Identification of the "active sites" of a surface-catalyzed reaction [J]. Science, 1996, 273: 1688-1690.

[25] Vang R. T., Honkala K., Dahl S., et al. Controlling the catalytic bond-breaking selectivity of Ni surfaces by step blocking [J]. Nat. Mater., 2005, 4: 160-162.

[26] Tsunoyama H., Ichikuni N., Sakurai H., et al. Effect of electronic structures of Au clusters stabilized by poly (N-vinyl-2-pyrrolidone) on aerobic oxidation catalysis [J]. J. Am. Chem. Soc., 2009, 131: 7086-7093.

[27] Chaki N. K., Tsunoyama H., Negishi Y., et al. Effect of Ag-doping on the catalytic activity of polymer-stabilized Au clusters in aerobic oxidation of alcohol [J]. J. Phys. Chem. C., 2007, 111: 4885-4888.

[28] Kim Y. D., Fischer M., Ganteför G. Origin of unusual catalytic activities of Au-based catalysts [J]. Chem. Phys. Lett., 2003, 377: 170-176.

[29] Beltrame P., Comotti M., Della Pina C., et al. Aerobic oxidation of glucose II. Catalysis by colloidal gold [J]. Appl. Catal. A, 2006, 297, 1-7.

(原载于 *Nature Materials*, 2012, 11: 49-52)

# X-Ray Photoelectron Spectrascopy Investigation of Ceria Doped with Lanthanum Oxide

Du Xueyan (杜雪岩)[1,2]　Li Wenchao (李文超)[2]
Liu Zhenxiang (刘振祥)[1]　Xie Kan (谢侃)[1]

(1. Institute of Physics, Chinese Academy of Sciences, Beijing 100080;
2. Department of Physical Chemistry, University of
Science and Technology, Beijing 100083)

The influence of doping lanthana on valent state of $CeO_2$ surface has been studied by using X-ray photoelectron spectroscopy. It was found that $La^{3+}$ doping in host $CeO_2$ lattice can promote to convert $Ce^{4+}$ into $Ce^{3+}$ and enhance adsorbing-carbonyl capacity of the materials. However, the inversion of $Ce^{4+}$ to $Ce^{3+}$ is depressed by dopant $La^{3+}$ during argon ion bombardment.

Much attentionhas been paid to cerium dioxide for its excellent catalysis property as a transient oxygen storage material in automobile exhaust gas catalytic converters[1,2]. It can store oxygen in the oxygen-rich condition and release oxygen in the oxygen-poor condition. Partial replcement of $Ce^{4+}$ ions with trivalent cations could create a corresponding number of anion vacancies and result in solid solution with high ionic conductivity[3-6]. The defective surface of these materials may enhance their catalytic properties. Kubsh et al. reported that lanthana can impede ceria crystallite growth in severe oxidizing environment as stabilizer. Moreover, the use of stabilized ceria containing lanthana resulted in poor three-way activity relative to unstabilized ceria.

Because the redox characteristics of La-doped ceria have been reported scarcely, the present paper is devotedto investigation of redox properties of La-doped ceria. Ascatalytic reactions occur at the surface of the catalyst, surface characterization is critical for understandingmaterial catalytic properties. In order to probe the surface various oxidation states that might occur in rare earth redox systems, X-ray photoelectron spectroscopy (XPS) was used to study La-doped ceria ceramic chips.

Samples were prepared by mixing ceria with 1 Lanthana different dopant levels of lanthana with magneticstirring apparatus for 30min. The samples were pressed into pellets (at $80kg \cdot cm^{-2}$), calcined in air at 1400℃ for three hours and mounted on a nickel sampleholder that allowed high temperature heat treatment.

The XPS were recorded in ESCALAB5 appatatus equipped with an Mg $K\alpha$ radiation. The background vacuum was better than $5 \times 10^{-7}$Pa during analysis. The samples were etched by using an

---

Supported by the National Natural Science Foundation of China under Grant No. 19874077, and Ph. D. Educational Fund. @ by the Chinese Physical Society

杜雪岩，1996~2000年于北京科技大学师从李文超教授攻读博士学位。目前在兰州理工大学工作，教授，博士生导师。发表论文60余篇，获省部级二等奖3项。

argon ion beam. The following ion gun parameters were used during sputtering: ion beam energy 2kV, emission current 20μA, sputtering time 5min.

The Ce $3d$ core level of pure ceria and the one doped with lanthana at room tempareture were shown in Fig. 1. Following convention[8-10], two sets of multiplets $u$ and $v$ ($3d_{3/2}$ and $3d_{5/2}$) have been labelled, respectively. Here $v$, $v''$ and $v'''$ can be attributed to $CeO_2$; $v$ and $v''$ are due to a mixture of $(5d\,6s)^0 4f^2\,O2p^4$ and $(5d\,6s)^0 4f^1 O2p^5$ configurations while $v'''$ is a pure $(5d\,6s)^0 4f^0 O2p^6$ final state. On the other hand, $v_0$ and $v'$ are due to a mixture of $(5d\,6s)^0 4f^2 O2p^4$ and $(5d\,6s)^0 4f^1 O2p^5$ configurations in $Ce_2O_3$; ustructures, due to the Ce $3d_{3/2}$ level, can be explainedin the same way. It can be seen that peaks $v'$, $v_0$, $u'$, and $u_0$ of curve $a$ are higher than those of curve $b$. This indicates that the $Ce^{3+}$ concentration of La-doped ceria is higher than that of pure ceria. Thus we can suggest on theoretical grounds that the most likely defect mechanism for ceria is through the production of trivalent cerium ions and anion vacancies[11]. It can be written as:

$$+2O_0^X \longrightarrow O_{2+} +2 \qquad (1)$$

$$La_2O_3 \xrightarrow{CeO_2} 2La'_{Ce+} +3O_0^X \qquad (2)$$

Considering Eqs. (1) and (2) gives

$$La_2O_3 +6Ce_{Ce}^X \longrightarrow 2La'_{Ce} +6Ce'_{Ce^{4+}} +O_2\,(g) \qquad (3)$$

These symbols are all conicdent with Kröger-Vink rule.

According to Eq. (3), it can be seen that the equation will move to the right as the lanthanum ions are incorprated into the lattice. That is, dopant lanthana can promote some $Ce^{4+}$ to convert into $Ce^{3+}$.

In order to investigate the influence of La-doping on the ruduction of cerium dioxide, the material has been sputtered with argon ion beams. The results are shown Fig. 2.

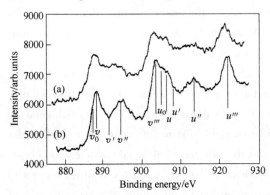

Fig. 1  Ce $3d$ spectra of pure ceria and ceria doped with lanthanum oxides: (a) $CeO_2 + La_2O_3$ (20% mass), (b) $CeO_2$

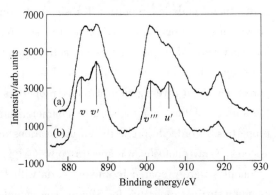

Fig. 2  Ce $3d$ spectra of ceria and ceria doped with lanthanum oxides after argon ion bombardment: (a) $CeO_2 + La_2O_3$ (10% mass), (b) $CeO_2$

It can be seen that two materials are both reduced during argon ion sputtering. However, before sputteringthe concentration of $Ce^{3+}$ in La-doped sample is higher than that of pure ceria. However, after sputteringthe concentration of $Ce^{3+}$ of the former one become lower than that of latter one, which seems to be that, on one hand, La doping promotes to convert $Ce^{4+}$ to $Ce^{3+}$, on the other hand, La doping prevents ceria being deep reduced. This phenomenon may be an important reason why the use of stabilized ceria containing lanthana results in poor three-way activity relative to unstabilized ceria. This phenomenon has also been found in other materials[12].

Fig. 3 presents the O 1s emission spectra for different lanthana levels. It can be seen that the high binding-energy shoulders rise as the lanthana levels increase. Praline et al[13]. attributed these features to hydroxyl or tosome hydroxyl-containing oxide, rather than to physisorbed oxygen. Laachir et al[14]. also found these peaks, who attributed to emission from hydroxyl and carbonate species. For our samples, C1s emission also gets higher as dopants levels increase, so it can be confirmed that the shoulder comes from carbonate species. These groups were mostly eliminated after heat treatment at 500℃.

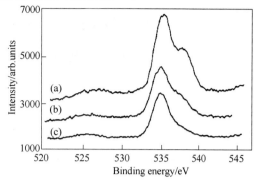

Fig. 3 Oxygen 1s XPS spctra of ceria and ceria doped with lanthanum oxide: (a) $CeO_2 + La_2O_3$ (10% mass), (b) $CeO_2 + La_2O_3$ (5% mass), (c) $CeO_2$

In conclusion, XPS spectra of pure $CeO_2$ and $CeO_2$ doped with lanthana have been obtained. Detailed analysis shows that the addition of a relatively small quantity of $La^{3+}$ to cerium dioxide results in dramatic effects as follows: (1) promoting to convert some $Ce^{4+}$ into $Ce^{3+}$; (2) enhancing the adsorbing-carbonyl capacity of sample surface; (3) depressing the inversion of $Ce^{4+}$ to $Ce^{3+}$ during argon ion sputtering.

## References

[1] H. C. Yao, Y. F. Yao. J. Catal, 1984, 86: 254.
[2] A. D. Longan, M. Shelef. J. Mater. Res., 1994, 9: 468.
[3] A. K. Bhattacharya, et al. J. Mater. Sci., 1996, 31: 5005.
[4] A. Cocco. Powder Metal. Int, 1990, 22: 25.
[5] R. Gerhardt-Anderson, A. S. Nowick. Solid State Ionics, 1981, 5: 547.
[6] V. Butler, et al. Solid State Ionics, 1983, 8: 109.
[7] A. Crucq. Catalysis and Automotive Pollution Control II [M]. Amsterdam: Elsevier Science Publishers B. V., 1991.
[8] A. Pfau, K. D. Schierbaum. Surf. Sci., 1994, 321: 71.
[9] A. Dauscher, et al. Surf. Interface Anal., 1990, 16: 341.
[10] M. Romeo, et al. Surf. Interface Anal., 1993, 20: 508.
[11] Mats Hillert, Bo Jansson. J. Am. Ceram. Soc., 1986, 69: 732.
[12] Zhenxiang Liu, et al. Applied Catalysis, 1989, 56: 207.
[13] G. Praline, et al. J. Electron Spectrosc. Relat. Phenom., 1980, 21: 17.
[14] Ahmidou Laachir, et al. J. Chem. Soc. Faraday Trans., 1991, 87: 1601.

(原载于 Chinese Physics Letters, 1999, 16 (5): 376-377)

# A New Highly Selective $H_2$ Sensor Based on $TiO_2$/PtO-Pt Dual-Layer Films

Du Xueyan[1]   Wang Yuan[1]   Mu Yongyan[2]
Gui Linlin[1]   Wang Ping[2]   Tang Youqi[1]

(1. State Key Laboratory for Structural Chemistry of Unstable and Stable Species, College of Chemistry and Molecular Engineering, Peking University, Beijing 100871; 2. Faculty of Science, Beijing University of Chemical Technology, Beijing 100029)

**Abstract:** A new highly selective hydrogen sensor based on $TiO_2$/PtO-Pt dual-layer films has been prepared. At 180-200℃, the prepared nanostructured sensor exhibits an excellent selectivity and good sensitivity to $H_2$ in air but is immune to many other kinds of reductive gases such as CO, $NH_3$, and $CH_4$. The sensor can give a faithful response to 1% $H_2$ in air, while the limitation for detecting $H_2$ in nitrogen is less than 1000 ppm. Influences of operation temperature and humidity on the sensing performance of the sensor were also investigated. The mechanism of the present sensor was attributed to the partial reduction of $Ti^{4+}$ to $Ti^{3+}$ by hydrogen catalyzed by a thin film of surface-oxidized Pt nanoparticles.

## 1 Introduction

Intensive studies have been made on gas sensors based on $TiO_2$ or $SnO_2$. Gas selectivity is a very important indicator that measures the ability of a sensor to precisely identify a specific gas in a gaseous mixture, which is a necessary characteristic for developing integrated gas sensor arrays. As an important chemical for many industrial processes, hydrogen leaks easily from systems and is one of the most explosive gases. Therefore, a lot of effort has been put into investigating hydrogen sensors and improving their selectivity[1-12]. It has been reported that covering a metal oxide sensor with a layer of organic or silica membrane as a molecular sieve renders the sensor an improved selectivity to hydrogen, but the decorated sensor still gives responses to the other reductive gases[8-12]. Recently, several selective hydrogen sensors have been reported which are insensitive to some reductive gases such as CO and $CH_4$ at a low concentration[1,8]. However, $NH_3$ even at a low concentration usually severely interferes with the detection of $H_2$[8]. On the other hand, selective detection of hydrogen in air is a more difficult task than that in an inert ambience because $O_2$ in air usually restrains a sensor from detecting $H_2$ with a high selectivity and sensitivity.

The gas-sensing performance and structure of a $Pt/TiO_{2-x}$ thin film prepared by evaporating a

---

杜雪岩，1996~2000 年于北京科技大学师从李文超教授攻读博士学位。目前在兰州理工大学工作，教授，博士生导师。发表论文60余篇，获省部级二等奖3项。

thin layer of Ti covered with a Pt film 6.5nm in thickness and oxidation treatment at 900K has been studied[13,14]. The sensor so prepared is capable of detecting reductive gases at 575K in a vacuum, but did not exhibit obvious selectivity to $H_2$. The $Pt/TiO_{2-x}$ film in its activated state is composed of metallic platinum and titania with a discontinuous island structure.

In this paper, we report the preparation and properties of a new highly selective sensor for detecting hydrogen in air. The present gas sensor consists of a continuous porous film of surface-oxidized Pt nanoparticles (PtO-Pt) on a glass substrate and a titania thin film covering the PtO-Pt film. The nanostructured dual-layer films are designed to utilize a process of partial reduction of $TiO_2$ with hydrogen, which is catalyzed by the PtO-Pt porous film, to induce a change in the concentration of charge carriers in the titania film at relatively low temperatures, which may increase the sensing selectivity for $H_2$ because $H_2$ as a reductant is more active than other usual reductive gases. Operated at 180-200℃, the $TiO_2$/PtO-Pt dual-layer film sensor exhibits a very high selectivity and good sensitivity toward hydrogen in air, which is a remarkable progress compared with the performance of previously reported titania-based hydrogen sensors[5-8].

## 2 Experimental section

(1) Materials. Tetraisopropyl orthotitanate ($Ti(OC_3H_7)_4$) and $PPh_3$ were purchased from Aldrich Chemical Co. Hexahydrated hexachloroplatinic acid ($H_2PtCl_6 \cdot 6H_2O$, 99%) was purchased from Beijing Hongke Chemical Products Co. Organic solvents and other chemicals of AR grade were used as received. $H_2$, CO, and $NH_3$ gases with purities higher than 99.99% were supplied by Beijing Analysis Equipment Co.

(2) Preparation of the PtO-Pt thin film. A toluene colloidal solution of $PPh_3$-modified Pt nanoclusters ($PPh_3$-Pt) with an average Pt particle size of 1.3nm was prepared as we reported previously[15]. A piece of glass or quartz substrate ($10 \times 10mm^2$) was immersed into this colloidal solution and kept there for 2 days, during which a homogenous selfassembly film of $PPh_3$-Pt nanoclusters with a thickness of about 100nm as measured by SEM was formed on the surface of the glass or quartz substrate. The $PPh_3$-Pt film on the substrate was washed with toluene, dried at 100℃ and then annealed in air at 400℃ for 30min to give a porous PtO-Pt film.

(3) Fabrication of the sensor. A colloidal solution of titanium oxide was prepared by hydrolyzing $Ti(OC_3H_7)_4$ in an aqueous solution of $HNO_3$ (0.1mol/L)[16]. Nanoparticles of titanium oxide in the obtained colloidal solution have a mean diameter of 4.1 nm with a size distribution from 3.4 to 5.4nm as determined by TEM measurements[17]. The colloidal solution (0.5mL) was dropped onto the PtO-Pt film and spun at 1500rpm for 20s to form a titanium oxide film on the PtO-Pt film. After being dried in air at 100℃, the films on the substrate were annealed at 400℃ for 30min to finish the preparation of a dual-layer film sensor.

(4) Characterization of the thin films. Microstructures of the prepared films were studied with a transmission electron microscope (JEM-200CX) and a field emission scanning electron microscope (JSM-6700F). XPS analyses of the prepared films were carried out on a VG ESCA LAB-5 (VG Co.) spectrometer. An $AlK\alpha$ X-ray source was used and the anode operated at 9kV and

18.5mA. The pass energy of the analyzer was fixed at 50eV. The vacuum of the analysis chamber was about $10^{-7}$ Pa, and all spectra were measured at a constant analyzer energy mode. The binding energies were calibrated with reference to that of C 1s at 284.8eV for adventitious hydrocarbon contamination.

(5) Gas sensitivity measurements. After two gold wires were attached as electrodes with Ag paste on the surface of a titania film with a distance of 5mm, the electric current passing through the sensor in different ambiences was measured at demanded temperatures using an electrochemical analyzer at a dc bias of 10V. The compositions of the sample gases were controlled by mass flow meters. In this work, gas sensitivity is defined as $S = (R_0 - R)/R_0$, where $R_0$ and $R$ are the measured resistances in air and in the sample gases, respectively.

## 3 Results and discussion

### 3.1 Structure of the dual-layer sensor

To investigate the microstructure of the self-assembly $PPh_3$-Pt film and the annealed film by TEM, a NaCl crystal substrate instead of a glass substrate was used in the preparation of a $PPh_3$-Pt thin film using the method described in the Experimental Section. After NaCl was dissolved in water, a piece of the film was transferred onto a TEM grid. Fig. 1 shows the TEM photograph of a self-assembly $PPh_3$-Pt thin film. Pt nanoparticles in the film have an average particle size of 1.3nm that is the same as the average particle size of Pt nanoparticles in the colloidal solution. The average distance between the close-stacked Pt nanoparticles in the film is less than 1nm. Large metal aggregations formed by directly contacting the Pt nanocores cannot be observed. XPS measurements revealed that the P species in the $PPh_3$-Pt film has a P 2p binding energy of 131.4eV, which is only a little higher than that of $PPh_3$

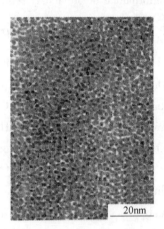

Fig. 1  TEM micrograph of a self-assembly thin film of $PPh_3$-Pt nanoclusters

(131.0eV), implying that the Pt nanoparticles are still surrounded by $PPh_3$ molecules. After the $PPh_3$-Pt film was heated in air at 400°C for 0.5h, $PPh_3$ in the film decomposed and the $PPh_3$-Pt film was transformed into a conductive and light-transparent film. Pt particles with a size up to 10nm formed by the aggregation of the small Pt nanoparticles during the heat treatment can be clearly observed in the TEM photograph of the annealed film (Fig. 2 (a)). The electron diffraction pattern of the annealed film (Fig. 2 (b)) shows the sharp diffraction points with $d$ spacings of 0.226, 0.196, 0.137, and 0.116nm, respectively, which can be indexed as (111), (200), (220), and (311) reflections of fcc Pt. Diffraction signals from the platinum oxide crystal cannot be observed in the electron diffraction pattern. However, XPS measurements revealed that the Pt $4f_{7/2}$ binding energy of the Pt species in the annealed film has a value of 72.4eV that is 1.5eV higher than that of metallic Pt (70.9eV) but is lower than that of PtO (73.8eV) by

1.4eV. From the TEM and XPS experiment results, it can be concluded that the annealed metal film is composed of Pt nanoparticles covered with a thin layer of platinum oxide.

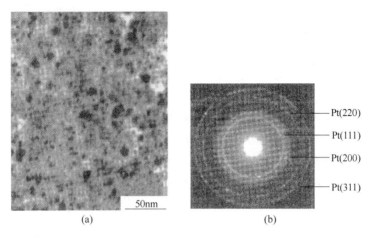

Fig. 2  TEM micrograph (a) and electron diffraction pattern (b) of a PtO-Pt film annealed in air at 400℃ for 0.5h

Fig. 3 and Fig. 4 show an SEM photograph and a scheme of the cross-section of the prepared $TiO_2$/PtO-Pt dual-layer sensor, respectively. From Fig. 3, it can be seen that the titania and PtO-Pt films are 110 and 100 nm in thickness, respectively. Holes with diameters of several tens of nanometers can be clearly observed in the PtO-Pt film, whereas the tiny holes in the titania film cannot be observed in the SEM micrograph since the resolution of SEM is not high enough.

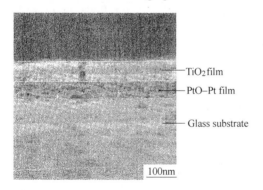
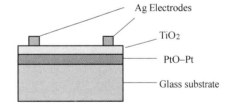

Fig. 3  HRSEM photograph of the cross-section of the $TiO_2$/PtO-Pt dual-layer films supported on a glass substrate

Fig. 4  Scheme of the cross-section of a prepared duallayer sensor

### 3.2  Performance of the prepared gas sensor

Fig. 5 shows the response ability to $H_2$ in air of the prepared $TiO_2$/PtO-Pt dual-layer sensor. When an air flow containing 2.0% $H_2$ was introduced into the system at 200℃, the measured resistance of the dual-layer sensor decreased from about 180 to 100MΩ, whereas after the hydrogen gas was turned off, the resistance reverted to the initial value as shown in Fig. 5. This response cycle can be repeated faithfully again and again. The sensing process of the prepared hydrogen sensor is therefore reversible and repeatable.

Sensitivities of the dual-layer sensor to different reductive gases at 200 ℃ are shown in Fig. 6. It can be observed from Fig. 6 that the sensing sensitivities toward CO, $NH_3$, and $CH_4$ are quite low and independent of the gas concentrations. In contrast, the sensor is quite sensitive to $H_2$, and the sensitivity to hydrogen strongly depends on the hydrogen concentration in air, indicating that the sensor has a very high selectivity to hydrogen in air, and is capable of semiquantitatively measuring the concentration of $H_2$ in air within a range from 1% to 10%. When the sensor is operated in $N_2$ ambience, the detecting limitation for $H_2$ is less than 1000ppm. It should be mentioned that, using a usual Pt metal film instead of the nanostructured platinum film, we could not prepare a dual-layer sensor due to the failure of preparing a $TiO_2$ thin film without cracks on the Pt metal film, which caused a direct contact between the Ag electrodes and the Pt film.

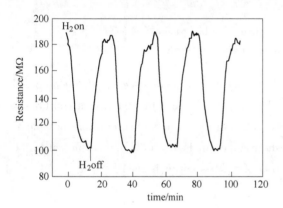

Fig. 5  Sensing resistance vs time for exposures of air containing 2.0% hydrogen. The operating temperature is 200 ℃

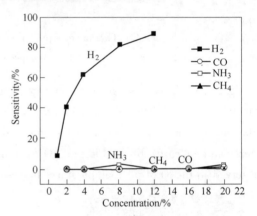

Fig. 6  Gas sensitivities of the $TiO_2$/PtO-Pt dual-layer sensor to $H_2$, $NH_3$, CO, and $CH_4$ in air

The influence of humidity on the sensing properties of the present sensor was investigated by comparing the response abilities of a prepared dual-layer sensor to $H_2$ in dry air and wet air obtained by passing an air flow through water in a vessel at room temperature. The influence of humidity on the sensing process is negligible at this condition as shown in Fig. 7.

Fig. 8 shows the temperature influence on the sensitivities to different reductive gases. The sensitivity to hydrogen increases with operating temperature up to 180 ℃; however, when the temperature is higher than 200 ℃, the sensitivity decreases gradually with the increasing operating temperature. Responses to $NH_3$, CO, and $CH_4$ in air cannot be observed at temperatures lower than 200 ℃; however, when the operating temperature is increased to 240 ℃, responses to $NH_3$ and CO with a sensitivity of about 20% can be observed. Therefore, the optimum operating temperature of the present sensor for detecting $H_2$ with a high selectivity is in the range from 180 to 200 ℃. The working temperature of the prepared sensor is much lower than that of hydrogen sensors based on $TiO_2$ reported previously[5,7], although it is higher than that of a hydrogen sensor made from Pd mesowires[1]. It should be mentioned that, in many real applications, a previous heat treatment at a temperature above 200℃ is necessary for removing adsorbed contaminants. Therefore, the stability to heat is also an important factor for a reliable sensor. For the present dual-layer sensor, a previous heat treatment at 400℃ in air does not influence its sensing properties, suggesting an excellent stability to heat.

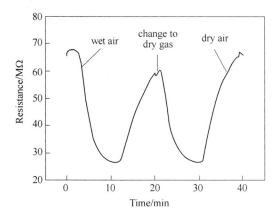

Fig. 7　Sensitivity to 4.0% $H_2$ in dry and wet air at 200℃. The thickness of the titania film is about 50nm

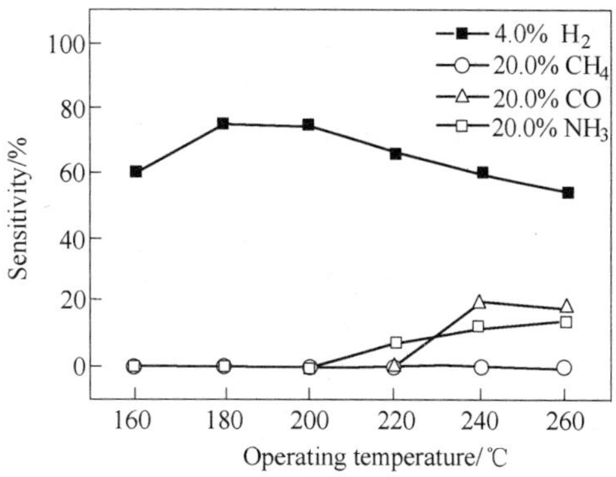

Fig. 8　Influences of the operating temperature on the sensitivities to different reductive gases in air

The present $H_2$ sensor exhibits satisfactory selectivity, sensitivity, and stability. However, it seems that further attempts are needed to improve the response speed several to 10 times before the sensor can be put into real application. We believe that embedding suitable catalysts in the layer of $TiO_2$ or optimizing the structure of the new type of sensors may improve the response speed remarkably.

## 3.3 Sensing mechanism of the $TiO_2$/PtO-Pt dual-layer films

The resistance of the PtO-Pt film is much lower than that of the titania film. At 200℃, the measured surface resistance of the PtO-Pt film is 1.4MΩ, whereas the surface resistance of a titania film directly coated on a glass substrate measured by the same method is about 310MΩ at 200℃ in air containing 4% $H_2$. The sensor resistance in the same condition is less than 100MΩ, implying that the measured electric current passes through the sensor in a way perpendicular to the surface of the sensor. To confirm this speculation, we interrupt the conductive path in the PtO-Pt film by making an insulated space of about 0.1mm width between the electrodes before coating the titania film on the PtO-Pt film. In this case, the measured sensor resistance is increased 10 times, proving that the conductive path of the $TiO_2$/PtO-Pt sensor, during its operation, is double the thickness of the titania film.

The sensing mechanism of the present $TiO_2$/PtO-Pt sensor can be attributed to a partial reduction of $Ti^{4+}$ in the titania film to $Ti^{3+}$ by hydrogen in the presence of the PtO-Pt thin film as a catalyst at 200℃ as we designed. Direct evidence for this mechanism was found from XPS measurements on the sensor before and after heating at 200℃ in air containing 10% $H_2$ (Fig. 9). After being treated with air containing hydrogen at 200℃ for 10min, the full width at half-maximum (fwhm) of the Ti $2p_{3/2}$ peak centered at 457.8eV increased from 1.7 to 2.2eV, indicating the creation of $Ti^{3+}$ ions and formation of surface oxygen vacancies in the film of titania nanoparticles[18,19]. The fwhm value of the Ti $2p_{3/2}$ peak reverted to 1.7eV after reoxidation in air at 200℃ for 10min, suggesting that the formed surface oxygen vacancies by the reduction of $H_2$ in the film of titania nanoparticles had been exhausted. In a real detection process, the sensor resistance is determined by an equilibrium between the formation and elimination of the charger carriers.

Fig. 10 shows the responses to $H_2$ (4%) in air of a titania film and the $TiO_2$/PtO-Pt film at 200℃. The $TiO_2$/PtO-Pt film exhibited a quite good sensitivity, and in contrast, the titania film did not show any observable response, indicating that the PtO-Pt film has a catalytic function for the partial reduction of the covered titania film.

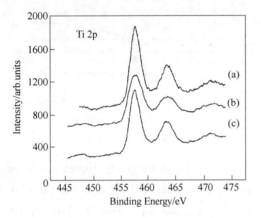

Fig. 9  Ti 2p XPS spectra of the sensor treated at different conditions: (a) untreated; (b) treated at 200℃ in air containing 10.0% hydrogen for 10min; (c) reoxidized in air at 200℃ for 10min

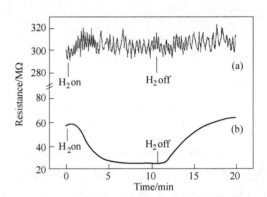

Fig. 10  Sensitivity of $TiO_2$(a) and $TiO_2$/PtO-Pt (b) films to 4.0% hydrogen in air at 200℃

The mechanism proposed above can well explain the effects of temperature and ambiences on the sensing properties of the prepared $TiO_2$/PtO-Pt sensor and is helpful for improving the response speed of the new type of sensors in a further study.

## 4　Conclusion

A hydrogen sensor composed of dual-layer films of titania and PtO-Pt nanoparticles was prepared for the first time, and exhibits a high selectivity and good sensitivity to hydrogen in air. The sensor can be used to semiquantitatively measure the concentration of $H_2$ in air within a range from 1% to 10% but does not respond to CO, $NH_3$, and $CH_4$. The optimum working temperature for detecting $H_2$ with an excellent selectivity is in the range from 180 to 200℃. XPS studies on the prepared sensor proved that the partial reduction of $Ti^{4+}$ to $Ti^{3+}$ and formation of surface oxygen vacancies in the film of titania nanoparticles is an important factor in the sensing mechanism. The PtO-Pt porous film prepared by annealing the self-assembly film of $PPh_3$-Pt nanoclusters plays the roles of an inserted electrode and a catalyst for the partial reduction of the titania film.

**Acknowledgements:** This work was financially supported by the Major State Basic Research Development Program (Project G2000077503) from the Chinese Ministry of Science and Technology and by grants from the NSFC (Project 29925308) and Chinese Ministry of Education (No. 99000). We express our thanks to Mr. Hua Shen at the laboratory for his help in the XPS measurements.

## References

[1] Favier F., Walter E. C., Zach M. P., et al. Science, 2001, 2227: 293.

[2] Kim C. K., Lee J. H., Lee Y. H., et al. Sensors and Actuators B, 2000116: 66:

[3] Tan O. K., Zhu W., Tse M. S., et al. Mater. Sci. Eng. B, 1999, 221: 58.

[4] Bévenot X., Trouillet A., Veillas C., et al. Sensors and Actuators B, 2000, 57: 67.

[5] Tang H., Prasad K., Sanjinés R., et al. Sensors and Actuators B, 1995, 71: 26-27.

[6] Birkefeld L. D., Azad A. M., Akbar S. A. J. Am. Ceram. Soc., 1992, 2694: 75.

[7] Mather G. C., F. Marques M. B., Frade J. R. J. Europ. Ceram. Soc., 1999, 887: 19.

[8] Hayakawa I., Iwamoto Y., Kikuta K., et al. Sensors and Actuators B, 2000, 55: 62.

[9] Katsuki A., Fukui K. Sensors and Actuators B, 1998, 30: 52.

[10] Fleischer M., Kornely S., Weh T., et al. Sensors and Actuators B, 2000, 205: 69.

[11] Weh T., Fleischer M., Meixner H. Sensors and Actuators B, 2000, 146: 68.

[12] Sharma R. K., Chan P. C. H., Tang Z., et al. Sensors and Actuators B, 2001, 160: 72.

[13] Walton R. M., D. Dwyer J., Schwank J. W., et al. Appl. Surf. Sci., 1998, 187: 125.

[14] Walton R. M., D. Dwyer J., Schwank J. W., et al. Appl. Surf. Sci. 1998, 125: 199.

[15] Wang Y., Ren J-W, Deng K., et al. Mater., 2000, 12: 1622.

[16] Shklover V., Nazeeruddin M.-K., Zakeeruddin S. M., et al. Mater., 1997, 9: 430.

[17] Liu W., Wang Y., Gui L., et al. Langmuir, 1999, 15: 2130.

[18] A. Turković, D. Šokčević. Appl. Surf. Sci., 1993, 68: 477.

[19] M. Z. Atashbar, H. T. Sun, B. Gong, et al. Thin Solid Film, 1998, 326: 238.

（原载于 *Chemistry of Malerials*, 2002, 14: 3953-3957）

# Mesoporous TiO$_2$ Thin Films Exhibiting Enhanced Thermal Stability and Controllable Pore Size: Preparation and Photocatalysed Destruction of Cationic Dyes

Wang Jinshu  Li Hui  Li Hongyi  Zou Chen  Wang Hong  Li Dasheng

(School of Materials Science and Engineering,
Beijing University of Technology, Beijing 100124)

**Abstract:** Ordered mesostructured TiO$_2$ thin films were constructed through the sol-gel combining with evaporation-induced self-assembly (EISA) method. It was found that the calcinations temperature, the kinds of block copolymer could vary the TiO$_2$ mesoporous structure. Based on tension stress calculated by the surface energy of crystallites and the compression calculated by interface energy between the crystallites, the thermodynamic study for the sample had been carried out and the critical crystallite size expression of the mesoporous film was presented for the prediction of the thermal stability of the mesoporous structure at high temperature. It was also found that varying the mass ratio of templating agent to inorganic precursor could adjust the pore size of mesoporous TiO$_2$. The pore size regulating mechanism had been discussed. The sample calcined at 450-500℃ which had higher specific surface area and larger pore size exhibited higher photocatalysed destruction capability of methylene blue.

**Keywords:** TiO$_2$; mesoporous structure; film; photocatalysed destruction

## 1 Introduction

There is intense interest in the synthesis of porous solids possessing ordered pore structures with large specific surface area owing to their wide applications including catalysis, chemical and biological separation, gas sensing, molecular electronic devices, and drug delivery[1,2]. Up to now, many efforts have been focused on the rational synthesis of metal oxide porous structures[3-5]. TiO$_2$ aroused great attention because of its outstanding chemical and physical properties[6-8]. The assembly of TiO$_2$ porous networks with well-controlled solid and pore morphologies is an active research area. Brinker et al. proposed the EISA method for the construction of well-defined TiO$_2$ porous materials[9,10]. However, the calcinations at the temperature over 450℃ would destroy the as-prepared mesostructure due to the crystallization of anatase frameworks[11,12]. Currently, many efforts including chemical doping (e.g. La, Pd, SiO$_2$, P dopants)[13-15], templating growth[16,17] and chemical modification of precursor[18-20], special treatment[21-23] have been carried out to improve the thermal stability of mesoporous TiO$_2$ framework. For example, Clement Sanchez's group reported a delayed rapid crystallization route to fabricate a TiO$_2$ mesostructure with highly thermal stability at a short period at 700℃ and extended period at 500℃[21,22]. Zhao et al. found that carbonizing the copolymer surfactant template can be used to support $TiO_2$ framework during high temperature crystallization[23]. Despite these great successes, most of these mentioned post-processing strategies are involved in some complicated procedures or introducing impurities, limiting the application of the obtained metal oxides. Recently, we have found that thick wall was favorable for the improvement of the thermal stability, which can be formed by inhibiting the hydrolysis and condensation of the inorganic precursor and controlling the aging process of the film[24].

Suitable Ti source for the preparation of highly ordered mesoporous structured $TiO_2$ films is quite limited. Compared with inorganic sources such as $TiF_4$, $TiCl_4$, and $Ti(SO_4)_2$, titanium alkoxides possess important advantages in tuning the homogeneity of initial solution and in controlling final size of nanoscaled $TiO_2$ crystallites[25]. Typically, titanium butoxide $Ti(OBu^n)_4$ has been proved to be a good Ti source for the construction of porous $TiO_2$ structures[26,27]. It has a higher linearity and larger size, and the in-situ released short-chain n-alkyl alcohol (n-butanol, BuOH) is very versatile to act as a co-surfactant or a swelling agent by solublizing hydrophobic additives inside the core of the micelles during the crystallization of porous $TiO_2$[28,29].

In this work, we studied the effect of the kinds of templating agent, concentration of block copolymer, calcinations temperature on the pore size, crystallite size of mesoporous $TiO_2$ prepared by using $Ti(OBu^n)_4$, and established a collapse criterion for the obtained $TiO_2$ mesoporous structure. It was found that the calculation results for the stability of mesostructure agreed well with the experimental results. Compared with the criterion reported in our previous work[30], this criterion is more easy to be carried out. Furthermore, the pore size regulation mechanism was investigated, and the photocatalytic activity of the as-prepared mesoporous $TiO_2$ materials was also studied.

## 2 Experimental

Pluronic F127 ($EO_{106}PO_{70}EO_{106}$), F108 ($EO_{133}PO_{50}EO_{133}$), P123 ($EO_{20}PO_{70}EO_{20}$), Brij58 ($EO_{20}CH_{16}H$) were used as the block copolymers and $Ti(OBu^n)_4$ was used as inorganic precursor. Mixture of F127, F108, P123, Brij58/Ethanol/HCl was added dropwise to $Ti(OBu^n)_4$/Ethanol/AcAc solution. The mole ratio of each component was F127, F108, P123, Brij58/$Ti(OBu^n)_4$/HCl/EtOH/AcAc = 0.0008:0.0726:0.5036:19.928:0.1476. In order to study the effect of the amount of the block copolymers on the pore size of $TiO_2$ mesostructure, the mass ratio of F127 to TBT, EtOH and AcAc was designed as $X$g:5g:100g:2.5g, where $X$ = 0.5g, 1.0g, 1.5g, 2.0g, 2.5g, which were assigned as S-1, S-2, S-3, S-4 and S-5, respectively. The films preparation method was similar to the $TiO_2$ thin films which were described in our previous paper[31]. The films were heated to different temperatures (350, 450, 500, 600, and 700℃) and then held for 2.5h, respectively.

## 3 Results and discussion

### 3.1 Effect of the kinds of block copolymer on the porous structure of $TiO_2$ films

It was reported that careful adjustment of the precursor and polymer molecular weight (MW) could vary the $TiO_2$ structure[31]. Fig. 1 shows the SEM micrographs together with the $N_2$ adsorption-desorption isotherms and the evolving pore size distribution plots of $TiO_2$ films prepared with different block copolymers followed by calcination at 500℃. As shown in Fig. 1 (a-1), the film prepared with F127 had uniform mesoporous structure. Fig. 1 (a-2) indicated that the sample had even cage-like pores since the hysteresis loops were not similar to Type H1 and Type H2 but in the intermediate between these types, which appeared that adsorption isotherms had broad hysteresis loops and no obvious difference in the steepness of the branches of adsorption and desorption[32]. The corresponding pore size distribution results indicated that the film had narrow pore size distribution with large pore size of 7.5nm. On the other hand, as shown in Fig. 1 (b), the $TiO_2$ film prepared with F108 also exhibited orderly mesoporous structure (Fig. 1 (b-1)) and the samples displayed Type IV adsorption isotherms (Fig. 1 (b-2)), which are representative of the bottle-like structure. The sample had smaller pore size compared with that prepared with F127, i.e., the pore size was only 4.3nm. However, the samples prepared with P123 had inhomogenous porous structure. As shown in Fig. 1 (c), the pores in the film had irregular shape and the $N_2$ adsorption-desorption isotherms also indicated the irregularity. As displayed in the insert of Fig. 1 (c-2), the film had two kinds of pores with the average pore size of 5.3nm and 12.5nm. As for the film prepared with Brji58, although some mesopores with the average pore size of 5.3nm still existed,

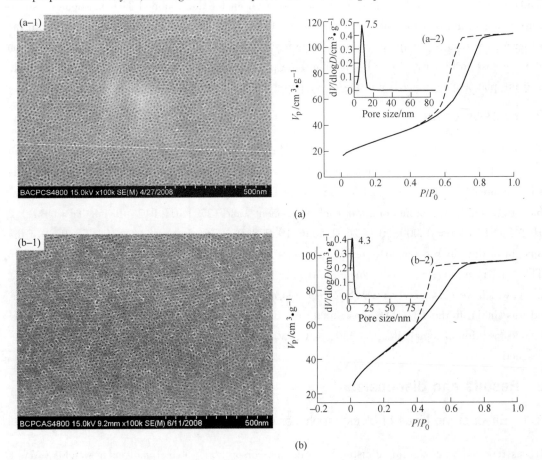

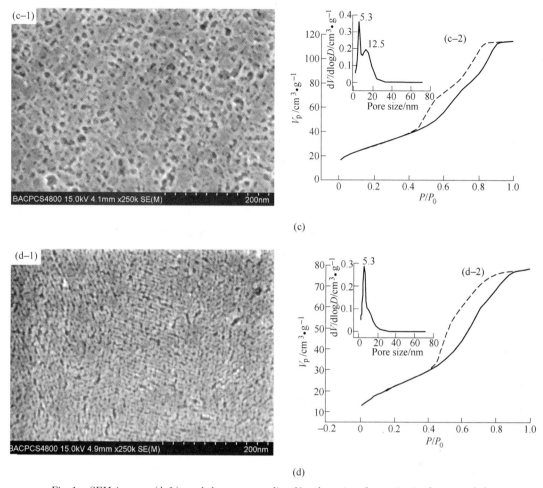

Fig. 1　SEM images (left) and the corresponding $N_2$ adsorption-desorption isotherms and the pore size distribution curves (right) of the $TiO_2$ thin fims synthesized with (a) F127, (b) F108, (c) P123 and (d) Brij58 followed by 500℃ calcinations

as shown in Fig. 1 (d-1) and (d-2), it was evident that the mesoporous structure deformed and collapsed. From Fig. 1, it could be seen that among these four kinds of block copolymer, F127 took an important role in the ordered mesostructure and bigger pores, which could be attributed to its long hydrophilic PEO and hydrophobic PPO segments together with high molecular weight. Therefore, in the following studies, we adopted F127 as the block copolymer.

## 3.2　Effect of calcinations temperature on the structure of $TiO_2$ films

Table 1 shows that the mesoporous structure could be kept up to the temperature of 600℃ and the crystallite grew with the increase of calcinations temperature. The crystallite growth of anatase had been widely studied[33,34]. Praserthdam et al. studied the relationship between the $TiO_2$ crystallite size and the temperature, which was described as[35]:

$$\ln \frac{d}{d_0} = k + n\ln \frac{T}{\sqrt{d_0}} \qquad (1)$$

Where, $d_0$ and $d$ are the initial TiO$_2$ crystallite diameter and that at the temperature of $T$, respectively, $T$ is the calcinations temperature, $n$ is the slope and $k$ is the constant. Taking the crystallite diameter at 350℃ as the original diameter and applying equation (1), the relationship between the crystallite diameter and the temperature was obtained, which is illustrated in Fig. 2. Fig. 2 demonstrated that the average crystallite size and temperature exhibited linear relation as described in equation (1) in the temperature range of 450-600℃. However, the average crystallite size of the sample calcined at 700℃ deviated from the linear segment. Kirsch studied the crystallization of mesoporous TiO$_2$ film by in-situ XRD method and found that the nucleation and growth were controlled by the surface crystallization, namely, the crystallite grew in the inside of the wall in the initial calcinations period and the growth rate decreased with the further calcinations[36].

Table 1  Pore size, crystallite size and specific surface area of the TiO$_2$ mesoporous films calcined at different temperatures

| $T/℃$ | $D_p$/nm | $D_e$/nm | $S_{BET}/m^2 \cdot g^{-1}$ | $D_{ec}$/nm |
|---|---|---|---|---|
| 350 | 4.8 | 7.6 | 145 | |
| 450 | 7.8 | 9.0 | 119 | |
| 500 | 7.5 | 10.3 | 101 | 14.4 |
| 600 | 7.4 | 12.7 | 70 | |
| 700 | — | 29.6 | 19 | |

$D_p$, $D_e$ and $D_{ec}$ are the pore size, crystallite size and critical crystallite size of TiO$_2$ mesoporous film, respectively. The sample calcined at 700℃ lost its mesoporous structure, so no pore size was provided in the Table.

The mesostructural evolution of the synthesized TiO$_2$ thin films with temperature is shown in Fig. 3. As shown in Fig. 3, ordered mesoporous structure was found for the TiO$_2$ films except for the sample prepared by calcinations at 700℃. When the sample was calcined at 350℃, as shown in Fig. 3 (a), many organic compounds still remained on the surface of the films, resulting in parts of pores had been blocked. When the calcinations temperature was increased to 500℃, as shown in Fig. 3 (b), uniform and ordered mesoporous structure was clearly shown in the sample without any substance absorbed on the film surface. Increasing the temperature to 600℃, uniformly ordered mesoporous structure was still maintained without the collapse of the mesoporous structure (Fig. 3 (c)). In order to investigate the mesoporous structure clearly, TEM observation was carried out for the films. As shown in Fig. 3 (e), the sample calcined at 600℃ had a well-organized mesostructure with the wall thickness

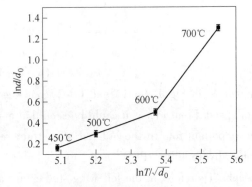

Fig. 2  The relation between $\ln d/d_0$ and $\ln T/\sqrt{d_0}$ of the synthesized mesoporous TiO$_2$ samples calcined at different temperatures

around 10-13nm. Further increasing the temperature to 700℃, the sample lost its initial mesostructure and wormlike particles existed in the film because of the grains' overgrowth, as shown in Fig. 3 (d), (f). From Fig. 3 (e), we could deduce that the thick wall of the framework played an important role in the highly thermal stability of the sample which exhibited as keeping its mesoporous structure at 600℃ at least for 2.5h.

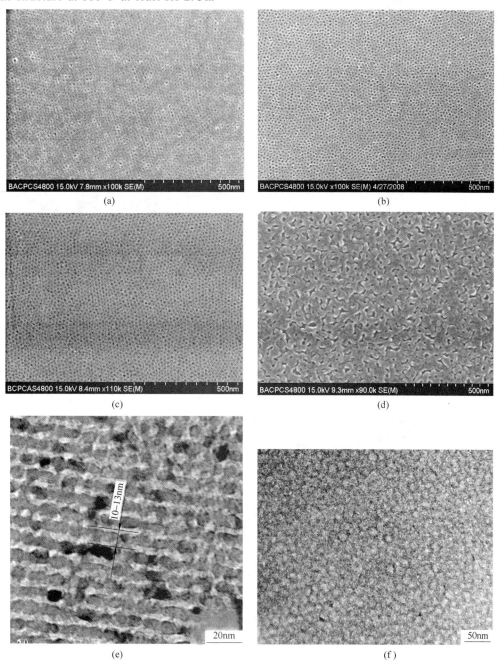

Fig. 3  The SEM images (a-d) and TEM images (e, f) of the synthesized mesoporous
TiO$_2$ thin films calcined at 350℃ (a), 500℃ (b), 600℃ (c,e), 700℃ (d,f)

Further analysis of the microstructure of the synthesized thin films has been carried out through the cross-sectional TEM images of the film calcined at 500℃, as shown in Fig. 4. The testing sample was prepared by the copper-coated method. As indicated by TEM image (Fig. 4 (a)), the thickness of the calcined mesoporous $TiO_2$ thin film was about 220nm. No obvious boundaries between the layers were found. A more detailed structural information could be obtained by HRTEM images (Fig. 4 (b) and (c)). The honeycomb structure of the framework could be clearly observed throughout the thickness of the film (Fig. 4 (b)). Unlike the well-aligned mesostructure of the top-view, the mesopores (white area) distributed randomly in the framework. The entire wall structure displayed anatase lattice fringe, indicating the highly crystalline nature of the film. The anatase crystallites which embedded in the framework (Fig. 4 (c)) oriented randomly, which allowed the crystallites to nucleate and grow within the limited space of the pore wall, avoiding the destruction of the mesostructure. The fast Fourier transform (FFT) of the image (inset Fig. 4 (c)) indicated the presence of a 4-fold symmetry[37].

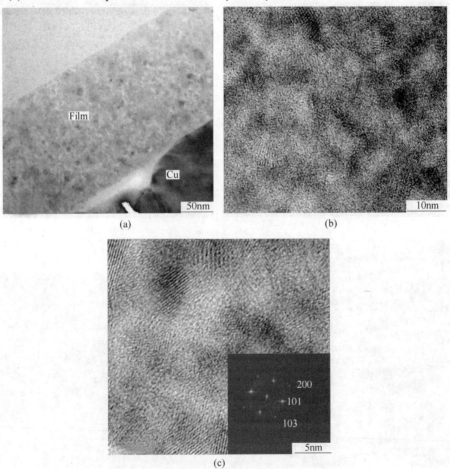

Fig. 4  Cross-sectional TEM images of the as-made mesostructured $TiO_2$ film calcined at 500℃ with different magnifications. (a) TEM; (b) and (c) HRTEM images. Inset (c) is the corresponding two-dimensional fast Fourier transformation (FFT) patterns of the full image

The Barrett-Joyner-Halenda (BJH) pore size analysis (Fig. 5) performed on the adsorption branches indicated that the samples calcined below 600℃ had a narrow pore size distribution. The structural parameters of the films at different calcination temperatures were listed in Table 1. The sample calcined at 350℃ possessed a small mean pore size of 4.8nm, whereas the sample calcined at 450℃ was 7.8nm. Furthermore, although the calcination temperatures increased from 450℃ to 500℃ and then 600℃, no much difference was found for the pore diameter, indicating the highly thermal stability of the films. The BET specific surface area was lower than those reported for mesoporous $TiO_2$ materials synthesized by templating F127[38,39]. This might be attributed to the thick inorganic walls of our samples. From the adsorption isotherms measured at 700℃, it could be seen that the sample had a broad pore size distribution due to the collapse of most pore walls, which was consistent with the SEM and TEM observations in Fig. 3 (d, f).

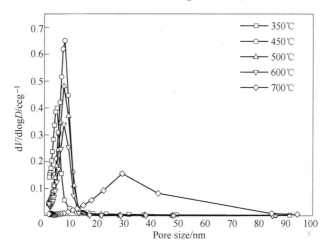

Fig. 5 The pore size distribution curves for the synthesized samples calcined at 350, 450, 500, 600, and 700℃

Raman method had been carried out for the analysis of crystallites since it was more suitable for the nanosized particle. The Raman spectra of $TiO_2$ films prepared at different temperatures was shown in Fig. 6. The wave shift at 146, 199, 398, 517 and 641 $cm^{-1}$ were caused by the vibration of anatase[40,41], in which wave shift at 146, 199 and 641 $cm^{-1}$ could be attributed to $E_g$ vibration model and 398 $cm^{-1}$ and 517 $cm^{-1}$ were designed as $B_{1g}$ model.

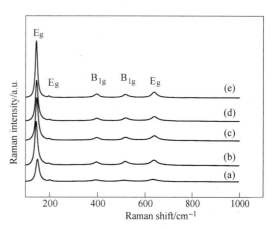

Fig. 6 Raman patterns of prepared mesoporous samples calcined at (a) 350℃, (b) 450℃, (c) 500℃, (d) 600℃, (e) 700℃

As shown in Fig. 6, five obvious Raman peaks existed in the five samples. The Raman vibration spectra displayed that anatase nanparticles had been obtained in the sample calcined at

350℃. The Raman vibration frequency of the nano $TiO_2$ was different from the ordinary $TiO_2$, i. e., the lowest frequencies $E_g$ of mesoporous $TiO_2$ calcined at 350℃ and 700℃ were 145 $cm^{-1}$ and 177 $cm^{-1}$, respectively, indicating that the Raman peak moved to the higher frequency side, which might be caused by the following two reasons: one was caused by the nanosized particles due to the blue shift of Raman frequency, and the another was caused by the oxygen vacancies[42]. As for our sample which was calcined in oxygen, the oxygen vacancies in the samples might be neglected, therefore, the blue shift of Raman frequency might be caused by the smaller crystallite size. The crystallite size increased with the calcinations temperature. Raman results also indicated that $TiO_2$ exhibited as anatase, even under high temperature of 700℃. The retardation of anatase-rutile transformation was probably due to the presence of acetyl acetone, which can make anatase stable to 800℃[43].

## 3.3 Thermal stability of mesoporus $TiO_2$ film

The thermal stability of the film is correlated with the temperature. The prediction of the mesoporous structure at different temperatures is important. A four-coordinate channel mode was proposed and the equation based on thermodynamic calculation has been established[30].

$$J = \frac{4D_{eff}}{kT(D_p + 0.87D_e)}\left(\gamma_{interf}\frac{-D_p + \sqrt{0.90D_e^2 - D_p^2}}{\pi D_p} - \gamma_{surf}\right) \quad (2)$$

Where, $D_p$ is the size of mesopores, $D_e$ is the size of crystallites, $D_{eff}$ is the effective coefficient, $\gamma_{interf}$ and $\gamma_{surf}$ are the interface energy and surface energy, respectively, $T$ is the temperature.

In the previous work, we presented that the mesoporous structure stability could be predicted by the value of $J$. When $J > 0$, the particles would diffuse toward the pores, causing the collapse of the mesoporous structure. Otherwise, the mesoporous structure would be stable. However, this method is relatively complicated since $J$ value must be calculated based on the measured data. Therefore, in this work, we deduced another method for the prediction of the thermal stability of the mesoporous structure.

From the discussion above, it could be found that the critical condition for the thermally stable mesoporous structure could be obtained at $J = 0$. Therefore, the critical crystallite $D_{ec}$ could be calculated and expressed as:

$$D_{ec} = \sqrt{\frac{\left(\frac{\gamma_{surf}}{\gamma_{interf}}D_p + \frac{1}{\pi}\right)^2 + D_p^2}{0.90}} \quad (3)$$

The surface energy and interface energy of $TiO_2$ could be calculated based on the equations (4) and (5), respectively[44].

$$\gamma_{surf} = 1.91 - 1.48 \times 10^{-4}(T - 298) \quad (4)$$

$$\gamma_{interf} = 1.32 - 1.48 \times 10^{-4}(T - 298) \quad (5)$$

The $TiO_2$ critical crystallite size ($D_{ec}$) of the mesoporous film could be calculated based on equations (3)–(5) and was shown in equation (6).

$$D_{ec} = \sqrt{\frac{\left[\frac{1.91 - 1.48 \times 10^{-4}(T - 298)}{1.32 - 1.48 \times 10^{-4}(T - 298)}D_p + \frac{1}{\pi}\right]^2 + D_p^2}{0.90}} \approx \sqrt{\frac{(1.45D_p + 0.32)^2 + D_p^2}{0.90}}$$

$$(6)$$

When $D_e > D_{ec}$, the crystallites exceeded the critical crystallite size and the substance diffused into the mesopores, leading to the collapse of the mesoporus structure. When $D_e < D_{ec}$, the TiO$_2$ film kept its mesoporous structure.

As shown in Table 1, the mesoporous film had the pore size in the range of 7.4-7.8nm in the temperature range of 450-600℃. Supposing that the average pore diameter of the film is 7.6nm (the mean value between 7.4nm and 7.8nm), based on the equation (6), the critical crystallite size $D_{ec}$ of the TiO$_2$ mesoporous film could be calculated as 14.4nm. The crystallite size $D_e$ of the films calcined at the temperature lower than 600℃ was smaller than $D_{ec}$, indicating that the films kept their mesoporous structure. $D_e$ of the film calcined at 700℃ was 29.6nm, larger than the critical value, therefore, the mesoporous structure collapsed. The above results indicate that the experimental results were consistent with the calculation results for the prediction of the thermal stability.

## 3.4 Regulation of pore size

The XRD patterns of mesoporous TiO$_2$ samples synthesized using F127 as the templating agent and Ti(OBu$^n$)$_4$ as the precursor with different mass ratios followed by calcinations at 500℃ are shown in Fig. 7. The small angel XRD results indicated that the thin films had highly organized mesoporous framework. Table 2 shows the pore size and crystallite size of the mesoporous film prepared with different mass ratio. As presented in Table 2, the crystallite size decreased and the pore size increased with the increase of mass ratio of F127/Ti(OBu$^n$)$_4$, i.e., when the F127/Ti(OBu$^n$)$_4$ mass ratio increased from 0.1 to 0.5, the crystallite size decreased from 14.3nm to 8.9nm and pore size increased from 5.4nm to 9.1nm, respectively, indicating that the wall thickness decreased.

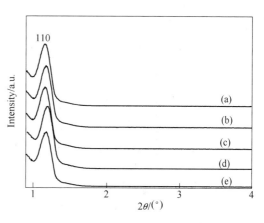

Fig. 7 The Small angel XRD patterns of the synthesized thin films
(a) S-1; (b) S-2;
(c) S-3; (d) S-4; (e) S-5

Table 2 The textural parameters of the synthesized mesoporous TiO$_2$ materials prepared with different mass ratios of F127/Ti(OBu$^n$)$_4$

| Sample | F127/Ti(OBu$^n$)$_4$ | $D_p^{①}$/nm | $D^{②}$/nm |
| --- | --- | --- | --- |
| S-1 | 0.1 | 5.4 | 14.3 |
| S-2 | 0.2 | 6.2 | 13.6 |
| S-3 | 0.3 | 7.1 | 11.2 |
| S-4 | 0.4 | 7.4 | 10.2 |
| S-5 | 0.5 | 9.1 | 8.9 |

① BJH average pore size.
② Average crystallite size.

The above results show that the pore size of the samples could be regulated by the mass ratio of

F127/Ti$(OBu^n)_4$. The pore size regulation mechanism could be explained as follows. Taking triblock-copolymer Pluronic F127 as the templating agent was the premise for the formation of the mesoporous TiO$_2$ with different pore size. F127 is a unique surfactant owing to its high molecular weight with long hydrophilic PEO and hydrophobic PPO segments, which offers the possibilities of the formation of hydrophobic core (PPO) of the micelles and hydrophilic corona with a certain thickness. According to the "three-phase model" which was presented for silica mesoporous structure[42], Ti-oxo species might react with micelles to form a three phase structure. Fig. 8 shows the schematic diagram of the formation of the synthesized mesoporous TiO$_2$ materials with different pore size. As shown in Fig. 8, hydrophobic phase Ⅰ (PPO) was located in the center, and the PEO chains Ⅱ existed in two parts, namely, embedding in the Ti-oxo matrix and situating between the hydrophobic core (phase Ⅰ) and the Ti-oxo (this part of PEO is called as phase Ⅱ). In other words, a part of the PEO chains might form a separate "phase" between the hydrophobic block and the Ti-oxo species. This part of PEO also contributed to the mesoporosity. In other words, since the templating agent could form micelles with a fixed thickness of PEO under a certain condition, the change of the organic-inorganic hybrid material (Phase Ⅲ) would change the thickness of the rest of PEO (Ⅱ). Therefore, the thickness of the organic-inorganic hybrid material could be changed by adjusting the mass ratio of templating agent to inorganic precursor, resulting in the change of pore size. It is known that the pore size is, to a large extent, dependent on the effective volume of the hydrophobic core (PPO) of the micelles. As shown in Table 2, when the concentration of templating agent F127 was low, more number of oligomeric Ti-oxo species could assembly with PEO chains to form inorganic-organic hybrid (phase Ⅲ), therefore, only a small number of PEO chains (phase Ⅱ) contributed to the microposity with the result of small pores in the framework. Since the concentration of precursor was fixed, the number of Ti-oxo species which could assembly with ethylene oxide decreased with the increase of the concentration of F127. Therefore, more number of PEO chains would contribute to the microposity, resulting in the formation of large pores and the decrease of the wall thickness of mesoporous TiO$_2$. Since the pore size of the mesoporous network was primarily dependent on the hydrophobic core diameter, the $n$-Butanol released by Ti$(OBu^n)_4$ inclined to penetrate the interface between the hydrophilic and hydrophobic domains of micelles, and located at the hydrophobic core due to the nonpolar nature, therefore, $n$-Butanol in situ also played a pronounced role in the construction of large pore mesostructured TiO$_2$

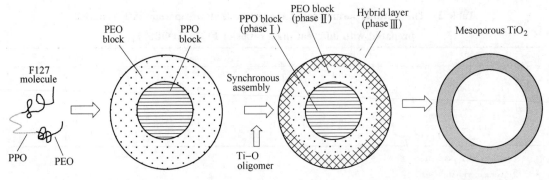

Fig. 8　Schematic diagram of the formation of the synthesized mesoporous TiO$_2$ materials with different pore size

framework.

### 3.5 Photocatalytic activity

The photocatalytic activity was evaluated by photodegradation of methylene blue, and the detailed measurement method has been described in our previous paper[45]. Fig. 9 shows the illumination time dependence of photodegradation property of methylene blue by using different mesoporous $TiO_2$ under irradiating high pressure mercury arc. It was clear that the sample calcined at 350℃ exhibited low photocatalytic activity because of its amorphous phase and remained organic compounds. Apart from the low crystallinity, the defects in the sample might become the recombination center of photogenerated holes and electrons, resulting in worse photocatalytic activity of the semiconductor. Only 29% MB was degraded after light irradiation for 180min. Increasing the calcinations temperature, the photocatalytic activity had greatly improved. Samples calcined at 450℃ and 500℃ photodegraded more than 70% MB (75% for the sample calcined at 500℃, 71% for that calcined at 450℃) in the same duration, which were superior to that of P25 (69% MB was photodegraded). However, the sample calcined at 600℃ showed relatively poor photocatalytic activity. 55% MB was photodegradated after being irradiated for 120min and kept almost stable with the irradiation time up to 180min. As shown in Table 1, the samples calcined at 450℃ and 500℃ had smaller crystallite size, higher specific surface area, larger pore size and porosity. It is well known that the larger specific surface area results in higher adsorption ability of MB. In addition, photo-generated carriers inside the nanoparticles are easier to diffuse to their surface, decreasing the recombination ratio of photo-generated electrons and holes. Furthermore, the high porosity can increase adsorption capability of the organic pollutant and the amount of hydroxyl radicals. Consequently, the mesoporous $TiO_2$ calcined at 500℃ had higher photocatalysed destruction capability of MB due to its larger specific surface area and mesopore volume. Increasing temperature to 600℃, the photocatalytic activity decreased due to the growth of the crystallites and decrease of the specific surface area. In addition, it was also notable that the mesoporous $TiO_2$ calcined at 450℃ and 500℃ exhibited higher photocatalytic capability than P-25 in ultraviolet light range.

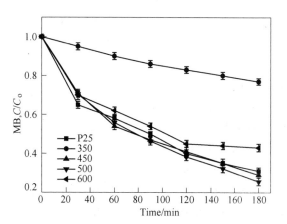

Fig. 9 Photocatalysed destruction of methylene blue of different samples

## 4 Conclusions

In summary, $TiO_2$ mesoporous films had been obtained by using $Ti(OBu^n)_4$-EtOH-HCl system with different templating agents through the sol-gel and EISA method. It was found that among F127, F108, P123 and Brij58, using F127 as the templating agent was easy for preparing mesoporous $TiO_2$ which exhibited an excellent thermal stability. The $TiO_2$ film could keep the mesoporous structure up to 600℃ but the structure would collapse with further increasing the calcinations temperature. The thermal stability of the $TiO_2$ mesoporous film could be predicated by thermodynamic method, and mesoporous structure collapse criterion was obtained as:

$$D_{ec} = \sqrt{\frac{(1.45D_p + 0.32)^2 + D_p^2}{0.90}}$$

Where, $D_p$ and $D_{ec}$ were the sizes of mesopores and critical crystallites, respectively.

When crystallite size $D_e$ at a certain temperature was smaller than $D_{ec}$, mesoporous structure remained stable; otherwise, the mesoporous structure would collapse. This criterion could be applied to determine the thermal stability of the available $TiO_2$ films with provided mesopore size and crystallite size at a certain temperature. Furthermore, the equation (3) could be a universal one for the prediction of the thermal stability of mesoporous films composed of other substance. The pore size could be adjusted by varying the mass ratio of templating agent to inorganic precursor. The mesoporous $TiO_2$ calcined at proper temperature which had higher specific surface area and smaller crystallite size had higher capability of photocatalysed destruction of methylene blue.

**Acknowledgments:** This work was supported by National Outstanding Young Investigator Grant of China (NO. 51225402), National Natural Science Foundation of China (NSFC: 51071005, 51002004), Beijing Municipal Commission of Education Foundation (KZ201010005001, KM201110005003), Beijing Innovation Talent Project.

## References

[1] G. J. D. A. A. Soler-Illia, C. Sanchez, B. Lebeau, et al. Chem. Rev., 2002, 102: 4093-4138.

[2] E. C. Goethals, A. Elbaz, A. L. Lopata, et al. Langmuir., 2013, 29 (2): 658-666.

[3] X. Lin, F. Rong, X. Ji, et al. Micropor. Mesopor. Mat., 2011, 142: 276-281.

[4] J. Pan, X. S. Zhao, Wan InLee. Chem. Eng. J., 2010, 2-3: 363-380.

[5] E. L. Crepaldi, G. D. A. A. Soler-Illia, A. Bouchara, et al. Angew. Chem. Int. Ed., 2003, 42: 347-351.

[6] T. Leshuk, R. Parviz, P. Everett, et al. ACS Appl. Mater. Interfaces, 2013, 5 (6): 1892-1895.

[7] A. Natoli, A. Cabeza, A. G. De la Torre, et al. J. Am. Ceram. Soc., 2012, 95: 502-508.

[8] X. Pan, Y. Zhao, S. Liu, et al. ACS Appl. Mater. Interfaces, 2012, 4 (8): 3944-3950.

[9] C. J. Brinker, Y. Lu, A. Sellinger, et al. Adv. Mater., 1999, 11: 579-585.

[10] Y. Lu, R. Ganguli, C. A. Drewien, et al. Nature, 1997, 389: 364-368.

[11] C. X. Lei, H. Zhou, C. Wang, et al. Electrochim. Acta, 2013, 87 (1): 245-249.

[12] T. Hongo, A. Yamazaki. Microporous Mesoporous Mater., 2011, 142: 316-321.

[13] S. Yuan, Q. R. Sheng, J. L. Zhang, et al. Micropor. Mesopor. Mat., 2008, 110: 501-507.

[14] C. X. He, B. Z. Tian, J. L. Zhang. J. Colloid. Inter. Sci, 2010, 344: 382-389.

[15] H. F. Yu. J. Phys. Chem. Solids, 2007, 68: 600-607.

[16] M. B. Zakaria, N. Suzuki, N. L. Torad, et al. Eur. J. Inorg. Chem., 2013: 2330-2335.

[17] E. Ortel, A. Fischer, L. Chuenchom, et al. Small, 2012, 8: 298-309.

[18] N. N. Khimich, B. I. Venzel, I. A. Drozdova, et al. Russ. J. Appl. Chem., 2002, 75: 1108-1112.

[19] M. Lackhoff, X. Prieto, N. Nestle, et al. Appl. Catal. B: Environ. 2003, 43: 205-216.
[20] V. Etacheri, M. K. Seery, S. J. Hinder, et al. Adv. Funct. Mater. , 2011, 21: 3744-3752.
[21] D. Grosso, G. J. A. A. Soler-Illia, E. o. L. Crepaldi, et al. Chem. Mater. , 2003, 15: 4562-4570.
[22] E. M. Ferrero, D. Grosso, C. Boissiere, et al. J. Mater. Chem. , 2006, 16: 3762-3767.
[23] R. Y. Zhang, B. Tu, D. Y. Zhao. Chem. Eur. J. , 2010, 16: 9977-9981.
[24] H. Li, J. S. Wang, H. Y. Li, et al. Mater. Lett. , 2009, 63: 1583-1585.
[25] J. H. Pan, X. Zhang, A. J. Du, et al. J. Am. Chem. Soc. , 2008, 130: 11256-11257.
[26] G. Calleja, D. P. Serrano, R. Sanz, et al. Ind. Eng. Chem. Res. , 2004, 43: 2485-2492.
[27] K. S. Liu, H. G. Fu, K. Y. Shi, et al. Nanotechnology, 2006, 17 (15): 3641-3648.
[28] P. Feng, X. Bu, D. J. Pine. Langmuir, 2000, 16: 5304-5310.
[29] F. Kleitz, J. Blanchard, B. Zibrowius, et al. Langmuir, 2002, 18: 4963-4971.
[30] J. S. Wang, H. Li, H. Y. Li, et al. J. Phys. Chem. C, 2012, 116 (17): 9517-9525.
[31] D. J. Kim, S. J. Kim, D. K. Roh, et al. Phys. Chem. Chem. Phys. , 2013, 15: 7345-7353.
[32] M. Kruk, M. Jaroniec. Chem. Mater. , 2001, 13 (10): 3169-3183.
[33] D. Fattakhova-Rohlfing, M. Wark, T. Brezesinski, et al. Adv. Funct. Mater. , 2007, 17: 123-132.
[34] K. M. Coakley, Y. X. Liu, M. D. McGehee, et al. Adv. Funct. Mater. , 2003, 13 (4): 301-306.
[35] P. Praserthdam, P. L. Silveston, O. Mekasuwandumrong, et al. Crystal Growth & Design, 2004, 4 (1): 39-43.
[36] B. L. Kirsch, E. K. Richman, A. E. Riley, et al. J. Phys. Chem. B, 2004, 108 (34): 12698-12706.
[37] S. Yoo, D. M. Ford, D. F. Shantz. Langmuir, 2006, 22: 1839-1845.
[38] E. L. Crepaldi, G J d A A Soler-Illia, D. Grosso, et al. New J. Chem. , 2003, 27: 9-13.
[39] E. Lancelle-Beltran, P. Prené, C. Boscher, et al. Chem. Mater. 2006, 18: 6152-6156.
[40] W. F. Zhang, Y. L. He, M. S. Zhang, et al. J. Phys. D: Appl. Phys, 2000, 33: 912-916.
[41] J. C. Parker, R. W. Siegel. Appl. Phy. Lett. , 1990, 57 (9): 943-945.
[42] B. Smarsly, S. Polarz, M. Antonietti. J. Phys. Chem. B, 2001, 105 (43): 10473-10483.
[43] Y. Djaoued. J. Sol-Gel Sci. Tech, 2002, 24: 255-264.
[44] L. Yin, Q. F. Zhou. Functional Mater. , 2000, 2: 186-189.
[45] J. S. Wang, H. Li, H. Y. Li, et al. Solid state Sci. , 2010, 4: 490-497.

(原载于 *ACS Applied Materials & Interfaces*, 2014, 6 (3): 1623-1631)

# Evaluation of $La_{0.3}Sr_{0.7}Ti_{1-x}Co_xO_3$ as Potential Cathode Material for Solid Oxide Fuel Cells

Du Zhihong[1]  Zhao Hailei[1,3]  Shen Yongna[1]  Wang Lu[1]
Fang Mengya[1]  Konrad Świerczek[2]  Zheng Kun[2]

(1. School of Materials Science and Engineering, University of Science and Technology Beijing, Beijing 100083, China;
2. AGH University of Science and Technology, Faculty of Energy and Fuels, Department of Hydrogen Energy, al. A. Mickiewicza 30, Krakow 30-059, Poland;
3. Beijing Key Lab of New Energy Material and Technology, Beijing 100083, China)

**Abstract:** Perovskites $La_{0.3}Sr_{0.7}Ti_{1-x}Co_xO_3$ (LSTCs, $x = 0.3 - 0.6$) are systematically evaluated as potential cathode materials for solid oxide fuel cells. The effects of Co substitution for Ti on structural characteristics, thermal expansion coefficients (TECs), electrical conductivity, and electrochemical performance are investigated. All of the synthesized LSTCs exhibit a cubic structure. With Rietveld refinement on the hightemperature X-ray diffraction data, the TECs of LSTCs are calculated to be (20-26) $\times 10^{-6}$ $K^{-1}$. LSTC shows good thermal cycling stability and is chemically compatible with the LSGM electrolyte below 1250℃. The substitution of Co for Ti increases significantly the electrical conductivity of LSTC. The role of doping on the conduction behavior is discussed based on defect chemistry theory and first principles calculation. The electrochemical performances of LSTC are remarkably improved with Co substitution. The area specific resistance of sample $La_{0.3}Sr_{0.7}Ti_{0.4}Co_{0.6}O_3$ on the $La_{0.8}Sr_{0.2}Ga_{0.8}Mg_{0.2}O_{3-\delta}$ (LSGM) electrolyte in symmetrical cells is 0.0145, 0.0233, 0.0409, 0.0930 $\Omega$ $cm^2$ at 850, 800, 750 and 700℃, respectively, and the maximum power density of the LSGM electrolyte (400μm) -supported single cell with the Ni/GDC anode, LDC buffer layer and LSTC cathode reaches 464.5, 648, and 775mW·$cm^{-2}$ at 850℃ for $x = 0.3, 0.45$, and 0.6, respectively. All these results suggest that LSTC are promising candidate cathode materials for SOFCs.

**Keywords:** $La_{0.3}Sr_{0.7}Ti_{1-x}Co_xO_3$; Cathode; Structural properties; electrical conductivity; electrochemical performance

## 1 Introduction

As electrochemical device that directly convert the chemical energy of fuels into electricity, solid oxide fuel cells (SOFCs) have attracted lots of attentions due to their high efficiency, low emission, noise-free nature and fuel flexibility[1,2]. The SOFC usually works in the temperature range of 900-1000℃. Lowering the operating temperature can reduce the cost and promote the commercialization

---

赵海雷，1989~1993年于北京科技大学师从李文超教授攻读博士学位。目前在北京科技大学工作，教授，博士生导师。获得教育部新世纪人才（2007）、北京市师德先进个人（2012）、宝钢优秀教师奖（2013）等荣誉称号，发表论文220余篇。

of SOFCs[3-5]. Many efforts have been made to develop intermediate-temperature SOFCs (IT-SOFCs)[6]. However, in the conventional case of the YSZ electrolyte and $La_{1-x}Sr_xMnO_3$ cathode, decreasing the operating temperature leads to a significant increase in both ohmic resistance mainly coming from the electrolyte and polarization resistance primarily from the cathode. For the ohmic resistance, this problem has been gradually solved by reducing the thickness of the electrolyte or employing alternative electrolyte materials like $Gd_{0.1}Ce_{0.9}O_{2-\delta}$ (GDC) and $La(Sr)Ga(Mg)O_3$ (LSGM)[7,8]. To address the cathodic polarization, the development of high-performance cathode materials has become critical. Mixed ionic-electronic conductors (MIEC) like doped $LaCoO_{3-\delta}$[9-11], $La_2NiO_4$[12-14] and $Ba_{0.5}Sr_{0.5}Co_{0.8}Fe_{0.2}O_{3-\delta}$ (BSCF)[15] have been extensively studied as potential cathode materials considering their high catalytic activity for oxygen reduction at lower and intermediate temperatures.

The cobalt-based perovskites have attracted increasing attention as alternative cathode materials for IT-SOFCs due to their mixed-conducting characteristics and high electrocatalytic activity towards oxygen reduction[15-17]. However, their practical application in IT-SOFCs is severely limited because of their large thermal expansion coefficient (TEC). The previous study reveals that introducing a stable $Ti^{4+}$ ion to replace part of Co ions of $Ba_{0.6}Sr_{0.4}CoO_{3-\delta}$ can successfully decrease the TEC[18]. Recently, $La_{0.5}Sr_{0.5}Co_{0.5}Ti_{0.5}O_3$ is reported to be a good candidate as a symmetrical electrode in IT-SOFCs. It shows good catalytic activity towards oxygen reduction in the cathode side and hydrogen oxidation in the anode side, and keeps good structural stability in both oxidizing and reducing atmospheres[19].

La-doped $SrTiO_3$ has been reported as a potential anode material for SOFCs due to its high electronic conductivity in reducing atmospheres and good chemical and structural stability upon redox cycling[20-23]. Substituting Ti with Co for $La_{0.3}Sr_{0.7}TiO_3$ (LST) could improve the oxygen ionic conductivity[24]. Taking into account that Co ions usually possess high redox ability, more Co substitution for Ti has the potential to enhance the catalytic activity of LST towards oxygen reduction, and thus can turn LST into a cathode material. The stable oxidation state of Ti ions in air ensures good structural stability and a lower TEC value for LST As symmetrical electrode materials, $La_{0.5}Sr_{0.5}Co_{0.5}Ti_{0.5}O_3$[19], $La_{0.4}Sr_{0.6}Ti_{1-y}Co_yO_{3-\delta}$[25] and $La_{2-x}Sr_xCoTiO_6$[26], have been investigated recently. These stdies focus mainly on the lattice structure evolution of $(LaSr)(TiCo)O_3$ with chemical composition and environmental atmosphere, but less on the electrochemical performance, except for the work of $La_{0.5}Sr_{0.5}Co_{0.5}Ti_{0.5}O_3$.

As B-site elements, the different electronic structures of Ti and Co ions endow perovskite oxides with much different properties, in terms of structural stability, lattice defect, electronic conductivity, and catalytic activity. The Co/Ti ratio will have a strong impact on the electrode performance of $(LaSr)(TiCo)O_3$ materials. In this work, $La_{0.3}Sr_{0.7}Ti_{1-x}Co_xO_3$ (LSTC, $x = 0.3 - 0.6$) materials are prepared and characterized as cathode materials, with the aim to get a deep understanding of the effect of Co content on the crystal structure, electrical conductivity and electrochemical properties. First principles calculation is employed to elucidate the electronic conduction behavior of

LSTC.

## 2 Experimental

### 2.1 Sample preparation

$La_{0.3}Sr_{0.7}Ti_{1-x}Co_xO_3$ ($x = 0.3 - 0.6$, LSTC) powders were synthesized by a citric acid-nitrate combustion process, with $Sr(NO_3)_2$ (99.9%, Sinopharm), $La_2O_3$ (99.9%, Sinopharm), $Co(NH_3)_2 \cdot 6H_2O$ (99.9%, Sinopharm), and $C_{16}H_{36}O_4Ti$ (Tetra-$n$-butyl Titanate, 99.9%, Sinopharm) as starting materials and citric acid monohydrate ($C_6H_8O_7 \cdot H_2O$, Guangdong Xilong) as complexant. Before being used, $La_2O_3$ was annealed at 1000℃ for 4h to remove possibly absorbed water and $CO_2$. First, $C_{16}H_{36}O_4Ti$ was dissolved in ethanol at a molar ratio of 1:30 to form a transparent solution, which was then added to a citric acid solution with a pH value of 5 to produce a stable $Ti^{4+}$ aqueous solution, followed by treatment in a water bath at 80℃ for 1h. The $La_2O_3$ and nitrate salts were dissolved in dilute nitric acid solution to form metal ion solution, the pH value of which was also adjusted to 5 with ammonia (AR, Sinopharm). Then the two resultant solutions were mixed together thoroughly. The amount of citric acid was fixed at 2:1 in molar ratio to the total amount of metal ions.

The obtained final solution was heated in a water bath at 80℃ until a gel was formed. The gel was transferred into an oven and heated at 250℃ to get a fluffy precursor, which was ground and subsequently calcined at 800℃ for 6h with an interval at 400℃ for 2h to obtain LSTC powders. The prepared LSTC powders were uniaxially pressed into pellets (19mm in diameter) and rectangular bars (40mm × 7mm × 3mm) with an appropriate amount of polyvingakohol (PVA, 1 wt%) as a binder, followed by sintering at 1200℃ ($x = 0.3, 0.45$) and 1150℃ ($x = 0.6$) for 10h in air to get dense samples for electrical conductivity measurement.

### 2.2 Cell preparation

The $La_{0.8}Sr_{0.2}Ga_{0.8}Mg_{0.2}O_{3-\delta}$ (LSGM), $Gd_{0.1}Ce_{0.9}O_{2-\delta}$ (GDC) and $La_{0.4}Ce_{0.6}O_{2-\delta}$ (LDC) powders were synthesized by the same citric acid-nitrate method[12]. The dense LSGM electrolyte (~400μm in thickness) was obtained by pressing the calcined powder into disks with a diameter of 19mm and then sintering at 1450℃ for 8h. NiO-GDC (6:4 in weight ratio) was used as the anode and the prepared LSTC as the cathode. LDC was employed as the buffer layer to prevent the reaction between Ni in the anode and LSGM electrolyte. All the electrolyte-supported cells were fabricated by screen-printing technique. The electrode slurries were prepared by mixing homogeneously the electrode powders with α-terpineol solution of 6 wt% ethylene cellulose in a weight ratio of 2:1. For the LSTC cathode inks, a little amount of flour was used as pore former.

For symmetrical cells of LSTC | LSGM | LSTC, LSTC slurries were screen-printed on both sides of the LSGM electrolyte symmetrically (active area 0.5cm$^2$), followed by calcining at 1200℃ for 2h. For a single cell with the configuration of Ni-GDC | LDC | LSGM | LSTC, the LDC slurry was deposited on one side of the LSGM electrolyte (active area 0.78cm$^2$) and fired at 1400℃ for

2h. Subsequently, the Ni-GDC anode slurry was screen-printed on the LDC layer and fired at 1300℃ for 2h. Finally LSTC ($x = 0.3 - 0.6$) was applied on the other side of the LSGM electrolyte and calcined at 1200℃ for 2h. The active area of both anode and cathode was 0.5cm$^2$. Ag paste was used as the current collector, which was painted in a grid structure on both sides of the cells and fired at 650℃ for 0.5h. The final cells were sealed on an alumina tube with a ceramic-based sealant (Cerama-bond 552-VFG, Aremco). Humidified $H_2$ (~3% $H_2O$) was fed as fuel to the anode with a flow rate of 40ml/min$^{-1}$, and pure $O_2$ or air (100ml·min$^{-1}$) as an oxidant to the cathode.

## 2.3 Structural characterization

The phases and crystal structure of LSTC ($x = 0.3 - 0.6$) samples were identified by X-ray diffraction (XRD, Rigaku D/max-AX-ray diffractometer) with CuKα radiation ($\lambda = 1.5406$Å). High Temperature XRD (HT-XRD) measurements were carried out on a PANalytical X'Pert Pro diffractometer (operated at 45kV and 40mA, with CuKα radiation) with an Anton Paar HTK 1200N oven-chamber over a $2\theta$ range of 10°-110° with a step size of 0.013° from room temperature to 900℃ in air. The Rietveld refinements of the XRD patterns were performed with GSAS/EXPGUI software[27,28]. The thermal expansion coefficient of LSTC was calculated with HT-XRD data. Thermogravimetric (TG) measurement was performed on a NETZSCH STA 449F3 between 50 and 835℃ in air with a heating rate of 10℃·min$^{-1}$. To examine the chemical compatibility of the LSTC cathode with the LSGM electrolyte, the LSTC ($x = 0.6$) powder was mixed uniformly with the LSGM powder in 1:1 mass ratio, and then pressed and calcined at 1250℃ for 5h in air. The calcined pellets were crushed and examined by XRD. X-ray photoelectron spectroscopy (XPS) was used to identify the oxidation state of Co ions in the synthesized LSTC ($x = 0.45$ and 0.6) on a RBD upgraded PHI-5000C ESCA system (Perkin Elmer) with Mg Kα radiation ($h\nu = 1253.6$eV). The data analysis was carried out by using XPSPeak4.1 provided by Raymund W. M. Kwok (The Chinese University of Hongkong, China). A scanning electron microscope (SEM, LEO-1450) was employed to observe the microstructure of the sintered dense pellets and cell electrodes.

## 2.4 First principls calculation

To get insight into the effect of Co-doping on the electronic structure of the LSTC samples, first-principles calculation was performed to get the density of states (DOS) of the material based on density functional theory using Materials Studio (MS) software. The detailed parameter setting was described elsewhere[29]. The structure model of the LSTC material was based on a cubic cell with 40 atoms ($La_2Sr_6Ti_8O_{24}$ and $La_2Sr_6Ti_4Co_4O_{24}$) that corresponded to a $2 \times 2 \times 2$ super cell of the ideal cubic simple $ABO_3$ perovskite ($Pm\bar{3}m$). $La_2Sr_6Ti_8O_{24}$ and $La_2Sr_6Ti_4Co_4O_{24}$ were used to approximately represent $La_{0.3}Sr_{0.7}TiO_3$ and Co-doped $La_{0.3}Sr_{0.7}Ti_{0.5}Co_{0.5}O_3$, respectively.

## 2.5 Electrical and electrochemical characterization

The total electrical conductivity of the samples was measured in static air and in Ar, respectively, using a standard four-terminal dc method in the temperature range of 300-850℃ with a step of 50℃. All the conductivity data were collected after equilibrium was achieved. For the thermal cycling test, the sample ($x = 0.6$) was heated to high temperature at 5℃·min$^{-1}$ and then cooled down naturally with furnace to RT for the first cycle, which was followed by heating again for the second and third cycles. During the heating process in each cycle, the electrical conductivity of the sample was collected. Impedance measurement of the symmetrical cells and single-cells was performed on a Solartron 1260 impedance gain/phase analyzer in combination with a 1287 electrochemical interface in the 0.1 to 10$^6$Hz frequency range with a perturbation amplitude of 10mV in air (20mL·min$^{-1}$). The current-voltage ($I$-$V$) curves were recorded in the range of 700-850℃ by using a Solartron 1287 Electrochemical interface controlled by CorrWare software, where the cell voltage was varied from OCV (open circuit voltage) to 0.3V.

## 3 Results and discussion

### 3.1 Lattice structure

Fig. 1 shows the XRD patterns of samples $La_{0.3}Sr_{0.7}Ti_{1-x}Co_xO_3$ (LSTC, $x = 0.3 - 0.6$), which were sintered for 10h at 1200℃ for $x = 0.3$, and 0.45 and at 1150℃ for $x = 0.6$. All samples exhibit a single phase with a $Pm\bar{3}m$ cubic perovskite structure and no impurities can be observed in the detection limit. With increasing content of Co, the diffraction peaks of LSTC shift progressively to a high-angle direction, indicating the shrinkage of lattice parameters. This is associated with the size difference between Ti and Co ions.

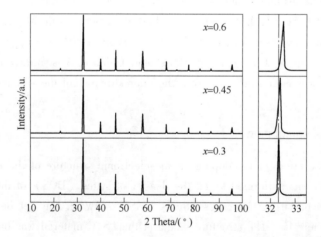

Fig. 1  XRD patterns of $La_{0.3}Sr_{0.7}Ti_{1-x}Co_xO_3$ samples sintered for
10h at 1200℃ for $x = 0.3$ and 0.45 and at 1150℃ for $x = 0.6$

In order to characterize the structure stability of LSTC during the heating process, the samples are subjected to HT-XRD examination. The typical HT-XRD results for $x = 0.6$ are shown in Fig. 2. The results reveal that the synthesized LSTC ($x = 0.3 - 0.6$) keep their cubic structure over the temperature range of RT-900℃, and no structure change or phase segregation is detected, indicating the good structural stability of the synthesized LSTC samples. To obtain detailed information about the crystal structure of the synthesized LSTC at different temperatures, Rietveld refinement is performed on the HT-XRD data using the GSAS/EXPGUI program. For all the LSTC samples, the $Pm\bar{3}m$ cubic perovskite structure is employed as an initial model, where La/Sr is located at 1a (0, 0, 0) site, Ti/Co at 1b (1/2, 1/2, 1/2) site, and O at 3c (0, 1/2, 1/2) site. The typical refinement results for sample with $x = 0.45$ at 25℃ and 900℃ are illustrated in Fig. 3. The refinement shows a good agreement between the calculated and observed profiles. The refined structure parameters for samples at 25 and 900℃ are summarized in Table 1. The lattice parameter of samples decreases with increasing content of Co for both at 25 and 900℃, which is consistent with the variation of the diffraction peaks in Fig. 1. This implies that the content of $Co^{4+}$ ions increases in LSTC, taking into account of the ionic radius of $Co^{4+}$ (0.67Å), $Co^{3+}$ (LS: 0.685Å, HS: 0.75Å) and $Ti^{4+}$ (0.745Å)[30], and the fact that the $Co^{3+}$ ion tends to take a higher spin state at high temperatures[31-33].

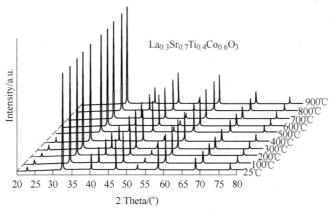

Fig. 2  XRD patterns of $La_{0.3}Sr_{0.7}Ti_{0.4}Co_{0.6}O_3$ at different temperatures

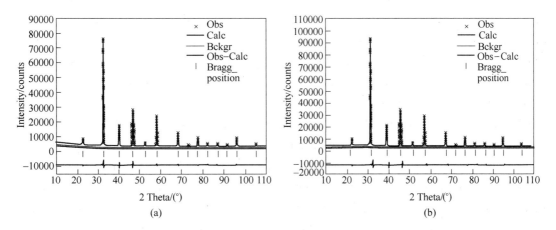

Fig. 3  XRD patterns and Rietveld refinement results of the sample with $x = 0.45$ at 25℃ (a) and 900℃ (b), observed (cross symbols) and calculated (continuous line)

Table 1 Structural parameters based on the Rietveld refinement of HT-XRD data of
LSTC at 25 and 900℃. La/Sr is located at 1a (0, 0, 0) position,
Co/Ti at 1b (1/2, 1/2, 1/2), and O at 3c (0, 1/2, 1/2) sites

| Item | $x=0.3$ | | $x=0.45$ | | $x=0.60$ | |
|---|---|---|---|---|---|---|
| | 25℃ | 900℃ | 25℃ | 900℃ | 25℃ | 900℃ |
| $a=b=c$ (Å) | 3.8783 (22) | 3.9451 (6) | 3.8667 (18) | 3.9391 (19) | 3.8591 (9) | 3.9366 (9) |
| La/Sr Uiso | 0.0150 (2) | 0.0356 (2) | 0.0150 (2) | 0.0348 (3) | 0.0176 (1) | 0.0361 (2) |
| Ti/Co Uiso | 0.0077 (3) | 0.0206 (3) | 0.0074 (3) | 0.0182 (4) | 0.0105 (3) | 0.0281 (4) |
| O Uiso | 0.0257 (6) | 0.0458 (7) | 0.0217 (6) | 0.0511 (17) | 0.0291 (7) | 0.0677 (9) |
| $\chi^2$ | 2.408 | 3.251 | 2.479 | 2.919 | 4.010 | 4.814 |
| $R_p$ (%) | 1.75 | 1.83 | 1.62 | 1.56 | 1.49 | 1.39 |
| $R_{wp}$ (%) | 2.46 | 2.83 | 2.35 | 2.50 | 2.42 | 2.56 |

With HT-XRD data, the lattice parameter of samples at different temperatures can be calculated by Rietveld refinement. The obtained lattice parameter variations of the investigated samples LSTC at different temperatures are depicted in Fig. 4. With the fitted linear slope, the thermal expansion coefficients of samples are derived, which are shown in the inset of Fig. 4. Two slopes with a bending at 300℃ can be observed for all three samples. The increased slope in the high temperature range, corresponding to a larger TEC value, is associated with the so-called chemical expansion, which is caused by the loss of lattice oxygen, reduction of B-site ions and/or transition from low-spin to high-spin state of partial Co ions[31,34,35]. In a low temperature range (RT-300℃), all samples show almost the same slope. However, in a high temperature range, the slope increases significantly with the Co content, corresponding to the increased TEC. The TEC is 20.7 (7), 23.6 (3) and 26.3 (2) $\times 10^{-6}$ $K^{-1}$ for the LSTC samples with $x=0.3, 0.45$ and 0.6, respectively. The increased TEC with increasing Co content is related to the weak Co-O bond compared to the Ti-O bond. The high content of Co in the sample will induce more lattice oxygen loss and thus more oxygen vacancy generation and more Co ion reduction, which thus result in large lattice expansion.

This assumption is supported by TG results. As shown in Fig. 5, there is a slow weight loss in the range RT-300℃, corresponding to a mass change ca. 0.09%, which probably originated from the evaporation of absorbed water. Above 300℃, a significant weight loss is observed for the sample with $x=0.6$, while only a slight weight loss is detected for sample with $x=0.3$, demonstrating the large lattice oxygen loss of the sample with a high Co content. This is consistent with the TEC results. Although high Co content leads to high TEC value, it can also give rise to more oxygen vacancies, which are beneficial to the electrode reaction process. It is worth noting that the TEC derived from the lattice parameter variation upon temperature is different from that obtained by directly measuring the dense sample. Considering that the dense sample usually contains more or less pores, the calculated TEC from lattice parameter change is higher than the practically measured value[17]. Nevertheless, the calculated TEC of LSTC ($x=0.3-0.6$) prepared in this work, (20-26) $\times 10^{-6}$ $K^{-1}$, is lower than the tested TEC of $La_{0.3}Sr_{0.7}CoO_{3-\delta}$ (28.8 $\times 10^{-6}$ $K^{-1}$), $SrCo_{0.8}Ti_{0.2}O_{3-\delta}$ (28.3 $\times 10^{-6}$ $K^{-1}$)[36] and BSCF((24.9-27.3) $\times 10^{-6}$ $K^{-1}$)[35,37].

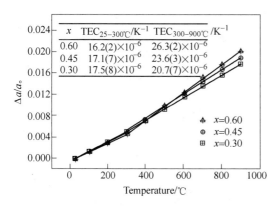 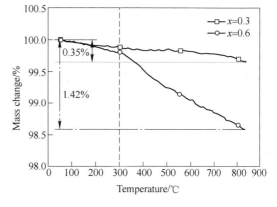

Fig. 4  Temperature dependence of the refined lattice parameter variations of different samples. The inset is the calculated thermal expansion coefficients

Fig. 5  TG results of LSTC ($x = 0.3$ and $0.6$), indicating the weight loss (%) of samples in air

## 3.2 Chemical compatibility

The chemical compatibility of cathode materials with an electrolyte is an important factor affecting the long-term performance of SOFC stacks. To evaluate the chemical compatibility of LSTC with electrolyte LSGM, LSTC ($x = 0.6$) and LSGM powders are mixed in 50:50 weight ratio, pressed to a pellet and then sintered at 1250℃ for 5h. The sintered pellet is crushed and ground to powders, followed by XRD examination. The XRD pattern is shown in Fig. 6. No impurity is detected in this situation, indicating a good chemical compatibility between LSTC cathode and LSGM electrolyte below 1250℃.

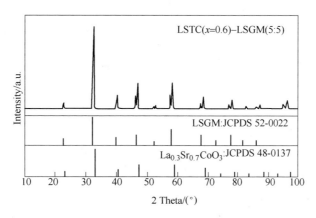

Fig. 6  XRD pattern of the LSTC-LSGM mixture calcined at 1250℃ for 5h

## 3.3 Electrical conductivity

Dense pellets with relative density above 95% are employed for electrical conductivity measurement. Fig. 7 presents the electrical conductivity of samples tested under various conditions. As shown in Fig. 7 (a), the conductivity of samples with $x = 0.6$ in air is higher than that in argon, showing typical p-type conduction behavior. This means that the electron-hole is the charge carrier in this material. With respect to the different samples, as displayed in Fig. 7 (b), the conductivity increases with increasing Co contents in air. For samples with $x = 0.45$ and $0.6$, the conductivity shows a maximum value with temperature, it increases initially and then decreases. While, the sample with $x = 0.3$ shows very low conductivity up to 600 °C and then presents a slight increase at high temperature. In Arrhenius plot (Fig. 7(c)), all samples exhibit a linear relationship in a low temperature range, indicating small polaron conduction behavior. However, at high temperature there is a deviation from the linearity, corresponding to the inflection of the conductivity curve in Fig. 7 (b), which is attributed to the loss of lattice oxygen from LSTC. This process can be expressed in eqn (1) and (2). The loss of lattice oxygen leads to the simultaneous generation of oxygen vacancies and free electrons. Although the former can promote the oxygen ion transport and accelerate the electrode reaction, while the latter will cause the annihilation of partial electron-holes, and thus result in the decrease of electronic conductivity. With increasing Co content, the inflection temperature of the conductivity curve decreases, which is mainly related to the weaker Co-O bond compared to the Ti-O bond.

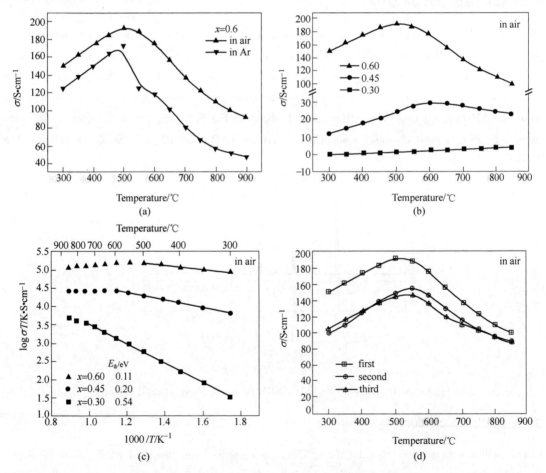

Fig. 7 (a) Temperature dependence of electrical conductivity of $La_{0.3}Sr_{0.7}Ti_{0.4}Co_{0.6}O_{3-\delta}$ in air and argon; (b) Temperature dependence of electrical conductivity and (c) its Arrhenius plot of $La_{0.3}Sr_{0.7}Ti_{1-x}Co_xO_{3-\delta}$; (d) Electrical conductivity of $La_{0.3}Sr_{0.7}Ti_{0.4}Co_{0.6}O_{3-\delta}$ under different thermal cycles in air

$$O_O^x \longrightarrow 2e' + V_O^{\cdot\cdot} + \frac{1}{2}O_2 \quad (1)$$

$$e' + h^{\cdot} \longrightarrow nil \quad (2)$$

At 800℃, the conductivity of LSTC reaches 24 and 110 S·cm$^{-1}$ for $x = 0.45$ and 0.6, respectively, which are comparable to or even higher than that of BSCF[35].

For charge compensation of $La_{0.3}Sr_{0.7}TiO_3$ without Co-doping, it is believed that the excessive positive charge produced by La-doping at Sr-site is balanced by the formation of A-site vacancies[25,38]. With respect to the system of $La_{0.3}Sr_{0.7}Ti_{1-x}Co_xO_3$, when $x = 0.3$, the excessive positive charge of La can be balanced by the substitution of Co for Ti, as expressed in eqn (3), provided that the oxygen vacancy concentration is limited in air. Because both $Co^{3+}$ and $Ti^{4+}$ ions are in the stable oxidation state and no electron defect is generated, the sample $La_{0.3}Sr_{0.7}Ti_{0.7}Co_{0.3}O_3$ shows a much lower conductivity.

$$0.3 [La_{Sr}^{\cdot}] = 0.3 [Co_{Ti^{4+}}^{3+'}] \quad (3)$$

For further substitution of Co for Ti ($x = 0.45$ and 0.6), the extra Co ions will take the oxidation state of $Co^{4+}$ in order to keep the neutrality of the material. Considering that considerable oxygen vacancies may produce in Co-rich samples, which will cause the generation of free electrons as shown in eqn. (1), the solid solution formula of $La_{0.3}Sr_{0.7}Ti_{1-x}Co_xO_{3-\delta}$ can be written as:

$$La_{0.3}^{3+}Sr_{0.7}^{2+}Ti_{1-x}^{4+}Co_{0.3+2\delta}^{3+}Co_{x-0.3-2\delta}^{4+}O_{3-\delta}$$

Accordingly, the concentration [$Co^{4+}$], corresponding to the concentration of electron-holes [$h^{\cdot}$], is proportional to the Co content $x$. As a result, the electronic conductivity increases with the Co content in LSTC materials, which is consistent with the results shown in Fig. 7 (b).

To confirm the assumption discussed above, XPS examination is performed to identify the oxidation state of Co ions in LSTC. The evolution of the Co 2p photoelectron spectra as a function of doping amount is shown in Fig. 8. Two broad peaks, belonging to Co $2p_{3/2}$ and Co $2p_{1/2}$ electrons, were observed, both of which can be deconvoluted into two peaks, assignable to $Co^{3+}$ and $Co^{4+}$, respectively[39-41]. With the integrated area of the peaks, the contents of each oxidation state of Co ions are calculated, which are listed in Fig. 8. Besides $Co^{3+}$, a considerable amount of $Co^{4+}$ is detected in both samples and the content of $Co^{4+}$ increases with increasing Co doping level, from 24% ($x = 0.45$) to 48% ($x = 0.6$). This result is consistent with the assumption of charge compensation. A satellite peak at around 786eV was evident, suggesting a small amount of $Co^{2+}$ coexisting on the surface of LSTC particles[40,42].

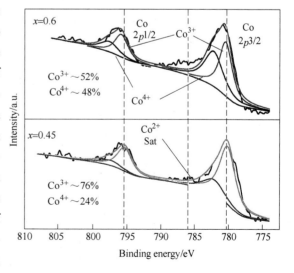

Fig. 8 XPS spectra of Co 2p of the synthesized $La_{0.3}Sr_{0.7}Ti_{1-x}Co_xO_3$ ($x = 0.45$ and 0.6)

In order to evaluate the performance stability of LSTC during the thermal cycling process, the conductivity the of sample with high Co content ($x = 0.6$) is tested under different cycles. The result is shown in Fig. 7 (d). The sample shows similar conductivity during the second and third cycles but much lower than that in the first cycle. The conductivity of mixed conductors depends strongly on the thermal history they experienced. As mentioned above, some lattice oxygen may lose at high temperature in the sintering process, which can lead to the decrease in conductivity. During the cooling process, these oxygen vacancies can be incorporated into the bulk of the sample again in order to maintain the thermodynamic equilibrium. In the case of fast cooling, however, this process cannot be thoroughly completed due to kinetic reasons, and more oxygen vacancies will remain, leading to reduced electrical conductivity. Because the thermal insulation system of the furnace for sample preparation in our lab is much better than that for electrical conductivity measurement, the sample, which has ever experienced the conductivity test and thus been subjected to the fast cooling process, usually presents a much lower conductivity value than the initial one. Fortunately, the sample displays similar electrical conductivity values in the subsequent cycles, demonstrating the highly reversibility of the structure and the good thermal cycling performance of LSTC as the cathode for SOFCs.

To get insight into the effect of Co-doping on the electronic conductivity of the LSTC, first-principle calculations are performed to get the density of states (DOS) based on density functional theory using Materials Studio software. For simplicity, $La_2Sr_6Ti_8O_{24}$ and $La_2Sr_6Ti_4Co_4O_{24}$ are approximately used as the calculation models for $La_{0.3}Sr_{0.7}TiO_3$ with and without Co-doping (Fig. 9). The calculated results are shown in Fig. 10. Significantly, Co-doping changes the density of states of electrons in LSTC. LSTC has transformed from n-type conductor (LST) to p-type conductor, since the Fermi level shifts from the conduction band (Fig. 10 (a)) to the valence band (Fig. 10 (b)). This finding is in good agreement with the experimental results (Fig. 7) and the previous reports that LST exnibits n-type conduction behavior under reducing atmosphere while Sr-doped $LaCoO_3$ shows p-type conduction characteristic[43-45]. The remarkably high conductivity values of LSTC ($x = 0.45$ and 0.6) are mainly attributed to the narrow bandgap and the observed strong hybridization of the Co and O states at the valence band edge and conduction band (Fig. 10 (b) and (c)). Both of them lead to low activation energy for electron-holes jumping along the Co-O-Co bonding network.

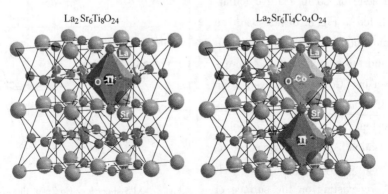

Fig. 9　Structure models of $La_2Sr_6Ti_8O_{24}$ and $La_2Sr_6Ti_4Co_4O_{24}$ for first-principles calculation

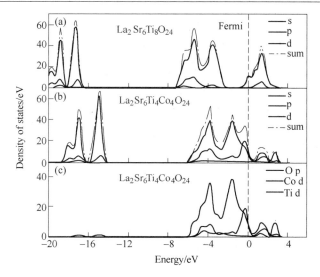

Fig. 10  Density of states of (a) $La_2Sr_6Ti_8O_{24}$ and (b) $La_2Sr_6Ti_4Co_4O_{24}$, (c) detailed view of the states of Op, Cod and Tid. Dashed lines represents the Fermi energy

## 3.4 Electrochemical performance

The cathode performance of LSTC is evaluated by symmetrical cells LSTC | LSGM | LSTC with AC impedance in the temperature range of 700-850 ℃. The typical Nyquist plots and fitting results with equivalent circuit $LR_0(Q_HR_H)(Q_LR_L)$ are shown in Fig. 11, where $L$ is the inductance, $R_0$ stands for the ohmic resistance, and $(Q_HR_H)$ and $(Q_LR_L)$ represent the constant phase element and resistance of the processes at high and low frequency, respectively. Because only one arc is observed for the sample with $x = 0.3$, $LR_0(QR)$ is used for data fitting. The area specific resist-

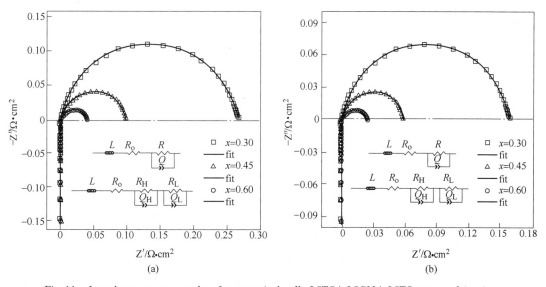

Fig. 11  Impedance spectra results of symmetrical cells LSTC | LSGM | LSTC measured in air at 750 ℃ (a) and 800 ℃ (b). The ohmic resistance has been subtracted from the impedance for direct comparison. Insert are the equivalent circuits adopted for data fitting

ances ($ASRs$) of the LSTC electrode, the characteristic capacitances and frequencies ($C = (QR)^{1/n}/R, f = 1/2\pi\ (QR)^{1/n}$) of the arcs are summarized in Table 2.

Table 2  Polarization resistances as a function of temperature for the symmetrical cell LSTC | LSGM | LSTC in air under the zero dc condition obtained from the data fitting with equivalent circuit $LR_O\ (Q_H R_H)\ (Q_L R_L)$ and $LR_O\ (QR)$

| | $T/℃$ | 700 | 750 | 800 | 850 |
|---|---|---|---|---|---|
| 0.3 | $R/\Omega \cdot cm^2$ | 0.498 | 0.268 | 0.162 | 0.104 |
| | $C/F \cdot cm^{-2}$ | 0.073 | 0.061 | 0.055 | 0.052 |
| | $F/Hz$ | 4.38 | 9.76 | 18.01 | 29.36 |
| | $ASR/\Omega \cdot cm^2$ | 0.498 | 0.268 | 0.162 | 0.104 |
| 0.45 | $R_H/\Omega \cdot cm^2$ | 0.0195 | 0.0090 | 0.0045 | 0.0023 |
| | $C_H/F \cdot cm^{-2}$ | 0.167 | 0.160 | 0.111 | 0.116 |
| | $F_H/Hz$ | 48.91 | 109.97 | 317.34 | 612.19 |
| | $R_L/\Omega \cdot cm^2$ | 0.220 | 0.091 | 0.053 | 0.035 |
| | $C_L/F \cdot cm^{-2}$ | 0.202 | 0.211 | 0.191 | 0.168 |
| | $F_L/Hz$ | 3.57 | 8.27 | 15.73 | 26.84 |
| | $ASR/\Omega \cdot cm^2$ | 0.2395 | 0.1000 | 0.0575 | 0.0373 |
| 0.6 | $R_H/\Omega \cdot cm^2$ | 0.0150 | 0.0079 | 0.0043 | 0.0025 |
| | $C_H/F \cdot cm^{-2}$ | 0.102 | 0.086 | 0.085 | 0.101 |
| | $F_H/Hz$ | 102.44 | 236.01 | 430.75 | 637.36 |
| | $R_L/\Omega \cdot cm^2$ | 0.078 | 0.033 | 0.019 | 0.012 |
| | $C_L/F \cdot cm^{-2}$ | 0.575 | 0.475 | 0.393 | 0.339 |
| | $F_L/Hz$ | 3.56 | 10.08 | 21.30 | 37.66 |
| | $ASR/\Omega \cdot cm^2$ | 0.0930 | 0.0409 | 0.0233 | 0.0145 |

It is suggested that the high frequency arc is mainly related to the charge transfer process while the low frequency arc is associated with the molecular oxygen dissociation process[46-49], as expressed in eqns. (4) and (5). Since the low frequency arc is much larger than the high frequency arc, the rate-limiting step of the electrode reaction should be the molecular oxygen dissociation processes.

$$O_{ad} + e' \rightleftharpoons O'_{ad} \quad (4)$$

$$O_{2,ad} \rightleftharpoons 2O_{ad} \quad (5)$$

The various resistances versus reciprocal temperature, accompanying with activation energy, are shown in Fig. 12. All the resistances ($R_H$ and $R_L$) decrease noticeably with increasing temperature, indicative of thermal activation behavior of the electrode reaction process. With increasing Co content, all the $R_H$ and $R_L$ decrease significantly, demonstrating that Co-doping improves the molecular oxygen dissociation and charge transfer processes of the electrode. The sample with $x = 0.6$ yields the lowest area-specific resistance of 0.0145 $\Omega \cdot cm^2$ at 850℃, followed by $x = 0.45$ and 0.3 with $ASR$ of 0.0373 and 0.104 $\Omega \cdot cm^2$, respectively, indicating that the electrode performance is significantly enhanced by Co substitution. The *ASR* of the LSTC electrode with $x = 0.6$ is superior to and comparable with the reported typical cathode materials based on the LSGM electrolyte, as summarized in Table 3.

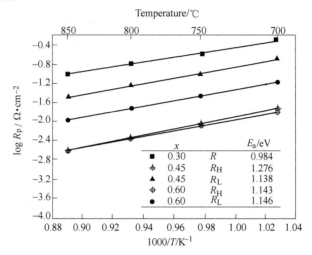

Fig. 12　Polarization resistances versus reciprocal temperature for the $La_{0.3}Sr_{0.7}Ti_{1-x}Co_xO_{3-\delta}$ cathode obtained in air

Table 3　ASR of the typical reported cathode materials based on the LSGM electrolyte and the prepared $La_{0.3}Sr_{0.7}Ti_{0.4}Co_{0.6}O_{3-\delta}$ in this work

| Composition | Temperature/℃ | $ASR/\Omega \cdot cm^2$ | Reference |
| --- | --- | --- | --- |
| $SmBaCo_2O_{5+x}$ | 800 | 0.031 | 50 |
| $PrBa_{0.5}Sr_{0.5}Co_2O_{5+x}$ | 800 | 0.027 | 51 |
| $GdBaCo_2O_{5+\delta}$ | 800 | ~0.138 | 52 |
| $Ba_{0.5}Sr_{0.5}Co_{0.8}Fe_{0.2}O_{3-\delta}$ | 800 | ~0.075 | 52 |
| $Sr_{0.7}Y_{0.3}CoO_{2.65-\delta}$ | 800 | 0.11 | 53 |
| $Sm_{0.5}Sr_{0.5}CoO_{3-\delta}$ | 800 | 1.34 | 54 |
| $La_{0.6}Sr_{0.4}Fe_{0.8}Co_{0.2}O_{3-\delta}$ | 700 | 0.1 | 55 |
| $Pr_2NiO_{4+\delta}$ | 700 | 0.23 | 55 |
| $La_{1.7}Ca_{0.3}Ni_{0.7}Cu_{0.3}O_{4+\delta}$ | 800 | 0.099 | 56 |
| $SrCo_{0.9}Nb_{0.1}O_{3-\delta}$ | 800 | 0.029 | 57 |
| $BaCo_{0.7}Fe_{0.2}Nb_{0.1}O_{3-\delta}$ | 750 | 0.06 | 58 |
| $Ba_{0.9}Co_{0.7}Fe_{0.2}Nb_{0.1}O_{3-\delta}$ | 800 | 0.02 | 59 |
| $La_{0.3}Sr_{0.7}Ti_{0.4}Co_{0.6}O_{3-\delta}$ | 700 | 0.0930 | This work |
|  | 750 | 0.0409 | This work |
|  | 800 | 0.0233 | This work |

In order to evaluate the cathode properties of LSTC, the LSGM electrolyte-supported single cell with the configuration of LSTC/LSGM/LDC/(Ni-GDC) is constructed and the performance is examined (Fig. 13). The open circuit voltage (OCV) of the cells is about 1.10V at 850℃, which is very close to the theoretical value (~1.13V), indicating that the gas leakage is very small. The cell performance of LSTC is remarkably enhanced by Co substitution. At 850℃, the

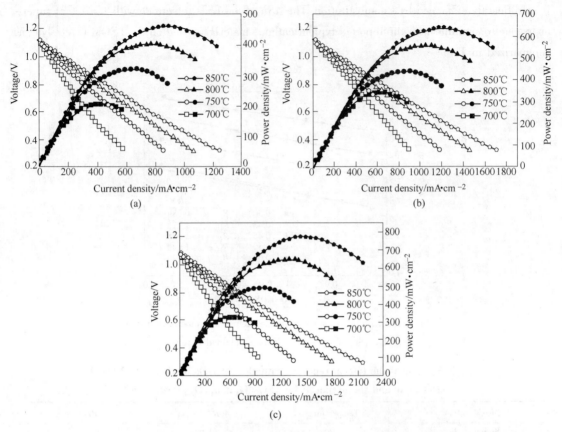

Fig. 13 Voltage and power density versus current density plots for single-cell LSTC/LSGM/LDC/(Ni/GDC) with humidified pure hydrogen as fuel and pure oxygen as an oxidant

(a) $x=0.3$; (b) $x=0.45$; (c) $x=0.6$.

maximum power density with $O_2$ as oxidant is 464.5, 648 and 775 mW·cm$^{-2}$ for $x=$ 0.3, 0.45 and 0.6, respectively, revealing the LSTC as a quite potential cathode material. With air as oxidant, the cell with LSTC ($x=0.45$) delivers the maximum power density of 597 mW·cm$^{-2}$ at 850 ℃ (Fig. 14), just slightly lower than that in pure $O_2$. Considering that the thickness of the LSGM electrolyte used in the single-cell is about 400 μm, the cell performance is acceptable, which can be further enhanced by using a thinner LSGM electrolyte and optimizing the electrode structure. The cell microstructure after the test is provided in Fig. 15. The electrodes still maintain the porous structure and good connection with the LSGM elec-

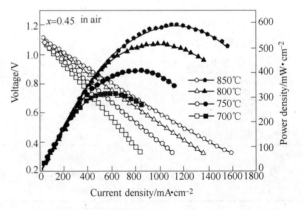

Fig. 14 Voltage and power density versus current density plots for single-cell LSTC ($x=0.45$)/LSGM/LDC/(Ni/GDC) with humidified pure hydrogen as fuel and air as an oxidant

trolyte. The excellent performance of electrolyte-supported single cells with the LSCT cathode is a strong indication of its potential as a cathode material for SOFCs.

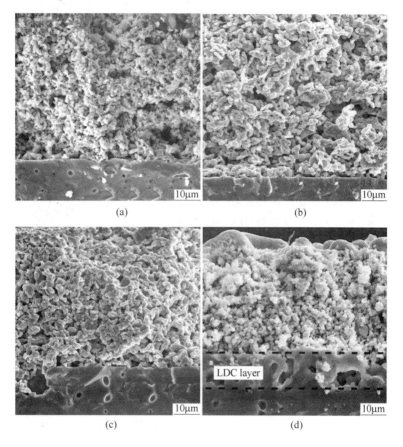

Fig. 15  SEM micrographs of the cross-section of the tested cells: LSTC cathode on LSGM electrolyte with (a) $x=0.3$, (b) $x=0.45$, (c) $x=0.6$; (d) Ni-GDC anode with an LDC buffer layer

## 4  Conclusions

The perovskites $La_{0.3}Sr_{0.7}Ti_{1-x}Co_xO_3$ ($x=0.3, 0.45$ and $0.6$) are prepared and investigated as potential cathode material for IT-SOFC. All the LSTC compounds show a cubic structure with lattice parameter decreasing upon Co substitution. The TECs of LSTC, derived from HT-XRD data, increases with the Co content. The LSTC samples with $x=0.3-0.6$ exhibit the TEC of $(20.7-26.3) \times 10^{-6}$ $K^{-1}$ in the temperature range of 300-900℃. Co substitution enhances significantly the electrical conductivity of LSTC due to the increased concentration of electronic holes and the reduced bandgap energy. LSTC has good chemical compatibility with the LSGM electrolyte below 1250℃. The Co substitution promotes the molecular oxygen dissociation and electron charge transfer processes occurring on the electrode/gas interface, sequentially decreases the *ASR* of the LSTC electrode and increases the power density of single-cells. The maximum power density of the LSGM electrolyte-supported single-cell with LSTC as the cathode increases with the Co content and can reach the value of 464.5, 648, and 775 mW·cm$^{-2}$ at 850℃ in pure $O_2$ for $x = 0.3$, 0.45, and 0.6, respectively. With air as an oxidant, the maximum power density is 597 mW·cm$^{-2}$ for $x = 0.45$. The high performance makes the LSTC ($x = 0.45$ and 0.6) promising candidates as cathode material for IT-SOFCs.

**Acknowledgements:** This work was financially supported by the National Basic Research Program of China (2013CB934003), Guangdong Industry-Academy-Research Alliance (2012B091100129) and the Fundamental Research Funds for the Central Universities (FRF-MP-12-006B).

## References

[1] S. C. Singhal. Solid State Ionics, 2000, 135: 305-313.
[2] S. C. Singhal. Solid State Ionics, 2002, 152-153: 405-410.
[3] S. McIntosh, R. J. Gorte. Chem. Rev., 2004, 104: 4845-4865.
[4] B. C. H. Steele. Nature, 1999, 400: 619.
[5] E. D. Wachsman, K. T. Lee. Science, 2011, 334: 935-939.
[6] B. C. H. Steele, A. Heinzel. Nature, 2001, 414: 345-352.
[7] Z. Tianshu, P. Hing, H. Huang, et al. Solid State Ionics, 2002, 148, 567-573.
[8] T. Ishihara, H. Matsuda, Y. Takita. J. Am. Chem. Soc., 1994, 116: 3801-3803.
[9] M. S. D. Read, M. S. Islam, G. W. Watson, et al. J. Mater. Chem., 2000, 10: 2298-2305.
[10] K. Huang, M. Feng, J. B. Goodenough, et al. J. Electrochem. Soc., 1996, 143 (11): 3630-3636.
[11] S. Khan, R. J. Oldman, F. Corà, et al. Phys. Chem. Chem. Phys., 2006, 8: 5207-5222.
[12] Y. Shen, H. Zhao, K. Świerczek, et al. J. Power Sources, 2013, 240: 759-765.
[13] A. Tarancón, M. Burriel, J. Santiso, et al. J. Mater. Chem., 2010, 20: 3799-3813.
[14] C. Sun, R. Hui, J. Roller. J. Solid State Electrochem, 2010, 14: 1125-1144.
[15] Z. Shao, S. M. Haile. Nature, 2004, 431: 170-173.
[16] A. Bieberle-Hütter, M. Søgaard, H. L. Tuller. Solid State Ionics, 2006, 177 (19): 1969-1975.
[17] C. Kuroda, K. Zheng, K. Świerczek, Int. J. Hydrogen Energy, 2013, 38: 1027-1038.
[18] H. Zhao, D. Teng, X. Zhang, et al. J. Power Sources, 2009, 186: 305-310.
[19] R. Martínez-Coronado, A. Aguadero, D. Pérez-Coll, et al. Int. J. Hydrogen Energy, 2012, 37: 18310-18318.
[20] X. Li, H. Zhao, X. Zhou, et al. Int. J. Hydrogen Energy, 2010, 35, 7913-7918.
[21] O. A. Marina, N. L. Canfield, J. W. Stevenson. Solid State Ionics, 2002, 149: 21-28.
[22] P. I. Cowin, C. T. G. Petit, R. Lan, et al. Adv. Energy Mater., 2011, 1: 314-332.
[23] X. Li, H. Zhao, F. Gao, et al. Electrochem. Commun., 2008, 10: 1567-1570.
[24] X. Li, H. Zhao, N. Xu, et al. Int. J. Hydrogen Energy, 2009, 34: 6407-6414.
[25] F. Napolitano, D. G. Lamas, A. Soldati, et al. Int. J. Hydrogen Energy, 2012, 37: 18302-18309.
[26] M. Yuste, J. C. Pérez-Flores, J. R. Paz, et al. Dalton Trans., 2011, 40: 7908-7915.
[27] A. C. Larson, R. B. Von Dreele. Los Alamos Natl. Lab. [Rep.] LA, 1994: 86-748, LAUR.
[28] B. H. Toby. J. Appl. Crystallogr., 2001, 34: 210-213.
[29] J. Wang, H. Zhao, Y. Shen, et al. ChemPlusChem, 2013, 78 (12): 1530-1535.
[30] R. D. Shannon. Acta Crystallogr., Sect. A: Cryst Phys., Diffr., Theor. Gem. Crystallogr., 1976, 32: 751.
[31] K. Huang, H. Y. Lee, J. B. Goodenough. J. Electrochem. Soc., 1998, 145: 3220.

[32] M. A. Korotin, S. Y. Ezhov, I. V. Solovyev, Phys. Rev. B: Condens. Matter Mater. Phys. , 1996, 54 (8): 5309.
[33] A. Podlesnyak, S. Streule, J. Mesot, et al. Phys. Rev. Lett. , 2006, 97 (24): 247208.
[34] Z. Shao, W. Yang, Y. Cong, et al. J. Membr. Sci. , 2000: 172; 177-188.
[35] B. Wei, Z. Lü, S. Li, et al. , Electrochem. Solid-State Lett. , 2005, 8 (8): A428-A431.
[36] V. V. Kharton, A. A. Yaremchenko, A. V. Kovalevsky, et al. J. Membr. Sci. , 1999, 163 (2): 307-317.
[37] P. Ried, P. Holtappels, A. Wichser, et al. J. Electrochem. Soc. , 2008, 155 (10): B1029-B1035.
[38] S. Hashimoto, L. Kindermann, F. W. Poulsen, et al. J. Alloys Compd. , 2005, 397: 245-249.
[39] E. Y. Konysheva, S. M. Francis, J. T. S. Irvine, et al. J. Mater. Chem. , 2011, 21: 15511-15520.
[40] Z. Cai, M. Kubicek, J. Fleig, et al. Chem. Mater. 2012, 24: 1116-1127.
[41] E. Konysheva, R. Blackley, J. T. S. Irvine. Chem. Mater. , 2010, 22: 4700-4711.
[42] Z. Cai, Y. Kuru, J. W. Han, et al. J. Am. Chem. Soc. 2011, 133: 17696-17704.
[43] P. R. Slater, D. P. Fagg, J. T. S. Irvine. J. Mater. Chem. , 1997, 7 (12): 2495-2498.
[44] C. D. Savaniu, J. T. S. Irvine. J. Mater. Chem. , 2009, 19: 8119-8128.
[45] J. Mizusaki, J. Tabuchi, T. Matsuura, et al. J. Electrochem. Soc. , 1989, 136 (7): 2082-2088.
[46] M. J. Escudero, A. Aguadero, J. A. Alonso, et al. J. Electroanal. Chem. , 2007, 611: 107-116.
[47] D. Chen, R. Ran, K. Zhang, et al. J. Power Sources, 2009, 188: 96-105.
[48] F. Mauvy, C. Lalanne, J. M. Bassat, et al. J. Electrochem. Soc. , 2006, 153 (8): A1547-A1553.
[49] J. Peñā-Martínez, D. Marrero-López, J. C. Ruiz-Morales, et al. Int. J. Hydrogen Energy, 2009, 34: 9486-9495.
[50] Q. Zhou, T. He, Y. Ji. J. Power Sources, 2008, 185: 754-758.
[51] S. Lü, G. Long, X. Meng, et al. Int. J. Hydrogen Energy, 2012, 37: 5914-5919.
[52] J. Peña-Martínez, A. Tarancón, D. Marrero-López, et al. Fuel Cells, 2008, 08 (5): 351-359.
[53] Y. Li, Y. N. Kim, J. Cheng, et al. Chem. Mater. , 2011, 23: 5037-5044.
[54] T. Duong, D. R. Mumm. J. Power Sources, 2013, 241: 281-287.
[55] B. Philippeau, F. Mauvy, C. Mazataud, et al. Solid State Ionics, 2013 249-250: 17-25.
[56] Y. Shen, H. Zhao, K. Świerczek, et al. J. Power Sources, 2013, 240: 759-765.
[57] F. Wang, Q. Zhou, T. He, et al. J. Power Sources, 2010, 195: 3772-3778.
[58] Z. Yang, C. Yang, B. Xiong, et al. J. Power Sources, 2011, 196: 9164-9168.
[59] Z. Yang, C. Yang, C. Jin, et al. Electrochem. Commun. , 2011, 13: 882-885.

(原载于 *Journal of Materials Chemistry A*, 2014, (2): 10290-10299)

# Platinum Decorated Aligned Carbon Nanotubes: Electrocatalyst for Improved Performance of Proton Exchange Membrane Fuel Cells

Yuan Yuan[1]  Joshua A. Smith[2]  Gabriel Goenaga[3]
Liu Dijia[4]  Luo Zhiping[5]  Liu Jingbo[2,6,7]

(1. Department of Environmental Engineering, Texas A&M University-Kingsville, MSC 213, Kingsville, TX 78363-8012, United States;
2. Department of Chemistry, Texas A&M University-Kingsville, MSC 161, Kingsville, TX 78363-8012, United States;
3. Chemical and Biomolecular Engineering, University of Tennessee, Knoxville, TN 37996, United States;
4. Chemical Sciences and Engineering Division, Argonne National Laboratory, Argonne, IL 60439-4837, United States;
5. Microscopy and Imaging Center and Materials Science and Engineering Program, Texas A&M University, 2257 TAMU, College Station, TX 77843, United States;
6. Department of Chemistry, Texas A&M University, P. O. Box 30012, College Station, TX 77842-3012, United States;
7. Department of Materials Science and Engineering, The University of Tokyo, 7-3-1 Hongo, Bunkyo-ku, Tokyo 113-8656, Japan)

**Abstract:** This study aims to improve the performance of proton exchange membrane fuel cells (PEMFCs) using carbon nanotubes as scaffolds to support nanocatalyst for power generation over prolonged time periods, compared to the current designs. The carbon nanotubes are prepared using chemical vapor deposition and decorated by platinum nanoparticles (Pt-NPs) using an amphiphilic approach. The PEMFC devices are then constructed using these aligned carbon nanotubes (ACNTs) decorated with Pt-NPs as the cathode. The electrochemical analyses of the PEMFC devices indicate the maximum power density reaches to $860 mW \cdot cm^{-2}$ and current density reaches $3200 mA \cdot cm^{-2}$ at 0.2V, respectively, when $O_2$ is introduced into cathode. Importantly, the Pt usage was decreased to less than $0.2 mg \cdot cm^{-2}$, determined by X-ray energy dispersive spectroscopy and X-ray photoelectron spectroscopy as complimentary tools. Electron microscopic analyses are employed to understand the morphology of Pt-ACNT catalyst (with diameter of 4-15 nm and length from 8 to 20μm), which affects PEMFC performance and durability. The Pt-ACNT arrays exhibit unique alignment, which allows for rapid gas diffusion and chemisorption on the catalyst surfaces.

**Keywords:** aligned carbon nanotubes; platinum nanoparticles; proton exchange membrane fuel cells; amphiphilic modification; nanostructural characterization; electrochemical performance

---

刘静波，于2001年获得北京科技大学材料科学与工程博士学位，目前任得克萨斯州农工大学—金斯维尔校区副教授。主持或参加由国家自然基金委（美国、中国）、加拿大自然科学和工程研究理事会及韦尔奇基金会资助项目15个，发表约60篇学术期刊文章。

## 1 Introduction

Proton exchange membrane fuel cells (PEMFCs) are one of the most important green energy conversion devices due to their high power density, energy conversion efficiency, flexible design and zero or near-zero emission to the environment[1-4]. The PEMFCs are composed of membrane electrode assembly (MEA), including a polymeric membrane electrolyte (commercially available Nafion 212 has been used in this study) in the center, and an anode and a cathode at each side[5-8]. On the anode compartment, fuel supplies, such as hydrogen ($H_2$) are oxidized to proton ($H^+$), whereas, oxidants, such as oxygen ($O_2$) are reduced on the cathode side[9-11]. The electrons travel from anode to cathode externally to provide direct current for stationary and portable application[12,13]. The $H^+$ ions transfer from the anode through the Nafion electrolyte and react with the $O_2$ to form water at the cathode surface through oxygen reduction reaction or ORR ($4H^+ + O_2 + 4e^- = 2H_2O$)[14,15].

Current design of the PEMFC encounters some problems, such as: materials compatibility, high material/manufacturing cost, and performance degradation over prolonged use, which collectively hinders widespread adaptation and commercialization[16,6,17]. To improve PEMFC performance and durability, and to lower catalyst cost, a great number of studies have focused on building new and robust anode and cathode catalysts[18-20]. The commonly used catalyst, Pt/C has been used for decades due to its high catalytical reactivity; however its expense is a major concern[21-23]. Recently, alternative supports have been applied to improve corrosion resistance, mass transfer and to reduce Pt loading. One of the most attractive catalyst-support materials is carbon nanotubes (CNTs) due to their large surface areas, high chemical stability, high electric conductivity and extraordinary mechanical strength[24-27]. In general, CNTs are coaxial cylinders of graphite sheets, which allow for the structural diversity. The most important structures of CNTs are single-walled nanotubes (SWNTs), multiwalled nanotubes (MWNTs), and aligned carbon nanotubes (ACNTs)[28-33]. The CNTs can be fabricated by carbon-arc discharge, laser ablation, and chemical vapor deposition[34-37]. Moreover, to increase the Pt specific surface area and lower its cost, highly dispersed Pt-NPs have been employed as cat-alysts supported by CNTs to facilitate the ORR process. ORR is the rate limiting step, which dominates the PEFMC efficiency due to its 4-electron involvement and large activation energy for reaction[38-41].

In the present study, Pt-ACNTs cathodic catalyst and MEA performance were investigated to optimize PEMFC catalyst fabrication methodology. This differs from other approaches, which have had issues with stability of the Pt-support[42] or large Pt size[43], which in turn can affect overall reactivity of cathode catalyst[44]. Therefore, ACNT supports have been fabricated to replace traditional carbon black-based catalysts[45]. The rationale to use CNTs lies on the facts that the $sp^2$ hybridization of carbon atoms provides high mechanical strength and also allows for the p electron in carbon atoms to be transferred to the unoccupied d-orbital of Pt. This phenomenon can further enhance the interactions between ACNT and Pt atoms, which reinforce the stability of the cathode materials under oxidative environment[26,46-48]. In addition to the stability of Pt-decorated ACNTs, the hydrophobicity of the ACNTs can prevent cathode flooding and favor water management, which consequently improves gas transport in the PEMFC electrode[49-51]. The Pt-ACNTs have orderly structure which can be tuned into the electrode support with high triple phase boundary (TPB) between cathode, electrolyte membrane and gas reactants. With the increase of TPB, the increased electrochemical behavior of the PEMFC devices is anticipated[52,53]. The contributions of this research are: (1) synthesis of ACNTs by the chemical vapor deposition (CVD) method to support Pt catalyst for ORR and to prevent cathode flooding; (2) decoration of ACNTs by Pt NPs through a novel and feasible amphiphilic chemical impregnation, followed by heat treatment; (3) integration of advanced analytical instrumentation techniques into PEMFCs electrochemical performance measurement.

## 2 Materials and methods

All reagents, chemicals and solvents were obtained from VWR International (West Chester, PA) and Sigma-Aldrich (St Louis, MO) to prepare the nanotubes and construct membrane electrode assemblies (MEAs), unless otherwise specified. The reagents were reagent grade and were used without further purification. Doubly distilled and 0.2μm filtered (Milli-Q) water was used in the dissolution of reactants. State-of-the-art facilities were used to characterize the nanostructure and electrochemical performance. The commercial MEA was procured from BASF Corporation (Port Arthur, TX) for comparative study, with loading of 0.5mg Pt · cm$^{-2}$ on both electrodes.

### 2.1 Nanocatalyst fabrication

The ACNTs were produced via the chemical vapor deposition (CVD) method. The main reactants were ferrocene (Fe($C_5H_5$)$_2$) and xylene ($C_8H_{10}$). The quartz glass slides (with 5cm$^2$ surface areas) were used as substrates for ACNT growth positioned inside of a quartz tube surrounded by a tube furnace with two independently controlled temperature zones. In the first zone, reactants (ferrocene and xylene solution) were injected at a rate of 0.225 or 0.250mL · min$^{-1}$ and vaporized at 225℃. In the second zone, the reactant molecules from the vaporized solution were decomposed over the seed of iron (Fe) particles to form nanotubes. Under the precise control of temperature, injection flow rate of reactants, and flow rate of inert gas (Ar) and hydrogen mixture, ACNTs with well-arranged arrays grew on the substrates. The Fe from ferrocene was continually fed as "seed", on which the carbon could accumulate into multiwalled tubes. ACNTs were then functionalized by Pt using wet impregnation. Several Pt-containing precursors, such as chloroplatinic acid hexahydrate ($H_2PtCl_6 \cdot 6H_2O$), platinum (II) acetylacetonate (Pt(acac)$_2$), platinum (IV) sulfite (Pt(SO$_3$)$_2$), and tetraammineplatinum (II) nitrate (Pt(NH$_3$)$_4$(NO$_3$)$_2$) were used to obtain mono-dispersed Pt nanoparticles. The Pt-modified ACNTs were activated by the heat treatment at 300℃ for 1h under reducing atmosphere ($H_2$). This process termed reduction, which also contributed to removal of the impurities in ACNTs. Next, ACNT catalyst was sprayed with Nafion solution with mass ratio of 1:1 (Nafion to ACNTs).

The ACNT-based membrane electrode assembly (MEA) fabrication has been previously reported[37,52] and its geographical surface area was kept at 5.0cm². The decal for anode consisted of a polymer substrate with a thin layer of teflon coating, painted with an ink composed of carbon-loaded Pt. The ink consisted of tetrabutylammonium hydroxide, Pt 20% (w/w), Nafion (w/w) 5%, and glycerol. The anode decal was placed into an oven heated to 140℃ for approximately an hour between each painting step. The painting-heating cycle was typically repeated up to five times or until the total weight of wet ink reached 8.00mg. The sample was then left in the oven overnight and usually yielded a desired weight of 7.00mg of anode catalyst, which corresponds to 0.2mg Pt $cm^{-2}$ loading. The anodic and cathodic substrates were then placed on two sides of the Na-form Nafion 212 membrane with the thickness averaged at 50μm. Hot pressing technique was then used to fabricate the MEA under temperature of 210℃ and pressure of 600 pounds-per-square-inch-gauge (psig) for 5-10min. The teflon (support for anode) and glass substrate (support for cathode) were then carefully removed from the MEA, which was subsequently treated in boiling $H_2SO_4$ (0.5M) aqueous solution for 1-2h. The $H_2SO_4$ acid treatment aims to exchange $Na^+$ by $H^+$ to increase the proton conductivity. The MEA was then washed in boiled distilled water for 1h to remove the $Na^+$, $Fe^{2+}$ and other impurities.

## 2.2 Nanocatalyst characterization

The nanostructure and elemental composition of the Pt-ACNT catalysts and MEA were characterized using several characterization techniques. The surface morphology, cross-sectional images and the thickness of the aligned carbon nanotubes were determined using a field emission scanning electron microscope (JSM-6701F) equipped with X-ray energy dispersive spectroscope (EDS, Department of Chemistry, Texas A&M University-Kingsville). An accelerating voltage of 20kV and high vacuum of about $1.0 \times 10^{-5}$m bar were generally employed. The substrate of cathode and MEA were mechanically fractured using a razor blade, enabling an easier estimation of the PEMFC thickness. The high resolution TEM (Tecnia F20-$G^2$) equipped with post-column Gatan Image Filter was applied, and magnifications were calibrated using standards of commercial cross-line grating replica and silicon carbide (SiC) lattice images to obtain images and electron diffraction patterns with high accuracy and resolution[54]. Raman spectroscopic analysis was conducted to evaluate the vibration mode and electron structure of Pt and ACNTs. Raman/Fourier transform infrared spectroscopy with confocal microscope (Hariba Jobin-Yvon LabRam IR system, Texas A&M University-College Station) was employed to obtain highly specific fingerprints to enable precise chemical and molecular characterization and identification. Raman module used was fiber optic coupled with the laser excitation lines of 632nm with the scan range of 100-3000$cm^{-1}$. The resolution was kept at 0.3$cm^{-1}$ per pixel for various lines.

## 2.3 Electrochemical performance measurement

The evaluation of PEMFC single cell performance with ACNT-based MEA was carried out using a commercially available test stand (Fuel Cell Technologies, Albuquerque, NM) and lab-assembled test stand. The pressurized and humidified $H_2$ and $O_2$ (air) were fed into the anode and cathode chambers as the fuel and oxidant under various flow rates. The cell operating conditions were controlled at various temperatures ranging from 70℃ to 90℃. Typically, the $H_2$ flow rate was set at 100mL·$min^{-1}$ and flow rate of $O_2$ at 300mL·$min^{-1}$ with the back pressure at 14.0 psig for both cathode and anode compartments. The potential sweeps were controlled at fast rate with 5s duration or slow rate with 30s duration per datum point. The data were collected using LabVIEW 7 software$^{TM}$. The cell voltage drop originating from the electrolyte resistance was not compensated in this study. The relationship between cell current density and power density as function of applied potential was established for estimation of PEMFC performance. The cell was generally purged by safe gas at anode and air at cathode at ambient temperature prior to the electrochemistry study. The conditioning of the PEMFC was conducted for 16 h before direct current (DC) polarization was performed.

## 3 Results and discussions

The Pt-decoration of ACNTs using amphiphilic wet chemical approach and their characterization using microscopic and spectro-scopic techniques are described. The electrochemical performance of the PEMFC devices constructed using Pt-ACNTs as cathode, Nafion as electrolyte membrane and Pt-loaded carbon graphite as anode is tested using fuel cell test station and the importance of the findings is summarized.

### 3.1 Pt-decoration of ACNT

In general, periodically aligned carbon nanotubes (ACNTs) can be synthesized with tunable pore diameters and rigid structural order using iron (Fe) as catalyst. The self assembly of multiwalled ACNTs with uniform ordered nanopores can be achieved. Importantly, the spacing between individual CNTs provides efficient channel for gas molecule transport. To improve the electrocatalytical activity, ACNTs must be modified using Pt-NPs. However, high dispersion and ultrafine Pt-NPs with high stability are extremely difficult to obtain due to poor wettability over hydrophobic ACNT by aqueous based precursor solution. Therefore, we focused on new catalyzing process to optimize the deposition of Pt-NPs on the surface of ACNTs. Conducting a series of experiments, we developed an amphiphilic wet-chemistry procedure (Fig. 1) to decorate and functionalize ACNTs. The two key points in this process are: (1) to maintain the distinctive alignment of well-channelled ACNT and (2) to dissolve the $H_2PtCl_6$ in aqueous solution for monodispersion and homogeneity. Therefore, hydrophilic and hydrophobic agents are needed to achieve the above goals. The critical step for this advancement is the modification of ACNT cathode catalyst using homogenous Pt-NPs, while the ACNTs alignment is well-maintained. This can be accomplished using a novel method, defined as "amphibious functionalization", in which the aromatic conjugate ring of the nanotube surfaces can then be functionalized by hydroxyl group, which forms ion-dipole intermolecular forces with $Pt^{4+}$ cations.

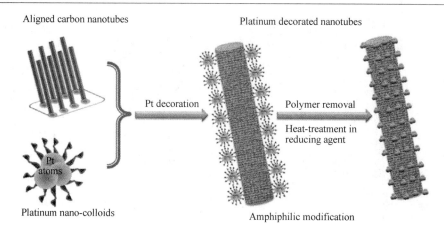

Fig. 1 The amphiphilic wet-chemistry approach to decorate ACNT using Pt nanoparticles, followed by reduction under $H_2$ atmosphere at 300℃

Two reagents, namely isopropanol ($C_3H_7OH$) and distilled water, which are miscible, were used to control the homogeneity from the molecular level. The focus centers on optimization of the Pt starting materials, dispersing agents, amount of the Pt mass, and Pt particle size. Firstly, the ACNT surfaces need to be functionalized using amphiphilic solvent (isopropanol containing 1 mol% of Nafion) to maintain its geometry and provide hydroxyl (OH) group. The water-soluble $Pt^{4+}$ compound (such as $H_2PtCl_6$) is then sprayed onto the surface of ACNT. Due to intermolecular forces (such as ion-dipole coupling), the $Pt^{4+}$ can then interact with ACNT and retain its alignment. Secondly, step-wise deposition of metallic cations onto charged ACNT surface assists self-assembly of amphiphilic block copolymers followed by formation of covalent bond with the ACNT "backbones". Lastly, graft copolymerization onto reactive seeded particles will allow cations to link with the ACNT backbones. Through polymer crafting and self assembly, the $Pt^{4+}$ can be uniformly distributed along the ACNT arrays. $Pt^{4+}$ cations were then reduced to Pt metal in $H_2$ atmosphere. It was found that Pt-NPs growth due to the Ostwald sintering has been successfully prevented and geometry of ACNTs has been well-maintained (also see Fig. 2 (a)).

## 3.2 Electron microscopic analyses of Pt-ACNTs

Scanning and transmission electron microscopic (SEM and TEM) analyses were conducted to understand the morphology and dispersion of Pt-decorated ACNTs due to the important role to secure the high performance and long-term stability of the electrocatalyst. The SEM cross-sectional image of Pt-ACNTs (Fig. 2 (a)) indicated that dense and uniform ACNTs were highly aligned when the seed of iron (Fe) catalyst was used. After Pt-decoration through the feasible amphiphilic wet-chemistry approach, the ACNT highly aligned geometry is well-maintained. This alignment is highly beneficial to channeling the gas flow more efficiently. SEM micrographs of the Pt-ACNT catalyst also show Pt-NPs uniformly embedded along the ACNT surface, with a diameter ranging from 2 to 4nm without further aggregation (Fig. 2 (b)). The highly dispersed Pt-NPs suggested a large fraction of Pt atoms located at the surface of the ACNTs supports. High dispersion of Pt-NPs draws

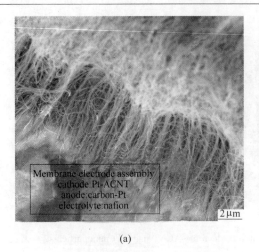 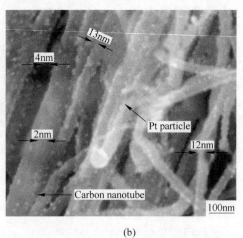

Fig. 2 Scanning electron microscopic analyses of the MEA composed of Pt-functionalized ACNT
(a) The Pt-ACNT aligned structure maintained well; (b) Pt nanoparticles averaged at size of
2-4nm before the electrochemical study is conducted

our attention for two reasons. First of all, high dispersion mitigates the usage of expensive metal (Pt). Secondly, the tunable particle sizes provide controllable structure, which in turn improves the PEMFC performance. The Pt-ACNT cathode catalysts are composed of multi-walled nanotubes with 5-19 layers. The SEM results also indicate that the oriented and uniform ACNTs arrays act as a support for cathodic catalyst to facilitate the $O_2$ gas diffusion. The improved specific surface areas of Pt-ACNTs advanced $O_2$ chemisorption and then increased the PEMFCs intrinsic reaction rates.

The high-resolution TEM images of Pt-ACNTs cathode catalyst were also obtained (Fig. 3 (a) – (d)) to evaluate the geometry and alignment of Pt-ACNT, and the Pt nanoparticles and their size distribution. Selected TEM images (Fig. 3 (a) and (b)) show the ACNTs have an average length of about 10m and a diameter of outer layer varying from 3.5 to 12.5nm. The results indicated that the thickness of ACNT layer thickness was comparable to that of a conventional membrane electrode assembly (MEA). From our experimental data, it can be concluded that morphology of PEM-FCs catalyst can efficiently channel the gas ($H_2$ and $O_2$) distribution, and in turn improve the electrochemical kinetics of PEMFC devices. The CVD-derived ACNTs consist of multiple concentric tubes with the interlayer distance approximately 0.396nm (Fig. 3 (c)), slightly larger than the interlayer distance in graphite. This observation evidently suggested that decoration of the ACNT occurred on the surface without insertion of the Pt-NPs (averaged size 2.7nm) into the inner ACNT layers due to the size effect. These SEM/TEM images corroborate that the ACNTs provide ordered inter-tube spacing, thereby, greatly facilitating gas diffusion. Under carefully controlled fabrication conditions, the length and diameter of the Pt-ACNT can be adjusted to yield optimal catalytic surface area[55,56]. This observation corroborates that the high electrochemical activity of the PEMFCs devices can be achieved (see Fig. 6 and related discussion). From the ring pattern indexing, it can be seen that crystalline phase of Pt-NPs and ACNTs were obtained. The Pt ring patterns (Fig. 3 (d)) were very well aligned with the standard face-centered cubic of Pt (PDF 00-004-0802, 3.924Å and 90°).

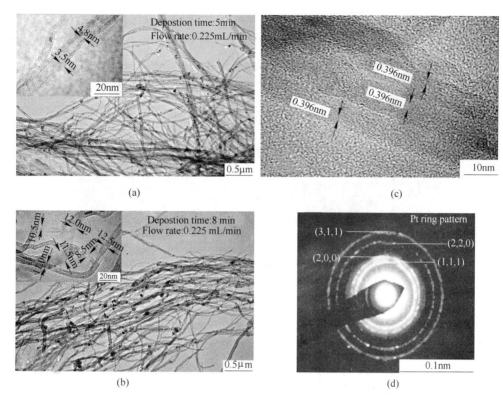

Fig. 3 Transmission electron microscopic analyses of the MEA composed of Pt-functionalized ACNT
(a) ACNT display narrower diameter when deposition was controlled at 5min; (b) ACNT display wider diameter when deposition was controlled at 8min; (c) the distance between multiwall nanotubes; (d) Pt ring pattern showing crystalline Pt formed

## 3.3 Elemental composition analyses of Pt-ACNTs

X-ray energy dispersive spectroscopy (EDS) and X-ray photo-electron spectroscopy (XPS) have been used as complementary techniques to confirm the stoichiometric Pt loading, ranging from 0.14 to 0.17mg · cm$^{-2}$, accordingly (Fig. 4 (a) and (b)). EDS results indicated the principle emission of Pt occurs at $M_\alpha$ at 2.048keV and $L_\alpha$ at 9.441keV, correspondingly. The XPS full spectra indicated the presence of the Pt and C elements, well indexed by their standard spectra for 4f and 1s electron configurations, respectively. The emissions at 283.6eV and 290.6eV were attributed to the C=C double bonds in the ACNT arrays. Metallic Pt can be characterized by asymmetric emission lines with peak tailing to the binding energy (BE) corresponding to its innermost 4f electron excitation (BEs of $4f_{7/2}$ located at 72.1eV and $4f_{5/2}$ at 76.4eV) with peak splitting 4.3eV. The composition analyses indicate that Pt loading can be decreased to 0.2mg · cm$^{-2}$ while the high power density of the PEM-FCs was maintained nicely. From the characteristic binding energy for C and Pt, it can be concluded that functionalization of ACNT by Pt is carried out through van der Waals intermolecular forces, without breaking C=C double bonds. The mono-dispersion of ultrafine Pt-NPs proves that decreased amount of Pt usage (<0.2mg · cm$^{-2}$) provides many electrochemically active sites, which consequently advance the PEMFC electrochemical performance.

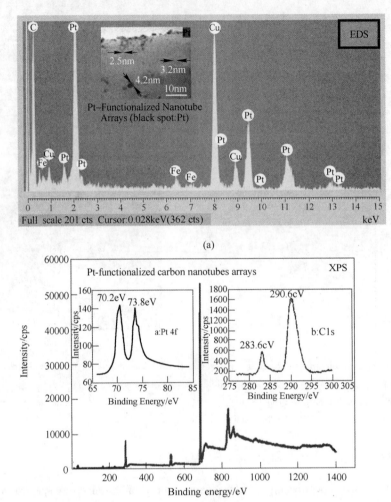

Fig. 4 Elemental composition analyses of the MEA composed of Pt-functionalized ACNT
(a) EDS analyses; (b) XPS analyses

## 3.4 Raman spectroscopic analyses of Pt-ACNTs

The characteristic spectra, radial breathing mode (RMB), D-band and G-band of a series of Pt-ACNT catalysts were evaluated by Raman spectroscopy. The results for six selected MEAs made from the Pt-ACNTs are consistent with standards of Pt and carbon nanotubes (Fig. 5 (a) and (b)). The Pt metal displays two characteristic spectra at frequency of 483.24 $cm^{-1}$ and 2084.98 $cm^{-1}$. The intensity of Pt is significantly low, confirming its small amount of loading. The main Raman features for carbon nanotubes are D band at 1250-1350 $cm^{-1}$ and G band at about 1600 $cm^{-1}$ (Fig. 5(a)). The Pt-decorated ACNTs displayed high intensity D band, suggest-ing that the strengthening of the D band is enhanced by Pt-NPs. This observation reveals the incorpora-

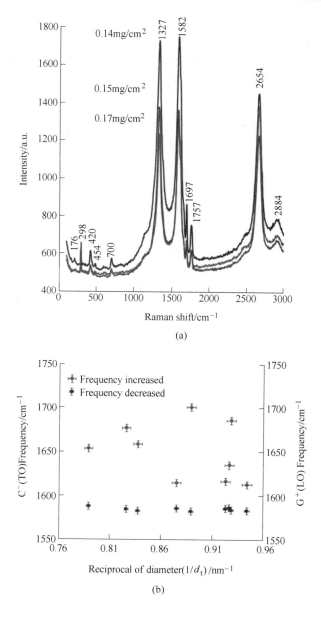

Fig. 5 Raman spectroscopic analyses of the Pt-functionalized ACNT at 633nm excitation
(a) The spectroscopy to demonstrating the molecular vibration;
(b) the G-band splitting due to the curvature effect

tion of platinum into carbon nanotubes. Meanwhile, the blue shift of D band to about 1315.10-1325.77cm$^{-1}$ occurred, suggesting the presence of crystalline graphite-like CNT was formed since it has a typical position of 1305-1330cm$^{-1}$[40]. This measurement is in agreement to the reported data[57,58] that these second order Raman features are resulting from defect-induced and double resonance of Pt-ACNTs. The Pt-decorated ACNTs exhibited tangential vibrational modes, which results from the in-plane optic phonon mode. This phenomenon corresponds to the stretching mode of sp² hybridization in carbon nanotubes [59,60]. The G band was split into a lower frequency (G⁻, 1584.17cm⁻¹) and higher frequency (G⁺, 1609.28cm⁻¹), and this splitting is corresponding to the reciprocal of the interlayer of nanotubes as shown in Fig. 5(b). The softening of G⁻ (TO) band, whose vibration perpendicular to the ACNT tubes, occurs due to the curvature in the pt-ACNT. On the other hand. the strengthening of the G⁺(TO) band was observed due to its vibration along the ACNT tubes. The intermediate frequency mode(IFM) also occurred at 702.73cm⁻¹ due to the defect in the Pt-ACNTs. Importantly. it can be seen the"radial breathing mode"(RBM) of pt-ACNT occurred at a lower frequency. ranging from 102.4cm⁻¹ to 298.98cm⁻¹. From the RBM determination, more accurate diameter calculation on CNTs can be obtained [61]. It was found that the diameter of CNT is inversely proportional to the Raman shift as demonstrated in Eq. (1):

$$\omega_{RBM} = \frac{A}{d_t} + B \tag{1}$$

where $\omega_{RBM}$ represents the radial breathing mode in cm⁻¹, $d_t$ is the diameter of the ACNT in nm, $A$ is a dimensionless constant (commonly used value is 218.2) and $B$ is also an empirical constant (19.6cm⁻¹), corresponding to the Raman shift.

Generally, an increase in the CNT diameter would result in a decrease in the RBM frequency [55,58]. It has to be noted that some satellite spectra were also seen due to the complexity of the MEA elemental composition and structure. Most importantly, the mode frequency of ACNTs and Pt depends on the laser excitation and the dimensional confinement.

## 3.5 Electrochemical performance of PEMFCs devices

The electrochemical performances of both Pt-ACNT and commercial ink-based PEMFCs were evaluated using cell current-potential polarization curves under identical operating condition. The DC polarization was conducted to identify the current density when potentials (0.2 – 1.0V) were applied.

The current generally varies for different MEA specimens with the low Pt loading (about 0.2mg · cm⁻²). The average current density at 0.2V from six selected Pt-ACNT MEA specimens was 3200mA · cm⁻² when $O_2$ is introduced into the cathode chamber (Fig. 6 (a)). The PEMFC power density can be calculated by multiplying the cell potential with associated cell current density. The Pt-ACNT MEA displays a respectable peak power density of 625mW · cm⁻² in air[62] and 860mW · cm⁻² in oxygen (Fig. 6 (b)). The various Pt-ACNT MEAs were tested and we found that the reproducibility of polarization over Pt-ACNT MEAs can be improved via tuning the geometry and structure of the nanotubes and Pt catalyst, although the variation is not as high as in some studies (reviewed by Gasteiger et al.)[21]. The stability of the same Pt-ACNT MEAs was determined at potential ranging from 0.2V to 1.0V with a sweep rate at 5s/cycle (Fig. 6 (c)) over about 7200 cycles. Over the measured cycles, the relative deviations in the polarization curves were negligible, suggesting high PEMFC MEA stability. In this study, the polarization curves within 600-1200 cycles were arbitrarily selected to demonstrate stability. The current density at 0.2V was maintained at 2860-2950mA · cm⁻² ($\sigma < \pm 1.5\%$) and peak powder density was maintained within 850-860mW·cm$^{-2}$ ($\sigma < \pm 0.6\%$), suggesting high stability of the PEMFC devices when operating in the $O_2$ atmosphere[27]. High stability and power/current densities are necessary when the fuel cell is required to withstand high dynamic load. It is important to point out that the Pt usage in this study was decreased to lower than 0.2mg·cm$^{-2}$ determined by XPS analyses. The mass activity $i_{m(0.9V)}$ and specific activity $i_{s(0.9V)}$ at 0.9V, though not measured here, have been previously reported by Liu et al. over Pt-ACNT cathode with the values being 101 A·$g_{Pt}^{-1}$ and 374μA·$cm_{Pt}^{-2}$, respectively[62].

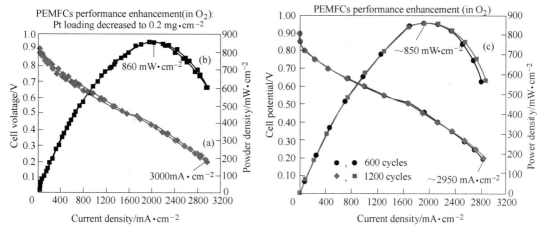

Fig. 6 Electrochemical performance of PEMFC device using Pt-ACNT as cathode catalyst
(a) current density in $O_2$; (b) power density in $O_2$; (c) stability of PEMFC devices

During this study, we also found a key drawback of ACNT prepared by flowing catalyst method, which produce high level of iron (Fe) over the CNTs. Fe could be oxidized and affected Nafion membrane conductivity during fuel cell operation. Such shortcoming, however, can be mitigated by using ACNT prepared by a different method.

For comparison purposes, a side-by-side single-cell polarization study was carried out over Pt-ACNT PEMFCs and conventional ink-based PEMFCs. The operating conditions were the same as previously described with operating temperature at 75℃, using humidified $H_2$ as fuel supply, and $O_2$ and air as oxidant. When air is used as the oxidant, the peak power density for commercial ink-based PEMFC reaches 281mW·cm$^{-2}$ at $V = 0.35V$ (Fig. 7 (a)). The Pt-ACNT PEMFCs, however, displayed a peak power density of 625mW·cm$^{-2}$ under the same condition. In oxygen atmosphere, the peak power density for commercial ink-based PEMFC reaches 470mW·cm$^{-2}$ at $V = 0.4V$ (Fig. 7 (b)), where the Pt-ACNT PEMFCs displayed a peak power density of 860mW·cm$^{-2}$ under the same condition.

From the performance comparison, it can be concluded that the Pt-ACNT PEMFCs exhibited similar performance to that of conventional ink-based PEMFCs in the low current region, where PEMFCs are generally designed to operate for the optimal energy conversion efficiency. However, the current and power densities have exceeded that of the conventional ink-based MEA at low voltage, and high output region. This observation further confirms our design goal that the alignment of

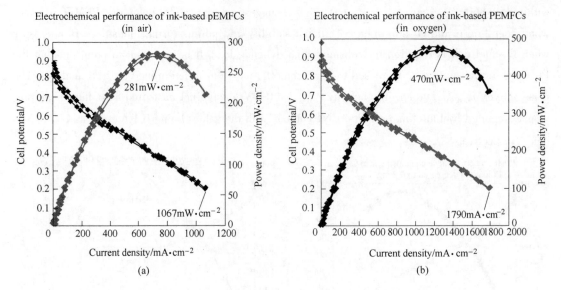

Fig. 7　Electrochemical performance of PEMFC device using conventional ink-based as cathode catalyst
(a) current density and power density in air; (b) current density and power density $O_2$

the ACNT would lead to an enhancement in current and power densities through improved mass transfer. We pointed out, however, that our ink-based MEA was limited only by the one from the commercial source. Others have reported better performances under various conditions, as can be found in Refs[21,63].

From the results comparison, it can safely be concluded that the alignment of carbon nanotubes contributed to performance improvement of the PEMFCs. Such gains are primarily from improved mass-transfer of gas flow and better conductance between the MEA and current collector by Pt-ACNTs being in aligned geometry. In addition, the alignment of ACNT also allows for channeling gas diffusion more efficiently compared with the traditional Pt-C catalysts. These inherent design features allow for substantial increase in reaction rates at the catalytic surface of the Pt-ACNT PEMFCs towards meeting the high power/current density requirements, in addition to the twin benefits of high utilization of the available Pt based catalyst due to the large specific surface area, aiding fast kinetics. Collectively, these benefits allow for the PEMFCs electrochemical behavior to be advanced and the practical realization of PEMFC based devices in automobiles to be furthered along the roadmap.

## 4　Conclusions

The Pt-ACNT cathodic catalyst was obtained by chemical vapor deposition, followed by wet chemistry impregnation to decorate the catalyst, and heat treatment to activate the catalyst. The catalyst was characterized through SEM, TEM, and Raman spectroscopy. The results show that uniform ACNTs coated with Pt mono-dispersive NPs have been directly grown on cathode materials. The observed diameter of the ACNTs ranges from 5 to 20nm and their lengths averaged 10μm, depending on the growth time period. The Pt NPs are well dispersed on the surface of ACNTs and their size ranges from 2 to 4nm. Additionally, the PEMFC performance tests indicate the maximum power density of PEMFC at 860mW · cm$^{-2}$ and current density at 3200mA · cm$^{-2}$ with the Pt NPs loading less than 0.2mg · cm$^{-2}$ when the $O_2$ was used as the oxidant.

**Acknowledgements:** The authors are grateful to the National Science Foundation (NSF) Centers of Research Excellence in Science and Technology (NSF CREST, HRD-0734850) at the Texas A&M University-Kingsville, U. S. Department of Energy (DOE), the Office of Science, the Divisions of Educational Programs, Chemical Sciences and Engineering at the Argonne National Laboratory (ANL), and China Scholarship Council. The Faculty and Student Team (FaST) program collectively funded by NSF and DOE is duly acknowledged. The ACNT-MEA research at ANL was supported by DOE Office of Energy Efficiency and Renewable Energy, Fuel Cell Technologies Program. The assistance and support from Argonne National Fuel Cell group members, the discussion with Dr. Yang from the ANL, Dr. Liang and Dr. Young's from Texas A&M University (TAMU) -College Station and access to the facility at Materials Characterization Center at TAMU are also duly acknowledged. Finally, Dr. Hayes was acknowledged for copy-editing this manuscript.

**Author contributions:** Y. Yuan undertook SEM and RAMAN data analyses. She also co-wrote the manuscript. J. Smith participated in this project as a FaST student fellow. He was trained to prepare carbon nanotubes and modify the ACNTs using platinum precursors. G. Goenaga conducted the student training and optimized the ACNTs fabrication variables. D.-J. Liu was the principal investigator of a DOE funded project of using ACNT as fuel cell electrode support. He is responsible for the concept development, design of experiment and project execution. He hosted the FaST fellows dur-ing their research at Argonne National Laboratory. Z. Luo collected the TEM images and EDS elemental compositions. J. Liu was temporarily employed at the Argonne National Laboratory as a FaST faculty fellow. All authors edited and proof-read this manuscript. D.-J. Liu and J. Liu also oversaw the project progress. Finally, J. Liu is responsible for submitting the manuscript and related figures.

## References

[1] S. Moghaddam, E. Pengwang, Y. Jiang, et al. Nat. Nanotechnol, 2010, 5: 230-236.
[2] R. Esposito, A. Conti. Polymer Electrolyte Membrane Fuel Cells and Electrocatalysts [M]. New York, Nova Science Publisher, 2009.
[3] H. W. Cooper. Chem. Eng. Prog., 2007, 103: 34-43.
[4] S. K. Buratto. Nat. Nanotechnol., 2010, 5: 176.
[5] G. J. O' La, H. J. In, E. Crumlin, et al. Int. J. Energy Res, 2007, 31: 548-575.
[6] J. D. Morse. Int. J. Energy Res., 2007, 31: 576-602.
[7] J. Jagur-Grodzinski. Polym. Adv. Technol., 2007, 18: 785-799.
[8] J. Pan, S. Lu, Y. Li, et al. Adv. Funct. Mater., 2010, 20: 312-319.
[9] J. Larminie, A. Dicks. Fuel Cell Systems Explained, second ed., [M]. London: John Wiley & Sons Ltd., 2003.

[10] S. Gamburzev, A. J. Appleby. J. Power Sources, 2002, 107: 5-7.
[11] J. J. Baschuk, X. Li. Int. J. Energy Res. , 2003, 27: 1095-1116.
[12] A. Kazim, P. Forges, H. T. Liu. Int. J. Energy Res. , 2003, 27: 401-414.
[13] M. A. Hickner, B. S. Pivovar, Fuel Cells, 2005, 5: 213-229.
[14] H. Yu, B. Yi. Fuel Cells, 2004, 4: 96-100.
[15] C. Wang, M. Waje, X. Wang, et al. Nano Lett. , 2004, 4: 345-348.
[16] G. Hörmandinger, N. D. Lucas. Int. J. Energy Res. , 1997, 21: 495-526.
[17] P. M. Gomadam, J. W. Weidner. Int. J. Energy Res. , 2005, 29: 1133-1151.
[18] J. F. Lin, V. Kamavaram, A. M. Kannan. J. Power Sources, 2010, 195: 466-470.
[19] V. Kamavaram, V. Veedu, A. M. Kannan. J. Power Sources, 2009, 188: 51-56.
[20] M. B. Fischback, J. K. Youn, X. Zhao, et al. Electroanalysis, 2006, 18: 2016-2022.
[21] H. A. Gasteiger, S. S. Kocha, B. Sompalli, et al. Appl. Catal. B, 2005, 56: 9-35.
[22] V. Mazumder, Y. Lee, S. Sun. Adv. Funct. Mater. , 2010, 20: 1224-1231.
[23] V. Di Noto, E. Negro, R. Gliubitszzi, et al. Adv. Funct. Mater. , 2007, 17: 3626-3638.
[24] X. Huang. R. Solasi, Y. Zou, et al. J. Polym. Sci Part B: Polym. Phys, 2006, 44: 2346-2357.
[25] Y. Shao, G. Yin, Z. Wang, et al. J. Power Sources, 2007, 167: 235-242.
[26] H. Chu, Y. Shen, L. Lin, et al. Adv. Funct. Mater. , 2010, 20: 3747-3752.
[27] J. Yang, G. Goenaga, A. Call, et al. Electrochem. Solid State Lett. , 2010, 13B: 55-57.
[28] A. L. Dicks. The role of carbon in fue cells [J] . Power Sources, 2006, 156: 128-141.
[29] A. M. Yashchenok, D. N. Bratashov, D. A. Gorin, et al. Adv. Funct. Mater, 2010, 20: 3136-3142.
[30] X. Yu. S. Ye. J. Power Sources, 2007, 172: 133-154.
[31] J. Sung. J. Huh, J. Choi, et al. Adv. Funct. Mater. , 2010, 20: 4305-4313.
[32] T. W. Ebbesen, P. M. Ajayan. Nature, 1992, 358: 220-222.
[33] Y. Chen, Z. lqbal, S. Mitra. Adv. Funct. Mater. , 2007, 17: 3946-3951.
[34] C. C. Chen, C. F. Chen, C. H. Hsu, et al. Diamond Relat. Mater. , 2005, 14: 770-773.
[35] M. Paradise, T. Goswami. Mater. Des. , 2007, 28: 1477-1489.
[36] F. N. Büchi, S. Srinivasan. J. Electrochem. Soc. , 1997, 144: 2767-2772.
[37] J. Yang, D. -J. Liu. Carbon, 2007, 45: 2842-2854.
[38] J. J. Baschuk, X. Li. J. Power Sources, 2000, 86: 181-196.
[39] S. M. Mlia, G. Zhang, D. Kisailus, et al. Adv. Funct. Mater. , 2010, 20: 3742-3746.
[40] R. Yu, L. Chen, Q. Liu, et al. Chem. Mater. , 1998, 10: 718-722.
[41] P. R. Poulsen, J. Borggreen, J. Nygard, et al. American Institute of physics (AIP) Conference Proceedings, Kirchberg, Tirol (Austria), 4-11 March, 2000 (Available from: http: //faculty. washington. edu/cobden/DHCpapers/Poulsen00. pdf) .
[42] Z. He, J. Chen, D. Liu et al. Diamond Relat. Mater. , 2004, 13: 1764-1770.
[43] K. Kinoshita. J. Electrochem. Soc. , 1990, 137: 845-848.
[44] Y. Y. Shao, G. P. Yin, Y. Z. Gao. J. Power Sources, 2007, 171: 558-566.
[45] J. G. Liu, Z. H. Zhou, X. X. Zhao, et al. Phys. Chem. Chem. Phys. , 2004, 6: 134-137.
[46] C. P. Deck, K. Vecchio. Carbon, 2005, 43: 2608-2617.
[47] S. B. Sinnott, R. Andrews, D. Qian. et al. Chem. Phys. Lett. , 1999, 315: 35-30.
[48] R. T. K. Baker, P. S. Harris. in: P. L. Walker. P. A. Thrower (Eds. ) Chem. Phys. Carbon, 1978, 14: 83-165.
[49] S. D. Miguel, O. Scelza, M. Romanmartinez, et al. Appl. Catal. A: Gen. , 1998, 170: 93-103.
[50] N. Giordano, E. Passalacqua, L. Pino, et al. Electrochim. Acta, 1991, 36: 1979-1984.

[51] T. Belin, F. Epron. Mater. Eng. B, 2005, 119: 105-118.
[52] D. -J. Liu, J. Yang, D. J. Gosztola. ECS Trans. , 2007, 5: 147-154.
[53] G. G. Wildgoose, C. E. Banks, R. G. Compton. Small, 2006, 2: 182-193.
[54] Z. P. Luo. Acta. Mater. , 2006, 54: 47-58.
[55] K. Lee, J. J. Zhang, H. J. Wang. et al. J. Appl. Electrochem. , 2006, 36: 507-522.
[56] X. Teng, X. Liang, S. N. Maksimuk, et al. Small, 2006, 2: 249-253.
[57] M. S. Dresselhaus, G. Dresselhaus, R. Saito, et al. Phys. Rep. 2005, 409: 47-100.
[58] L. M. Malard, M. A. Pimenta, G. Dresselhaus, et al. Phys. Rep. , 2009, 473: 51-87.
[59] M. S. Dresselhaus, A. Jorio, M. Hofmann, et al. Nano Lett. , 2010, 10: 751-758.
[60] L. G. Cancado, A. Hartschuh, L. Novotny. J. Raman Spectrosc. , 2009, 40: 1420-1426.
[61] A. A. Green, M. Hersam. Nat. Nanotechnol. , 2009, 4: 64-70.
[62] D. -J. Liu. J. Yang, N. Kariuki et al. ECS Trans. , 2008, 16（2）: 1123-1129.
[63] H. A. Gasteiger, J. E. Panels, S. G. Yan. J. Power Sources, 2004, 127: 162-171.

（原载于 *Journal of Power Sources*, 2011, 196: 6160-6167）

# Preparation and Characterization of $Li^+$-modified $Ca_xPb_{1-x}TiO_3$ Film for Humidity Sensor

Liu Jingbo[1]   Li Wenchao[1]   Zhang Yanxi[2]   Wang Zhimin[2]

(1. University of Science and Technology Beijing, Beijing 100083;
2. Heilongjiang University, Harbin 150080)

**Abstract:** A novel active element for a humidity sensor based on $Ca^{2+}$-doped $PbTiO_3$ perovskite ceramic film has been prepared by the sol-gel technique. The addition of alkali ions, especially, $Li^+$ dopant was found to be beneficial to the improvement of the humidity-sensitive characteristic. The resulting film modified by $Li^+$ calcined at 850℃ for 1h was characterized by X-ray diffraction (XRD) and transmission electronic microscope (TEM). The results obtained show that the crystalline size and the lattice distortion are the key factors that influence the humidity properties. At room temperature, and 100Hz frequency, the variations of sensor's resistance values ($R$) are higher than three orders of magnitude over the working relative humidity ($\varphi$) range, from 8 to 93.5% RH. The curves log $R$ versus $\varphi$ of the designed sensor show excellent linearity, high sensitivity ($R_{8\%RH}/R_{93\%RH} > 10^3$), minimal hysteresis ($< \pm 2\%$ RH) and rapid response ($<8$ s).

**Keywords:** $Li^+$-doped $CaPbTiO_3$; film; characterization; humidity sensor; hysteresis

## 1 Introduction

The advantages of non-stoichiometric perovskite material sensors have been highlighted in many papers[1-4] and many of this family of materials in the form of thin film are well-suited to integration into microsensor devices[5,6]. Among them, lead titanate is the most promising due to its piezoelectric, pyroelectric, mechanical properties[7,8]. Through modification, $PbTiO_3$-based materials can be utilized in many fields, such as surface wave probe, photo-conducting devices, field-effect transistor, permanent memory permanent storage, and ultrasonic transducer[9-12]. Up to now, the $Li^+$-modified $Ca^{2+}$-doped $PbTiO_3$ system humidity-sensitive materials have not been reported. It is known that pure $PbTiO_3$ is less sensitive especially in the low relative humidity range. In the experiment, we have found although $Ca^{2+}$-doped $PbTiO_3$ film sensor has a higher humidity-sensitive characteristic, its poor linearity prevents it from being applied for the purpose of sensor. The $Li^+$ dopant is helpful in the improvement of both linearity and sensitivity of $Ca^{2+}$-doped $PbTiO_3$ sensor.

The fabrication of a miniature sensor, a trend in sensor development, is also feasible with the

---

刘静波，于2001年获得北京科技大学材料科学与工程博士学位，目前任得克萨斯州农工大学—金斯维尔校区副教授。主持或参加由国家自然基金委（美国，中国）、加拿大自然科学和工程研究理事会及韦尔奇基金会资助项目15个，发表约60篇学术期刊文章。

$Li^+$-doped $PbTiO_3$ films. The films are expected to have a rapid response to the analyte due to the relatively short diffusion distance[13]. In this present investigation, the $Li^+$-modified $Ca^{2+}$-doped $PbTiO_3$ humidity-sensitive film is prepared by sol-gel technique which appears to be a very promising method and a cost-effective route[14], because it offers significant advantages such as high purity, chemical homogeneity[15], controlled particle size, and lower processing temperature and requirement of considerably less equipment. In the framework, the preparation parameters of film by sol-gel for a stable humidity sensor with high sensitivity have been optimized. To ensure the desired humidity-sensitive response, the control of nano-particle size and size distribution were carried out. For this purpose, pure $PbTiO_3$, $Ca^{2+}$-doped, $Li^+$-modified sensors have been investigated.

## 2 Preparation and characterization of the sensitive element

The sensor consisted of $Li^+$-modified $Ca_{0.35}Pb_{0.65}TiO_3$ as a sensitive film material, Cu-electrode and $Al_2O_3$-substrate as shown in Fig. 1. A series of films with different compositions in Table 1 were prepared by sol-gel technique using various parameters. The $Ca(OAc)_2$, $Pb(OAc)_2$ and $Ti(O^nBu)_4$ were employed as starting materials. The flowchart for the preparation of $Ca_xPb_{1-x}TiO_3$-based film is shown in Fig. 2. The matrix sol was prepared by mixing a solution of titanium tetrabutoxide monomer in butylalcohol (17mL/20mL) with $Ca(OAc)_2$ and $Pb(OAc)_2$ which

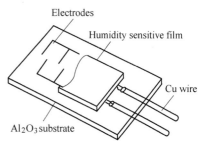

Fig. 1 Illustration of the $Ca_xPb_{1-x}TiO_3$ sensor

had been dissolved in the mixture of water and alcohol (0.1mol/40mL). In order to improve formation of film, acidity (pH) should be controlled in the range of 3-3.5. By spin-coating, the film was then deposited on the $Al_2O_3$ of 12mm×8mm substrate which contained comb-shaped interdigitated six pairs of electrodes, separated by gaps which have the same width as the figures. The films were allowed to preheat at 50℃ for 0.5h and heat-treated incrementally to crystallize at a temperature with a heating rate of 5℃/min. The heat-treatment was carried out at lower temperature range 550–900℃ compared with solid-reaction at >1200℃. At room temperature, the resistance of $Li^+$-modified $Ca_xPb_{1-x}TiO_3$ sensing element with the pores decreases with an increase of relative humidity by three or four orders of magnitude. Our study focused on its sensitivity, hysteresis, response time, etc.

Table 1 Relative humidity by various saturated salt solution

| Saturated salt solution | Relative humidity, RH/% |
|---|---|
| NaOH | 8 |
| $MgCl_2$ | 32.8 |
| $Mg(NO_3)_2$ | 52.9 |
| NaCl | 75.3 |
| $KNO_3$ | 93.6 |

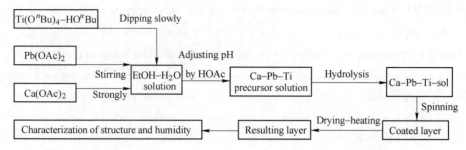

Fig. 2  Flow chart for preparation of $Ca_xPb_{1-x}TiO_3$-based thick film

The synthesized samples were characterized by X-ray diffraction (XRD) (Model, Dmax-ⅢB, Japan Rutaki) using Cu K$\alpha_1$ in order to determine the crystalline phase and the average crystallite size as well as lattice distortion. Crystalline morphology, crystallite size, and size distribution were observed by transmission electronic microscope (TEM) (Model 1200Ex, Japan Hitachi).

The humidity sensor illustrated in Fig. 1 was introduced through a small hole into an otherwise sealed receptacle, and suspended in air above five different saturated salt solutions for the relative humidity change from 8 to 94% RH (shown in Table 1). The standard resistor and the alternative voltage were employed in order to determine the resistance of the humidity sensor and the measuring circuit[16] shown in Fig. 3. At room temperature, the resistance measurement was carried out by impedimeter at the testing parameters of $R_s$ = 100kΩ, $V_{out}$ = 4V, 100Hz, and room temperature (Model 1658, The USA GenRad).

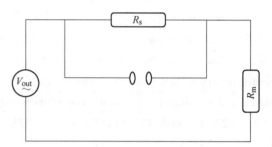

Fig. 3  Measurement circuit of sensor resistance

## 3  Results and discussion

### 3.1  Characterization of film

The XRD spectra (Fig. 4) of the different $Ca_xPb_{1-x}TiO_3$ films indicate that their crystalline structure is mainly per-ovskite $ATiO_3$ and does not show any evidence of any second phase. The main feature of the spectra is the narrowing and intensifying of the XRD peaks with the increase of the calcination temperature which are resulting in grain size increase and higher crystallinity. The average crystalline size as shown in Table 2 was deduced from the half-width of the XRD peak at about $2\theta = 30$ according to Scherrer formula[17].

$$D = \frac{k\lambda}{\beta\cos\theta}$$

where $k$ is a constant, $\lambda$ the wavelength of Cu K$\alpha$, $\beta$ the line broadening and $\theta$ the diffraction angle.

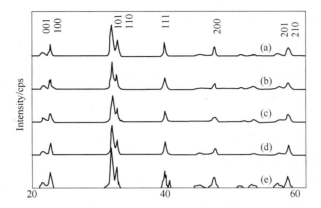

Fig. 4　The XRD spectra of different samples

Table 2　The firing temperature ($T$), preliminary size ($D$),
lattice distortion ($\delta$) of different samples

| | Different samples | $T$/℃ | $D$/nm | $\delta$ |
|---|---|---|---|---|
| (a) | PTO | 800 | 39.21 | 1.0639 |
| (b) | PTO:Ca (0.15) | 700 | 25.69 | 1.0510 |
| (c) | PTO:Ca (0.35) | 850 | 22.61 | 1.0456 |
| (d) | PTO:Ca/Li (0.35/0.05) | 850 | 21.89 | 1.0414 |
| (e) | PTO:Ca/Li (0.35/0.10) | 850 | 22.92 | 1.0456 |

Fig. 5 indicates the lattice distortion $\delta$ monitored through the variation of the lattice parameters $a$ and $c$, which are evaluated from the XRD spectra. The STEM morphological analysis in Fig. 6 shows the film is compacted and continuous with crystalline grain. The crystalline size is around 30-50nm and size distribution is narrow (illustrated in Fig. 6).

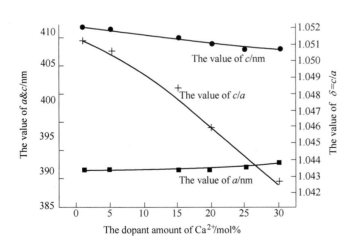

Fig. 5　Lattice distortion of $Ca^{2+}$-doped $PbTiO_3$ samples

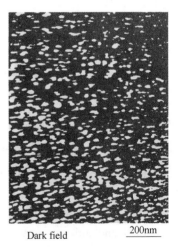

Fig. 6　The TEM microscopy of $Li^+$-modified $CaPbTiO_3$ in a dark field

## 3.2 Humidity characteristic measurement

### 3.2.1 Relative humidity-resistance curves

The dependence of resistance $R$ for different samples on the relative humidity ($\varphi$) at room temperature is shown in Figs. 7 and 8, where the humidity-resistance curves are plotted on a semi-logarithmic graph. From the results, it has been found that pure $PbTiO_3$ only showed the sensitivity in the high humidity above 80% RH. Although the $Ca^{2+}$-doped sample sintered at 950℃/1h showed high sensitivity in the wide range, it can not still be used for sensor purposes due to its poor linear characteristic. Therefore, lithium salt has been doped to enhance the sensitivity and linearity. The doping of 0.5-1 mol% $Li^+$ in $CaPbTiO_3$ film sensor has enhanced the sensitivity by two orders of magnitude, $R_{8\% RH}/R_{93\% RH} > 10^3$. Fig. 8 indicates the relationship between $\log R$ and $\varphi$, $\log R = \alpha + \beta\varphi$, where $\alpha$ and $\beta$ are constants. The correlation factors are 0.9949 and 0.9937, respectively, and show the curves are nearly linear from 8 to 93% RH. There is reasonable assumption that the increase of the sensitivity with $Li^+$ doping may be due to charge density of $Li^+$ ions which highly polarize the absorbed water molecules. This in turn will provide more free $H^+$ or $H_3O^+$ ions for conduction, thus, decreasing resistance[18,19]. The highest sensitivity and best linear behavior has been achieved in the samples at 850℃ for 1h with 0.5% and 1.0mol% $Li^+$ doping.

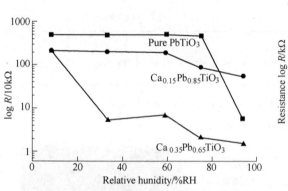
Fig. 7  The curves of $R$-$\varphi$ of pure and $Ca^{2+}$-doped $PbTiO_3$

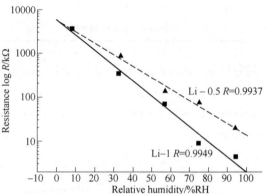
Fig. 8  The curves of $R$ and $\varphi$ of $Li^+$-modified sensor

### 3.2.2 Hysteresis and response characteristic

The typical curves of humidity-resistance of the thick film sensor during adsorption and desorption processing are shown in Fig. 9. The maximum hysteresis is within ±2% RH. It reveals that the sensor has the minimal hysteresis. It is known that Li-O bonding is a hydrophilic group. The $OH^-$ hydroxyl ions are readily absorbed on the sensing element surface. There exists attractive force between water and surface if water vapor is absorbed on the surface of the crystal grain. In order to be desorbed, the force must be eliminated and then hysteresis is caused[16].

The response characteristic of the sensor at room temperature is revealed in Fig. 10. The sensor responds rapidly to the change of relative humidity. The response time which is up to the pores and hydrophilic bonding Li-O is about 8 and 15s for moisture adsorption and desorption processing, respectively. Although the response time of desorption is longer, there is no negative effect on the application of the sensor.

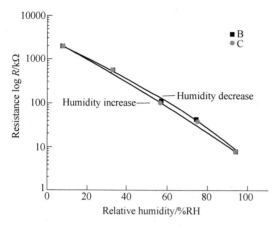

Fig. 9　Humidity hysteresis of Li$^+$-modified sensor

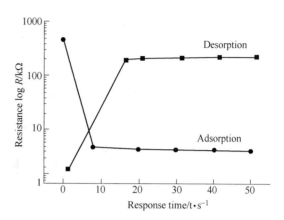

Fig. 10　The frequency characteristic of Li$^+$-modified sensor

### 3.2.3　The frequency characteristic

The sensitivity depends on the frequency at which the resistances are measured. Fig. 11 shows the sensitivity results obtained at 10,000, 1000 and 100Hz, respectively. At 10000Hz, the resistance decreases slightly ($R_{8\%\,RH}/R_{93\%\,RH} > 10$) with an increase of the relative humidity. At 1000Hz, the linearity is poor. At 100Hz, log $R$ is linear with $\varphi$, with a high sensitivity of more than four orders of magnitude over the whole relative humidity range tested. The higher frequency, the lower sensitivity in the humidity ranges. In order to gain high relative humidity sensitivity over the entire range, and good linearity of the curve, low frequency should be selected.

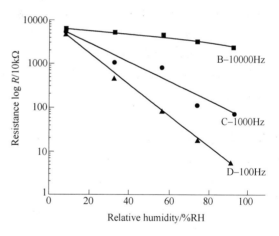

Fig. 11　Response characteristic Li$^+$-modified sensor

### 3.2.4　Stability

The resistance of the humidity sensor was measured after storage in a natural environment for 1 month. It has been found that there exists a little drift of resistance even after a long time, showing the properties of the prepared elements are very stable over time. The results are showed in Fig. 12. The drift occurrence is strictly related to the ion-type sensing mechanism. The drift is limited because the charge carriers are Li$^+$ ions, which are not significantly affected by the formation of stable surface hydroxyls over time. The sensitive properties can also be ascribed to the structure of the material. It is assumed that the stability can be related to Jahn-Teller effect.

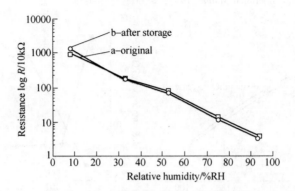

Fig. 12 Stability of the Li$^+$-modified sensor

## 4 Conclusion

(1) Li$^+$-modified CaPbTiO$_3$-based homogeneous thin film for humidity sensor has been prepared by sol-gel processing, and the heat treatmens is lower than that of solid reaction.

(2) Li$^+$ dopant can improve the sensitivity and linearity of the Ca$_x$Pb$_{1-x}$TiO$_3$ sensor and its appropriate content is 0.5 – 1 mol%.

(3) The humidity sensor has exhibited excellent characteristics of good linearity, high sensitivity, minimal hysteresis, rapid response time, and wide working range.

**Acknowledgements:** The studies were supported by Natural Science Foundation of Heilongjiang Province (No. E9919). The authors are thankful to Dr. Jierong YING of Tsinghua University and Keton Kersting of Beijing Information Technology Institute for revision of this paper.

### References

[1] J. Riegel, K. H. Hardtl. Analysis of combustible gases in air with calorimetric gas sensors bases on semiconducting BaTiO$_3$ ceramics [J]. Sens. Actuators B1, 1990: 54-57.

[2] Q. Fabin, L. Xi, G. Kui, et al. High pressure effect on nanocrystalline BaTiO$_3$ and its humidity sensitive properties [J]. J. Functional Mater. , 1999, 30 (2): 186-187.

[3] T. J. Hwang, G. M. Choi. Humidity response characteristic of barium titanate [J]. J. Am. Ceram. Soc. , 1993, 76 (3): 766-768.

[4] G. Gerlach, K. Sager. A piezoresistive humidity sensor [J]. Sens. Actuators A43, 1994: 181-184.

[5] B. Morten, G. De Cicco, M. Prudenziati. A thick-film resonant sensor for humidity measurements [J]. Sens. Actuators A, 1993, 37/38: 337-342.

[6] Z. Xiaohua, O. Toft, C. Quanxi, et al. Study on sensing mechanism of Mg-SrTiO$_3$ thick film [J]. J. Functional Mater. , 1999, 30 (1): 43-45.

[7] R. S. Nasar, M. Cerqueira, E. Longo, et al. Experimental and theoretical study on the piezoeletric behaviour of

barium doped PZT [J]. J. Mater. Sci. , 1999, 34: 3659-3667.

[8] D. S. Paik, S. E. Park, T. R. Shrout, et al. Dielectric and piezoeletric properties of perovskite materials a cryogenic temperature [J]. J. Mater. Sci. , 1999, 34: 469-475.

[9] Z. Tao. PbTiO$_3$ [J]. J. Functional Mater. , 1997, 28 (6): 604-606.

[10] Y. -H. Jeong, S. -H. Lee, J. -H. Yoo, et al. Voltage gain characteristic of using PbTiO$_3$ system ceramics [J]. Sens. Actuators, 1999, 77: 126-130.

[11] G. De Cicco, B. Morten, D. Dalmengo, et al. Pyroeletricity of PZT-based thick films [J]. Sens. Actuators, 1999, 76: 409-415.

[12] J. Wang, B. Xu, G. Liu, et al. Influence of doping on humidity sensing properties of nanocrystalline BaTiO$_3$ [J]. J. Mater. Sci. Lett. , 1998, 17: 857-859.

[13] R. Radhouane, T. Ban, Y. Ohya, et al. Humidity-sensing characteristic of divalent-metal-doped indium oxide thin films [J]. J. Am. Ceram. Soc. , 1998, 81 (2): 321-327.

[14] W. Qu, W. Wbodarski, J. -U. Meger. Comparative study on micromorphology and humidity sensitive properties of thin-film and thick-film humidity sensors based on semiconducting MnWO$_4$ [J]. Sens. Actuators, 2000, B64: 76-82.

[15] S. Kazaoui, J. Ravez. Dielectric relaxation in Ba(Ti$_{0.8}$Zr$_{0.2}$)O$_3$ ceramics prepared from sol-gel and solid state reaction powders [J]. J. Mater. Sci. , 1993: 1211-1219.

[16] Y. Xin, S. Wang. An investigation of sulfonated polysulfone humidity-sensitive materials [J]. Sens. Actuators, 1994, A1: 147-149.

[17] B. D. Cullity. Elementary of X-Ray Diffractions, Second Eddition [M]. MA: Addison Alesley, 1978.

[18] M. K. Jain, M. C. Bhatnagar, G. L. Sharma. Effect of Li$^+$ doping on ZrO$_2$-TiO$_2$ humidity sensor [J]. Sens. Actuators, 1999, B55: 180-185.

[19] K. -S. Chen, T. -K. Lee, F. -J. Liu. Sensing mechanism of a porous ceramic as humidity sensor [J]. Sens. Actuators, 1999, B55: 96-111.

(原载于 *Sensors and Actuators B*, 2001, 75: 11-17)

# 摩托车尾气催化净化技术原理与应用

秦建武

(北京信诚联合科技咨询有限公司,北京 100080)

**摘 要**:催化净化系统是摩托车尾气排放控制的最有效方法。摩托车催化器是氧化型的金属蜂窝载体催化剂。在催化剂涂层中,一般添加稀土 La、Ce 和 Ba、Zr 等基本金属(base metal)作为稳定剂和助催化剂,以改善涂层的热稳定性和增加储氧功能。所用贵金属活性组分多为 Pt 和 Rh。催化剂的性能劣化和失效主要是由热老化和中毒引起的。为了保证摩托车催化剂的催化效果,必须进行发动机化油器的技术匹配,以改善空燃比,降低原始排放。为了保证整车一致性达到国家标准要求的限值和耐久性,在应用催化器时,须考虑催化剂的工作特性和是否有必要配装二次空气产品。

**关键词**:摩托车;排放控制;催化剂

## Catalytic Technology and Application for Motorcycle Emission Control

Qin Jianwu

**Abstract**: Catalyst technology is one of the most effective ways for motorcycle exhaust control. Motorcycles are fitted with oxidation catalytic metal type. Thermal durability and oxygen storage of catalysts can be improved by adding various rare earth elements La, Ce and base metal Ba, Zr oxide stabilizers and promoters. Catalyst performance deteriorations are caused by thermal deactivation and catalyst poisoning. For ensuring catalyst effect, adjusting the carburetor is essential to improve Air/Fuel ratio and emission level. In order to meet limits and durability of the stricter standard identically, when the use of catalytic converters on motorcycle, properties of catalyst must be considered and second air injection (SAI) system are necessarily installed.

**Keywords**: motorcycle; emission control; catalyst

## 1 引言

随着我国经济的高速发展,机动车的生产量和保有量迅速增长。我国作为世界第一大摩托车生产国,摩托车的年产量已近 1500 万辆,摩托车的排放问题日益突出。2004 年,我国开始实行第二阶段摩托车排放标准(欧Ⅱ标准),从而在政策法规方面,为摩托车催

---

秦建武,1987~1990 年于北京科技大学师从李文超教授攻读硕士学位。目前在南京亿达高科环保技术有限公司工作,任执行董事。

化净化技术的实际应用提供了根本保证。

与汽车闭环电喷发动机（EFI）不同，摩托车发动机由化油器提供混合气。而大多数摩托车化油器结构简单，没有加速系统和加浓系统，为了保证摩托车在全工况范围下的稳定工作，发动机的混合气一般很浓[1]。此外摩托车发动机的转速远高于汽车发动机，其排气空速和热冲击均很大。因此，摩托车催化器产品和使用条件与汽车三元催化器相比有很大的差异。在解决摩托车排放问题时，需要了解摩托车发动机的特点和摩托车催化器的产品特性和技术原理，进行发动机与催化器的系统匹配，才能保证排放控制效果。本文综述了摩托车催化净化系统的产品技术和应用特点，在实际应用的试验基础上，对催化剂的老化及失效原因进行分析，对二次空气产品技术和整车系统匹配进行了技术讨论，以期为摩托车排放污染的治理工作提供参考。

## 2 摩托车催化系统的技术介绍

### 2.1 摩托车催化剂的催化反应原理

在影响摩托车排放污染物浓度的诸多因素中，进入发动机气缸的燃油（Fuel）和空气（Air）的比例，简称为空燃比（A/F），是最显著的因素。

A/F 为 14.7 时，为理论空燃比，此时混合气燃烧最完全；当 A/F < 14.7 时，混合气偏浓，称为贫氧燃烧，CO、HC 排放增加，而 $NO_x$ 较少；当 A/F > 14.7 时，混合气偏稀，为富氧燃烧，此时 CO、HC 排放减少，但 $NO_x$ 增加。在应用摩托车催化剂时，其转化率与尾气中的氧含量即 A/F 密切相关。

在催化剂的作用下，发动机的排气发生以下氧化和还原反应[2]：

CO、HC 的氧化反应：

$$C_m H_n + (m + n/4) O_2 = mCO + n/2 H_2O$$
$$CO + 1/2 O_2 = CO_2$$

NO（or $NO_2$）的还原反应：

$$NO(or\ NO_2) + CO = 1/2 N_2 + CO_2$$
$$NO(or\ NO_2) + H_2 = 1/2 N_2 + H_2O$$
$$(2 + n/2) NO(or\ NO_2) + C_m H_n = (1 + n/4) N_2 + mCO_2 + n/2 H_2O$$

当排气中的氧含量和 $NO_x$ 浓度比较低，不足以氧化 CO 和 HC 时，CO、HC 将进行以下反应：

水煤气反应（Water gas shift reaction）：

$$CO + H_2O = CO_2 + H_2$$

烃—水蒸气重整反应（Steam reforming reaction）：

$$C_m H_n + m H_2O = mCO + (m + n/2) H_2$$

由此可见，如果摩托车发动机混合气较浓，而排气中的氧含量又不足，会发生烃—水蒸气重整催化反应，则可能发生 HC 排放降低，而 CO 不减反增的现象，不利于摩托车的排放达标。

催化剂在以上化学反应中，对参与化学反应的分子起一种活化作用，使反应物分子的化学结构发生有利于化学反应的变化，催化剂本身并不参与最终反应产物，借助催化作用，使上述反应的活化能降低，从而加快反应速度。

因摩托车发动机的空燃比 A/F 一般小于 14.7，其排放以 CO、HC 为主，$NO_x$ 较少，用于化油器发动机的摩托车催化剂均为氧化型（二元催化剂），主要催化 CO 与 HC 的氧化反应。只有摩托车采用闭环控制电喷发动机后，三元催化剂才有可能在摩托车排放控制上得到应用。因为只有在发动机以理论空燃比（A/F 为 14.7）范围运行的条件下，三元催化剂才可能同时具有高效净化 CO、HC 和 $NO_x$ 三种有害气体的能力。而闭环控制发动机能精确控制空燃比在理论空燃比范围内。

## 2.2 摩托车催化器的产品技术

摩托车催化器由载体、高比表面积涂层和含贵金属活性组分的催化材料组成。根据载体材质，分有金属载体和陶瓷载体两种，其中金属载体又可按结构分为蜂窝式和热管式。

金属蜂窝载体催化剂因其较大的比表面积和较高的转化率，一般作为主催化剂在摩托车上得到广泛应用，其安装在摩托车消声器内，距发动机较远。热管式催化剂具有较低的排气背压，便于安装到靠近发动机排气管位置，可改善催化剂的起燃和减轻蜂窝催化剂的负担，成本低但转化效率也低，可配合蜂窝主催化剂使用（见图1）。

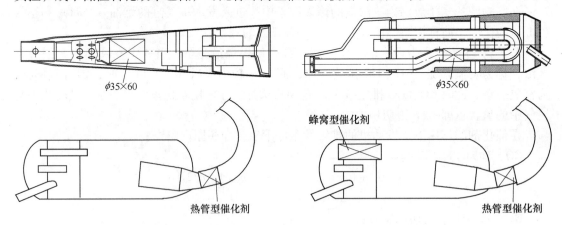

图 1 蜂窝型和热管型催化剂在几种摩托车排气消声系统中的安装图示

### 2.2.1 金属蜂窝载体

蜂窝载体是催化剂的骨架，催化剂分散在蜂窝载体的内孔表面上，可改善催化剂的热传导性和机械强度。与汽车三元催化器不同，摩托车催化剂主要采用金属蜂窝载体，其材质为 Fe、Cr、Al 和稀土添加剂 Ce、Y 的高温合金。与陶瓷载体相比，金属载体具有耐热冲击、排气阻力小、机械强度高、热容量小等优点（见表1）。

表 1 金属蜂窝载体与陶瓷蜂窝载体性能比较

| 性 能 参 数 | 陶瓷蜂窝载体 | 金属蜂窝载体 |
| --- | --- | --- |
| 孔密度/CPSI | 400 | 400 |
| 壁厚/mm | 0.15 | 0.04 |
| 开口率/% | 76.6 | 91.6 |
| 几何表面积/$m^2 \cdot L^{-1}$ | 2.8 | 3.5 |
| 热导率/$W \cdot m^{-1} \cdot K^{-1}$ | 0.8 ~ 1 | 14 ~ 22 |
| 热容量/$kJ \cdot kg^{-1} \cdot K^{-1}$ | 1.05 | 0.5 |
| 热膨胀系数（$\Delta L/L$）/$K^{-1}$ | $1 \times 10^{-6}$ | $15 \times 10^{-6}$ |
| 最高承受温度/℃ | 1360 | 1500 |

金属蜂窝载体对催化剂的性能有以下影响：
（1）几何参数：
1）几何表面积（GSA）越大，越有利于催化反应的进行；
2）开口率（OFA）越大，排气背压越小；
3）孔密度（CPSI）增大，可提升催化剂的空速特性和转化率，但排气背压加大。
（2）物理参数：
1）热容量：热容量越小，催化剂起燃越快；
2）热导率：热导率越大，催化剂越易被加热起燃。

根据发动机的性能参数和排放控制目标，在设计催化剂技术方案，选定金属载体的尺寸和孔密度时，既要考虑发动机的动力性和经济性，又要满足催化剂的转化率指标和空速特性。

### 2.2.2 涂层

涂层是把一种由多孔物质（$\gamma$-$Al_2O_3$）与稳定剂和助催化剂组成的混合物浆料（Slurry），涂布在蜂窝载体的内孔表面上，经烘干焙烧而成的高比表面积的衬底材料。其作用为[3]：
（1）增加催化剂的有效面积，提高催化剂的热稳定性；
（2）提供合适的微孔结构和催化活性中心；
（3）高度分散贵金属活性组分，节省贵金属用量，降低成本。

对氧化铝涂层的要求为：分布均匀，表面积大，与金属载体有很高的附着力，热稳定性好。

由于催化剂浆料（Slurry）与金属载体材质不同，热膨胀系数有很大的差异，若浆料配方和涂布工艺不合适，会影响催化剂与金属载体的结合强度，在催化剂使用过程，容易造成涂层脱落，从而影响催化剂的性能和寿命。在进行催化剂浆料涂布时，通常需要对金属载体进行预处理，并调整浆料配方。

涂层中的$\gamma$-$Al_2O_3$，在800℃以上，易发生相变形成$\alpha$相，使微孔烧结，晶粒变粗，导致涂层的比表面积降低，催化活性下降。大量研究表明[4-6]，在浆料中添加稀土La、Ce和贱金属Ba、Zr、Ca等氧化物稳定剂，可提高涂层的耐温能力，阻止晶粒团聚，可提高氧化铝涂层的相变温度。同时，稀土Ce通过氧化铈的变价反应，可作为储氧材料，在富氧气氛中储氧，在贫氧气氛中释氧，从而改善贫氧状态下CO、HC的氧化反应。另外，稀土Ce还促进水煤气反应，并对铂族贵金属具有助催化剂作用，可降低贵金属的用量。

### 2.2.3 含贵金属活性组分的催化材料

一般选用铂族金属铂（Pt）、铑（Rh）、钯（Pd）为催化剂的主要活性组分。可以根据排放标准的限值及耐久性要求、发动机的排放情况和成本要求，调整贵金属的比例和含量，如Pt/Rh为7/1～5/1、纯Pt、30～50g/ft³等。贵金属的比例和含量直接影响催化剂的性能和成本。

在贵金属活性组分中，Rh对CO和$NO_x$的反应活性好，Pt对CO和HC的氧化活性好，Pt、Rh抗硫（S）中毒性好；而Pd对HC的氧化活性好，但抗S中毒性差。另外，由于含钯催化剂对水煤气反应有较差的催化活性，而且对 HC 的吸附性较强，在贫氧条件下易促进 HC 的不完全氧化反应，可能会导致 CO 浓度增高。因此，摩托车催化剂仍以贵金属 Pt 和 Rh 为主，而很少使用 Pd[6,7]。

对于贱金属催化剂，其吸氧能力较强而且牢固，氧分子不易脱附，使催化剂的供氧速率受到制约。在发动机高污染、高空速和低温下，与贵金属催化剂相比，具有致命弱点，即易烧结失效、易发生硫中毒、空速特性差、需加大催化剂体积等。由此可见，贱金属催化材料一般只作为助催化剂，而不能作为主要的活性组分。

## 2.3 摩托车催化器的工作特性

在摩托车排放控制中应用催化器时，如何合理地安装使用催化剂和保证催化剂的正常工作，需要了解催化器的主要特性。

（1）温度特性。温度特性是指 CO、HC 在一定的浓度和空速条件下，净化率随温度变化的情况。CO、HC 的净化率为 50% 的温度，既为催化剂的起燃温度，也是催化剂开始有效工作的温度。起燃温度越低，催化剂发挥作用越快。一般新鲜催化剂的起燃温度为 250~300℃，老化后会有一定程度的升高。

（2）空速特性（SV）。所谓空速是 1 小时内通过催化剂的气体在标准状态下的体积与催化剂的体积之比。对于同一催化剂，在一定条件（温度、浓度、压力）下，由于排气空速不同，其净化效果也不同。摩托车的发动机转速越高，排气空速越大，排气在催化剂停留的时间越短，对催化反应不利，催化剂的转化效率降低。摩托车排放控制能否达到预期目标，需要综合考虑催化剂在各种工况下的排放净化效果。

（3）空燃比特性（A/F）。对同一催化剂，在相同的温度和空速条件下，排气中的氧含量不同，CO、HC 或 $NO_x$ 转化率随氧含量的变化而变化，富氧条件有助于提高 CO 和 HC 的转化率，而贫氧条件有助于提高 $NO_x$ 的转化率。摩托车催化剂主要催化转化 CO 和 HC，为了保证净化效果，摩托车排气系统需要有充足的氧含量，即较大的空气过量系数。可以通过化油器的优化匹配或引入二次空气系统补充氧气。

（4）热稳定性。即催化剂长期在高温下运行，引起涂层结构和活性组分化学状态的改变，导致催化剂的活性和净化率的变化情况。摩托车发动机的排气温度和流速变化很大，这就要求催化剂具有良好的耐热性能，防止催化剂的使用寿命下降（见图 2）。

（5）抗中毒性能。摩托车催化剂在使用过程，由于燃油和机油中的铅（Pb）、硫（S）、磷（P）等毒性化合物吸附在催化剂上或与活性组分发生化学反应，所引起的催化剂活性下降或失效现象，称为催化剂中毒。其中铅是很强的催化剂毒物，目前还没有任何催化剂可以避免铅中毒。尽管我国已推广使用无铅汽油，但仍有微量残余铅存在，而催化剂的铅中毒是累积性的，会明显破坏催化剂的性能。添加稀土化合物可改善催化剂的抗铅、硫中毒能力。

## 2.4 摩托车催化器老化及失效原因的分析讨论

催化器的老化实际上是催化剂的性能劣化，直接影响催化器的性能和耐久性，其过程是一个复杂的物理、化学变化过程。而催化剂的失效主要是由高温热老化、催化剂中毒及

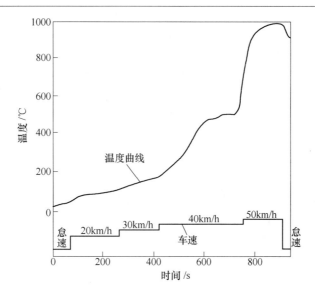

图 2　摩托车催化器中心温度随车速的变化关系

机械损伤造成的。其中高温热老化是摩托车催化剂的主要失活方式。

### 2.4.1　催化剂的热老化失活机理

催化剂在高温（850℃以上）下使用，高比表面积涂层材料中的氧化铝发生相变和微孔烧结，比表面积大幅下降，同时，贵金属活性组分的晶粒也发生烧结长大或形成合金，导致催化剂的性能下降或失活。通过添加稀土金属氧化物等稳定剂可以改善催化剂的耐高温性能。

此外，催化器的规格和外部使用条件不合适，也容易产生催化剂的高温老化。

（1）发动机的排放情况。若发动机燃烧不好，CO 和 HC 排放过高，催化剂负担过重，长期使用，会造成热老化，影响催化剂的使用寿命。若点火系统不良造成发动机持续失火，导致催化剂温度大幅升高，会引起严重的热老化，直接使催化剂高温失效。对于二冲程摩托车，由于排气中 HC 含量远高于四冲程，在氧化反应过程大量放热，导致催化剂升温过高（可达 1000℃），更易造成催化剂热老化失效[8]。

（2）催化器的规格。在设计摩托车排放控制方案时，应综合考虑各种影响催化剂耐久性的因素，以确定催化器的尺寸和孔密度。当摩托车发动机排放较差或有失火现象时，若采用高孔密度的催化剂，由于转化效果好，更易产生催化剂温升，反而比低孔密度的催化剂容易热老化，此外其抵抗发动机失火损坏能力也差。

### 2.4.2　催化剂的中毒失活

催化剂中毒可分为物理中毒和化学中毒。其中物理中毒也称物理覆盖，主要是由机油中的磷、锌、钙、燃油清洁剂中的硅和汽油添加剂 MMT 中的锰等燃烧后所产生的化合物灰烬以及含碳沉积物（碳结焦），覆盖在催化剂表面，引起气体扩散通道受阻，导致催化性能下降[3]。覆盖引起的催化剂中毒是暂时中毒，失活也是可逆的，在催化剂工作过程中可自动再生，但有可能影响催化剂性能和寿命，严重的表面覆盖会导致最高转化率大幅下降。而化学中毒主要由汽油中的铅、硫和卤化物与贵金属活性组分发生化学反应，导致催化剂的永久失活。

### 2.4.3 机械损伤引起的失效

尽管摩托车的减震系统较差，行驶时的振动较大，但由于采用金属载体，摩托车催化器有较高的抗震性能，一般的机械振动不会损伤催化器。但在摩托车工况变化很快时，金属载体处于很高的温度变化率的排气下，较薄的金属芯片和较厚的外壳之间会产生不同的热膨胀变形，引起较高的热负荷和机械负荷，出现金属蠕变疲劳现象。若金属载体制作不良，会发生芯体与外壳的脱离，造成催化器的失效。

## 3 摩托车催化净化技术的二次空气产品（SAI）

因摩托车尾气排放主要为 CO 和 HC，所用催化剂为氧化型。为了保证氧化反应的催化效果，除了调整匹配化油器外，许多情况需要引入二次空气，以调节空燃比（A/F）在偏稀的范围。二次空气的导入分为发动机前的电控补气技术和发动机后的二次空气单向补气阀方式。

### 3.1 电控补气（电控化油器）技术

电控补气产品是在化油器底座下加装进气管，并与旁通电磁阀相连，电磁阀进气端与空滤器连接，一般从点火线圈取发动机转速信号，反馈到电控单元 ECU，ECU 按设定的程序，根据发动机的转速情况，向电磁阀发出脉冲电信号指令，通过电磁阀的开启/闭合动作频率，控制二次空气的补入量；当发动机转速不高时，节气门开度不大，发动机的进气有压差，电磁阀开启后会吸气补充到进气歧管，与混合气一起进入发动机，达到调节空燃比的目的。而在发动机高速工况下，节气门开度很大或全开，此时进气压差很小或没有，电控补气效果不明显。

为了控制成本，电控补气产品一般为开环控制（无氧传感器），其补气量不能根据排气的实际氧含量进行自我调节。因此，为了精确控制混合气的空燃比，必须与不同的发动机进行良好的匹配，以确定 ECU 的控制程序和参数。若控制匹配合适，会很好地改善发动机的排放，并有节油效果。但因电控补气为脉冲进气，若匹配调整不当，也会发生发动机动力下降，摩托车怠速不稳，启动困难等不良现象。

### 3.2 机后二次空气技术（SAI）

机后二次空气产品包含单向进气阀（RV）和二次空气控制阀（AICV），图 3 为安装示意图。

（1）单向进气阀 RV。通过发动机排气的脉冲压差，当负压时，RV 从外界吸入空气到排气管内，为氧化催化剂的高效工作补充氧含量。通过调整设定 RV 的金属簧片的厚度、簧片止口和开口区的面积等参数，来控制补气量。此外，RV 的补气量要适当，当发动机的空燃比正常，此时若 RV 补气量过大，反而对 CO 和 HC 的催化氧化起负面影响，因为过量的外界空气会降低催化剂的温度，影响转化效率[9]。由于 RV 是机后补气，不会对发动机的性能产生影响，其成本也远低于电控补气产品，但是 RV 在使用中，存在进气噪声问题，而且若安装在跨骑式摩托车的排气管上，对车辆外观有影响。

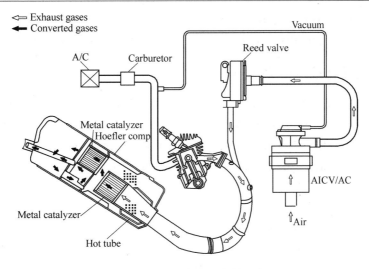

图 3  机后二次空气产品安装示意图

对于 RV 的使用，以下因素会影响 RV 的补气效果：

1）发动机工况。由于 RV 采用簧片式单向阀结构，从试验测量结果看，受簧片响应频率的制约，RV 在发动机转速中低速时具有一定的补气效果，而在高速情况下，补气量很小（见图 4）。

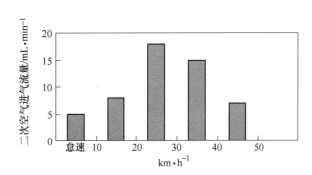

图 4  某轻便摩托车的二次空气进气流量与车速的变化关系

2）RV 的安装位置。单向进气阀 RV 一般安装在排气管上或发动机排气缸头；若安装在排气管的位置距消声器入口过近，发动机排气进入消声器时，会产生反射气流，导致排气脉冲压差的正压过大而负压过小，影响 RV 的进气效果。为了保证补气效果，可对发动机排气管的不同位置测试排气压差。一般情况下，RV 安装位置靠近发动机的补气效果较好。

3）进气连接管尺寸。连接管的直径过小和长度过长，会影响进气量，不利于 RV 的补气。

（2）二次空气控制阀 AICV。对四冲程摩托车，需加装 AICV。因为当摩托车急减速时，会有大量未燃高温混合气从发动机排入排气系统，若 RV 仍正常吸气，未燃高温混合气与吸入的空气相遇，会发生爆燃，出现消声器"放炮"现象。通过把 RV 与 AICV 连接，当摩托车急减速时，AICV 会阻断 RV 的进气通路，阻止外界空气的继续吸入，从而防止消声器的"放炮"。对二冲程摩托车，因其排气温度低，很少发生爆燃，一般不用加装 AICV。

因为二次空气（SAI）产品在发动机高速工况时的补气效果不佳，对于原机排放比较恶劣，特别是高速工况混合气很浓的发动机，还须对化油器进行匹配调整。

## 4 摩托车催化净化技术的系统匹配

发动机化油器的合理匹配，可降低发动机的原始排放，减轻催化剂的负担，为催化剂提供良好的使用条件（如合适的空燃比），保证摩托车催化器更有效持久地发挥作用（见图5）；另一方面，摩托车催化器安装在发动机的排气消声器内，必然会对发动机的性能和消声器造成影响，而对催化器与发动机及整车进行合理的系统匹配，可保证发动机的动力性和经济性，并保证消声器的正常工作。

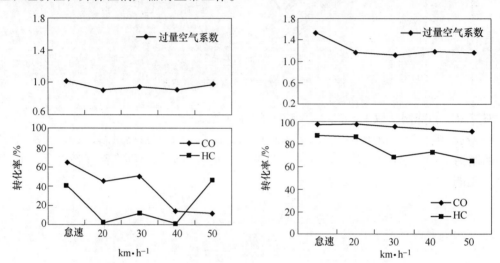

图 5　化油器调节前后过量空气系数与催化器净化效率的变化情况（未加二次补气装置）

### 4.1 发动机化油器的匹配

以前我国没有制订严格的摩托车排放标准，在设计化油器产品时，很少考虑对发动机排放的影响，而化油器对发动机的空燃比和排放水平有至关重要的影响。在对摩托车进行排放控制时，应首先在发动机台架或整车转鼓（底盘测功机）上进行化油器匹配试验，调节化油器至富氧状态，可明显改善尾气排放，尤其是 CO 排放（见表2）[1]；匹配合适的化油器应不能明显影响发动机的动力性（动力损失不高于3%）。一般从以下几方面匹配调整化油器：

表 2　调整摩托车化油器的整车排放工况测试结果

| 车　型 | 调整化油器 | $CO/g \cdot km^{-1}$ | $HC + NO_x/g \cdot km^{-1}$ |
| --- | --- | --- | --- |
| 骑式 125 摩托车（四冲程） | 化油器 A | 5.982 | 0.655 |
|  | 化油器 B | 11.742 | 1.440 |
|  | 化油器 C | 7.709 | 0.841 |
| 踏板 125 摩托车（四冲程） | 化油器 1 | 12.996 | 1.446 |
|  | 化油器 2 | 8.432 | 0.807 |
|  | 化油器 3 | 17.712 | 3.096 |

(1) 主量孔。主量孔的大小与通过的燃油流量成正比，调配主量孔后，在不同的发动机和负荷工况下，对空燃比有不同的影响。

(2) 油针和主喷嘴。对于摩托车用柱塞式化油器，调节油针结构和尺寸（直径、锥度及位置）、主喷嘴参数等，都会影响空燃比。

(3) 柱塞切角高度。切角高度直接影响喉管真空度和出油量，切角高度小，空气吸入阻力大，混合气变浓。

(4) 产品加工控制。尽管我国摩托车零部件制造水平有所提高，但化油器关键部件的加工精度差，离散性较大，为了保证优化匹配化油器后的效果，应严格控制量孔、油针等的加工精度和产品制造的一致性。

尽管匹配调整化油器可以明显降低发动机的原始排放，但难以保证效果的一致性，而且摩托车在使用过程，化油器会发生"漂移"，偏离匹配好的状态。为了整车一致性满足更严格排放标准的限值和耐久性要求，还须配装催化器，并建议加装二次空气产品（SAI）以保证催化器正常工作所需的氧含量（见图6）。

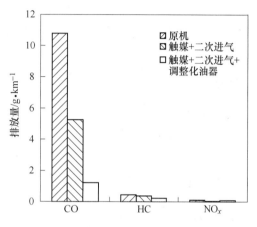

图6　骑式125摩托车不同状态下的工况法测试结果

### 4.2　催化器与整车的匹配

摩托车催化器与发动机及整车一般需要进行以下的匹配工作：

(1) 测试整车发动机的原始排放，若原机排放恶劣，应通过台架发动机和整车系统试验，重新匹配调整化油器，改善发动机的燃烧状况和空燃比。

(2) 根据发动机参数和优化匹配后的整车排放水平，考虑排放标准的限值及耐久性要求、消声器结构和成本因素，确定催化剂的贵金属含量和比例、载体的尺寸和孔密度。测试排气系统及消声器在不同工况下的温度分布，选择合适的催化器安装位置。

(3) 在台架发动机和整车转鼓上进行综合性能试验（包括发动机的外特性、净化效果等），若综合性能满意，则进行整车工况试验，以考核行驶性能；并可通过发动机台架的老化试验，对催化器进行耐久性评估。

### 4.3　催化器对摩托车排气消声器的影响

催化器一般安装在摩托车消声器内，因此会对消声器的装配制造和使用带来一些不良影响，特别是跨骑式摩托车；同时也对消声器的消音效果产生影响。

(1) 对消声器的温度影响。催化器通过净化CO和HC而放热，使催化器安装位置的温度比原消声器高许多（高达几百度），必须对净化消声器的表面温度进行控制，以保护乘骑者不被烫伤和防止跨骑式摩托车消声器表面电镀层的高温变色。可通过优化催化器结构及安装位置，以控制反应温度；使用耐温隔热材料或加装防护板，以降低消声器表面温度。

（2）对消声器装配生产的影响。要合理调整净化消声器的装配生产工艺，以避免酸洗、电镀等环节对催化剂的损坏，确保催化剂的质量。可采用催化器的组合装配或双层消声器外壳等办法，应对电镀问题。在焊装催化器时，还要注意消声器结构的密封，防止排气的泄漏，保证催化效果。

（3）对消声器材质，结构和噪声的影响。加装催化器后，改变了排气温度和空气动力性。由于排气温度的提高，要求消声器的材质能耐高温腐蚀，玻璃纤维棉吸音材料也应改为耐温隔热的陶瓷纤维棉。由于催化器载体的蜂窝孔道结构，其本身也是一种消声结构，可降低高频噪声。而摩托车消声器以抗性结构为主，在中低频消声效果好[10]。因此，若优化设计消声器和催化器的安装结构，可改善消声器的高频消声效果。

## 5 结语

（1）催化净化技术是摩托车排放控制广泛使用的成熟可靠技术，与其他净化方法相比，具有合理的性价比，更容易在摩托车上实施应用。

（2）面对未来日益严格的排放法规和摩托车产品的市场竞争，期望开发低成本高性能的摩托车催化剂，如降低贵金属含量或开发低成本的用于摩托车的含钯催化剂，并提高催化剂的活性和保证其耐久性。

（3）在应用催化净化技术控制摩托车排放时，除了要求催化净化效果外，还要综合考虑整车的动力性和经济性指标，以及对排气消声器的影响，对系统进行优化设计与匹配是必要的。

（4）为了保证摩托车产品一致性达到国家标准的限值和耐久性要求，在采用匹配发动机化油器和配装催化器的同时，建议加装二次空气产品。

## 参 考 文 献

[1] 国家环保总局标准司. 摩托车排放污染防治技术指南 [M]. 北京：中国环境出版社，2003.
[2] Kummer J. Catalysts for Automobile Emission Control [J]. Prog. Energy Combust Sci., 1980（6）：177-199.
[3] 胡逸民，李飞鹏. 内燃机废气净化 [J]. 北京：中国铁道出版社，1994.
[4] 王建昕，傅立新，黎维彬. 汽车排气污染治理及催化转化器 [M]. 北京：化学工业出版社，2000.
[5] J. C. Summers, S. A. Ausen. Interaction of Cerium Oxide with Noble Metals [J]. J. Catalysis, 1979（58）：131.
[6] 盐川二朗著，翟羽伸译. 稀土的最新应用技术 [M]. 北京：化学工业出版社，1993.
[7] G. Bickle, T. Yoshikawa, A. Schafer-Sindlinger, et al. Emission Control System for Two-Stroke Engine-A Challenge for Catalysis [J]. SAE Paper 982710, 1998.
[8] D. R. Palke, M. A. Tyo, J. E. Dillon. Catalytic Aftertreatment of Vehicle Exhausts from Two-Stroke Motorcycles [J]. SAE paper, 1996.
[9] H-C. Wu, S-M. Yang, A. Wang. Emission Control Technology for 50 and 125cc Motorcycles in Taiwan [J]. SAE Paper 980938, 1988.
[10] 黎志勤，黎苏. 汽车排气系统噪声与消声器设计 [M]. 北京：中国环境科学出版社，1991.

（原载于《摩托车技术》，2004，（11）：91-97）

# 基于 WEB 的人工神经网络材料设计系统

刘国华　包　宏　李文超

（北京科技大学，北京　100083）

**摘　要**：开发了一个基于 WEB 的人工神经网络材料设计系统。该系统用 Delphi MIDAS 技术结合 ActiveX 技术实现对数据库的维护；利用 MATLAB 的 WEB 技术结合其数据库工具箱和神经网络工具箱编写程序，从材料数据库中获得数据进行神经网络计算，并将结果通过 WEB 服务器返回给浏览器。

**关键词**：WEB；人工神经网络（ANN）；材料设计

## Artificial Neural Network System for Materials Design Based on WEB

Liu Guohua　Bao Hong　Li Wenchao

(University of Sciences and Technology Beijing, Beijing 100083)

**Abstract**: An artificial neural network system was built for materials design based on WEB. The Delphi MIDAS combined with ActiveX technology was used for realizing materials database management. The MATLAB WEB technology combined with MATLAB database toolbox and artificial neural network toolbox was employed to obtain data from materials database and to process the data. The results of computation can be send to browsers by web server.

**Keywords**: WEB; artificial neural network (ANN); materials design

## 1 引言

　　人工神经网络是试图模拟人脑神经系统的结构与功能特征的数学处理方法。它具有自学习能力，不需要任何先验函数的假设，即可从实验数据中自动总结规律。目前的新材料设计中，常是仅有组分、工艺和性能之间的相关实验数据，而其间内在规律尚不很清楚。对于一个材料设计系统而言，这可认为是仅有数据库而无知识库。用数据库中的数据对人工神经网络进行训练，并用训练好的网络充当"知识"来预测未知，从而建立材料设计智能系统是一种有效的手段[1]。现有的用于材料设计的人工神经网络智能系统一般都是单机版的，限制了它的推广和应用。近年来，由于 Internet 的发展，开发与之相适应的智能系统有着巨大的现实与潜在需求[2]。鉴于此，本文开发了一个基于 WEB 的人工神经网络材料设计系统，以提供 Internet 实时计算服务。

---

刘国华，1999~2002 年于北京科技大学师从李文超教授攻读博士学位。目前在广州白云区发改局工作。

## 2 系统构建

本文选用 Microsoft 公司的 NT Server4.0 作为网络操作系统，以 Internet Information Server 4.0（IIS 4.0）作为 WEB 服务器，以 SQL Server 7.0 作为数据库系统。整个材料设计系统分为两大模块，即数据库维护模块和人工神经网络模块。数据库维护模块用 Delphi 的 MIDAS 技术实现，首先编制数据应用程序服务器，它起到数据代理的作用，然后编制能下载到浏览器中的 Activeform 与数据库应用程序服务器通讯，实现对数据库查询、修改等操作。人工神经网络模块是利用 Matlab 的 WEB 技术来实现的，该技术提供了一个名为 Matweb.exe 的 CGI 程序和一个多线程 TCP/IP 服务器 Matlabserver。Matweb.exe 用于建立 WEB 服务器与 Matlabserver 之间的联结；而 Matlabserver 则负责解释并运行 Matlab 程序（.M 文件），本文中的 Matlab 程序是指利用 MATLAB 的数据库工具箱和神经网络工具箱编写的 WEB 神经网络程序。本系统的结构如图 1 所示。

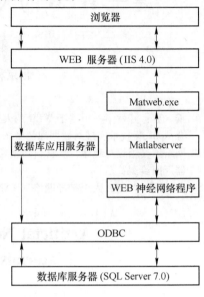

图 1　系统结构示意图

## 3 数据库维护的实现

我们用 Delphi 的多层分布式应用程序服务器技术（MIDAS）[3]来实现基于 WEB 的数据库维护模块。多层结构是在客户层和数据库层之间构造一层或更多层次的中间层。典型的多层结构分为三层：第一层是数据库服务器，提供数据的存储和管理功能；第二层是应用程序服务器，即中间层，它把业务逻辑单独提取出来，负责处理用户请求，并且把结果返回给用户；第三层是瘦客户机层，它只提供应用的用户界面，不直接返问后台数据库。中间层可根据需要分布在不同的计算机上，从而形成分布式应用系统。以这种方式构造整个系统有更大的灵活性。利用 Delphi 的 MIDAS 技术建立 WEB 数据库应用程序有两种方式[4]，一是将 MIDAS 技术与 ActiveX 技术相结合，由嵌入到网页中并下载到客户端的 ActiveX 控件或 Activeform 与 MIDAS 数据库应用程序服务器通讯并取得数据；二是将 MIDAS 技术与 WEB Broker 技术相结合，由使用 WEB Broker 技术编制的 WEB 应用程序从 MIDAS 数据库应用程序服务器取得数据，并在服务器端产生出 HTML 代码，传给客户端。第一种方式的功能更强大，所以在本文中采用第一种方式。

要想使下载到客户端的 Activeform 能对数据库进行操作，首先要构造数据库应用程序服务器。其关键是远程数据模块，该模块通过 Database 及 Table 等数据集组件与远程数据库建立连接，并利用 DatasetProvider 组件为 Activeform 中的 ClientDataset 组件留下数据供应接口。客户端就是通过数据供应接口来间接访问数据库的。构造好的数据库应用程序服务器可根据均衡负载、提高系统性能的需要，将其运行并注册到相应计算机上。在注册好数据库应用程序服务器后就可编制 Activeform。Activeform 需要一个通讯组件与应用程序服务器连接，Delphi 提供了四种不同的通讯协议，它们是 DCOM、TCP/IP、OLEnterprise、

CORBA。考虑到我们是以 WEB 方式开发,我们选用 WEB 方式下广泛使用的 TCP/IP 协议,其所用的通讯组件为 Socket Connection。使用 TCP/IP 协议,应用程序服务器端必须运行一个专门的运行期软件 ScktSrver.exe。客户的请求首先传递给 ScktSrver.exe,然后再创建数据模块的实例,在此基础上,通过 Activeform 中 ClientDataset 的数据接口从应用程序服务器存取数据库,并用 DBGrid 等数据库显示组件显示到 Activeform 上。如此获得与数据库的交互能力,实现数据库维护。下面是数据库更新程序的关键部分:

```
procedure Tclientform.ButtonUpdateclick (sender : Tobject);
begin
if clientdatasetANNsample.applyupdates (0) =0 then
    clientdatasetANNsample.refresh;
end;
```

## 4　WEB 神经网络程序的实现

要开发基于 WEB 的神经网络系统,要求编程语言能方便地编制神经网络程序、能访问数据库,并能支持 WEB 开发方式。我们选用 MATLAB 语言,这是因为 MATLAB 的基本元素是无需定义维数的矩阵,很适合于处理涉及大量矩阵运算的神经网络编程问题,而且它还提供了关于神经网络的工具箱,即用 MATLAB 语言构造出了神经网络理论中所涉及的公式运算、矩阵操作和方程求解等大部分子程序,以用于神经网络设计、训练。我们只要根据需要调用相关程序,免除了编写复杂而庞大的算法程序的困扰[5],所以 MATLAB 在编制神经网络程序和此类程序的执行效率上是其他传统计算语言无法比拟的。此外,MATLAB 还提供了数据库工具箱和 MATLAB WEB 技术,支持数据库操作和 WEB 开发。MATLAB 数据库工具箱允许访问 SQL Server、Oracle、Sybase、Informix 等大部分关系型数据库。MATLAB WEB 技术是采用混合 CGI 技术[6]对 WEB 服务器的扩展,也就是说该 WEB 服务器扩展程序分为两个部分:"瘦" CGI 程序 Matweb.exe 和"胖"伙伴进程 Matlabserver。Matweb.exe 只负责接收用户输入和把用户输入发送给伙伴程序而不做任何其他工作;Matlabserver 是在 Windows NT 的后台运行的"系统服务",它负责解释并运行后缀为.M 的 WEB 神经网络程序,该程序根据用户输入完成数据库连接、神经网络计算并获得计算结果等任务。

下面以网络训练为例说明其具体开发过程。

(1)制作收集用户输入信息的 HTML 文件。在 HTML 文件的 form 中用文本框或下拉列表等形式收集用户输入,并以 POST 方式将用户输入信息传给 CGI 程序 matweb.exe。其基本结构如下:

```
<form action = "/cgi-bin/matweb.exe"  method = "POST" >
  <input type = "hidden"  name = "mlmfile"  value = "anndbtrain" >
数据集名:
  <input type = "text"  name = "table" >
……
隐含层神经元数:
  <select name = "innum" >
      <option value = "5"  selected >5
      ……
```

```
       </select>
       ......
       <input type="submit" name="Submit">
       </form>
```

当用户提交这一表单时,Web 服务器会调用 CGI 程序 matweb.exe 来接收用户输入信息,然后,该 CGI 程序会把这些信息发送给其伙伴进程 Matlabserver,由它来运行在 form 中指定的 WEB 神经网络应用程序 anndbtrain.m。

(2) 编制 WEB 神经网络程序 anndbtrain.m。该程序首先根据用户输入通过 MATLAB 数据库工具箱连接到 ODBC 数据源上,并查询出所要操作的样本数据。将样本数据进行标准化处理和主成分分析后,把样本数据分成训练集和监控集两部分。训练集中的数据被用来计算梯度并更新网络权重。监控集用来防止网络过拟合的出现,它是通过在网络训练过程中计算自身的误差变化来监控网络训练的。通常情况下,在网络训练初期,监控集误差和训练集误差一样会不断减小,而当网络训练到一定时间后,监控集的误差开始增大,如果该误差一直增大,可以认为网络训练出现了过拟合,此时,训练会停止,返回监控集误差最小时的网络权重,并将结果以 (3) 中做好的模板返回给用户。

(3) 为上述 WEB 神经网络应用程序做一个 HTML 模板文件,以显示神经网络程序的返回结果,包括训练后的网络响应值 RESPONSE、网络响应值与实际值间的方差 $S$ 和相关系数 $R$ 等(返回结果放在两个$之间)。其基本结构如下:

```
       <html>
       数据集$table$的样本数据的训练结果</font>
       <table   autogenerate="$ACTUAL$">
       <caption align=center>实际目标值</caption>
         <tr>
         <td>    </td>
         </tr>
       </table>
       <table autogenerate="$RESPONSE$">
       <caption align=center>网络响应值</caption>
         <tr>
          <td>    </td>
         </tr>
       </table>
       <p>本次训练的网络响应值与实际目标值的关系如下:</p>
       方差:$S$
       相关系数:$R$
       </html>
```

## 5　系统的使用

当用户在浏览器中输入 WEB 神经网络材料设计系统主页的 URL 并且连接成功后,即可使用该系统。首先要用数据库中的样本数据对神经网络进行训练。在浏览器中根据提示选择合适的参数后开始训练网络（网络训练的界面如图 2 所示），直至网络的响应值与实际输出间的方差和相关系数满足要求后，网络训练结束。程序会将训练结果以文件的形式保存下来，作为从这些数据中获得的"知识"（这种知识是通过神经元间联结权重的隐式形式来实现的），以供下一步操作新数据预测未知之用，当用户使用预测程序时，程序会调用训练结果文件，应用已有"知识"对未知数据作出预测。我们将系统应用于合成陶瓷材料的工艺参数优化中，取得了很好的效果。

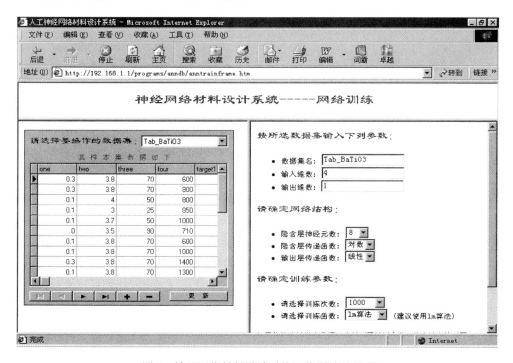

图 2　神经网络材料设计系统网络训练的界面

# 6　结束语

本工作实现了神经网络材料设计系统的 WEB 计算功能，将有助于推广人工神经网络在材料设计中的应用。

参 考 文 献

[1] 吕允文，夏宗宁，赖树纲，等．材料设计专家系统与人工神经网络的应用［J］．材料导报，1994，8（6）：1-4．
[2] 高全泉．关于人工智能技术在 Internet 上的应用与发展［J］．计算机科学，2000，27（6）：13-16．
[3] 张虹，甄青坡．基于 MIDAS 构建多层分布式结构及应用［J］．计算机科学，2000，27（2）：32-35．
[4] 徐新华．Delphi 5 高级编程- COM、CORBA 与 INTERNET 编程［M］．北京：人民邮电大学出版社，2000．
[5] 丛爽．面向 MATLAB 工具箱的神经网络理论与应用［M］．合肥：中国科学技术大学出版社，1998．
[6] 孙琨，奚舸．Web 数据库应用技术与主要产品（上）［J］．计算机与通信，1999，（3）：66-70．

（原载于《计算机工程与应用》，2001，（20）：141-142）